Dedication

First, to those closest to me: my parents, Roxanne, Barbara, Dan, and Juliann. And second, to the memory of Thomas L. Jacobs of UCLA, the person who inspired my life-long interest in organic chemistry. Without all of you, this book would never have come into being.

A COMPLETE INTRODUCTION TO MODERN NMR SPECTROSCOPY

ROGER S. MACOMBER
University of Cincinnati and
Pepperdine University

A Wiley-Interscience Publication

JOHN WILEY & SONS, INC.
New York • Chichester • Weinheim • Brisbane • Singapore • Toronto

Library of Congress Cataloging-in-Publication Data:

Macomber, Roger S.
  A complete introduction to modern NMR spectroscopy / Roger S.
  Macomber.
      p. cm.
    "A Wiley-Interscience publication."
    Includes bibliographical references and index.
    ISBN 0-471-15736-8 (alk. paper)
    1. Nuclear magnetic resonance spectroscopy. I. Title.
  QD96.N8M3 1997
  543′.0877--dc21   97-17106

Printed in the United States of America.

10 9 8 7 6 5 4 3 2 1

CONTENTS

PREFACE

In the decade since the first version of this book was written, the field of nuclear magnetic resonance (NMR) spectroscopy has made, quite literally, quantum leaps. Computer-controlled NMR spectrometers with high-field superconducting magnets, once available only to the most well-funded institutions, are now commonplace. Two-dimensional NMR techniques, in their infancy a decade ago, have blossomed into an indispensable array of tools to elucidate the molecular structure of compounds as complex as proteins. And who is unaware of the growing importance of NMR to medical diagnosis through the technique of magnetic resonance imaging (MRI)?

Nuclear magnetic resonance has become ubiquitous in such divergent fields as chemistry, physics, material science, biology, medicine, and forensic science, among others. In writing this book it has been my goal to provide a monograph aimed at anyone, not just chemists, with an interest in learning about NMR. Medical students, whose interest lie not so much in molecular structure as it does in the three-dimensional distribution of magnetic nuclei such as hydrogen, will find what they are looking for in the beginning three chapters, which discuss the physics of NMR signal generation, and Chapter 16, which describes MRI. If your interests are in biochemistry, Chapter 14 will be very useful to you. The majority of you will use NMR to elucidate molecular structures, so in Chapters 4–13 I have tried to give you everything you will need to get that job done efficiently.

A first-year college chemistry course is the only scientific background I have assumed the reader has. All the necessary details are developed from the most basic level. The approach is relatively nonmathematical, with only a few simple equations. But do not let this fool you. By the end of the book you will be well prepared to solve any molecular structure problem given a complete set of NMR data. And you will be able to proceed confidently to any advanced treatise on NMR.

Above all, I have tried to make the text clear, logical, and interesting to read. There are hundreds of figures and actual spectra to illustrate the many topics. The vast majority of spectra were obtained on modern high-field spectrometers. Because of the way I structured the book, I recommend that you proceed through it chapter by chapter, rather than skipping around. Nonetheless, I have tried to help those readers who skip around by adding liberal references to earlier sections that have supporting information on each topic.

Soon after starting Chapter 1 you will note that I have adopted a semiprogrammed approach. That is, there are frequent example problems (with solutions) to test your mastery of the topic at hand. To get the most from your reading, try to work each problem as you encounter it. They often contain important additional information about the material just covered. Then, at the end of each chapter there is a chapter summary and several review problems to see if you have mastered the concepts in that chapter. There are also two self-tests (after Chapters 7 and 13) that will help you assess your overall mastery of the subject. The answers to these review and self-test problems appear in Appendix 1.

I hope you enjoy this book and that it inspires you to learn all you can about modern NMR methods. I would also appreciate greatly any feedback you would like to offer.

Malibu, California

Roger S. Macomber

ACKNOWLEDGMENTS

Many individuals have contributed mightily to this book, and without their help I could not have gone forward with the project.

About two years ago Jeff Holtmeier, a consulting editor at John Wiley, approached me to see if I had any interest in writing an updated edition of my earlier book on NMR spectroscopy, published in early 1988. When I said yes, he set about getting the original edition reviewed by a half-dozen experts in the field. With these in hand, he single-handedly promoted the project through the appropriate Wiley channels, resulting in a contract in late 1995.

As the prospectus for the new edition was being developed, several of my colleagues agreed to contribute chapters in their particular areas of expertise. George Kreishman and Elwood Brooks (UC's staff NMR spectroscopist) collaborated to write Chapter 14, dealing with the applications of NMR to biochemistry. Jerry Ackerman, a one-time colleague of mine at the University of Cincinnati and now director of NMR spectroscopy at Massachusetts General Hospital and assistant professor of radiology at the Harvard Medical School, offered to contribute Chapters 15 and 16 on solid-state NMR and magnetic resonance imaging.

Fortuitously, at that time I signed the contract, I was just beginning a previously arranged two-quarter sabbatical leave, during which I planned to make major progress on the new edition. My first stop was the Chemistry Department at the University of Hawaii, hosted by my friend and colleague, Karl Seff. Here I spent a most enjoyable month writing drafts of several chapters, delivering and attending seminars, surfing, and biking. Then, after six weeks back in Cincinnati, I headed to the University of Utah, where I was hosted by my friends Pete Stang and Wes Bentrude. Before I arrived, the winter in Utah had been exceptionally mild, with absolutely no snow on the slopes, so I expected to get a lot of writing done. Before I left Utah, over 120 inches of fresh powder had fallen. Need I say more?

While at Utah a number of individuals offered to contribute example spectra for inclusion in the book. I have mentioned the source of each contributed spectrum in the text or figure legend, but let me introduce them here, as well: Bobby L. Williamson (with Peter Stang's group), Alan Sopchik (with Wes Bentrude's group), Dhileepkumar Krishnamurthy and Stan McHardy (with Gary Keck's group), and John Bender and Soren Giese (with Fred West's group). I also wish to thank Steve Fetherston and Rosemary Laufer, able staff members who made my visit even more productive and enjoyable.

Back again in Cincinnati, Elwood Brooks, Marshall Wilson, and graduate students Pat Hutchins, Mark Guttaduaro, and Sheela Venkitachalam contributed additional spectra for the book, as did David Watt of the University of Kentucky. But there are two major contributors of example spectra who deserve special thanks. Sadtler Research Laboratories (Division of Bio-Rad Laboratories, Inc.), through staff scientist Marie Scandone, provided many of the one-dimensional ^{1}H and ^{13}C spectra. And David Lankin, a graduate of this department and currently a research scientist and NMR expert at Searle, provided the bulk of the special 2D and related spectra in Chapters 12 and 13, with the help of Geoffrey Cordell of the University of Illinois School of Pharmacy.

Finally, three more of my faculty colleagues at UC deserve special thanks. Albert Bobst advised me on the EPR section of Chapter 11, Frank Meeks provided mathematical advice, and Allan Pinhas helped edit the entire set of proofs. To all these individuals, I give a sincere thank you!

Roger S. Macomber

FREQUENTLY USED SYMBOLS AND ABBREVIATIONS

●, an unpaired electron

↑ or ↓, possible orientations (with respect to $\mathbf{B}_0$) of the magnetic moment of an electron or an $I = \frac{1}{2}$ nucleus

[ ], concentration in moles per liter

{ }, indicates the nucleus being irradiated in a double-resonance experiment

2D, NMR spectrum where signal intensity is a function of two frequencies

A, (1) mass number of a nucleus; (2) net CIDNP absorption

a, hyperfine coupling constant in an EPR spectrum

α, (1) one possible orientation (with respect to $\mathbf{B}_0$) of the magnetic moment of an electron or an $I = \frac{1}{2}$ nucleus; (2) the flip angle in a pulsed NMR experiment

$AA'BB'$, Pople designation for a coupled four-spin system consisting of two chemically but not magnetically equivalent nuclei in each of two sets

A/E, absorption–emission CIDNP net effect

APT, attached proton test

$\mathbf{B}_0$, external (applied) magnetic field vector

$\mathbf{B}_1$, oscillating magnetic field vector of the observing channel

$\mathbf{B}_2$, oscillating magnetic field vector of the "irradiating" channel

β, possible orientation (with respect to $\mathbf{B}_0$) of the magnetic moment of an electron or an $I = \frac{1}{2}$ nucleus

β_e, Bohr magneton

C_n, n-fold axis of symmetry

C(A)T, computed (axial) tomography

COSY, 2D correlation spectroscopy

CSR, chemical shift reagent

d, doublet (signal multiplicity)

D, deuterium (^{2}H)

Δ, a change (e.g., ΔE, a change in energy)

δ, chemical shift signal position (ppm)

$\delta_{11}, \delta_{22}, \delta_{33}$, chemical shift tensors

$\Delta\delta_X$, substituent shielding parameter

$\Delta\nu$, frequency difference (Hz) between the signal of interest and the operating frequency of the spectrometer; also used for electric quadrupole interaction

$\Delta\nu_0$, frequency difference (Hz) between two sites at the slow exchange limit

$\delta\nu$, frequency difference (Hz) between the signal of interest and the frequency of the internal standard signal; also a measure of spectral resolution

δ_x, spatial resolution along x axis

Δ_x, field of view along x axis

DEPT, distortionless enhancement by polarization transfer

DMF, dimethylformamide

E, (1) energy; (2) net CIDNP emission

E/A, emission–absorption CIDNP net effect

EPI, echo planar imaging

EPR, electron paramagnetic resonance

ESR, electron spin resonance (same as EPR)

exp, exponentiation on the base *e*

F$_1$, the FT of the evolution/mixing time parameter in a 2D NMR experiment

F$_2$, the FT of the detection time parameter in a 2D NMR experiment

FID, free induction decay time-domain signal

FOV, field of view

f$_p$, fraction of *p*-orbital character

f$_s$, fraction of *s*-orbital character

FT, Fourier transformation

g (factor), signal position parameter in an EPR spectrum

G°, standard-state free energy

γ, magnetogyric ratio of a nucleus

Γ$_n$, the type of CIDNP net effect

Γ$_m$, the type of CIDNP multiplet effect

G$_x$, magnetic field gradient

GHz, gigahertz (10^9 Hz)

h, Planck's constant

η, signal intensity enhancement due to NOE

HSC, 2D heteroscalar shift-correlated NMR spectrum

HET2DJ, 2D heteronuclear *J*-resolved experiment

HETCOR, same as HSC

HOM2DJ, 2D homonuclear *J*-resolved experiment

HSC, 2D heteronuclear shift correlation

Hz, unit of frequency (cycles per second)

I, magnetic spin of a nucleus

INADEQUATE, 1D and 2D NMR experiments to demonstrate direct C–C connectivity

INEPT, insensitive nuclei enhanced by polarization transfer

IOU, index of unsaturation

J, joules, a unit of energy

J, internuclear coupling constant (Hz)

J$_r$, residual internuclear coupling constant (Hz) observed in an off-resonance decoupled spectrum

k, rate constant for exchange

k$_c$, rate constant at coalescence

K, equilibrium constant

L, number of lines (multiplicity) of a signal

m, magnetic spin quantum number of a nucleus

m$_s$, magnetic spin quantum number of an electron

M, net (macroscopic) magnetization vector

M$_0$, net (macroscopic) magnetization vector at time zero

M$_t$, net (macroscopic) magnetization vector at time *t*

M$_{x,y}$, component of the net (macroscopic) magnetization vector in the *x, y* plane

μ, (1) magnetic moment of a nucleus; (2) parameter used to calculate the CIDNP net effect

MHz, megahertz (10^6 Hz)

MRI, magnetic resonance imaging

n, (1) number of nuclei in an equivalent set; (2) mixing coefficient of an *sp*^{n} hybrid orbital

N, number of neutrons in a nucleus

ν, frequency (Hz)

ν$_{av}$, average position (Hz) of two or more signals undergoing exchange

ν$_0$, operating frequency (Hz) of an NMR spectrometer

ν$_0'$, operating frequency (MHz) of an NMR spectrometer

ν$_1$, observing frequency in a double-resonance experiment

ν$_2$, "irradiating" frequency in a double-resonance experiment

ν$_{1/2}$, signal halfwidth

ν$_{1/2}^0$, signal halfwidth at the fast exchange limit

NMR, nuclear magnetic resonance

NOE, nuclear Overhauser effect

NOESY, a 2D COSY spectrum showing NOE interactions

ω, angular frequency (rad s^{-1})

P, population of a given state or energy level

ppm, parts per million

Pro-R, configuration of an atom about a prochiral center

Pro-S, configuration of an atom about a prochiral center

PSR, paramagnetic shift reagent

π_x, $\mathbf{B}_1$ pulse causing a 180° rotation around the x' axis

$(\pi/2)_x$, $\mathbf{B}_1$ pulse causing a 90° rotation around the x' axis

q, quartet (signal multiplicity)

q, electric quadrupole

r, internuclear distance

R, (1) exchange ratio, $k/\Delta\nu_0$, (2) absolute configuration at a chiral center; (3) spectroscopic resolution; (4) ideal gas constant

ROI, region of interest

s, singlet (signal multiplicity)

s, (1) same as m_s, (2) an atomic orbital

S, absolute configuration at a chiral center

σ, (1) representing a plane of symmetry; (2) shielding constant; (3) parameter used in describing the CIDNP multiplet effect; (4) cylindrically symmetric bond or molecular orbital

S/N, signal-to-noise ratio

SPI, selective population inversion

sp^n, hybridized atomic orbital with mixing coefficient n

SW, sweep (or spectral) width (Hz)

t, time (s)

t, triplet (signal multiplicity)

t_{acq}, acquisition time

TE, echo time

t_d, dwell time (s)

t_p, duration (μs) of a pulse

TR, repetition time

t_w, delay time

T (Tesla), unit of magnetic field strength

T, absolute temperature (K)

T_ρ, absolute rotating frame temperature

T_1, spin–lattice (longitudinal) relaxation time of a nucleus

$T_{1\rho}$, rotating-frame spin–lattice relaxation time

T_2, spin–spin (transverse) relaxation time of a nucleus

$T_{2\rho}$, rotating-frame spin–spin relaxation time

T_2^*, *effective* spin–spin (transverse) relaxation time of a nucleus

T_{CP}, cross-polarization time constant

τ, (1) old chemical shift scale; (2) lifetime (s)

τ_c, correlation time of a nucleus

θ, (1) angle between the internuclear vector and $\mathbf{B}_0$, (2) dihedral (torsional) angle

TMS, tetramethylsilane (internal standard)

TOCSY, total correlation 2D NMR spectroscopy

W, watt (measure of rf power)

X, mole fraction

x,y,z, Cartesian coordinate system in the laboratory frame of reference

x',y',z, Cartesian coordinate system in the rotating frame of reference

z axis, the axis of $\mathbf{B}_0$, the external (applied) magnetic field

Z, atomic number

1

SPECTROSCOPY: SOME PRELIMINARY CONSIDERATIONS

1.1 WHAT IS NMR SPECTROSCOPY?

Nuclear magnetic resonance (NMR) spectroscopy is the study of molecular structure through measurement of the interaction of radio-frequency electromagnetic radiation with a collection of nuclei immersed in a strong magnetic field. These nuclei are parts of atoms that, in turn, are assembled into molecules. An NMR spectrum, therefore, can provide detailed information about molecular structure and dynamics, information that would be difficult, if not impossible, to obtain by any other method.

It was in 1902 that physicist P. Zeeman shared a Nobel Prize for discovering that the nuclei of certain atoms behave strangely when subjected to a strong external magnetic field. And it was exactly 50 years later that physicists F. Bloch and E. Purcell shared a Nobel Prize for putting the so-called **nuclear Zeeman effect** to practical use by constructing the first crude NMR spectrometer. It would be an understatement to say that, during the succeeding years, NMR has completely revolutionized the study of chemistry and biochemistry, not to mention having a significant impact on a host of other areas. Nuclear magnetic resonance has become arguably the single most widely used technique for elucidation of molecular structure. But before we can begin our foray into NMR, we need to review a few fundamental principles from physics.

1.2 PROPERTIES OF ELECTROMAGNETIC RADIATION

All spectroscopic techniques involve the interaction of matter with **electromagnetic radiation**, so we should begin with a description of the properties of such radiation. The light rays that allow our eyes to see this page constitute electromagnetic radiation in the visible region of the electromagnetic spectrum. Each electromagnetic ray can be pictured as shown in

Figure 1.1. Notice that the wave is actually composed of two **orthogonal** (mutually perpendicular) waves that oscillate exactly *in phase* with each other. That is, they both reach peaks, nodes, and troughs at the same points. One of these waves describes the electric field vector (**E**) of the radiation, oscillating in one plane (e.g., the plane of the page); the other describes the magnetic field vector (**B**) oscillating in a plane perpendicular to the electric field. Thus, both these fields exhibit *uniform periodic* (e.g., sinusoidal) *motion*. The axis along which the wave propagates (the abscissa in Figure 1.1) can have dimensions of either time or length.

The wave(s) pictured in Figure 1.1 can be characterized by two independent quantities, **wavelength** (λ) and maximum **amplitude** (E_0 and B_0 in the figure). The *intensity* of a wave is proportional to the square of its amplitude. Knowing that electromagnetic radiation travels with a fixed velocity c (3.00×10^{10} cm s^{-1} in a vacuum), we can alternatively describe the wave as having a **frequency** ν, which is the inverse of the peak-to-peak time t_0 in the figure:

$$\nu = \frac{1}{t_0} \qquad (1.1)$$

where t_0 is measured in seconds and ν has units of cycle per second (cps or s^{-1}), now called **hertz** (Hz) in honor of the physicist H. Hertz.

Recognizing that the wave travels a distance λ in time t_0, we can derive a second relationship:

$$c = \frac{\lambda}{t_0} = \lambda\nu \qquad (1.2)$$

Thus, the wavelength and frequency of a given wave are not independent quantities; rather, they are inversely propor-

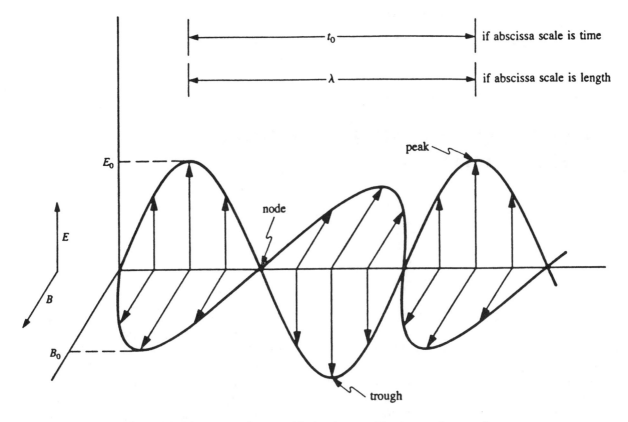

Figure 1.1. Electromagnetic wave with electric vector **E** and magnetic vector **B**.

tional. Radiation of high frequency has a short wavelength, while radiation of low frequency has a long wavelength.

The known electromagnetic spectrum (Table 1.1) ranges from cosmic rays of extremely high frequency (and short wavelength) to **rf** (radio-frequency) radiation of low frequency (and long wavelength). The narrow visible region in the middle of the electromagnetic spectrum corresponds to radiation of wavelength 380–780 nm (1 nm = 10^{-9} m = 10^{-7} cm) and frequency 4×10^{14}–8×10^{14} Hz. Our optic nerves do not respond to electromagnetic radiation outside this region.

In addition to its wave properties, electromagnetic radiation also exhibits certain behavior characteristic of particles. A particle, or **quantum**, of radiation is called a **photon**. For our purposes the most important particlelike property of a photon is its *energy* (E). Each photon possesses a discrete amount of energy that is directly proportional to its frequency (if we regard it as a wave). This relationship can be written

$$E = h\nu \tag{1.3}$$

where h, **Planck's constant**, has values of 6.63×10^{-34} J s per photon. Alternatively, h can be expressed on a *per-mole* basis through multiplication by Avogadro's number (6.02×10^{23} mol^{-1}) and division by 10^3 J(kJ)$^{-1}$ to give $h = 3.99 \times 10^{-13}$ kJ s mol^{-1}. (A mole of photons is referred to as an **Einstein**.)

Since the strength of a chemical bond is typically around 400 kJ mol^{-1}, radiation above the visible region in Table 1.1 has sufficient energy to **photodissociate** (break) chemical bonds, while radiation below the visible region does not (see Table 1.1). Of particular interest to us for NMR purposes is rf radiation, the same frequency range that carries communication signals to our radios and televisions. We will normally be using radiation with frequencies of 200–750 MHz (1 MHz = 10^6 Hz), at the low end of the energy scale in Table 1.1. This, it will turn out, is exactly the amount of energy we will need to perform NMR experiments.

To summarize, if two photons possess the same energy, they correspond to waves (or *wavelets*) of the same frequency and the same maximum amplitude. The total intensity of an electromagnetic beam is therefore the number of photons delivered per second.

■ **EXAMPLE 1.1** Derive the relationship between the energy of a photon and its wavelength.

□ *Solution:* We can rearrange Eq. (1.2) to $\nu = c/\lambda$. Substituting for ν in Eq. (1.3) gives

$$E = h\nu = \frac{hc}{\lambda}$$

□

TABLE 1.1   Electromagnetic Spectrum

Radiation	Wavelength, λ (nm)	Frequency, ν (Hz)	Energy (kJ mol^{-1})
Cosmic rays	$<10^{-3}$	$>3 \times 10^{20}$	$>1.2 \times 10^8$
Gamma rays	10^{-1}–10^{-3}	3×10^{18}–3×10^{20}	1.2×10^6–1.2×10^8
X-rays	10–10^{-1}	3×10^{16}–3×10^{18}	1.2×10^4–1.2×10^6
Far ultraviolet	200–10	1.5×10^{15}–3×10^{16}	6×10^2–1.2×10^4
Ultraviolet	380–200	8×10^{14}–1.5×10^{15}	3.2×10^2–6×10^2
Visible	780–380	4×10^{14}–8×10^{14}	1.6×10^2–3.2×10^2
Infrared	3×10^4–780	10^{13}–4×10^{14}	4–1.6×10^2
Far infrared	3×10^5–3×10^4	10^{12}–10^{13}	0.4–4
Microwaves	3×10^7–3×10^5	10^{10}–10^{12}	4×10^{-3}–0.4
Radio frequency	10^{11}–3×10^7	10^6–10^{10}	4×10^{-7}–4×10^{-3}

It is perhaps worthwhile to mention that the velocity (v) of electromagnetic radiation *decreases* as it passes through a condensed medium (e.g., a liquid or solution). The ratio of its speed in a vacuum (*c*) to its velocity in the medium is called the **index of refraction** (*n*) of the medium:

$$n = \frac{c}{\text{v}} \qquad (1.4)$$

The magnitude of *n* for a given medium varies inversely with the wavelength of radiation, but it is always greater than unity. The *energy* of a photon (unless it is absorbed) is unaffected by passage through the medium, so its frequency must also be unchanged [Eq. (1.3)]. Therefore, its *wavelength* must have decreased (to λ') in order to preserve the relationship in Eq. (1.2):

$$\text{v} = \lambda'\nu$$

where $\lambda' = \lambda/n = \lambda\text{v}/c$.

1.3   INTERACTION OF RADIATION WITH MATTER: THE CLASSICAL PICTURE

Now that we know something about electromagnetic radiation, let us turn to the question of what factors control the interaction of such radiation with particles of matter. The three main types of interactions of interest to spectroscopists are absorption, emission, and scattering. When **absorption** occurs, the photon disappears and its entire energy is transferred to the particle that absorbed it. The resulting particle with this excess energy is said to be in an **excited state**. It can relax back to its **ground state** by **emitting** a photon, which carries off the excess energy.

Radiation is **scattered** when the direction of propagation of the photon is shifted by some angle, the result of passing close to a perturbing particle. If the frequency of the radiation is unchanged, the scattering is described as **elastic**. However, if the frequency has changed (**inelastic** scattering), this indicates that there was a partial exchange of energy between the photon and the particle.

In the case of NMR spectroscopy we will be concerned only with absorption and emission of rf radiation. **Quantum mechanics**, the field of physics that deals with energy at the microscopic (atomic) level, allows us to define **selection rules** that describe the probability for a photon to be absorbed or emitted under a given set of circumstances. But even classical (i.e., pre-quantum-mechanical) physics tells us there is one requirement shared by all forms of absorption and emission spectroscopy: For a particle to absorb (or emit) a photon, the particle itself must first be in some sort of uniform periodic motion with a characteristic fixed frequency. Most important, the frequency of that motion must *exactly* match the frequency of the absorbed (or emitted) photon:

$$\nu_{\text{motion}} = \nu_{\text{photon}} \qquad (1.5)$$

This fact, which at first glance might appear to be an incredible coincidence, is actually quite logical. If a photon is to be absorbed, its energy, which is originally in the form of the oscillating electric and magnetic fields, must be transformed into energy of the particle's motion. This transfer of energy can take place only if the oscillations of the electric and/or magnetic fields of the photon can constructively interfere with the "oscillations" (uniform periodic motion) of the particle's electric and/or magnetic fields. When such a condition exists, the system is said to be in **resonance**, and only then can the act of absorption take place. By the way, do not confuse the term *resonance* in this context with the concept of resonance (conjugation) of electrons used to describe the structure of molecules.

■ **EXAMPLE 1.2**   The C=O bond in formaldehyde vibrates (stretches, then contracts) with a frequency of 5.13×10^{13} Hz. (a) What frequency of radiation could be absorbed by this vibrating bond? (b) How much energy would each photon deliver? (c) To which region of the electromagnetic spectrum does this radiation belong? (d) Are photons of this region capable of breaking bonds?

☐ *Solution:* (a) From Eq. (1.5) we know the frequencies must match; therefore, $\nu_{photon} = 5.13 \times 10^{13}$ Hz. (b) From Eq. (1.3),

$$E_{photon} = h\nu = (6.63 \times 10^{-34} \text{ J s})(5.13 \times 10^{13} \text{ s}^{-1})$$

$$= 3.40 \times 10^{-20} \text{ J} = 20.5 \text{ kJ mol}^{-1}$$

(c) From Table 1.1 we see that radiation of this frequency and energy falls in the infrared region. (d) No. This amount of energy is less than half that required to break even the weakest chemical bond. However, absorption of such a photon does create a vibrationally excited bond, which is more likely to undergo certain chemical reactions than is the same bond in its ground state. ☐

At this point you might think that the frequency-matching requirement places a heavy constraint on the types of absorption processes that can occur. After all, how many kinds of periodic motion can a particle have? The answer is that even a small molecule is constantly undergoing many types of periodic motion. Each of its bonds is constantly vibrating; the molecule as a whole and some of its individual parts are rotating in all three dimensions; the electrons are circulating through their orbitals. And each of these processes has its own characteristic frequency and its own set of selection rules governing absorption!

All of the above forms of microscopic motion are what we might describe as *intrinsic*. That is, the motion takes place all by itself, without intervention by any external agent. However, it is possible under certain circumstances to *induce* particles to engage in additional forms of periodic motion. Still, to achieve resonance, we need to match the frequency of this induced motion with that of the incident radiation [Eq. (1.5)].

For example, an ion (or any charged particle, for that matter) follows a curved path as it moves through a magnetic field. If we carefully adjust the strength of the magnetic field, the ion will follow a perfectly circular path, with a characteristic fixed frequency that depends on its mass, charge, velocity, and strength of the magnetic field. Matching this characteristic **cyclotron** frequency with incident electromagnetic radiation of the same frequency can lead to absorption, and this is the basis of a technique known as **ion cyclotron resonance (ICR) spectroscopy**. We will discover in Chapter 2 that a strong magnetic field can also be used to induce

certain nuclei to move with uniform periodic motion of a different type.

1.4   UNCERTAINTY AND THE QUESTION OF TIME SCALE

If you have ever tried to take a photograph of a moving object, you know that the shutter speed of the camera must be adjusted to avoid blurring the image. And, of course, the faster the object is moving, the shorter must be the exposure time to "freeze" the motion. We have very similar considerations in spectroscopy.

Suppose you owned a collection of very extraordinary chameleons that were able to change colors instantaneously from white to black or black to white every 1 s. If you took a picture of them with a shutter speed of 10 s, each of the little critters would appear to be gray. But if you decreased the exposure time to 0.01 s, the photograph would show black ones and white ones in roughly equal numbers but no gray ones! Thus, to capture the individual colors, your exposure time must be significantly shorter than the lifetimes of the species, in this case the 1-s lifetime of each colored form.

There are many types of molecular chameleons, that is, molecules that constantly undergo some sort of reversible reorganization of their structures. If absorption of the photon is fast enough, we will detect both the "black" and "white" forms of the molecule. But if the absorption process is slower than the interconversion, we will detect only some sort of *time-averaged* structure. The situation therefore boils down to the question: How long does it take for a particle to absorb a photon? Unfortunately, such a question is impossible to answer with complete precision.

In 1927, W. Heisenberg, a pioneer of quantum mechanics, stated his **uncertainty principle:** There will always be a limit to the precision with which we can *simultaneously* determine the energy and time scale of an event. Stated mathematically, the product of the uncertainties of energy (ΔE) and time (Δt) can never be less than h (our old friend, Planck's constant):

$$\Delta E \, \Delta t \geq h \qquad (1.6)$$

Thus, if we know the energy of a given photon to a high order of precision, we would be unable to measure precisely how long it takes for the photon to be absorbed. Nonetheless, there is a useful generalization we *can* make. Using Eq. (1.3), we can substitute $h \, \Delta \nu$ for the ΔE in Eq. (1.6), giving

$$\Delta t \geq \frac{1}{\Delta \nu}$$

where $\Delta \nu$ is the uncertainty in frequency. As a result, the time required for a photon to be absorbed (Δt) must be approximately as long as it takes one "cycle" of the wave to pass the particle. That length of time, t_0 in Figure 1.1, is nothing more than $1/\nu$. This result stands to reason if we consider that the particle would have to wait through at least one cycle before it could sense what the radiation frequency was. At least we now have an order-of-magnitude idea of how fast our shutter speed must be in order to "freeze" a given molecular event. We will encounter the uncertainty principle at several points along our voyage through NMR spectroscopy.

■ **EXAMPLE 1.3** Suppose our NMR experiment required the use of rf radiation with a frequency of 250 MHz to examine formaldehyde (see Example 1.2). Will this NMR experiment enable us to see the various individual lengths of the C=O bond as it vibrates, or will we detect only a time-averaged bond length?

☐ *Solution:* The vibrational time scale ($1/\nu = 1/(5.13 \times 10^{13}$ Hz) $= 1.9 \times 10^{-14}$ s) is much shorter (faster) than the NMR time scale [$1/\nu = 1/(2.5 \times 10^8$ Hz) $= 4 \times 10^{-9}$ s]. Therefore, NMR can only detect a time-averaged C=O bond length. ☐

Equipped with this knowledge about electromagnetic radiation, periodic motion, resonance, and time scale, we are now ready to enter the intriguing world of the atomic nucleus.

CHAPTER SUMMARY

1. Nuclear magnetic resonance spectroscopy involves the interaction of certain nuclei with radio-frequency (rf) electromagnetic radiation when the nuclei are immersed in a strong magnetic field.

2. Electromagnetic radiation is characterized by its frequency (ν) or wavelength (λ), which are inversely proportional [Eq. (1.2)]. Radio-frequency radiation used in NMR spectroscopy typically has frequencies on the order of 200–750 MHz.

3. The energy of a photon (a quantum of radiation) is directly proportional to its frequency [Eq. (1.3)].

4. For radiation to be absorbed by a particle, the frequency of the radiation must equal the frequency of the particle's periodic motion.

5. The Heisenberg uncertainty principle [Eq. (1.6)] defines the time scale of radiation absorption event as inversely proportional to the radiation's frequency. Processes that occur faster than the spectroscopic time scale are time averaged during the absorption process.

REVIEW PROBLEMS (Answers in Appendix 1)

1.1. The linear HCN molecule rotates around an imaginary axis through its center of mass and perpendicular to the molecular axis. The frequency of this rotation is 4.431598×10^{10} Hz. (a) What frequency of radiation could be absorbed by this rotating molecule? (b) To which region of the electromagnetic spectrum does such radiation belong?

1.2. When laser light with $\lambda = 1064$ nm impinges on a sample of formaldehyde (Example 1.2), most of the light is scattered elastically. But a small number of scattered photons emerge with $\lambda = 1301$ nm. Account for this exact wavelength. (This is an example of Raman spectroscopy.)

1.3. What is the shortest lifetime a species could have and still be detectable with visible light having $\lambda = 500$ nm?

2

MAGNETIC PROPERTIES OF NUCLEI

2.1. THE STRUCTURE OF AN ATOM

The compounds we examine by NMR are composed of molecules, which are themselves aggregates of atoms. Each atom has some number of negatively charged **electrons** whizzing around a tiny, dense bit of positively charged matter called the **nucleus**. The *size* of an atom is the volume of space that the electron cloud occupies. However, >99.9% of the *mass* of an atom is concentrated in its nucleus, though the nucleus occupies only one trillionth (10^{-12}) of the atom's volume. Even the nucleus can be further dissected into other fundamental particles, including **protons** and **neutrons**, not to mention a host of other subnuclear particles that help hold the nucleus together and give nuclear physicists something to wonder about.

2.1.1 The Composition of the Nucleus

It is the number of protons in an atom's nucleus (Z, the atomic number) that determines both the atom's identity and the charge on its nucleus. In the **periodic table** of the elements (Appendix 2) the atomic number of each element is shown to the right of its chemical symbol. Every nucleus with just one proton is a hydrogen nucleus, every nucleus with six protons is a carbon nucleus, and so on. Yet, if we carefully examine a large sample of hydrogen atoms, we find that not all their nuclei are identical. It is true that all have just one proton, but they differ in the number of neutrons. Most hydrogen atoms in nature (99.985%, to be exact) have no neutrons ($N = 0$), but a small fraction (0.015%) have one neutron ($N = 1$) in addition to the proton. These two forms are the naturally occurring stable **isotopes** of hydrogen, and they are given the symbols ^{1}H and ^{2}H, respectively. The leading superscript is the **mass number** (A) of the isotope, which is the integer sum of Z and N:

$$A = Z + N \qquad (2.1)$$

The isotope ^{2}H is usually referred to as deuterium (D), or heavy hydrogen, but most isotopes of other elements are identified simply by their mass number. The **atomic mass** listed for each element in the periodic table is a weighted average, the fractional abundance of each isotope times its exact mass, summed over all naturally occurring isotopes.

■ **EXAMPLE 2.1** Tritium, ^{3}H, is a radioactive (unstable) isotope of hydrogen. What is the composition of its nucleus?

□ *Solution:* Since the atom is an isotope of hydrogen, $Z = 1$. The mass number A is 3 and therefore, from Eq. (2.1), $N = 2$. Thus, the nucleus consists of one proton and two neutrons. □

■ **EXAMPLE 2.2** Natural chlorine ($Z = 17$) is composed of two isotopes, ^{35}Cl and ^{37}Cl. The atomic mass listed for chlorine in the periodic table is 35.5. (a) What is the composition of each nucleus? (b) What is the natural abundance of each isotope? (You may assume for the purposes of this question that the exact mass of each isotope is exactly equal to its mass number, though in general this is not the case.)

□ *Solution:* (a) Chlorine-35 has $Z = 17$ (17 protons), $A = 35$, and $N = 18$ (18 neutrons); ^{37}Cl has $Z = 17$, $A = 37$, and $N = 20$. (b) Since the atomic mass of 35.5 is a weighted average of a mixture of ^{35}Cl and ^{37}Cl, we can use a little algebra to calculate the fraction (f) of each isotope:

$$(f_{35} \cdot 35) + (f_{37} \cdot 37) = 35.5$$

and since only the two isotopes are present,

$$f_{35} + f_{37} = 1.00$$

Therefore,

$$(f_{35} \cdot 35) + (1.00 - f_{35})(37) = 35.5$$

$$35f_{35} - 37f_{35} = 35.5 - 37$$

$$f_{35} = 0.75(75\%) \quad \text{and} \quad f_{37} = 0.25(25\%) \qquad \square$$

2.1.2 Electron Spin

Before we delve further into the properties of the nucleus, let us momentarily shift our attention back to one of the electrons zooming around the nucleus. Just like photons, electrons exhibit both wave and particle properties. Each electron wave in an atom is characterized by four **quantum numbers**. The first three of these numbers can be taken as the electron's address and describe the energy, shape, and orientation of the volume the electron occupies in the atom. This volume is called an **orbital**. The fourth quantum number is the **electron spin quantum number** s, which can assume only two values, $+\frac{1}{2}$ or $-\frac{1}{2}$. (Why $\pm\frac{1}{2}$ was selected rather than, say, ±1 will be described a little later.) The **Pauli exclusion principle** tells us that no two electrons in an atom can have exactly the same set of four quantum numbers. Therefore, if two electrons occupy the *same* orbital (and thus possess the same first three quantum numbers), they must have *different* spin quantum numbers. Therefore, no orbital can possess more than two electrons, and then only if their spins are *paired* (opposite).

Is there any other significance to the spin quantum number? Yes, indeed! Because the electron can be regarded as a particle spinning on its axis, it has a property called **spin angular momentum**. Further, because the electron is a *charged* particle ($Z = -1$), this spinning gives rise to a *magnetic moment* (symbol μ) represented by the boldface vector arrows in Figure 2.1. We describe such a species as having a **magnetic dipole**. The two possible values of s correspond to the two possible orientations of the magnetic moment vector in an external magnetic field, "up" (in the same direction as the external field) or "down" (in the opposite direction to the external field). These two **spin states** are *degenerate* (i.e., have the same energy) in the absence of an external magnetic field. Moreover, if all the electrons in an atom are *paired* (i.e., each orbital contains two electrons), all up spins are canceled by down spins, so the atom as a whole has zero magnetic moment.

However, when *unpaired* electrons are immersed in an external magnetic field, the two states are no longer degenerate. An electron oriented *opposite* to the field ($s = -\frac{1}{2}$ in Figure 2.1) has lower energy (and greater stability) than an electron oriented *with* the field ($s = +\frac{1}{2}$). It is the interconversion of these two spin states that is centrally important to the technique known as **electron paramagnetic resonance spectroscopy** (Chapter 11). But for now, we return to the nucleus.

2.1.3 Nuclear Spin

The proton is a spinning charged ($Z = 1$) particle too, so it should not surprise us to learn that it also exhibits a magnetic moment. And as with the electron, its magnetic moment has only two possible orientations that are degenerate in the absence of an external magnetic field. To differentiate **nuclear spin states** from electronic spin states, we will adopt the convention of labeling nuclear spin states with the **nuclear spin quantum number** m. Thus, for a proton, m can assume values of only $+\frac{1}{2}$ or $-\frac{1}{2}$. We describe such a nucleus as having a **nuclear spin** (I) of $\frac{1}{2}$. Because nuclear charge is the opposite of electron charge, a nucleus whose magnetic moment is aligned with the magnetic field ($m = +\frac{1}{2}$) has the lower energy (Figure 2.2). (If an isotope has a negative μ, the lowest energy state is the one with the most negative m value.)

Perhaps surprisingly, neutrons also exhibit a magnetic moment and a nuclear spin of $I = \frac{1}{2}$, even though they are uncharged. Therefore, they too can adopt two different orientations in a magnetic field. But because the sign of μ for a neutron is negative (Table 2.1), the more stable orientation corresponds to $m = -\frac{1}{2}$.

So, we have established that ^{1}H nuclei (i.e., protons) exhibit two possible magnetic spin orientations. What about other isotopes? From Chapter 1 you might remember that

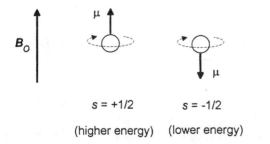

$$s = +1/2 \qquad s = -1/2$$
(higher energy)   (lower energy)

Figure 2.1. Two possible orientations of the magnetic moment (μ) of a spinning electron in an external magnetic field (**B$_0$**).

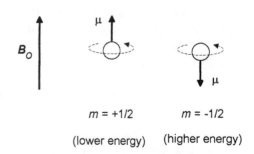

$$m = +1/2 \qquad m = -1/2$$
(lower energy)   (higher energy)

Figure 2.2. Two possible orientations of the magnetic moment (μ) of a spinning proton in an external magnetic field (**B$_0$**).

TABLE 2.1 Magnetic Properties of Selected Particles[a]

Isotope	%[b]	Z	N	A	I	γ[c]	ν[d]	μ[e]	Q[f]	S[g]
n	—	0	1	1	$\frac{1}{2}$	−183.26	29.167	−1.91315	0	0.322
^{1}H	99.985	1	0	1	$\frac{1}{2}$	267.512	42.5759	2.79268	0	1.00
^{2}H	0.015	1	1	2	1	41.0648	6.53566	0.857387	2.73×10^{-3}	9.65×10^{-3}
^{7}Li	92.58	3	4	7	$\frac{3}{2}$	103.96	16.546	3.2560	-3×10^{-2}	0.293
^{10}B	19.58	5	5	10	3	28.748	4.5754	1.8007	7.4×10^{-2}	1.99×10^{-2}
^{11}B	80.42	5	6	11	$\frac{3}{2}$	85.828	13.660	2.6880	3.55×10^{-2}	0.165
^{13}C	1.108	6	7	13	$\frac{1}{2}$	67.2640	10.7054	0.702199	0	1.59×10^{-2}
^{14}N	99.63	7	7	14	1	19.325	3.0756	0.40347	1.6×10^{-2}	1.01×10^{-3}
^{15}N	0.37	7	8	15	$\frac{1}{2}$	−27.107	4.3142	−0.28298	0	1.04×10^{-3}
^{17}O	0.037	8	9	17	$\frac{5}{2}$	−36.27	5.772	−1.8930	-2.6×10^{-2}	2.91×10^{-2}
^{19}F	100	9	10	19	$\frac{1}{2}$	251.667	40.0541	2.62727	0	0.833
^{23}Na	100	11	12	23	$\frac{3}{2}$	70.761	11.262	2.2161	0.14	9.25×10^{-2}
^{27}Al	100	13	14	27	$\frac{5}{2}$	69.706	11.094	3.6385	0.149	0.206
^{29}Si	4.70	14	15	29	$\frac{1}{2}$	−53.142	8.4578	−0.55477	0	7.84×10^{-3}
^{31}P	100	15	16	31	$\frac{1}{2}$	108.29	17.235	1.1305	0	6.63×10^{-2}
^{33}S	0.76	16	17	33	$\frac{3}{2}$	20.517	3.2654	0.64257	-6.4×10^{-2}	2.26×10^{-3}
^{35}Cl	75.53	17	18	35	$\frac{3}{2}$	26.212	4.1717	0.82091	-7.89×10^{-2}	4.70×10^{-3}
^{37}Cl	24.47	17	20	37	$\frac{3}{2}$	21.82	3.472	0.6833	-6.21×10^{-2}	2.71×10^{-3}

[a]Abstracted in part from *The 64th CRC Handbook of Chemistry and Physics*, CRC Press, Boca Raton, FL, 1984.
[b]Natural abundance.
[c]Magnetogyric ratio in units of 10^6 rad T^{-1} s^{-1}.
[d]Resonance frequency in megahertz in a 1-T field.
[e]Magnetic moment in nuclear magnetons.
[f]Electric quadrupole moment in barns.
[g]Sensitivity (relative to proton) for equal numbers of nuclei at constant field; $S = 7.652 \times 10^{-3} \mu^3 (I+1)/I^2$.

Zeeman found only certain isotopes give rise to multiple nuclear spin states when immersed in an external magnetic field. This is because *only isotopes with an odd number of protons (odd Z) and/or an odd number of neutrons (odd N) possess nonzero nuclear spin.* Nuclei with zero nuclear spin (those with an even Z and even N) have zero nuclear magnetic moment and *cannot be detected by NMR methods.*

Here is the reason that the *parity* (odd or even number) of protons and neutrons is so important: A proton spin can only pair with (cancel) another proton spin, but not a neutron spin, and vice versa. This rule allows us to assign every isotope to one of three groups.

Group 1: Nuclei with Both Z and N even (and Therefore A Even).

In such nuclei all proton spins are paired and all neutron spins are paired, resulting in a net nuclear spin of zero ($I = 0$). Such nuclei are invisible to NMR. Some examples include the abundant isotopes ^{12}C, ^{16}O, ^{18}O, and ^{32}S.

Group 2: Nuclei with Both Z and N Odd (and Therefore A Even).

Such nuclei must have an odd number of unpaired proton ($I = \frac{1}{2}$) spins, and an odd number of unpaired neutron ($I = \frac{1}{2}$) spins, so the net magnetic spin must be a nonzero integer [i.e., an integer multiple of $2(\frac{1}{2})$]. Such nuclei are detectable by NMR. A few common examples are ^{2}H ($I = 1$), ^{10}B ($I = 3$), ^{14}N ($I = 1$), and ^{50}V ($I = 6$).

Group 3: Nuclei with Even Z and Odd N, or Odd Z and Even N (In Either Case Odd A).

These nuclei must have an even number of proton spins (all paired) and an odd number of unpaired neutron spins, or vice versa. Therefore, the net magnetic spin is an odd integer multiple of $\frac{1}{2}$, and these nuclei can be detected by NMR. Here are some examples: ^{1}H ($I = \frac{1}{2}$), ^{11}B ($I = \frac{3}{2}$), ^{13}C ($I = \frac{1}{2}$), ^{15}N ($I = \frac{1}{2}$), ^{17}O ($I = \frac{5}{2}$), ^{19}F ($I = \frac{1}{2}$), ^{29}Si ($I = \frac{1}{2}$), ^{31}P ($I = \frac{1}{2}$).

Finally, remember that different isotopes of the same element can have different nuclear spins, some of which *are* detectable by NMR, others of which are not.

■ **EXAMPLE 2.3** Predict the nuclear spin I of ^{4}He, ^{6}Li, and ^{7}Li, and indicate which are detectable by NMR.

☐ *Solution:* Assign each nucleus to one of the three groups described above.

Nucleus	A	Z	N	Group	Predicted I	Detectable?
^{4}He	4	2	2	1	0	No
^{6}Li	6	3	3	2	Integer	Yes
^{7}Li	7	3	4	3	n/2 (*n* odd)	Yes

The actual values of I are $I = 0$ for ^{4}He, $I = 1$ for ^{6}Li, and $I = \frac{3}{2}$ for ^{7}Li. ☐

Just when we begin to understand that a nuclear spin number of $\frac{1}{2}$ gives rise to two spin orientations in a magnetic field, we are confronted with nuclear spin values of $\frac{3}{2}$, $\frac{5}{2}$, and even 6. What is the significance of this? The explanation is actually quite straightforward. Although *single* atomic particles such as protons, neutrons, and electrons can adopt only two magnetic spin orientations, complex nuclei can adopt more than two. In fact, the total number (**multiplicity**) of possible spin states (i.e., the different values of m) is determined solely by the value of I:

$$\text{Multiplicity} = 2I + 1 \qquad (2.2)$$

Each of these $2I + 1$ states has its own spin quantum number m in the range $m = -I, -I + 1, \ldots, I - 1, I$ (listed in the order of decreasing energy and increasing stability). Thus, for nuclei with $I = \frac{1}{2}$, the multiplicity is 2, and these two states are $m = +\frac{1}{2}$ and $m = -\frac{1}{2}$.

■ **EXAMPLE 2.4** Calculate the spin state multiplicity for each of the nuclei below, and list the value of m for each state from highest to lowest energy:

$$^{11}\text{B}, \,^{12}\text{C}, \,^{14}\text{N}, \,^{17}\text{O}, \,^{31}\text{P}$$

☐ *Solution:* The I values for these nuclei appear among the above list of three groups. Using Eq. (2.2), we first calculate the multiplicity of spin states and then their m values.

Nucleus	I	Multiplicity	m Values
^{11}B	$\frac{3}{2}$	4	$-\frac{3}{2}, -\frac{1}{2}, \frac{1}{2}, \frac{3}{2}$
^{12}C	0	1	0
^{14}N	1	3	$-1, 0, 1$
^{17}O	$\frac{5}{2}$	6	$-\frac{5}{2}, -\frac{3}{2}, -\frac{1}{2}, \frac{1}{2}, \frac{3}{2}, \frac{5}{2}$
^{31}P	$\frac{1}{2}$	2	$-\frac{1}{2}, \frac{1}{2}$

☐

All nuclei with $I = \frac{1}{2}$ have a spherical (symmetrical) distribution of spinning charge, so the electric and magnetic fields surrounding such nuclei are spherical, homogeneous, and isotropic in all directions. By contrast, nuclei with $I > \frac{1}{2}$ have a nonspherical distribution of spinning charge, resulting in nonsymmetrical electric and magnetic fields. This imparts an **electric quadrupole** (Q) to the nucleus, a property that can complicate their NMR behavior. As a result, the most commonly studied nuclei are those with a nuclear spin of $\frac{1}{2}$.

2.2 THE NUCLEUS IN A MAGNETIC FIELD

2.2.1 More about the Nuclear Zeeman Effect

As we have said, a nucleus with nuclear spin I adopts $2I + 1$ nondegenerate spin orientations in a magnetic field. The states separate in energy, with the largest positive m value corresponding to the lowest energy (most stable) state. It is this separation of states in a magnetic field that is the essence of the nuclear Zeeman effect.

The energy of the ith spin state (E_i) is directly proportional to the value of m_i and the magnetic field strength $\mathbf{B}_0$ (that is, energy is *quantized* in units of $\gamma h\mathbf{B}_0/2\pi$)

$$E_i = -m_i \frac{\gamma\, h\, \mathbf{B}_0}{2\pi} \qquad (2.3)$$

In this equation h (Planck's constant) and π have their usual meanings, while γ is called the **magnetogyric ratio**, a proportionality constant characteristic of the isotope being examined (more on this a little later). The minus sign in the equation follows from the convention of making a positive m correspond to a lower (negative) energy. Figure 2.3 graphically depicts the variation of spin state energy as a function of magnetic field strength for two different nuclei, one with $I = \frac{1}{2}$, the other with $I = 1$. Notice that as field strength increases, the *difference* in energy (ΔE) between any two spin states also increases proportionally. For a nucleus with $I = \frac{1}{2}$, this difference is

$$\Delta E = E_{(m = -1/2)} - E_{(m = 1/2)}$$

higher   lower

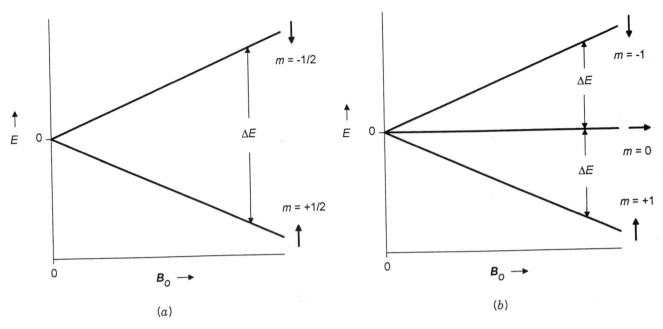

Figure 2.3. Nuclear Zeeman effect. (*a*) A nucleus with $I = \frac{1}{2}$. (*b*) A nucleus with $I = 1$. The arrow beside each spin state line indicates the orientation of the magnetic moment in a vertical magnetic field.

$$= -\left[\left(-\frac{1}{2}\right) - \left(\frac{1}{2}\right)\right]\frac{\gamma\,h\,\mathbf{B}_0}{2\pi}$$

$$= \frac{\gamma\,h\,\mathbf{B}_0}{2\pi} \tag{2.4}$$

And now you realize why values of $\pm\frac{1}{2}$ were picked for m (and s too, for that matter). It is so that the difference in energy between two neighboring spin states will always be an integer multiple of $\gamma\,h\,\mathbf{B}_0/2\pi$.

The magnetogyric ratio γ describes how much the spin state energies of a given nucleus vary with changes in the external magnetic field. Each isotope with nonzero nuclear spin has its own unique value of γ, though the magnitude of γ depends on the units selected for $\mathbf{B}_0$. We will use the unit tesla (T) for magnetic field strength so that γ has units of radians per tesla per second (2π radians in one cycle of 360°). As we will see in Chapter 3, modern commercial NMR spectrometers are equipped with magnets that generate fields ranging from ca. 5 to 16 T. For comparison, the earth's magnetic field is a mere 6×10^{-5} T. (*Note:* Earlier books on NMR used the gauss or kilogauss for magnetic field strength; $1\,\text{T} = 10^4\,\text{G} = 10\,\text{kG}$.)

In Table 2.1 are listed many of the common isotopes examined by NMR techniques, together with their nuclear constants. Notice that a bare proton has the largest γ value of any nuclear particle, while heavier nuclei, surrounded by many subvalence electrons, tend toward lower values. This will become significant later. Relative sensitivity is the strength of the NMR signal that is generated by a fixed number of nuclei of a given isotope relative to the signal obtained from an equal number of ^{1}H nuclei. If we compare the data on natural abundance and sensitivity, we see why historically the most easily studied nuclei were ^{1}H, ^{19}F, and ^{31}P. Indeed, prior to the 1970s these three were the only nuclei routinely studied with commercially available instrumentation. More recently, however, instruments have become available that can routinely examine a wide variety of other ubiquitous elements, including ^{13}C (which is of immense importance to organic chemists), ^{15}N, ^{23}Na, and ^{29}Si, to mention just a few.

■ **EXAMPLE 2.5** (a) What is the energy difference between the two spin states of ^{1}H in a magnetic field of 5.87 T? (b) Of ^{13}C?

□ *Solution:* (a) Use Eq. (2.4) and the γ values in Table 2.1.

$$\Delta E = \frac{\gamma h \mathbf{B}_0}{2\pi}$$

$$= \frac{(267.512 \times 10^6 \text{ rad T}^{-1}\text{s}^{-1})(6.63 \times 10^{-34} \text{ J s})(5.87 \text{ T})}{2(3.14 \text{ rad})}$$

$$= 1.66 \times 10^{-25} \text{ J}$$

(b) For ^{13}C, $\gamma = 67.2640 \times 10^6$, so $\Delta E = 4.18 \times 10^{-26}$ J, about one-fourth the difference for ^{1}H. ☐

2.2.2 Precession and the Larmor Frequency

We now know that nuclei with $I \neq 0$, when immersed in a magnetic field, adopt $2I + 1$ spin orientations, each with a different energy. But before these nuclei can absorb photons, they must be oscillating in some sort of uniform periodic motion (Section 1.3). Fortunately, quantum mechanics requires that the magnetic moments are actually *not* statically aligned exactly parallel or antiparallel to the external magnetic field, as Figure 2.2 implied. Instead, they are forced to remain at a certain angle to $\mathbf{B}_0$, and this causes them to "wobble" around the axis of the field at a fixed frequency. Why is this so?

If you have ever played with a spinning top, you may know that it is the spin angular momentum of the top that prevents it from falling over and also causes it to wobble in addition to spinning. This periodic wobbling motion that the top assumes in a *gravitational* field is called **precession**. The earth precesses on its axis in much the same way, though much more slowly. In an exactly analogous way, the magnetic moment vector of a nucleus in a magnetic field also precesses with a characteristic *angular* frequency called the **Larmor frequency** (ω), which is a function solely of γ and $\mathbf{B}_0$:

$$\omega = \gamma \, \mathbf{B}_0 \qquad (2.5)$$

The angular Larmor frequency, in units of radians per second, can be transformed into linear frequency ν (in reciprocal seconds or hertz) by division by 2π:

$$\nu_{\text{precession}} = \frac{\omega}{2\pi} = \frac{\gamma \mathbf{B}_0}{2\pi} \qquad (2.6)$$

This precessional motion causes the tip of the magnetic moment vectors (either up or down) to trace out a circular path, as shown in Figure 2.4. Note also that the precession fre-

quency is *independent* of m, so that all spin orientations of a given nucleus precess at the same frequency in a fixed magnetic field.

■ **EXAMPLE 2.6** (a) At 5.87 T, what is the precession frequency ν of a ^{1}H nucleus? A ^{13}C nucleus? (b) In what region of the electromagnetic spectrum does radiation of these frequencies occur?

☐ *Solution:* (a) For ^{1}H, using the γ value from Table 2.1, we find

$$\nu = \frac{\gamma \mathbf{B}_0}{2\pi} = \frac{(267.512 \times 10^6 \, \mathrm{rad\,T^{-1}s^{-1}})(5.78 \, \mathrm{T})}{2(3.14 \, \mathrm{rad})}$$

$$= 2.50 \times 10^8 \, \mathrm{s^{-1}} = 250 \, \mathrm{MHz}$$

Similarly, for ^{13}C, $\nu = 62.9$ MHz.

After you have completed these calculations the "long" way, using $\mathbf{B}_0$ and γ values from Table 2.1, try this easier way. In Table 2.1, the data in the column labeled ν are the precession frequencies (in megahertz) of each nucleus in a 1.00-T magnetic field. Simply multiplying these numbers by the actual field strength (in tesla), directly gives the value of ν at any other field strength. Thus, for ^{1}H,

$$\nu = (42.5759 \, \mathrm{MHz\,T^{-1}})(5.87 \, \mathrm{T}) = 250 \, \mathrm{MHz}$$

(b) From Table 1.1 we note that these frequencies fall in the rf region. ☐

■ **EXAMPLE 2.7** At what *magnetic field strength* do protons (^{1}H nuclei) precess at a frequency of 300 MHz?

☐ *Solution:* Rearranging Eq. (2.6), we find

$$\mathbf{B}_0 = \frac{2\pi\nu}{\gamma}$$

$$= \frac{2(3.14 \, \mathrm{rad})(300 \times 10^6 \, \mathrm{s^{-1}})}{267.512 \times 10^6 \, \mathrm{rad\,T^{-1}\,s^{-1}}}$$

$$= 7.04 \, \mathrm{T} \qquad \qquad ☐$$

We are almost ready to perform our NMR experiment. We have immersed the collection of nuclei in a magnetic field; each nucleus is precessing with a characteristic frequency. To observe resonance, all we have to do is irradiate them with electromagnetic radiation of the appropriate frequency, right? Well, almost!

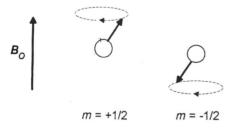

Figure 2.4. Precession of the magnetic moment in each of the two possible spin states of an $I = \frac{1}{2}$ nucleus in external magnetic field $\mathbf{B}_0$.

2.3   NUCLEAR ENERGY LEVELS AND RELAXATION TIMES

2.3.1   Boltzmann Distribution and Saturation

In Chapter 1 we hinted that once a particle absorbs a photon, the energy originally associated with the electromagnetic radiation appears somehow in the particle's motion. Where does the energy go in the case of precessing ^{1}H nuclei? Because there are only two spin states possible, the energy goes into a spin *flip*. That is, the photon's energy is absorbed by a nucleus in the lower energy spin state ($m = +\frac{1}{2}$), and the nucleus is flipped into its higher energy spin state ($m = -\frac{1}{2}$). This situation is depicted in Figure 2.5. And remember that this spin flip does *not* change the precessional frequency of the nucleus.

We have already calculated the energy gap between these two spin states [Eq. (2.4)], and this must equal the energy of the absorbed photon [Eq. (1.3)]. Combining these with Eq. (2.6) gives us

$$\Delta E = \frac{\gamma h \mathbf{B}_0}{2\pi} = h\nu_{\text{precession}} = E_{\text{photon}} = h\nu_{\text{photon}} \tag{2.7}$$

Thus, as we expected from Chapter 1, for resonance to occur, the radiation frequency must exactly match the precessional frequency.

But there is a fly in the ointment. Quantum mechanics tells us that, for net absorption of radiation to occur, there must be more particles in the lower energy state than in the higher one. If the two populations happen to be equal, Einstein predicted theoretically that transition from the upper ($m = -\frac{1}{2}$) state to the lower ($m = +\frac{1}{2}$) state (a process called **stimulated emis-**

sion) is exactly as likely to occur as absorption. In such a case, no *net* absorption is possible, a condition called **saturation**.

Is there any reason to expect that there will be an excess of nuclei in the lower spin state? The answer is a qualified yes. For any system of energy levels *at thermal equilibrium*, there will always be more particles in the lower state(s) than in the upper state(s). However, there will always be *some* particles in the upper state(s). What we really need is an equation relating the energy gap (ΔE) between the states to the relative populations of (numbers of particles in) each of those states. This time, quantum mechanics comes to our rescue in the form of the **Boltzmann distribution**:

$$\frac{P_{(m=-1/2)}}{P_{(m=1/2)}} = e^{-\Delta E/kT} = \exp\left(\frac{-\Delta E}{kT}\right) \tag{2.8}$$

where P is the population (or fraction of the particles) in each state, T is the absolute temperature in Kelvin (not to be confused with nonitalicized T for Tesla), and k (the Boltzmann constant) has a value of 1.381×10^{-23} J K^{-1}.

■ **EXAMPLE 2.8**   At 25°C (298 K) what fraction of ^{1}H nuclei in 5.87 T field are in the upper and lower states? See Example 2.5 for the value of ΔE.

□ *Solution*: Use Eq. (2.8) and the results of Example 2.5:

$$\frac{P_{(m=-1/2)}}{P_{(m=+1/2)}} = \exp\left(-\frac{\Delta E}{kT}\right)$$

$$= \exp\left(-\frac{1.66 \times 10^{-25}\ \text{J}}{1.381 \times 10^{-23}\ \text{J K}^{-1} \times 298\ \text{K}}\right) = 0.99996$$

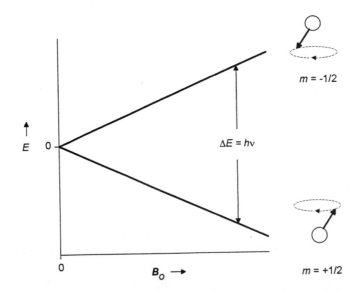

Figure 2.5.   Relative energy of both spin states of an $I = \frac{1}{2}$ nucleus as a function of the strength of the external magnetic field $\mathbf{B}_0$.

Since there are only two spin states, $P_{(m=-1/2)} = 1 - P_{(m=+1/2)}$, so $P_{(m=-1/2)} = 0.49999$ and $P_{(m=+1/2)} = 0.50001$ □

As you can see from the above example, the difference in populations of the two ^{1}H spin states is exceedingly small, on the order of 20 ppm. And the difference for other elements is even smaller because of their smaller γ values. But do not despair. This difference is sufficient to allow an NMR signal to be detected. It is this small difference, however, that accounts in part for the relatively low sensitivity of NMR spectroscopy compared to other absorption techniques such as infrared and ultraviolet spectroscopy. Remember, factors such as a stronger magnetic field (**B**$_0$), a larger magnetogyric ratio, or a lower temperature all contribute to a larger population difference, reduce the likelihood of saturation, and lead to a more intense NMR signal.

2.3.2 Relaxation Processes

Our NMR theory is almost complete, but there is one more thing to consider before we set about designing a spectrometer. We indicated previously that at equilibrium in the *absence* of an external magnetic field, all nuclear spin states are degenerate and, therefore, of equal probability and population. Then, when immersed in a magnetic field, the spin states establish a new (Boltzmann) equilibrium distribution with a slight excess of nuclei in the lower energy state.

A relevant question is this: How long after immersion in the external field does it take for a collection of nuclei to reequilibrate? This process is *not* infinitely fast. In fact, the rate at which the new equilibrium is established is governed by a quantity called the **spin–lattice** (or **longitudinal**) **relaxation time**, T_1. The exact relation involves *exponential decay*:

$$\frac{P_{eq} - P_t}{P_{eq} - P_0} = \exp\left(-\frac{t}{T_1}\right) \qquad (2.9)$$

where $P_{eq} - P_t$ is the difference between the equilibrium population of a given state (for example, the $m = +\frac{1}{2}$ state) and the population after time t and the subscript zero refers to $t = 0$. (P_{eq} is thus the population at $t = \infty$.)

■ **EXAMPLE 2.9**   (a) Suppose that for a certain set of ^{1}H nuclei at 25°C, the value of T_1 is 0.20 s. How long after immersion in a 5.87-T magnetic field will it take for an initially equal distribution of ^{1}H spin states to progress 95% of the way toward equilibrium? (b) What would happen if the magnet were turned off at this point?

□ *Solution:*   (a) From Example 2.8, we know that the final equilibrium population of the $m = +\frac{1}{2}$ state will be

0.50001. At 95% of equilibrium $P_{eq} - P_t = 0.05(P_{eq} - P_0)$. Use Eq. (2.9) and solve for t:

$$\frac{P_{eq} - P_t}{P_{eq} - P_0} = \frac{0.05(P_{eq} - P_0)}{P_{eq} - P_0} = \exp\left(-\frac{t}{T_1}\right)$$

$$0.05 = \exp\left(-\frac{t}{T_1}\right)$$

Taking the natural logarithm (ln) of both sides of the equation, we find

$$-3.00 = -\frac{t}{T_1} \quad \text{or} \quad t = 3.00 T_1$$

Since $T_1 = 0.20$ s, $t = 3(0.20 \text{ s}) = 0.60$ s.

(b) When the field strength returns to zero, the collection of nuclei will decay exponentially toward the original equal populations of spin states at a rate still governed by the same value of T_1. □

Figure 2.6 graphically depicts the situation in Example 2.9. The arrows in the three diagrams below the graph represent the distribution of individual precessing ^{1}H magnetic moments, either up or down. Initially there are equal numbers of nuclei in each spin state. But at equilibrium in the magnetic field, there is a 20 ppm excess of up spins (exaggerated in the middle diagram). When the magnetic field is turned off, the collection decays back to the original equal distribution.

The values of T_1 range broadly, depending on the particular type of nucleus, the location of the nucleus (atom) within a molecule, the size of the molecule, the physical state of the sample (solid or liquid), and the temperature. For liquids or solutions, values of 10^{-2}–10^2 s are typical, though some quadrupolar nuclei have (faster) relaxation times of the order of 10^{-4} s. For crystalline solids, T_1 values are much longer (Section 2.5). For now, just remember that the larger the value of T_1, the longer it takes for a collection of nuclei to reach (or return to) equilibrium.

There is another reason why the magnitude of T_1 is important. Suppose we have a Boltzmann distribution of nuclei precessing in a magnetic field, and we irradiate the collection with photons of precisely the correct frequency (and energy) to cause transitions (spin flips) between the lower ($m = +\frac{1}{2}$) level and the upper ($m = -\frac{1}{2}$) states. Because there is initially such a small difference between the populations of the two states, it will not be long before the populations are equalized through the absorption of the photons! This, of course, means the spin system has become saturated and no further net absorption is possible. However, if we turn *off* the source of rf radiation, the system can relax back to the Boltzmann distribution (at a rate controlled by T_1) and absorption can

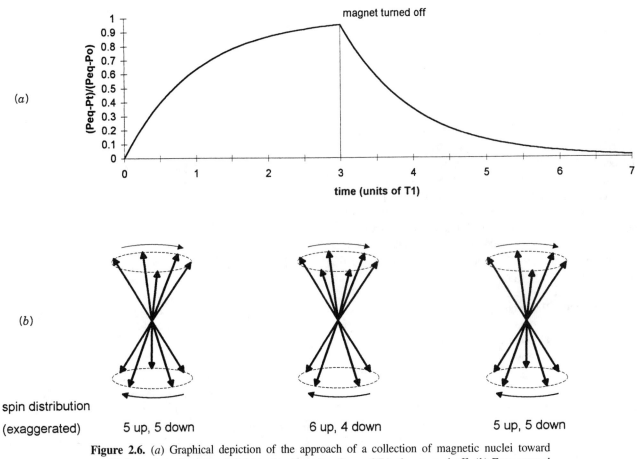

(a)

(b)

spin distribution

(exaggerated)

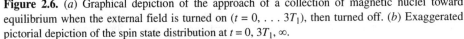

Figure 2.6. (*a*) Graphical depiction of the approach of a collection of magnetic nuclei toward equilibrium when the external field is turned on ($t = 0, \ldots 3T_1$), then turned off. (*b*) Exaggerated pictorial depiction of the spin state distribution at $t = 0, 3T_1, \infty$.

resume. This fact presents us with a paradox. The spin system must absorb enough photons for us to be able to detect the signal instrumentally, but not so much as to cause saturation.

This brings us to the question (to be covered more fully in Section 3.1) of how an NMR signal is actually generated. Figure 2.7 depicts a collection of $I = \frac{1}{2}$ nuclei at equilibrium, in a magnetic field aligned with the +z axis. (By convention, the static magnetic field, $\mathbf{B}_0$, always defines the *z* axis.) Before irradiation begins (Figure 2.7*a*), the nuclei in both spin states are precessing with the characteristic frequency, but they are completely *out of phase,* that is, randomly oriented around the *z* axis. The *net* nuclear magnetization **M** is the vector sum of all the individual nuclear magnetic moments, and its magnitude is determined by the excess of up spins over down spins. Here, **M** is aligned parallel to $\mathbf{B}_0$; it has no precessional motion and no component in the *x,y* plane.

Now suppose there were a way to produce another magnetic field *perpendicular* to $\mathbf{B}_0$. However, this new field ($\mathbf{B}_1$) will be much weaker than $\mathbf{B}_0$, and it will *precess* in the *x,y* plane, oscillating at exactly the same frequency as the nuclear magnetic moments (Figure 2.7*b*). (In Section 3.1 we will describe how to generate such an oscillating magnetic field, but it should not surprise you to learn that it involves electro-*magnetic* radiation of the same frequency.) At any rate, a rather strange thing happens when this irradiation by $\mathbf{B}_1$ begins: All of the individual nuclear magnetic moments become *phase coherent.* That is, they *focus,* tracking the oscillating magnetic field, and form a precessing "bundle" as shown in Figure 2.7*c*. Provided we have not saturated the system by absorbing too may photons, this phase coherence also requires that **M** tip away from the *z* axis and begin to precess around the *z* axis, again with the characteristic Larmor frequency. As such, **M** now has a component in the *x,y* plane ($\mathbf{M}_{x,y}$) oscillating with the same frequency. The flip angle α that **M** makes with the *z* axis controls the magnitude of $\mathbf{M}_{x,y}$ by the relation

$$\mathbf{M}_{x,y} = \mathbf{M} \sin \alpha \qquad (2.10)$$

The angle α is, in turn, determined by the power and duration of the irradiation by $\mathbf{B}_1$. Ultimately, the actual NMR signal

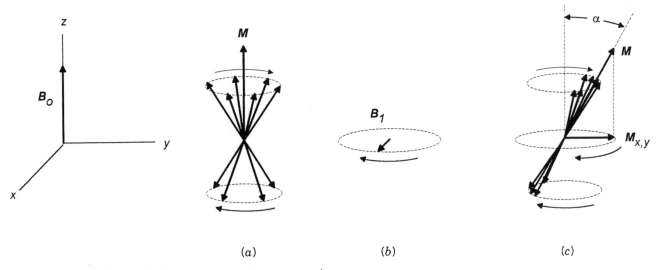

Figure 2.7. Precession of a collection of $I = \frac{1}{2}$ nuclei around external magnetic field $\mathbf{B}_0$. Here, **M** represents the net nuclear magnetization, the vector sum of all the individual nuclear magnetic moments. (*a*) Before irradiation by $\mathbf{B}_1$. (*b*)The orientation of the rotating magnetic field $\mathbf{B}_1$. (*c*) During irradiation by $\mathbf{B}_1$.

will be generated from the oscillation of $\mathbf{M}_{x,y}$ (Section 3.1), so the maximum signal intensity will occur when α equals 90° (Why?).

There is one other type of relaxation process that must be mentioned at this point. After irradiation ceases and $\mathbf{B}_1$ disappears, not only do the populations of the $m = +\frac{1}{2}$ and $m = -\frac{1}{2}$ states revert to the Boltzmann distribution, but also the individual nuclear magnetic moments begin to lose their phase coherence and return to a random arrangement around the z axis (Figure 2.7*a*). This latter process, called **spin–spin** (or **transverse**) **relaxation**, causes decay of $\mathbf{M}_{x,y}$ at a rate controlled by the **spin–spin relaxation time** T_2. Normally, T_2 is much shorter than T_1. A little thought should convince you that if $T_2 < T_1$, then spin–spin (dephasing) relaxation takes place much faster than spin–lattice (Boltzmann distribution) relaxation.

2.4   THE ROTATING FRAME OF REFERENCE

Frequently in this book we will want to depict nuclear spin orientations like those shown in Figure 2.7. More often than not, we will focus our attention on the net nuclear magnetic moment (**M**) rather than on the individual nuclear spins. Because **M** will sometimes precess around $\mathbf{B}_0$ (i.e., the z axis), we need a more convenient way than the dashed ellipses used so far to depict **M** as it precesses and changes orientation. Henceforth we will use another convention to represent this precessional motion of **M**, the **rotating frame of reference**, which is designed to show the effects of $\mathbf{B}_1$ on **M**.

In Figure 2.8 are shown four representations of **M** in the "old" way, the so-called **laboratory frame of reference**, the normal x,y,z coordinate system as viewed by a stationary observer in the lab. In part (a) of the figure $\mathbf{B}_0$ and $\mathbf{B}_1$ are off, so the populations of individual up and down nuclear spins are equal, and the magnitude of **M** is zero. In (b) the equilibrium distribution of spins with $\mathbf{B}_0$ on has been achieved and **M** is aligned along the z axis, even though the *individual* nuclear magnetic moments still precess around the z axis. Part (c) shows **M** tipped to an α of 45° through its interaction with $\mathbf{B}_1$, and the resulting precession of **M** describes a cone. Finally in (d) the flip angle is 90° so the precession of **M** describes a disc in the x,y plane.

Instead, suppose the x and y axes were themselves precessing clockwise (when viewed from above) around the z axis at the same frequency the nuclear spins are precessing. Further suppose we, the observers, were precessing around the z axis at the same frequency. To differentiate this rotating coordinate system from the fixed (i.e., laboratory frame) system, we will use labels x',y', and z' to represent the three rotating axes (the z' axis is coincident on and equivalent to the z axis). To us rotating observers, the rotating axes and $\mathbf{B}_1$ appear stationary, and **M** will rotate in the plane perpendicular to $\mathbf{B}_1$. These relationships are shown in Figure 2.9.

Clearly, viewing the motion of **M** in the rotating frame simplifies our drawings. Still, it is important to remember this about rotating-frame diagrams: Whenever **M** is anywhere except directly along the z' axis, it has a component oscillating in the x,y plane (laboratory frame), and this is what gives rise to an NMR signal.

■ **EXAMPLE 2.10**   (a) Look back at Figure 2.6. Draw a laboratory frame diagram that shows **M** at $t = 0, T_1, 2T_1, \ldots, 6T_1$. (b) How would the diagram change in the rotating frame?

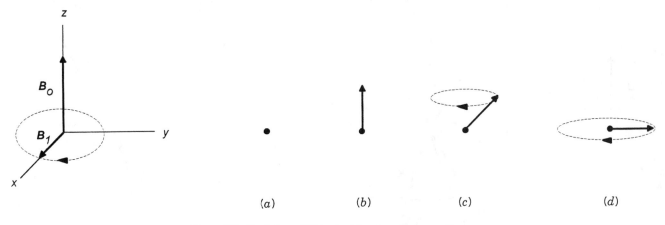

Figure 2.8. Depiction of **M** in the laboratory frame; see text.

□ *Solution:* (a) See Figure 2.10. (b) Since **M** is at all time aligned with the *z* axis, the diagram would look exactly the same in the rotating frame.   □

■ **EXAMPLE 2.11** Look back at Figure 2.7. Draw a rotating frame diagram that shows how **M** changes orientation when **B**₁ is turned on long enough to give an α of 90°, then turned off at $t = 0$. Show the orientation of **M** initially, then at α = 90°, then at t values of T_2, $2T_2$, $3T_2$, and ∞. Note that **B**₁ is positioned along the *x'* axis. (You may neglect the effects of spin flips and longitudinal relaxation for the purposes of this example.)

□ *Solution:* See Figure 2.11. Note that **M** remains in the *y',z'* plane.   □

   The result shown in Example 2.11 is very important. The rotating magnetic field **B**₁ (depicted initially in Figure 2.7*b*) now appears stationary (here, along the *x'* axis) in the rotating frame. The significance of this fact is shown in Figure 2.12. While the external field **B**₀ tends to keep **M** aligned with the *z* (*z'*) axis, **B**₁ causes **M** to precess around the *x'* axis, tilting **M** in the *y',z'* plane. (The overall motion of **M** in the laboratory frame would describe a complex spiral, but the component of precession around **B**₀ does not show up in the rotating frame.) Since **B**₁ is much weaker than **B**₀, the precession of **M** around **B**₁ is much slower than its precession around **B**₀. The stronger **B**₁ is and the longer it is on, the more **M** will precess around it, increasing the flip angle α. When **B**₁ turned off, **M** relaxes exponentially back toward the *z'* axis at a rate governed by T_2.

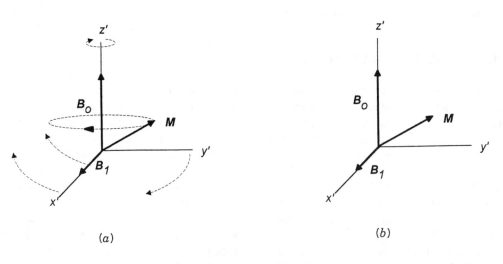

Figure 2.9. Axes of the rotating frame of reference, as viewed (*a*) by a stationary observer in the laboratory frame and (*b*) by an observer precessing in the rotating frame.

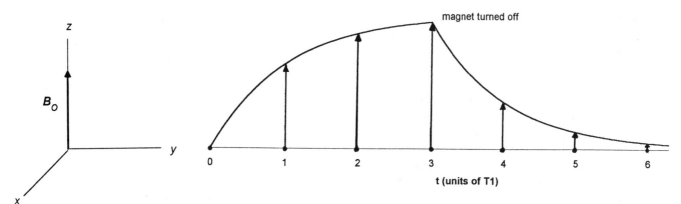

Figure 2.10. Pictorial solution to Example 2.10. Compare with Figure 2.6*a*.

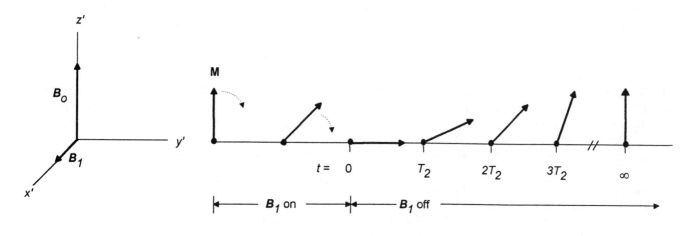

Figure 2.11. Pictorial solution to Example 2.11.

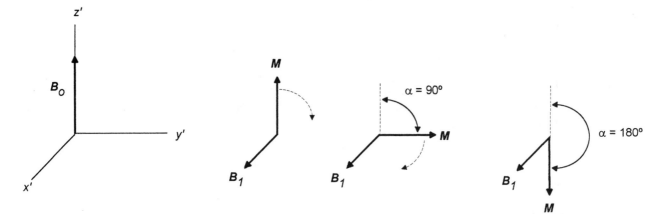

Figure 2.12. Precession of **M** around **B**$_1$ in the rotating frame.

2.5   RELAXATION MECHANISMS AND CORRELATION TIMES

The complete microscopic details of how longitudinal (spin–lattice, T_1) and transverse (spin–spin, T_2) relaxation occur is beyond the scope of this book. But a little further discussion might be profitable in order to provide us with at least a qualitative understanding of the subject.

Look again at Figure 2.7c. What causes the individual up and down nuclear spins in a "bundled" set of identical target (observed) nuclei to randomize their phasing (defocus) after B_1 is turned off? One mechanism for spin–spin relaxation can be pictured as follows. Suppose that one of the up spins and one of the down spins instantaneously exchange energy. In this way, the up spin is converted to a higher energy down spin, and vice versa, with no *net* change in energy. However, although the orientations (up or down) have exchanged, the exact phasing has not, as shown in Figure 2.13. Repetition of this process with other pairs of up/down spins will have the ultimate effect of randomizing their phasing, driving $\mathbf{M}_{x,y}$ to zero and with it the NMR signal.

Spin–spin relaxation can also occur when other nearby oscillating magnetic or electric fields interfere with the external field $\mathbf{B}_0$, causing some of the nuclei to experience a slightly augmented magnetic field while others experience a slightly diminished one. Those nuclei in the region of the augmented field will precess slightly faster, while those experiencing the diminished field will precess slightly slower (Figure 2.14). This will result in the "fanning out" of individual spin vectors again with no net energy change.

As we will see in Section 3.2, there is also a limit to the homogeneity of $\mathbf{B}_0$ itself. Even the finest magnets produce a field strength that varies ever so slightly around the region containing the sample. And this small range in field strength causes nuclei in one part of the sample to precess at very slightly different frequencies, again leading to dephasing of the nuclear spins once $\mathbf{B}_1$ is turned off. This mechanism for spin–spin relaxation is usually the dominant one and gives rise to an *effective* spin–spin relaxation time known as T_2^*, where $T_2^* < T_2$ (the "natural" spin–spin relaxation time). All three types of spin–spin relaxation are driven by the **second law of thermodynamics**: In the absence of other forces, a system will tend spontaneously to attain that arrangement with maximum *entropy* (disorder).

Thermodynamics also tells us that systems tend spontaneously toward equilibrium, which is characterized by a *minimum* energy (or **free energy**, to be exact). One component of free energy is the entropy mentioned above. Normally the dominant component of free energy is **enthalpy** (heat content). Within a magnetic field, the equilibrium (Boltzmann) distribution of nuclear spins is the one with minimum enthalpy (and maximum entropy). Any other distribution will have higher enthalpy (and free energy). For such a higher enthalpy distribution to relax back to equilibrium, it must dissipate its excess energy to the surroundings. In the context of NMR, these surroundings (the *lattice*) comprise other nearby nonidentical magnetic nuclei that can, but need not necessarily, be part of the same molecule as the nuclei of interest. The lattice can also be regarded as an infinite heat (energy) sink to or from which energy can be transferred without changing its temperature.

The most important mechanism for spin–lattice relaxation involves a direct (through space) interaction between the magnetic dipole of a target nucleus and that of lattice nuclei. Since lattice nuclei are undergoing constant periodic motion (e.g., rotation and translation), the local magnetic fields due to their magnetic moments will also be oscillating at the same frequencies. When the frequency of this motion is comparable to the frequency of precession of the target nucleus (e.g., 250 MHz), there can be a mutual spin flip. But since these nuclei are nonidentical, there will be a net change in energy accompanying the exchange. That is, energy will either be passed *to*

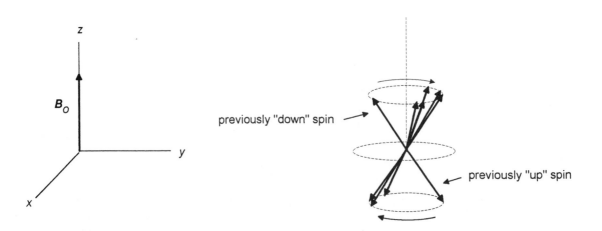

previously "down" spin

previously "up" spin

Figure 2.13. Result of one spin–spin exchange, shown in the laboratory frame. Compare with Figure 2.7c. Here, $\mathbf{M}$ is not shown.

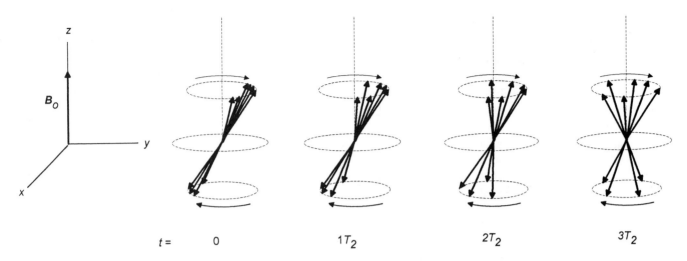

$t = \quad 0 \qquad\qquad 1T_2 \qquad\qquad 2T_2 \qquad\qquad 3T_2$

Figure 2.14. Laboratory frame diagram of "effective" spin-spin relaxation. Here, **M** is not shown.

the lattice nucleus (if the target nucleus drops to a lower energy level) or energy will be absorbed *from* the lattice nucleus (if the target nucleus is promoted to a higher energy level). These exchanges continue at a rate governed by T_1 until equilibrium is reestablished.

The above dipole–dipole mechanism for spin–lattice relaxation depends on the interaction of the target nucleus with the magnetic field $\mathbf{B}_l$ of a lattice nucleus with magnetic moment μ_l. The magnitude of $\mathbf{B}_l$ is governed by the equation

$$\mathbf{B}_l = \frac{\mu_l\,(3\cos^2\theta - 1)}{r^3} \qquad (2.11)$$

where θ is the angle between the external field $\mathbf{B}_0$ and a line of length r connecting the two nuclei. This equation shows that the effectiveness of spin–lattice relaxation (as measured by how short T_1 is) is increased by the lattice nucleus having a large μ and being as close to the target nucleus as possible to (i.e., in the same molecule or in high concentration in the bulk medium).

In addition to the direct interaction of magnetic dipoles, spin–lattice relaxation can proceed by way of interactions between the magnetic dipole of the target nucleus and fluctuating electric fields in the lattice. This is why neighboring quadrupolar nuclei (those with $I > \frac{1}{2}$, Section 2.1) can bring about very efficient spin–lattice relaxation (short T_1 values).

As mentioned above, the frequency of rotational or translational motion of magnetic and electric fields in the lattice nuclei is critical to the effectiveness of spin–lattice relaxation. It cannot be too fast or too slow. It is common to express the frequencies of these types of molecular motion in terms of a so-called **correlation time** τ_c. If the angular rotation frequency is ω (in radians per second), the rotational correlation time is $1/\omega$, the time required for a molecule (or part of a molecule) to rotate 1 rad. Similarly, the translational correla-

tion time can be equated to the time required for a molecule to move a distance equal to one molecular diameter. In both cases τ_c is an average measure of how long the two nuclear magnetic dipoles remain in the appropriate relative orientation to interact. Furthermore, it can be shown that

$$\frac{1}{T_1} \propto \frac{\tau_c}{1 + (2\pi\nu_0\tau_c)^2} \qquad (2.12)$$

where ν_0 is the precessional frequency of the target nucleus. This equations tells us that for very fast molecular motion (i.e., when $1/\tau_c \gg 2\pi\nu_0$), $1/T_1$ is proportional to τ_c (T_1 is inversely proportional to τ_c). That is, as correlation time *increases* (molecular motion slows), relaxation time *decreases* (the rate of relaxation increases). Conversely, for slow molecular motion (i.e., when $1/\tau_c \ll 2\pi\nu_0$), T_1 is directly proportional to τ_c; they both increase together. The minimum in T_1 (ca. 10^{-3} s), and hence the most efficient spin–lattice relaxation, occurs when $\tau_c = (2\pi\nu_0)^{-1}$.

As we will see later, the magnitudes of relaxation and correlation times are influenced by many factors, such as temperature, viscosity of the medium, and size of the molecules involved. For example, in crystalline solids where all translational and rotational motion has ceased, T_1 values are exceptionally large while T_2 values are exceptionally small. A major problem that results from inefficient spin–lattice relaxation is that the target nuclei are much more easily saturated (Section 2.3), making it difficult to obtain the desired NMR signal. And, as we will see in Section 3.5, highly efficient spin–spin relaxation gives rise to very broad signal peaks. On the other hand, by measuring relaxation times and correlation times, we are able to obtain detailed information about how the giant molecules (e.g., polymers and proteins) actually move (see Chapter 14).

CHAPTER SUMMARY

1. The nucleus of an atom consists of a number (Z) of protons and a number (N) of neutrons. The atomic number Z determines the identity of the nucleus, while the sum $Z + N$ determines the mass number (A) of the nucleus.

2. Isotopes of a given element have the same value of Z but different values of N and A.

3. Nuclear spin (I) is a property characteristic of each isotope and is a function of the parity of Z and N. The values of I can only be zero, n (an integer), or $n/2$ (where n is an odd integer). Only if $I \neq 0$ can the isotope be studied by NMR methods. The most frequently studied nuclei are those with $I = \frac{1}{2}$ (e.g., ^{1}H and ^{13}C).

4. Each isotope with $I \neq 0$ has a characteristic magnetogyric ratio (γ, Table 2.1) that determines the frequency of its precession in a magnetic field of strength $\mathbf{B}_0$ [Eq. (2.6)]. It is this frequency that must be matched by the incident electromagnetic radiation (actually, the oscillating magnetic field $\mathbf{B}_1$) for absorption to occur.

5. When a collection of nuclei with $I \neq 0$ is immersed in a strong magnetic field, the nuclei distribute themselves among $2I + 1$ spin states (orientations), each with its own value of magnetic spin quantum number m_i. The quantity $2I + 1$ is called the multiplicity of spin states. Nuclei in each spin state precess at the same frequency.

6. The energy of the ith nuclear spin state is given by Eq. (2.3).

7. The relative population of each spin state is determined by the Boltzmann distribution, Eq. (2.8). Under conditions of a typical NMR experiment the ratio of spin state populations is near unity, differing only by a few parts per million.

8. If the two (or more) spin state populations become equal, the system is said to be saturated and no net absorption can occur.

9. After $\mathbf{B}_1$ is turned off, nuclei can change their nuclear spin orientations through two types of relaxation processes. Spin–lattice (longitudinal) relaxation (governed by relaxation time T_1) involves the return of the nuclei to a Boltzmann distribution. Spin–spin (transverse) relaxation (governed by relaxation time T_2 or T_2^*) involves the dephasing of the bundled nuclear spins. Normally $T_2^* < T_2 < T_1$.

10. The rotating frame (of reference) is a Cartesian coordinate system where the x and y axes (designated x' and y') rotate around the z (z') axis at the precessional frequency of the target nuclei. The rotating frame is drawn as it would appear to an observer precessing at the same frequency. In the rotating frame $\mathbf{B}_1$ is statically aligned along (for example) the x' axis, and net nuclear magnetization $\mathbf{M}$ lies in the x', z' plane. No precession around $\mathbf{B}_0$ appears in the rotating frame.

11. Spin–spin relaxation can be accomplished either by mutual energy exchange between two target nuclei or by inhomogeneities in the local magnetic fields. In either case, the relaxation is entropy driven and involves no net change in energy of the system of target spins. Spin–lattice relaxation involves energy exchange between a target nuclear spin and fluctuating magnetic or electric fields in the lattice (the collection of neighboring nonidentical magnetic nuclei). The efficiencies of both types of relaxation depend critically on the similarity of the oscillation frequency (or correlation time) of the interacting nucleus compared to the precessional frequency of the target nucleus.

ADDITIONAL RESOURCES

1. There is an excellent computer tutorial entitled *The Basics of NMR Spectroscopy*, written by Joseph P. Hornak and available through him at the Department of Chemistry, Rochester Institute of Technology, Rochester, NY 14623. This software uses realistic graphic animations to show such processes as spin equilibration, absorption, and relaxation.

2. Becker, E. D. *High Resolution NMR*, 2nd ed., Academic, New York, 1980.

REVIEW PROBLEMS (Answers in Appendix 1)

2.1. How many protons and neutrons are there in a ^{10}B nucleus? *Hint*: B has atomic number 5.

2.2. To which of the three groups of nuclei does ^{10}B belong? Without looking at Table 2.1, what can you say about the I value for ^{10}B?

2.3. Using the data in Table 2.1, determine the spin multiplicity of ^{10}B and all possible values of m_i. Then draw a diagram (resembling Figure 2.3) showing how the energy of each spin state of ^{10}B varies with $\mathbf{B}_0$.

2.4. Calculate the precessional frequency of ^{10}B in a 5.87-T magnetic field.

2.5. Calculate the energy gap between adjacent spin states of ^{10}B in a 5.87 T magnetic field.

2.6. Calculate the population ratio of the $m = -3$ to the $m = +3$ states of ^{10}B in a 5.87-T magnetic field at 25°C.

2.7. Suppose we are studying ^{1}H nuclei at a field strength of 5.87-T. Assume that the oscillating field $\mathbf{B}_1$ is only 10^{-5} as strong as $\mathbf{B}_0$. (a) How fast will $\mathbf{M}$ precess around $\mathbf{B}_1$. (b) How much time is required for $\mathbf{M}$ to precess one full revolution around $\mathbf{B}_1$? (c) Calculate the flip angle α at the following times after irradiation with $\mathbf{B}_1$ begins:

0, 0.10, and 0.20 ms (1 ms = 10^{-3} s). (d) Draw a rotating-frame diagram that shows **M** at each of the above times. (e) Suppose B_1 is turned off 0.20 ms after it was turned on. Describe what happens to **M** in terms of T_1 and T_2 using a rotating-frame diagram.

2.8. A collection of ^{1}H nuclei is irradiated by B_1 to give a flip angle of 45°, during which about a fourth of the excess up spins absorb photons and flip to down spins. Then B_1 is turned off. (a) At this point, what is the magnitude (length) of the **M** vector, compared to its initial equilibrium magnitude? (b) Use a rotating-frame diagram to show what happens to **M** after B_1 is turned off.

2.9. Explain what effect dissolved oxygen (O_2) might have on longitudinal relaxation of ^{1}H nuclei. *Hint*: The oxygen molecule has two unpaired electrons with the same s value.

3

OBTAINING AN NMR SPECTRUM

3.1 ELECTRICITY AND MAGNETISM

From a physics course in your past you are probably aware that there is an intimate connection between electricity and magnetism. Let us review a few of the relevant physical principles.

3.1.1 Faraday Induction in the Receiver Coil

When a steady direct current of electricity (electrons) passes through a loop of wire [by attaching the ends of the wire to a battery or other source of direct-current (dc) voltage], a steady magnetic field is established along the axis of the loop (a line perpendicular to the loop and through its center); see Figure 3.1. The higher the current (amperes), the greater is the strength of the resulting magnetic field. The field is also strengthened by using a coil of wire made up of several loops or by coiling the wire around an iron bar. These principles are used in the construction of all electromagnets and superconducting magnets (Section 3.2).

If the direction of current flow in the coil is reversed, so is the direction of the magnetic field. And if the current is oscillating (i.e., alternating current), the resulting linearly polarized magnetic field will oscillate (change directions) at the same frequency (Figure 3.2).

Now let us take the same loop of wire and replace the voltage source with an ammeter. Initially, of course, no current registers in the ammeter. However, if we pass a bar magnet down into the coil, the needle of the ammeter deflects, indicating a current *as long as the bar magnet is moving*. Change the direction of the magnet's movement and you will see that the direction of the current reverses. This effect is called **Faraday induction:** A current is *induced* in the wire by the *movement* of the magnetic field near the wire.

Suppose that instead of a bar magnet we use the precessional motion of **M** (the net nuclear magnetization vector; Section 2.3) to induce an oscillating current [i.e., a radio-frequency alternating-current (ac) signal] in a coil of wire. We will orient the coil so that its axis lies anywhere in the x,y plane, for example, along the y axis and perpendicular to $\mathbf{B}_0$, as in Figure 3.3a. Henceforth, any time there is a component of **M** oscillating in the x,y plane ($\mathbf{M}_{x,y}$[Figure 2.7]), an alternating current of the same frequency will be induced in the coil's circuitry. We will label this loop the **receiver coil**, for it is here that the NMR signal is generated.

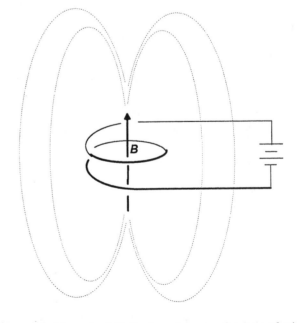

Figure 3.1. Magnetic field **B** along the axis of a loop of wire carrying direct current. The dotted lines indicate a few of the lines of magnetic flux.

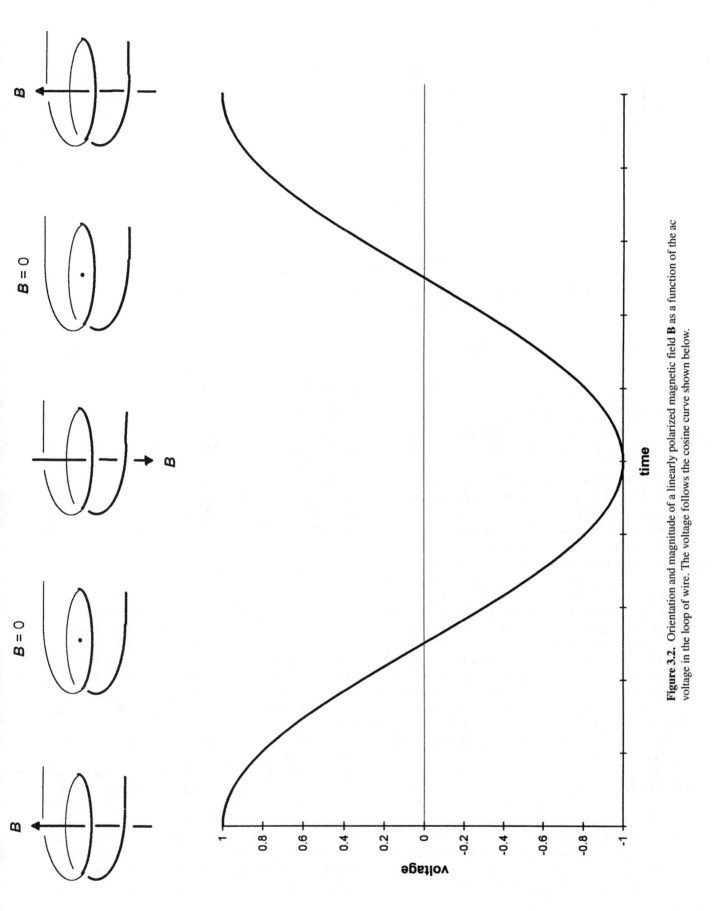

Figure 3.2. Orientation and magnitude of a linearly polarized magnetic field **B** as a function of the ac voltage in the loop of wire. The voltage follows the cosine curve shown below.

23

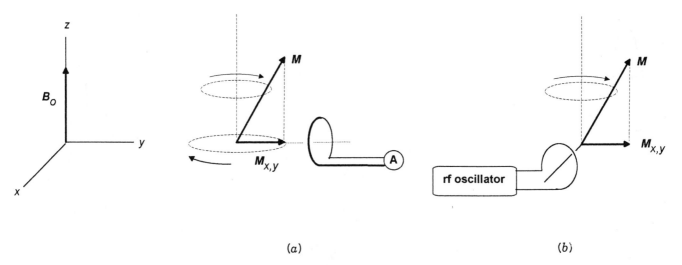

Figure 3.3. Orientation of (*a*) the receiver coil (attached to ammeter **A**) around the *y* axis and (*b*) the transmitter coil (attached to rf oscillator) around the *x* axis.

3.1.2   Generation of B_1 in the Transmitter Coil

Remember from Section 2.3 that to tip **M** off the *z* axis, so it has a component in the *x,y* plane, we need an "irradiating" magnetic field (**B**₁) that oscillates at exactly the precessional frequency of the nuclei of interest and is oriented perpendicular to **B**₀. How are we going to generate such a precessing magnetic field?

Suppose we orient a second loop of wire, henceforth called the **transmitter coil**, so that its axis is aligned with the *x* axis, perpendicular to both **B**₀ and the axis of the receiver coil; see Figure 3.3*b*. As we saw in Figure 3.2, passage of an rf alternating current through the transmitter coil will generate a magnetic field (**B**₁) that is *linearly polarized* (oscillates back and forth) along the *x* axis, as in Figure 3.4*a*. Now here is the important part. This linearly polarized field can be viewed as if it were the vector sum of two oppositely phased, *circularly polarized* rotating magnetic fields (**B**₁ and **B**₁′), as shown in Figure 3.4*b*. Note how the two circularly polarized vectors in Figure 3.4*b* add together to generate the linearly polarized vector in Figure 3.4*a*. One of these, **B**₁, is rotating clockwise, in the same direction as the nuclear moments precess around **B**₀ (Section 2.2.2). Therefore, when viewed in the rotating frame (Figure 3.4*c*), **B**₁ will always be aligned with the *x*′ axis, exactly where it is needed to bring about the precession of **M**; compare Figures 3.4*c* with 2.10.

3.2   THE NMR MAGNET

From the foregoing discussion we can list the basic components of an NMR spectrometer. There will be a magnet to generate **B**₀, an rf oscillator to generate **B**₁ in the transmitter coil, a receiver coil to pick up the signal, the electronics (including a computer and plotter) to turn the signal into a spectrum, and, of course, a sample to be analyzed. In this chapter we will refine our spectrometer design as we consider its performance and limitations.

3.2.1   The Magnet

The three most important characteristics of the magnet in any NMR spectrometer are the strength, stability, and homogeneity of its magnetic field **B**₀. Not only is the precessional (resonance) frequency of identical nuclei directly proportional to the strength of **B**₀ (Section 2.2), but so is the difference in precessional frequencies ($\Delta\nu$) of nonidentical nuclei:

$$\Delta\nu = \nu_1 - \nu_2 = \frac{\gamma_1\,\mathbf{B}_0}{2\pi} - \frac{\gamma_2\,\mathbf{B}_0}{2\pi} = \frac{(\gamma_1 - \gamma_2)\mathbf{B}_0}{2\pi} \tag{3.1}$$

Therefore, it is advantageous to use the strongest available magnet to obtain the greatest separation (i.e., resolution) between NMR signals. Remember also that the stronger field results in larger energy gaps between spin states [Eq. (2.4)] and hence greater populations in the lower energy states [Eq. (2.8)]. This serves to enhance the intensity of the NMR signal, which turns out to be approximately proportional to the square of **B**₀.

Magnets are of three general types: permanent magnets, electromagnets, and superconducting magnets. There are advantages and disadvantages with each type. Permanent magnets are less costly, they have relatively stable fixed magnetic fields, and they require no electric current to generate the field. Unfortunately the strength of their fields is so limited (ca. 1.4 T) that they were used only in the first generation of commercial NMR spectrometers. Electromagnets used for NMR applications, on the other hand, are huge and more costly to build and operate, but their field strengths range up to the strongest

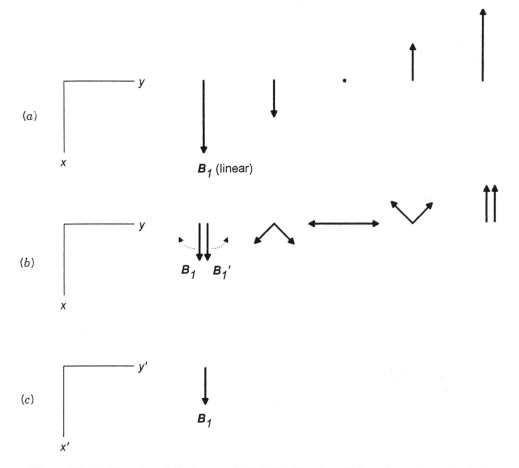

Figure 3.4. (*a*) Linearly polarized magnetic field **B**$_1$ (linear) oscillating along the *x* axis of the laboratory frame; compare with Figure 3.2. (*b*) Resolution of **B**$_1$ (linear) into oppositely rotating circularly polarized magnetic fields, **B**$_1$ and **B**$_1'$ both oscillating at ν_0. (*c*) Orientation of **B**$_1$ in the rotating frame.

ever achieved, in excess of 34 T (which corresponds to a ^{1}H resonance frequency of 1.45 GHz, where 1 GHz = 10^9 Hz). However, electrical resistance to the high current necessary to generate strong magnetic fields generates considerable heat that must be efficiently dissipated to assure a stable field.

The superconducting magnets used in NMR spectrometers constitute a special subcategory of electromagnets. Each superconducting magnet is designed to provide a specific nominal field strength, currently in the range of ca. 6–18 T, corresponding to ^{1}H operating frequencies in the range of 250–750 MHz. The cylindrical solenoid through which the large direct current flows is made of a unique niobium–tin alloy that becomes **superconducting** (that is, develops zero electrical resistance) when cooled to 4 K (4 degrees above absolute zero) by immersion into liquid helium in a cryostat that includes an outer jacket filled with liquid nitrogen. Figure 3.5*a* shows a complete Bruker DMX-500 NMR spectrometer. The internal configuration of the magnet and probe region is detailed in the cutaway diagram of Figure 3.5*b*.

The appropriate current is initially established in the cooled solenoid with the aid of a high-voltage dc source. Then the two ends of the solenoid loop are "short circuited," removing the dc source. However, because the solenoid circuit has zero resistance, the large direct current continues coursing through the solenoid indefinitely (as long as the helium holds out!), thereby generating a very strong and stable magnetic field.

As strong and stable as the magnetic field is, it is still necessary to provide some mechanism by which the stability of the field is monitored and controlled. This can be achieved through an electronic feedback technique known as **locking**. We first select a substance with nuclei that give rise to a strong NMR signal (the **lock signal**) at a different frequency from those of the nuclei of interest. If the lock substance is kept physically apart from (but close to!) the sample, it is referred to as an **external lock**. More commonly, the lock substance is used as the solvent for the sample and is termed an **internal lock**. In either case, the frequency of the lock signal is continuously monitored and electronically compared to a fixed

reference frequency rf oscillator. Any difference between the lock and reference frequencies causes a direct microcurrent to pass through a secondary coil (known as the Z^0 gradient coil, aligned with and inside the solenoid). This results in a small secondary magnetic field to be generated, aligned either *with* B_0 (increasing its magnitude slightly) or *against* B_0 (decreasing its magnitude slightly) until the lock frequency once again matches the reference frequency. At this point the magnet is said to be "locked."

The most common lock systems monitor the signal from deuterium (^2H, $I = 1$), so it is common in NMR to use deuteriated solvents such as D_2O or $CDCl_3$ (deuteriated chloroform). Many such deuteriated solvents are readily available.

■ **EXAMPLE 3.1**   If the spectrometer's magnetic field varied by +0.00001 T (that is, about 2 ppm), what magnitude of change would be introduced in the resonance frequency of ^1H nuclei at 5.87 T?

□ *Solution:* We can use a simple proportion:

$$\frac{0.00001 \text{ T}}{5.87 \text{ T}} = \frac{\Delta\nu}{250 \text{ MHz}}$$

$$\Delta\nu = 0.000430 \text{ MHz} = 430 \text{ Hz}$$

As we will see later, this ~2 ppm shift would be a very large shift indeed!   □

■ **EXAMPLE 3.2**   Suppose the lock signal frequency is found to be slightly less than the constant reference frequency of the rf oscillator. Should the magnetic field be increased or decreased to bring it back to the nominal value?

□ *Solution:* Remember [Eq. (3.1)] that the frequency of any signal increases in direct proportion to the field strength. Thus, to increase the lock signal frequency, we need to increase the field strength.   □

Once a stable field is established, the question remains as to whether that field is completely *homogeneous* (uniform) throughout the region of the sample. The level of homogeneity required for a given NMR experiment depends on the desired level of resolution, which in turn controls the precision of the measurement. In the case of ^1H nuclei at 5.87 T, for example, Example 3.1 suggests that to achieve a precision of ±1 Hz at a frequency of 250 MHz (four parts per billion!), the field must be homogeneous to the extent of 2.35×10^{-8} T!

Such phenomenal uniformity can be achieved by means of two additional instrumental techniques. First, the sample vessel (normally a precisely constructed glass tube) is positioned along the center (z axis) of the solenoid in a region called the **probe** (Figure 3.5b) that is separated from the region cooled by liquid helium and whose temperature can therefore be independently controlled. The tube is spun around its axis at ca. 100 Hz by means of an air stream that turns a small plastic turbine attached to the top of the tube.

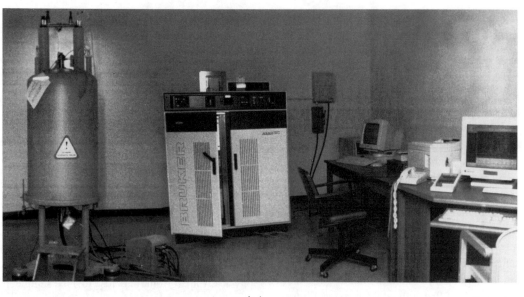

(a)

Figure 3.5. (*a*) Bruker model DMX-500 NMR spectrometer. The magnet cryostat is located on the left.

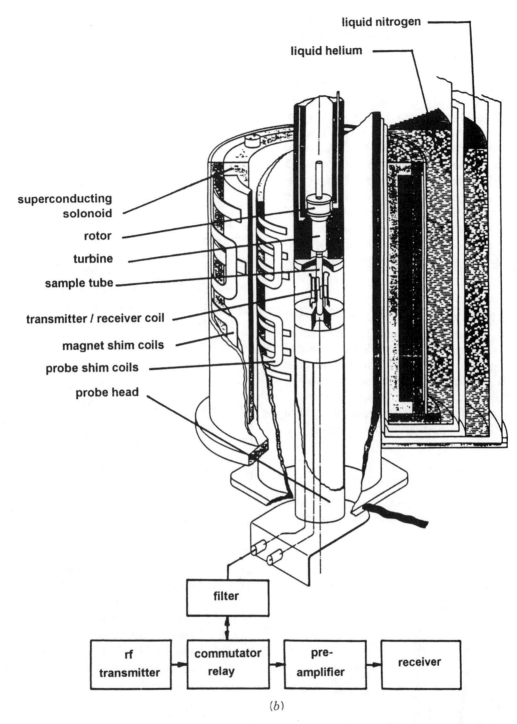

Figure 3.5. (*b*) Cutaway diagram of the cryostat, showing the essential parts of the magnet and probe. (Courtesy of Bruker Instruments, Inc.)

This spinning helps to "average out" any slight inhomogeneities of the field in the sample region.

Second, the contour (shape) of the field itself can be varied (within very narrow limits) by passing extremely small currents through a complex series of **shim coils** located around the probe cavity. The resulting small fields in these coils can be adjusted to further improve the homogeneity of the field. This process, called **shimming** or *tuning* the magnet, is accomplished by adjustment of a dizzying series of shim controls (labeled by the field axis most affected, e.g.,

Z^2, Z^3, X, Y, XZ, YZ, XY, X^2, $-Y^2$, etc.) before each spectrum is taken. With the advent of microprocessor-controlled spectrometers, the bulk of these adjustments are now accomplished automatically.

We are left with a paradox. To get an acceptable signal, we need to have as many nuclei as possible in the sample region (remember the minute difference between spin state populations?). But we must restrict our sample to the smallest possible volume in the interest of precision and resolution. The eventual compromise results in a sample cavity of ~0.05 cm^3 for instruments that accommodate narrow-bore (5-mm-o.d.) tubes to ~1.0 cm^3 for instruments using wide-bore (10–20-mm-o.d.) tubes. [However, for medical use, some low-resolution NMR instruments have been constructed that accommodate entire human bodies! (See Chapter 16)] Typically, ^{1}H and ^{13}C spectra require ca. 5 and 15 mg of sample, respectively.

■ **EXAMPLE 3.3**   A typical thin-wall narrow-bore NMR tube has an outside diameter of 5.00 mm (0.500 cm) and an inside diameter of 4.20 mm. It is usually filled to a height of ~1 in. (2.5 cm). (a) What is the wall thickness of the glass tube? (b) If the total volume of the sample cavity (including the tube) is 0.050 cm^3, what are the dimensions (radius r, height h, and volume) of the space that is actually occupied by sample? [*Hint*: The volume of the cylinder is $\pi r^2 h$.]

□ *Solution:*   Refer to Figure 3.6.

(a) Wall thickness = (5.00 mm – 4.20 mm)/2 = 0.40 mm.
(b) The diameter of the sample is the same as the inside

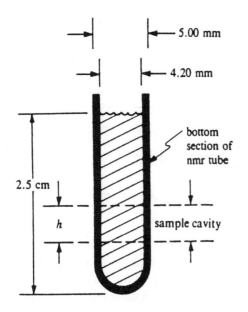

Figure 3.6. Bottom section of an NMR sample tube and the sample cavity where the NMR signal is generated.

diameter of the tube, 0.42 cm. The height of the sample cavity is given by

$$h = \frac{V_{\text{total}}}{\pi \, r_{\text{total}}^2} = \frac{0.050 \text{ cm}^3}{\pi (0.25 \text{ cm})^2} = 0.255 \text{ cm}$$

Since $r_{\text{sample}} = 0.42$ cm/2 = 0.21 cm,

$$V_{\text{sample}} = \pi \, r_{\text{sample}}^2$$

$$h = \pi (0.21 \text{ cm})^2 (0.255 \text{ cm}) = 0.035 \text{ cm}^3 = 35 \text{ } \mu\text{L}$$

Thus, only 70% of the sample cavity is filled with sample in this case. Clearly, the thinner the wall of the tube, the more sample actually gets into the sample cavity. Furthermore, for samples in very limited supply (<1 mg), precisely machined Teflon inserts for the tubes are available to both reduce the amount of dead volume in the tube below the sample cavity and prevent formation of vortices (whirlpools) of the solution within the tube when it is spun.   □

There is one more thing you should know about the powerful magnets used in today's NMR spectrometers. If you get too close to the magnet itself, you run the risk of demagnetizing anything on your person, such as credit cards, magnetic tapes, and computer diskettes. (This fact was central to the plot of a recent novel by Joseph Wambaugh.) But no need to worry; it will not affect your magnetic personality!

3.2.2   Assembling the Pieces

We are now ready. The magnetic field is locked at 5.87 T and a sample containing a substance with ^{1}H nuclei, dissolved in a deuterium-containing solvent, is spinning in the probe. We activate the rf oscillator in the transmitter coil, producing a **B**$_1$ field oscillating at 250 MHz radiation, and . . . yes! A weak signal is detected in the receiver circuit! But let us not celebrate yet. What has this told us, except that we have ^{1}H in the sample, precessing at 250 MHz? Certainly, we would like to be able to adjust parameters to test for nuclei precessing at other frequencies as well. How can we accomplish that?

3.3   SIGNAL GENERATION THE OLD WAY: THE CONTINUOUS-WAVE (CW) EXPERIMENT

Recall from Chapter 2 that, at a given field strength, each magnetic isotope ($I \neq 0$) precesses at a unique frequency governed by its magnetogyric ratio [Eq. (2.6)]; see Figure 3.7. Thus, if we intend to use our spectrometer to observe nuclei other than ^{1}H, there must be a way to vary either the operating (**B**$_1$) frequency or the magnetic field strength.

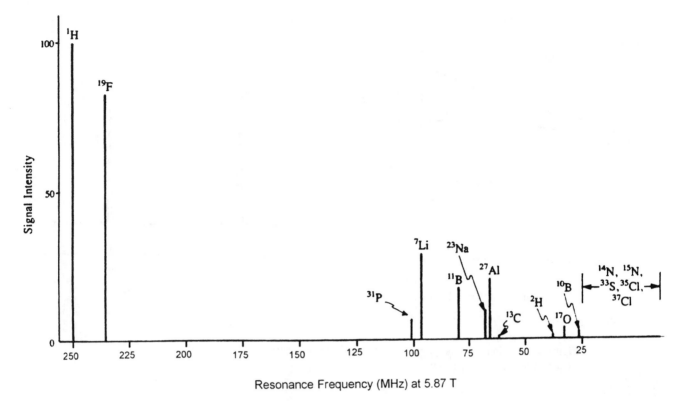

Figure 3.7. Resonance frequencies and relative sensitivities of several common isotopes at 5.87 T; compare with Table 2.1.

From the discussion in Section 3.2, it is clear that the field strength of any type of electromagnet can only be varied within an extremely narrow range. This means that the nominal operating frequency for, say, ^{1}H nuclei is also fixed. (In fact, the model number of most commercial NMR spectrometers is the same as the ^{1}H operating frequency in megahertz.) Since, for a given instrument, we cannot change $\mathbf{B}_0$, the only way to investigate nuclei other than ^{1}H is to switch to a different rf oscillator ($\mathbf{B}_1$ source) with a frequency appropriate for the specific nucleus of interest (e.g., ^{13}C or ^{31}P). In the old days, switching oscillator circuits could be a rather tedious process, but these days it requires little more than pushing a few buttons or typing a few commands into the computer that controls the instrument.

However, as we will see in Chapter 5, the real value of NMR does not come from generating one signal for each isotope present in the sample (e.g., one signal each for ^{1}H, ^{13}C, or ^{31}P). Rather, we will find that even the individual nuclei of the same isotope in different molecular environments can precess at *slightly different* frequencies (on the order of parts per million), and these small differences can tell us much about the details of the molecular structure involved. Thus, we normally examine only one isotope (e.g., ^{1}H) in each NMR spectrum; that is, each spectrum is **homonuclear**. Since we are examining only one isotope at a time, we only have to

vary the magnetic field (field sweep) or rf frequency (frequency sweep) over narrow limits to see all of the nuclei of that specific target isotope. All older NMR spectrometers, equipped with conventional electromagnets, used either field- or frequency-sweep technology. Because both involved continuous operation of the rf ($\mathbf{B}_1$) oscillator, they were both referred to as continuous-wave (cw) techniques.

3.3.1 Frequency-Sweep Mode

In the case of a frequency-sweep cw experiment, the magnetic field strength ($\mathbf{B}_0$) was fixed. Suppose our sample contained two different nuclei (A and B) characterized by slightly different resonance frequencies, with $\nu_A > \nu_B$. The spin state energies of these two nuclei are depicted in Figure 3.8a. In order for nuclei A and B to generate resonance signals, Eq. (2.7) requires they be irradiated with $\mathbf{B}_1$ frequencies given by

$$\nu_A = \frac{\Delta E_A}{h} \tag{3.2a}$$

$$\nu_B = \frac{\Delta E_B}{h} \tag{3.2b}$$

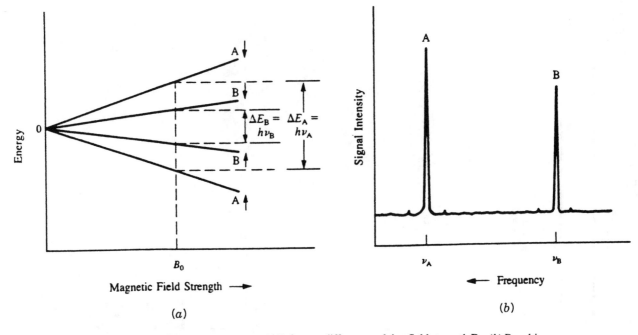

Figure 3.8. (*a*) Spin state energy gaps (ΔE) for two different nuclei at field strength $\mathbf{B}_0$. (*b*) Resulting frequency-sweep NMR spectrum.

Thus, it was necessary to sweep the operating frequency over the range ν_A to ν_B (or vice versa) using a *variable-frequency* rf transmitter. This *range* of frequencies ($\nu_A - \nu_B$) is called the **spectral width** (or sweep width), SW, of the spectrum and was limited to about 2000 Hz. Beginning at a frequency a little less than ν_B, the frequency of the continuous rf radiation was gradually increased to a value above ν_A while the receiver circuit was constantly monitored for a signal. As the frequency passed first through ν_B, then through ν_A, two separate signals were detected. A plot of the intensity of these signals versus frequency (Figure 3.8*b*) is our first **frequency-domain** NMR spectrum, with frequency increasing from *right to left* (as is the convention).

3.3.2 Field-Sweep Mode

Although some early NMR spectrometers did employ the frequency-sweep (at constant field) technique, it turned out to be electronically simpler to maintain the constant $\mathbf{B}_1$ oscillator frequency appropriate for the target isotope and sweep (vary) the magnetic field to achieve resonance for the nuclei in the sample. By passing a slowly increasing direct current though **sweep coils**, the magnetic field $\mathbf{B}_0$ could be varied through a limited range while maintaining its homogeneity in the sample area. Let us investigate how this change affected the spectrum.

Because the operating radio frequency (ν_0) was now constant, only photons of energy $h\nu_0$ were available in the experiment [see Eq. (1.3)]. Thus, the *energy gap* [ΔE, Eq. (2.4)] for *each* nucleus had to be adjusted by varying the magnetic field strength $\mathbf{B}_0$. By rearranging Eq. (2.7), we can solve for the field strength at which nuclei A and B entered resonance:

$$\mathbf{B}_A = \frac{2\pi\nu_0}{\gamma_A} \tag{3.3a}$$

$$\mathbf{B}_B = \frac{2\pi\nu_0}{\gamma_B} \tag{3.3b}$$

Figure 3.9*a* depicts the situation graphically. This time, instead of varying the $\mathbf{B}_1$ frequency until it matched the precessional frequencies of nuclei A and B, we varied the precessional frequencies (by varying the field strength) until they matched the $\mathbf{B}_1$ frequency. Most importantly, the resulting spectrum (Figure 3.9*b*) was *indistinguishable* from the frequency-sweep spectrum, except that the abscissa was calibrated in magnetic field units increasing from *left to right*.

Comparing Figure 3.8*b* with Figure 3.9*b* reveals that nucleus A enters resonance at a higher frequency (at constant field) or at a lower field (at constant frequency) than does nucleus B. For this reason, the right-hand (low-frequency) side of an NMR spectrum is often referred to as the *high-field* (or *upfield*) *side*, while the left-hand (high-frequency) side is called the *low-field* (or *downfield*) *side*.

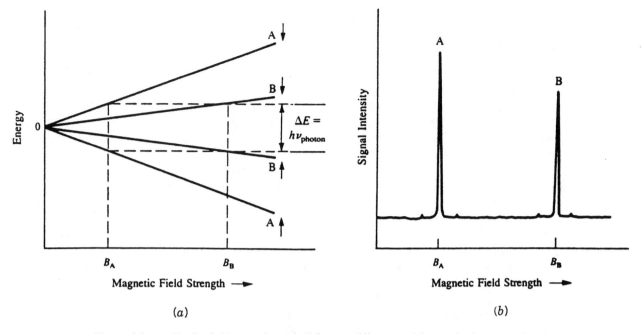

Figure 3.9. (*a*) The B_0 field strength required for two different nuclei to attain the same spin state energy gap ΔE. (*b*) Resulting field-sweep NMR spectrum.

■ **EXAMPLE 3.4** Suppose that, at a nominal field strength of 2.35 T, NMR signals were detected at 100 and 25.2 MHz. (a) What are the two isotopes giving rise to the signals? (b) If the spectrum were obtained by the field-sweep method, what field strength would be required to observe the latter signal using an operating frequency of 100 MHz?

□ *Solution:* (a) By referring back to the ν column in Table 2.1, we see that isotopes are ^1H and ^{13}C, respectively. (b) Use Eq. (3.3) and the value of γ for ^{13}C from Table 2.1:

$$\mathbf{B}_{C-13} = \frac{2\pi\nu_0}{\gamma_{C-13}} = \frac{2(3.14\ \text{rad})(100 \times 10^6\ \text{s}^{-1})}{67.3 \times 10^6\ \text{rad}\,T^{-1}\,\text{s}^{-1}} = 9.33\ \text{T}$$

It is impossible for an NMR electromagnet designed for 2.35 T to produce a 9.33-T field!   □

3.3.3 Spinning Sidebands and the Signal-to-Noise Ratio

A closer examination of the signals in Figures 3.8*b* and 3.9*b* brings up some additional points. Figure 3.10 is an enlargement of one of the peaks. Notice the two small signals symmetrically flanking the large signal. These so-called **sidebands** or **satellite peaks**, marked with asterisks in the figure, are an unavoidable result of spinning the sample to improve field homogeneity (Section 3.2.1). If the field has been well shimmed, such **spinning sidebands** rarely exceed 1% of the height of the main peak. Moreover, they will always be

separated from the main peak by *exactly* the spin rate in hertz. This is, in fact, the most direct way to measure the sample spin rate. One can, of course, completely eliminate spinning sidebands by turning off the spinner. But this would cause substantial broadening of the signal (with an accompanying decrease in signal height) because of the small inhomogeneities present in the magnetic field.

■ **EXAMPLE 3.5** Suppose the two sidebands in Figure 3.10 are each separated from the main peak by 58 Hz. (a) What

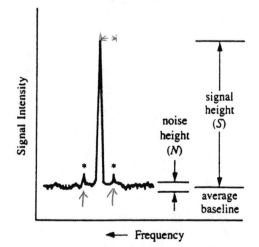

Figure 3.10. An NMR signal peak showing sidebands (*) and signal (*S*) to noise (*N*) ratio.

is the spin rate of the sample tube? (b) How could you confirm that these peaks were indeed spinning sidebands and not due to something else?

□ *Solution:* (a) Because the frequency separation of a sideband from the main peak is equal to the spin rate, the latter must also be 58 Hz. (b) Varying the spin rate (by adjusting an air flow directed at the spinner paddle) causes spinning sidebands to shift position accordingly. If the peaks do not shift position, they are not spinning sidebands. There are other types of satellite peaks that we will encounter later.   □

You will also notice in Figure 3.10 that the signal baseline is not perfectly flat. Instead, there is perceptible random background noise causing a continuous wiggle in the baseline. Even in the most carefully built and tuned spectrometers there will always be some electronic noise generated by the various circuits within the instrument, and the way the signal data are processed. Normally, this does not present a problem, provided the sample gives signals strong enough to be easily differentiated from the noise. But what about really weak signals? For example, a sample of natural hydrogen (Table 2.1) contains 99.985% ^{1}H with a relative sensitivity of 1.00, while a sample of natural carbon contains only 1.1% ^{13}C, which has a relative sensitivity of 1.59×10^{-2}. These two factors, low natural abundance and low relative sensitivity, combine to make the carbon signal only about 2×10^{-4} times as intense as a hydrogen signal for equal numbers of atoms of each element. Thus, a ^{13}C signal is much harder to distinguish from noise than is a ^{1}H signal.

To deal with this problem, we define a quantity called **signal-to-noise ratio** *(S/N)*, where *S* and *N* are approximated by their respective peak heights, as shown in Figure 3.10.

■ **EXAMPLE 3.6** Using a ruler, estimate *S/N* in Figure 3.10.

□ *Solution:* The exact values of *S* and *N* will depend on the scale of your ruler. But the ratio of the two numbers, *S/N*, should be about 20.   □

The goal in an NMR experiment is to maximize *S/N*, and this can be achieved in several ways. One method is to use higher rf power and a series of rf filters to remove some of the noise. While this technique results in some improvement, care must be taken to avoid saturation problems (see Section 2.3). A far better improvement can be obtained by relying on a result of the **central limit theorem** from information theory. This theorem tells us that if the spectrum is scanned *n* times and the resulting data are added together, signal intensity will increase directly with *n* while noise (being random) will only increase by $\sqrt{n}$:

$$\left(\frac{S}{N}\right)_n = \left(\frac{S}{N}\right)_1 \left(\frac{n}{\sqrt{n}}\right) = \left(\frac{S}{N}\right)_1 \sqrt{n} \qquad (3.4)$$

where $(S/N)_1$ is the ratio after a single scan. Thus, *S/N* improves linearly with $\sqrt{n}$.

To make use of this multiple-scan technique, we first need an extremely stable field. Then we need a computer to collect the signal data (digitally) from each scan, add the data together, and then divide the sum by *n*. The technique is known as **signal averaging** or **CAT** (computer-averaged transients, not to be confused with the diagnostic X-ray technique called *computer-aided axial tomography*). Because each scan of an older cw instrument could require 10 or more minutes, this process used to take hours, during which time the magnetic field had to be kept as perfectly stable and homogeneous as possible. A momentary flicker in the building power during an overnight experiment could cause all the data to be useless.

■ **EXAMPLE 3.7** A single 10-min cw NMR scan of a certain highly dilute sample exhibited *S/N* of 1.9. How much time would be required to generate a spectrum with *S/N* of 19?

□ *Solution:* We can rearrange Eq. (3.4) to calculate the required number of scans:

$$\sqrt{n} = \frac{(S/N)_n}{(S/N)_1} = \frac{19}{1.9} = 10$$

$$n = 10^2 = 100$$

The time required for 100 scans is

$$t = (100 \text{ scans})(10 \text{ min per scan})$$

$$= 1000 \text{ min} = 16 \text{ h } 40 \text{ min}$$

No sleep tonight!   □

3.4 THE MODERN PULSED MODE FOR SIGNAL ACQUISITION

Further improvement in *S/N* had to await the development of faster computer microprocessors, which was exactly what happened during the 1980s. Armed with very fast and efficient microcomputers with large memories, chemists discovered they could now generate NMR signals in an entirely new way.

Have you ever been sitting in a quiet room when suddenly a fixed pitch noise (sound waves) causes a nearby object to vibrate in sympathy? This is an example of acoustic resonance, in which the frequency of the sound waves exactly matches the frequency with which the object naturally vibrates. If the pitch of the sound changes, the resonance ceases.

This is because only when the two frequencies match can the sound waves *constructively* interfere (reinforce) the motion of the object. At any other frequency the interaction will be one of *destructive* interference and the net result will be no vibration by the object. This is exactly analogous to the continuous-wave NMR experiment. But have you not also heard how a single loud sound such as a sonic boom can set many different objects vibrating for a long period after the boom is over? This occurs because the brief sound pulse acts as though it is a mixture of all audible frequencies. Might it also be possible to generate NMR signals this way?

In a **pulsed-mode** NMR experiment, which is performed at both constant magnetic field *and* constant rf frequency, rf radiation is supplied by a brief but powerful computer-controlled pulse of rf current through the transmitter coil. This "monochromatic" (single-frequency) pulse, centered at the operating frequency ν_0, is characterized by a power (measured in watts and controlling the magnitude of $\mathbf{B}_1$) and a **pulse width** (t_p), the duration of the pulse measured in microseconds. However, as a direct consequence of the uncertainty principle [Eq. (1.6)], this brief pulse acts as if it covers a *range*

of frequencies from $\nu_0 - \Delta\nu$ to $\nu_0 + \Delta\nu$ (Figure 3.11), where $\Delta\nu$ is nothing more than the inverse of t_p (Section 1.4). Thus, the shorter the pulse, the greater the range of frequencies covered.

The exact value of ν_0 is designed to be slightly offset from the range of nuclear precession frequencies to be examined (the spectral width, SW, in hertz). Therefore, the SW can be no greater (and preferably less) than $\Delta\nu$, leading to the relationship

$$SW \leq (t_p)^{-1} \quad \text{or} \quad t_p \leq (SW)^{-1} \qquad (3.5)$$

The power of the pulse must be sufficient to cause all sets of nuclei precessing at frequencies within the SW to become phase coherent, each set tipping their $\mathbf{M}$ vectors toward the x',y' plane (Section 2.3). The nearer a nucleus's precession frequency is to ν_0, the more excitation energy the nucleus receives from the pulse, and the greater its flip angle α. However, the closer α is to 90°, the stronger will be the signal generated by a set of nuclei (Section 2.3). At the powers normally used, a pulse as long as 1/SW causes flip angles to

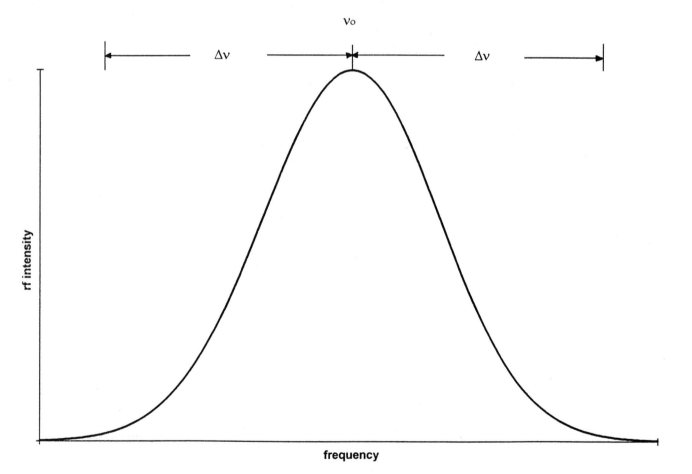

Figure 3.11. Frequency distribution profile for an rf pulse of duration (pulse width) t_p. The frequencies range from $\nu_0 - \Delta\nu$ to $\nu_0 + \Delta\nu$, where $\Delta\nu$ equals $1/t_p$. The sweep width can be either half of the curve.

exceed 90°, so t_p is generally set to approximately one-fourth of $(SW)^{-1}$. This also avoids using the outer "tails" of the pulse profile (Figure 3.11). The ultimate result of the pulse is a complex signal in the receiver circuit that is the superposition of the precession frequencies of *all* nuclei within the SW.

■ **EXAMPLE 3.8** At 5.87 T the required ¹H spectral width is normally on the order of 5000 Hz. (a) At what frequency should ν_0 be positioned? (b) What should be the value of t_p?

□ *Solution:* (a) At a field strength of 5.87 T, the nominal ¹H resonance frequency is 250 MHz, though the exact value of ν_0 should be slightly offset from all nuclei of interest. (b) $t_p \approx (4\,SW)^{-1} = (4 \times 5000\ \text{Hz})^{-1} = 5 \times 10^{-5}\ \text{s} = 50\ \mu\text{s}.$ □

The next step in a pulsed-mode experiment is to monitor the induced ac signal (voltage or current) in the receiver coil as a function of *time*. In order to carry out signal averaging, the data must be collected digitally. Thus, the computer instantaneously measures the voltage in the receiver coil at regular intervals (called **dwell time** t_d). A typical set of such data might look something like Figure 3.12. The pattern described by these 100 points is not obvious because data

points have not been collected often enough, at least for our eyes to recognize it. The experiment can be repeated with the dwell time reduced by a factor of 5, thus affording 500 data points in the same total time. Figure 3.13 shows the result: a more detailed wave function, made even more recognizable by connecting sequential dots (Figure 3.14).

Take a moment to look carefully at the wave function in Figure 3.14. You will see that the pattern is made up of two superimposed cosine waves, a lower frequency wave (which goes through 5 complete cycles during one "time unit") and a higher frequency wave (which goes through 25 complete cycles during one time unit). If the time unit were *seconds*, the two frequencies would be 5 and 25 Hz, respectively. Furthermore, the lower frequency component is about twice as intense as the higher frequency one. It turns out that, rather than measuring actual precessional frequencies (ν_i), it is electronically easier and more precise to measure the difference between ν_i and the operating frequency ν_0 ($\Delta\nu_i = \nu_i - \nu_0$) for each signal.

So, one important lesson is that the receiver circuit must be sampled often enough to make the pattern recognizable. The **Nyquist theorem** tells us that in order for a computer to determine the frequency ($\Delta\nu_i$) of each component of the wave function by this method, it must acquire at least two points

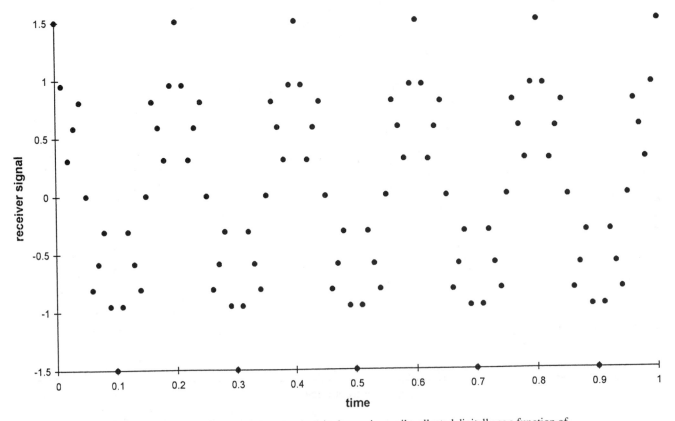

Figure 3.12. The ac signal (voltage or current) in the receiver coil, collected digitally as a function of time.

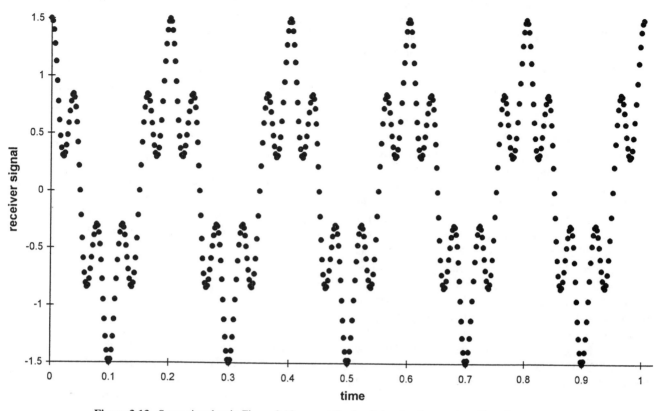

Figure 3.13. Same signal as in Figure 3.12, except the dwell time has been reduced by a factor of 5.

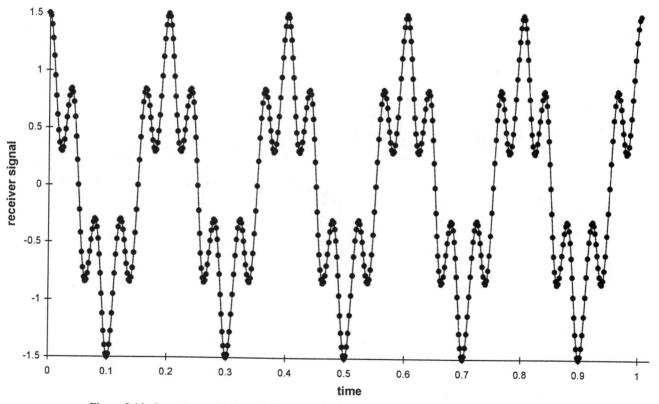

Figure 3.14. Same data as in Figure 3.13, except the dots are connected to make visualization easier.

per cycle for each contributing wave. Therefore, so that frequencies of Δv_i from 0 to SW hertz can be determined, the computer must collect $2\times$ SW data points per second, which is equivalent to a dwell time no larger than the inverse of twice SW:

$$t_d \leq (2\ SW)^{-1} \tag{3.6}$$

Actually, the spectroscopic data would more closely resemble the pattern in Figure 3.15, which is the same as the wave in Figure 3.14, except that the overall intensity of the signal decays exponentially with time. (Note that the decay does *not* affect the frequencies.) Such a pattern is called the **modulated free induction decay** (FID) signal (or **time-domain** spectrum). The decay is the result of spin–spin relaxation (Section 2.3.2), which reduces the net magnetization in the x, y plane. The envelope (see Section 3.6.2) of the damped wave is described by an exponential decay function whose decay time is T_2^*, the effective spin–spin relaxation time.

Most molecules examined by NMR have more than just two sets of nuclei, each set with a different frequency (Δv_i). Furthermore, each set has its own relaxation times (T_1 and T_2), and usually there are different numbers of nuclei within each set. These factors combine to give complex digital FID curves consisting of n data points (usually in the thousands), numbered from 0 to $n - 1$.

For example, Figure 3.16 shows the ^{1}H FID curve for toluene (**3-1**):

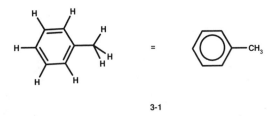

3-1

From the FID data the computer determines the frequency (Δv_i) and intensity of each component wave. How does it accomplish this magic? The computer performs a **Fourier transformation** (FT) of the FID data according to the equation

$$F_j = \sum_{k=0}^{n-1} T_k \exp\left(-\frac{2\pi i j k}{n}\right) \quad \text{for } j = 0, \dots, n-1 \tag{3.7}$$

where T_k represents the kth point of the time-domain (FID) data and F_j represents the jth point in the resulting frequency-domain spectrum. Thus, *each F_j point* (from 0 to $n - 1$) requires a summation over all n points in the FID curve, and the total calculation involves n^2 multiplications and additions.

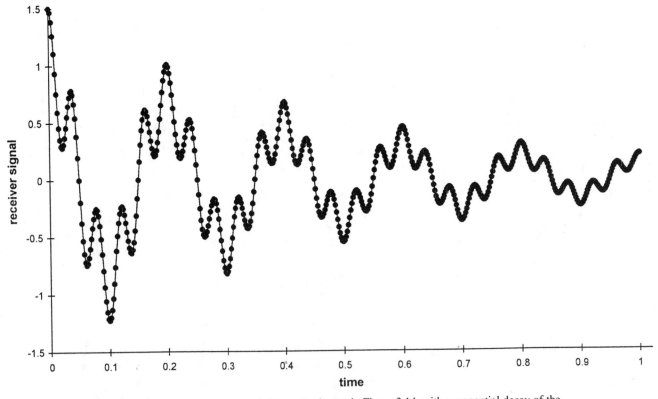

Figure 3.15. Simulated FID signal. The same data as in Figure 3.14, with exponential decay of the signal added.

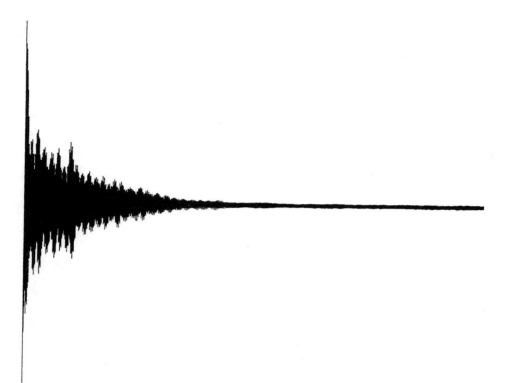

Figure 3.16. Actual ^{1}H FID curve for toluene (**3-1**).

An algorithm developed by Cooley and Tukey simplified this extremely time-consuming calculation, bringing it within the capability of modern microcomputers. Today, the transformation takes only a few seconds, after which the frequency-domain spectrum F_j can be plotted. (The frequency-domain spectrum corresponding to Figure 3.16 will be discussed in Chapter 5.)

During the FT of the FID data the computer will perform a series of additional manipulations on the time- and frequency-domain data in order to present the most accurate peak shape and spectrum appearance, including such computational techniques as **zero filling** and **apodization**. Zero filling is a process for retrieving the maximum resolution present in the FID data without increasing the RAM requirement of the computer. Apodization eliminates sinusoidal wiggles in the baseline around a frequency-domain signal peak caused by truncating the FID acquisition too soon (using a t_{acq} that is too short). Further details of these techniques are beyond the scope of this book, but the interested reader may wish to consult the Additional Resources section at the end of this chapter. With input from the operator, adjustments can also be made to the **phasing** of each signal, ensuring that the baseline on both sides is symmetrically horizontal.

In experiments with either dilute samples or insensitive nuclei (such as ^{13}C), one pulse usually does not give a sufficiently high S/N to allow the determination of the component frequencies and intensities accurately. The S/N can be improved by repeating the pulse–data acquisition sequence, then adding the new FID data to the previously acquired data, as in a CAT experiment. The required number of pulse sequences (scans) is determined by the desired S/N. But there is an additional consideration: To avoid saturation, one must allow enough time between pulses for the nuclei to return (or nearly return) to their original equilibrium (Boltzmann) distribution. The required **delay time** (t_w) is a function of the T_1 values for each set of nuclei of interest (Sections 2.3 and 2.5).

As a typical example of the pulse–data acquisition–delay sequence (hereafter referred to as the **pulse sequence**), we consider the parameters necessary when generating a typical pulsed-mode 250-MHz ^{1}H NMR spectrum. First, we select a spectral width that covers all the nuclei of interest, usually around 4000 Hz. Remember that the SW determines the pulse width $[t_p \approx (4 \text{ SW})^{-1}]$, which in this case would have to be 62.5 µs. Since the frequency (Δv_i) of each component wave can be uniquely determined by two points per cycle and our spectral width might include frequencies anywhere from 0 to 4000 Hz, we must sample points twice that fast, or 8000 Hz (a dwell time t_d of 125 µs), to ensure collecting all relevant frequencies [Eq. (3.6)]. Moreover, we are limited by our computer's memory capacity. Generating 8000 points per second will fill up 32K (32,768 locations) of the computer's RAM in 4.1 s. This fact determines the maximum **acquisition time** (t_{acq}), the length of time a given FID signal is actually monitored. Moreover, **resolution** (R), the ability to distinguish two nearby signals, is inversely proportional to acquisition time:

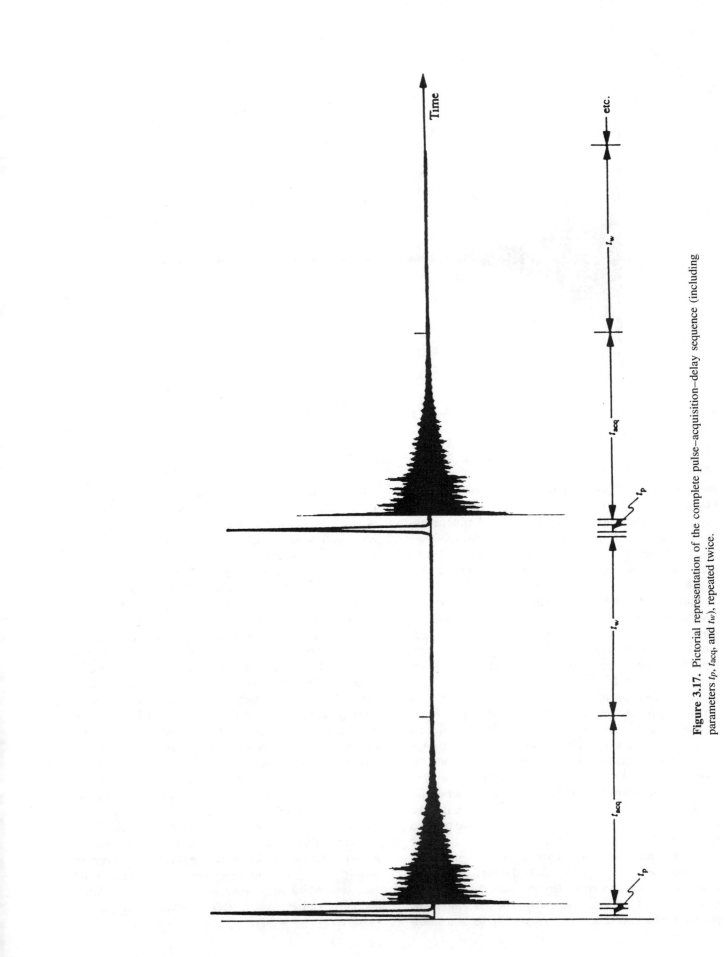

Figure 3.17. Pictorial representation of the complete pulse–acquisition–delay sequence (including parameters t_p, t_{acq}, and t_w), repeated twice.

$$R = (t_{acq})^{-1} \qquad (3.8)$$

Thus, the 4.1 s acquisition time would provide a resolution of $(4.1 \text{ s})^{-1} = 0.24$ Hz. If we require a resolution better (smaller) than 0.24 Hz, either we need more computer RAM to accommodate a longer acquisition time or we have to decrease our spectral width window.

Because the FID signal decays due to spin–spin relaxation, there is a time limit beyond which further monitoring of the FID provides more noise than signal. The usual compromise is to have a short enough *dwell time* to cover the spectral width and a long enough *acquisition time* to provide the desired resolution, consistent with computer memory limitations.

After collecting data from one pulse, we must wait for the nuclei to relax to equilibrium. A total of at least $3T_1$ is usually adequate, but part of this is spent as acquisition time. Thus, the additional *pulse delay time* (t_w), the time between the end of data acquisition and the next pulse, is given by

$$t_w = 3T_1 - t_{acq} \qquad (3.9)$$

Since ^{1}H nuclei normally exhibit T_1 values on the order of 1 s, we need essentially no additional delay after 4 s acquisition time. For ^{13}C and other slow-to-relax nuclei, however, substantial delay times are sometimes required. Following the pulse delay, the sample is irradiated with another pulse, and the data acquisition sequence begins anew.

Figure 3.17 depicts this sequence. After some number of pulse sequences (depending on sample concentration and isotope being studied), the data are subjected to a Fourier transformation, and the S/N of the resulting frequency-domain spectrum is ascertained. If it is not yet adequate, the pulse sequence is resumed until the desired S/N is attained.

To summarize, a pulsed-mode NMR experiment encompasses two steps: (1) the collection of FID data by a pulse–data acquisition–delay sequence, repeated enough times to yield a time-averaged signal possessing the desired S/N, and (2) a FT of the FID data followed by plotting of signal intensity as a function of frequency. The main thing to remember is that the frequency-domain spectrum generated in a pulsed-mode/Fourier transform experiment contains all the same information as the spectrum obtained in a cw experiment (e.g., Figure 3.8), but in principle all the relevant spectroscopic information can be generated in just a few seconds from a single rf pulse (S/N limitations notwithstanding). And there is one further incidental advantage to the pulse-mode technique. Because the receiver coil is off during the pulse and the transmitter coil is off during data acquisition, we can use just one coil to do both jobs (Figure 3.5b)!

■ **EXAMPLE 3.9** Suppose we are about to acquire a 62.5-MHz ^{13}C NMR spectrum. The desired spectral width window is 15,000 Hz. The nuclei have T_1 values up to 3 s and

the computer RAM has a capacity of 64K (65,536 locations). The desired resolution is 0.5 Hz. (a) Using Figure 3.17 as a guide, suggest values for the various data acquisition parameters. (b) How many data points can be collected from one pulse sequence with these values? (c) If 1000 pulse sequence repetitions (*scans*) are required for the desired S/N, how long will the total experiment take?

☐ *Solution:* (a) Use Eqs. (3.5), (3.6), (3.8), and (3.9):

$$t_p \approx (4 \text{ SW})^{-1} = (4 \times 15{,}000 \text{ Hz})^{-1} = 17 \text{ μs}$$

$$t_d = (2 \text{ SW})^{-1} = (30{,}000 \text{ Hz})^{-1}$$

$$= 33 \text{ μs} \quad (30{,}000 \text{ data points per second})$$

$$t_{acq} = R^{-1} = (0.5 \text{ Hz})^{-1} = 2 \text{ s}$$

$$t_w = 3(T_1) - t_{acq} = 3(2 \text{ s}) - 2 \text{ s} = 4 \text{ s}$$

b) Acquiring 30,000 data points per second for 2 s results in 60,000 points for each pulse, well within the computer's RAM limits. (c) Since the pulse time is negligible, the total pulse sequence time is essentially equal to $t_{acq} + t_w$ $(= 3T_1)$; thus, each pulse sequence requires 6 s. If we need 1000 scans, the total time required for the experiment is 6000 s (1 h 40 min). Although this may seem like a long time, remember from Example 3.7 that just 100 cw scans required nearly 17 h! ☐

Lest you worry that you will have to supply each parameter in every pulse sequence each time you take an NMR spectrum, you can relax. Modern spectrometers are menu driven. For most common nuclei all you have to do is select the nucleus and the computer assigns the appropriate pulse sequence parameters for most normal cases. However, when you are dealing with an unusual structure or an uncommon nucleus, you may have to do some experimenting with the parameters to get the best spectroscopic data.

3.5 LINE WIDTHS, LINESHAPE, AND SAMPLING CONSIDERATIONS

3.5.1 Line Widths and Peak Shape

A typical signal peak in a frequency-domain NMR spectrum is characterized by several features. Most important are its frequency (its position in the spectrum, Δv_i) and intensity (peak height, or more accurately, the area under the peak). But it is also characterized by a *shape* and a **halfwidth**. The halfwidth ($v_{1/2}$) of a signal is the width (in hertz) of the peak at half height (Figure 3.18). The shape, which is determined during the FT process, is described as **Lorentzian** (as opposed to, say, Gaussian; see review problem 3.3). A Lorentzian signal peak (L_v) conforms to the lineshape equation

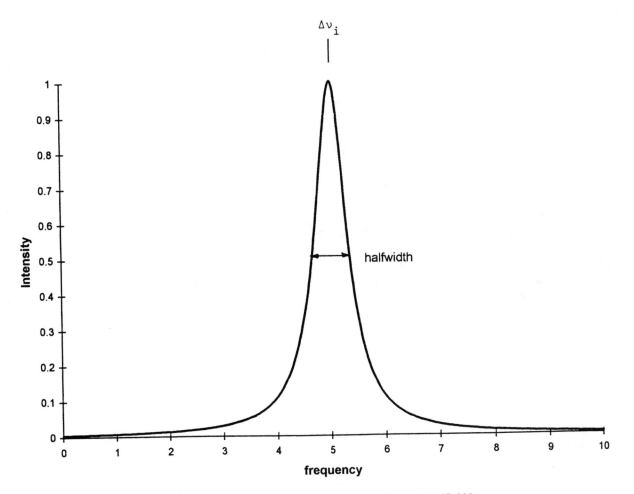

Figure 3.18. Lorentzian NMR signal peak centered at Δv_i, with halfwidth $v_{1/2}$.

$$L_v = \frac{a\,\Delta v_i^2}{\Delta v_i^2 + b(v - \Delta v_i)^2} \qquad (3.10)$$

where a is the maximum peak height at the center of the signal (Δv_i), v is any point along the frequency axis, and b (a dimensionless quantity) is inversely related to the square of the halfwidth by Eq. (3.11). In generating Figure 3.18 the parameters

$$v_{1/2} = 2\,\frac{\Delta v_i}{\sqrt{b}} \qquad (3.11)$$

were $a = 1.00$ intensity unit, $\Delta v_i = 5.00$ frequency units, and $b = 200$. Most of the signal peaks we encounter in NMR spectra will appear much sharper than the one in Figure 3.18, but this is only because the halfwidth of a typical NMR signal is 0.5 Hz out of a spectral width of several thousand hertz.

In Section 1.4 we found that the uncertainty principle establishes spectroscopic time scales. It also has a significant impact on the halfwidth of signals. One version of Eq. (1.6)

says that the uncertainty in precessional frequency of a nucleus (Δv) is inversely related to the uncertainty in lifetime of the spin state (Δt). If we use $v_{1/2}$ as a measure of the uncertainly in frequency, it must be inversely proportional to the lifetime of the spin state. What controls the lifetime of the spin state? How fast it relaxes, of course! So, for Δt we can substitute the relaxation time T:

$$v_{1/2} = \frac{1}{T} \qquad (3.12)$$

In other words, the halfwidth of a spectroscopic signal is inversely proportional to the lifetime (relaxation time) of the species giving rise to the signal. Nuclei that are slow to relax (large values of T) give sharp signals; nuclei that relax rapidly give broad signals. The halfwidth is therefore controlled by the *fastest* type of relaxation (smallest value of T). For liquids and solutions, this corresponds to T_2^* values (Section 2.5). Typical NMR halfwidths for solution samples range from about 0.1 to 0.5 Hz, except when dynamic chemical processes are occurring (Chapter 10). By the way, do not confuse halfwidth, which is controlled by the rate of spin–spin relaxation, with signal *intensity*. In a pulsed-mode spectrum, intensity *increases* as spin–lattice relaxation time (T_1) decreases.

3.5.2 Sampling Considerations

As we saw in Section 3.2, the intensity of an NMR signal is determined in part by the number of nuclei giving rise to the signal. We normally use solution samples to obtain reasonable relaxation times and halfwidths. It is typical to dissolve approximately 5–10 mg of a substance of average molecular weight (say, 200 daltons) in 0.4 mL of deuteriated solvent. Using special microtechniques, it is possible in some cases to obtain ^{1}H spectra on as little as 10 µg of sample!

When picking a solvent there are several considerations. First, and most obvious, the solvent must be inert, and it must dissolve the sample! Of course, if your sample is sparingly soluble or if there is only a fraction of a milligram of sample available, you can usually resort to signal-averaging techniques to obtain the desired *S/N*. Besides that, it is best if the solvent does *not* contain any nuclei of the isotope to be examined.

Make sure you use thoroughly cleaned and dried NMR tubes, and never assume that a new tube, fresh out of the box, is clean to NMR standards.

3.6 MEASUREMENT OF RELAXATION TIMES

At this point it would be useful to review many of the important topics covered in the first three chapters by discussing a few of the actual techniques used to measure relaxation times.

3.6.1 Review of Exponential Decay

Take a moment to look back at Figure 2.10. It shows how the magnitude of the *z*-component net magnetization vector **M**, when displaced from its Boltzmann equilibrium value, always recovers exponentially back to this equilibrium value at a rate determined by T_1, the spin–lattice (longitudinal) relaxation time. In this case, we will label this value $\mathbf{M}_0$, the equilibrium magnitude of **M** in the magnetic field. Since the magnitude of **M** at any time ($\mathbf{M}_t$) is solely dependent on the difference between the populations of the up and down spin states, we can recast Eq. (2.9) in the form

$$\mathbf{M}_t = \mathbf{M}_0 \left[1 - \exp\left(-\frac{t}{T_1}\right) \right] \quad \text{✳} \quad (3.13a)$$

which is exactly the equation for the first half of the curve in Figure 2.10, with $\mathbf{M}_\infty = \mathbf{M}_0$. Alternatively, this equation can be expressed in natural logarithmic form as

$$-\ln(\mathbf{M}_0 - \mathbf{M}_t) = \frac{t}{T_1} - \ln(\mathbf{M}_0) \quad (3.13b)$$

which shows that a plot of $-\ln(\mathbf{M}_0 - \mathbf{M}_t)$ versus time will be linear with slope $1/T_1$.

In an analogous manner, the decay toward zero of an NMR signal (i.e., the *x,y* component of **M**) by spin–spin (transverse) relaxation follows the equation

$$\mathbf{M}_t = \mathbf{M}_0 \left[\exp\left(-\frac{t}{T_2}\right) \right] \quad (3.14a) \quad \text{✳}$$

where $\mathbf{M}_0$ is now the magnitude of $\mathbf{M}_{x,y}$ the instant $\mathbf{B}_1$ is turned off. This equation describes a curve identical to the second half of Figure 2.10. And it too can be recast in logarithmic form:

$$\ln\left(\frac{\mathbf{M}_0}{\mathbf{M}_t}\right) = \frac{t}{T_2} \quad (3.14b)$$

showing that a plot of $\ln(\mathbf{M}_0/\mathbf{M}_t)$ versus time will be linear with slope $1/T_2$. (Here, $\mathbf{M}_\infty = 0$.)

3.6.2 Measurement of T_2^*

We have already hinted at two ways to measure T_2^*. In cases where the observed line width is due solely to *effective* (field-inhomogeneity-caused) spin–spin relaxation, the reciprocal of the halfwidth of the peak *is* T_2^* [Eq. (3.12)]. Also, in Section 3.4 we noted that the envelope of the FID signal follows exponential decay governed by T_2^*. In principle, we could measure the amount of time ($t_{1/2}$) required for the FID signal to decay to exactly half its initial magnitude ($\mathbf{M}_t = \frac{1}{2}\mathbf{M}_0$), at which point Eq. (3.14b) gives

$$T_2^* = \frac{t_{1/2}}{\ln 2} = \frac{t_{1/2}}{0.693}$$

In Figure 3.19 the envelope of the FID is shown by the dashed lines, and $t_{1/2}$ occurs at 0.33 time units.

3.6.3 Measuring T_2 by the Spin Echo Technique

In the usual case where T_2^* is less than T_2, the latter value cannot be determined from normal FID data. However, there is a way to measure "true" (as opposed to effective) T_2 values, and it involves a frequently encountered technique known as the spin echo.

Let us review what happens to $\mathbf{M}_0$ while and after it is irradiated by a $\mathbf{B}_1$ pulse. First, we set the power and pulse duration (pulse width t_p) to result in a flip angle α of 90° ($\pi/2$ rad). Since in the present case $\mathbf{B}_1$ is oriented along the *x′* axis of the rotating frame, we will refer to this as a $90_{x'}$ pulse. Figure 3.20*a* shows the initial orientation of **M** and Figure

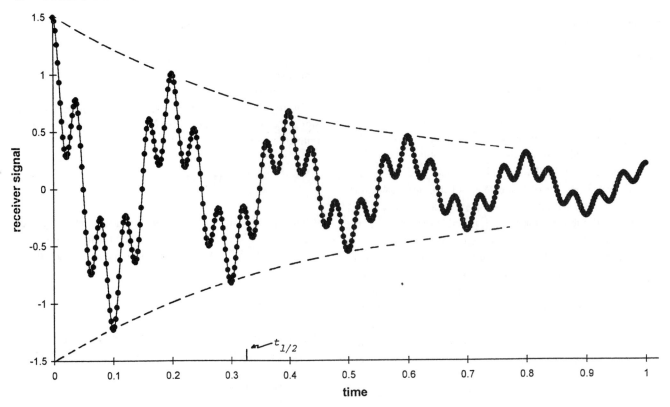

Figure 3.19. Data in Figure 3.15, with the signal envelope shown by the dashed lines. The half-life of the signal is $t_{1/2}$, about 0.33 time units.

3.20*b* the result of a $90_{x'}$ pulse: **M** has rotated (precessed) clockwise around the x' axis (as viewed along the x' axis toward the origin) and now lies along the $+y'$ axis.

At this point, with $\mathbf{B}_1$ off, the individual nuclear magnetic moments that comprise **M** begin to dephase (relax transversely) because inhomogeneities in the magnetic field $\mathbf{B}_0$ cause some of the nuclei to begin precessing slightly faster than ν_0, while others precess slightly slower (Figure 2.4). In the rotating frame, nuclei precessing faster than ν_0 will begin to move clockwise (when viewed from above), while the slower ones will move counterclockwise. We will depict this by a "fanning out" of **M** in the x',y' plane, as shown in Figure

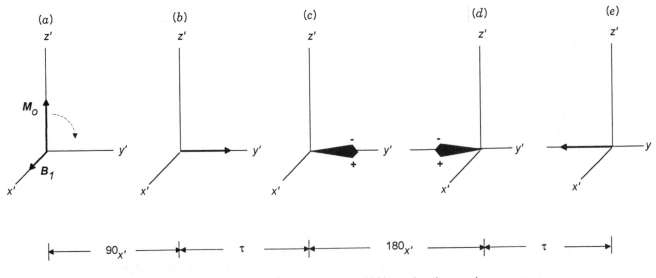

Figure 3.20. Pulse sequence and resulting behavior of **M** in a spin echo experiment; see text.

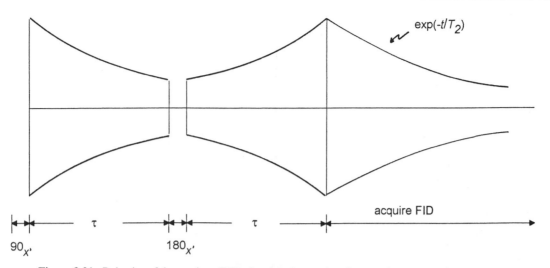

Figure 3.21. Behavior of the receiver (FID) signal during a spin echo experiment. Only the envelope of the FID is shown.

3.20c, where the "+" side of the "fan" indicates the faster moving nuclear moments and the "−" the slower ones.

We will let this fanning out evolve for a period τ, at which point we will deliver a $180_{x'}$ pulse. This second pulse will have the effect of rotating the entire fan 180° clockwise around the x' axis, leaving it centered on the $-y$ axis, as depicted in Figure 3.20d. Now think about what will occur next. The "+" nuclear moments will continue to move clockwise, the "−" ones counterclockwise, and if we again wait a period exactly equal to τ, all the individual nuclear moments will refocus on the $-y$ axis (Figure 3.20e), undoing all the effects of local inhomogeneities! This regeneration of **M** (and the corresponding FID signal) is termed the **spin echo**, and since **M** now lies in the x',y' plane, it generates an FID curve (Figure 3.21) that undergoes relaxation [Eq. (3.14b)] governed by the "true" spin–spin relaxation time T_2.

The spin echo technique, which has many uses in NMR spectroscopy, has introduced us to our first example of a multipulse sequence: $90_{x'}-\tau-180_{x'}-\tau-$acquire. We will see another example below and many more examples in Chapters 12 and 13.

3.6.4 Phase-Sensitive Detection and Spin Inversion–Recovery

Because of the way the FID signal is electronically sensed, there is information buried in the FID data other than simply the frequency, intensity, and relaxation time of the nuclei. The detector circuit also detects the *phase* of the signal. Back in Figure 1.1 we showed that the magnetic and electric vectors of an electromagnetic wave are "in phase," both reaching peaks and valleys at the same instant. Two waves are exactly out of phase when one reaches a peak at the same instant the other reaches a valley. The phase of an FID signal is always

compared to the phase of ν_0, which is taken to define the $+y'$ axis of the rotating frame. Thus, whenever **M** has a component on the $+y'$ axis, the resulting signal is said to have a positive phase (with respect to ν_0). Fourier transformation of such a signal results in a normal (upward) Lorentzian peak. If, instead, **M** is aligned with the $-y'$ axis, it is exactly 180° out of phase with ν_0. The FT of this FID will lead to a negative (downward) peak.

Take a moment to look at review problem 2.7 and Figures A.2 and A.3 in Appendix 1. In this problem the 0.20-ms pulse resulted in an *inversion* of **M** (a flip angle of 180° resulting from a $180_{x'}$ pulse). Then **M** decayed exponentially back to M_0 as described by the equation

$$\mathbf{M}_t = \mathbf{M}_0\left[1 - 2\exp\left(-\frac{t}{T_1}\right)\right] \qquad (3.15)$$

The extra 2 in this equation, compared to Eq. (3.13a), arises from the fact that, when fully inverted, **M** must relax from $-\mathbf{M}_0$, through zero, to $+\mathbf{M}_0$, or a total of $2\mathbf{M}_0$. If we could measure the exact time (t_z) required for **M** to reach zero (the asterisk in Figure A.3), we could then calculate T_1 by recasting Eq. (3.15) in the form

$$T_1 = \frac{t_z}{\ln 2} = \frac{t_z}{0.693} \qquad (3.16)$$

There is a problem, however. Since **M** is located along the z axis at all times in this experiment, it cannot generate an FID signal! So how can we tell when **M** reaches t_z? Take a moment to see if you can come up with a way.

Suppose that after the $180_{x'}$ pulse we allow **M** to begin to relax [by Eq. (3.15)] for a period τ that is *less* than t_z; then we apply a $90_{x'}$ pulse. This will rotate the residual **M** onto the $-y$

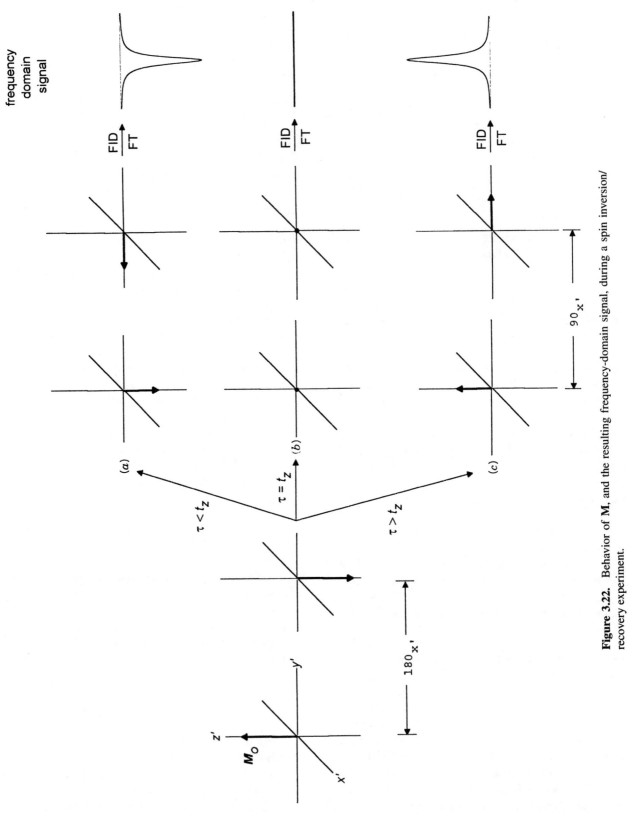

Figure 3.22. Behavior of **M**, and the resulting frequency-domain signal, during a spin inversion/recovery experiment.

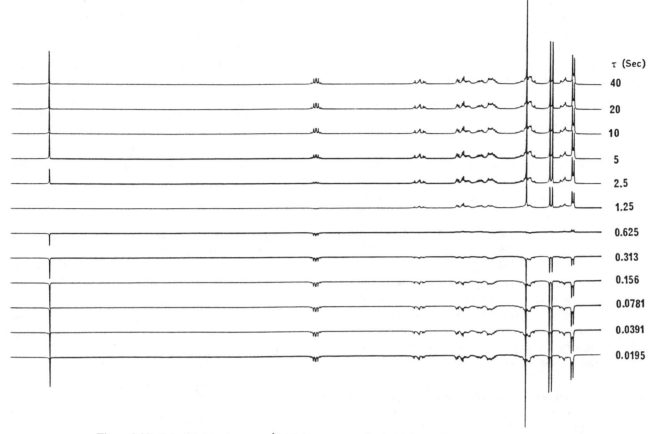

Figure 3.23. Spin inversion/recovery ^{1}H NMR spectrum of artimisinin as a function of delay time (τ). (Courtesy of David Lankin.)

axis, resulting in an FID signal whose FT is a *negative* frequency-domain peak (Figure 3.22a). If, on the other hand, τ is *greater* than t_z, **M** will have become positive during the delay, and the subsequent $90_{x'}$ pulse will put it on the $+y$ axis, leading ultimately to a positive peak (Figure 3.22c). However, if we happen to hit it just right and set τ *equal* to t_z, **M** will have vanished and the $90_{x'}$ pulse will have nothing to rotate—no FID, no peak (Figure 3.22b).

In practice this **inversion–recovery** experiment is done by programming the computer to carry out the following sequence of events automatically:

1. Set the value of τ to zero.

2. Carry out the pulse sequence $180_{x'}$–τ–$90_{x'}$–FID acquisition–FT (save the spectrum).

3. Increment the value of τ by $\Delta\tau$ ($<<t_z$) and repeat step 2.

4. Repeat step 3 until τ has been incremented through about $3t_z$.

When the frequency-domain data are plotted out, each peak in the spectrum will reach a minimum (flat line) when $\tau = 0.693T_1$ for the nuclei giving rise to that signal. Provided you have picked a small enough value for $\Delta\tau$, you will be able to determine the T_1 value for each set of nuclei in the molecule of interest in just this one experiment.

■ **EXAMPLE 3.10** In the complex molecule artemisinin (the structure of which can be seen in Chapter 13) each hydrogen exhibits its own signal. Figure 3.23 shows the ^{1}H spectrum of this compound, obtained by the inversion–recovery pulse sequence. Estimate the value of T_1 for each signal.

☐ *Solution:* Most of the signals go through zero intensity at a τ_z value of approximately 0.625 s. From Eq. (3.16), this corresponds to a T_1 value of $0.625/0.693 = 0.90$ s. The two-line signal furthest to the right-hand side of the spectrum relaxes slightly faster (becomes positive at a smaller value of τ_z), while the left most signal relaxes more slowly ($T_1 = 1.25/0.693 = 1.80$ s). ☐

The above multipulse inversion–recovery sequence is our first example of a 2D NMR spectrum. That is, it is a plot of signal intensity versus both the frequency of the signal ($\Delta\nu_i$)

and the length of the delay (τ). We will see many other examples of 2D pulse sequences in Chapter 13.

Before we can venture much further into the subject of NMR, we must be able to look at any molecular structure and decide how many signals of each type we can expect. To do this, we will need a little training in symmetry, and that is our next topic.

CHAPTER SUMMARY

1. A modern NMR spectrometer consists of the following components:

   a. A superconducting magnet capable of sustaining a strong, stable, homogeneous magnetic field. The nominal field strength of the magnet is maintained by electronic locking, and homogeneity is maintained by shimming.

   b. A probe within the magnetic field made up of a sample cavity, a transmitter coil (in which the oscillating magnetic field $\mathbf{B}_1$ is generated), and a receiver coil (in which the NMR signal is generated by Faraday induction). In modern pulsed-mode spectrometers, one coil serves both purposes.

   c. Appropriate electronic circuitry, a computer, and peripheral devices to detect, process, and display the NMR signals.

2. Older continuous wave (cw) NMR spectrometers were designed to operate in one of two modes:

   a. Frequency sweep, where the magnetic field strength was fixed and the $\mathbf{B}_1$ irradiation frequency was varied (swept).

   b. Field sweep, where the irradiation frequency was fixed and the field strength was varied by means of sweep coils.

   In either case, only one isotope at a time can be examined.

3. The signal-to-noise ratio (S/N) of an NMR signal improves with $\sqrt{n}$, where n is the number of times (scans) the signal is measured and averaged [Eq. (3.4)].

4. The pulsed-mode technique for acquiring an NMR spectrum involves the following steps once the sample is immersed in the magnetic field:

   a. Irradiation of the sample with a short but powerful $\mathbf{B}_1$ pulse of rf radiation. The pulse width (t_p) determines the range of frequencies covered [Eq. (3.5)].

   b. Digital acquisition of the free induction decay (FID) signal induced in the receiver coil as a function of time. The length of time between acquisition of each data point is called dwell time [t_d, Eq. (3.6)]. The total length of time that the FID signal is monitored is called the acquisition time (t_{acq}), which is inversely related to the spectroscopic resolution [Eq. (3.8)].

   c. If signal-to-noise considerations require FID signal averaging, multiple pulse sequences are carried out. There is a delay (t_w) within each pulse sequence to allow the nuclei to relax back to spin equilibrium (at a rate governed by T_1).

   d. Fourier transformation of the FID (time-domain) signal into a frequency-domain spectrum [Eq. (3.7)], including certain curve-smoothing procedures.

5. The NMR signal peaks are Lorentzian in shape. The halfwidth ($v_{1/2}$) of the peak is normally a function of T_2^*, the effective spin–spin relaxation time, which in turn is mainly dependent on inhomogeneities in the magnetic field.

6. An NMR spectrometer can only be configured to examine one isotope at a time, resulting in a homonuclear spectrum. Because the magnetic field strength $\mathbf{B}_0$ is fixed, each different isotope requires a different $\mathbf{B}_1$ oscillator (operating) frequency v_0.

7. When preparing solution samples for NMR, care should be taken in the choice of a solvent. It is best if the solvent is deuterated, inert, and lacks any nuclei of the target isotope.

8. The spin echo technique is a multipulse sequence used (among other things) to measure true T_2 values. The spin inversion–recovery techniques is a multipulse 2D technique for measuring T_1 values.

ADDITIONAL RESOURCES

1. Derome, A. E., *Modern NMR Techniques for Chemical Research*, Pergamon, Oxford, 1987.

2. King, R. W., and Williams, K. R., "The Fourier Transform in Chemistry," a four-part series in the *Journal of Chemical Education*: Part 1, *66,* A213 (1989); Part 2, *66,* A243 (1989); Part 3 (which also includes a useful glossary of terms), *67,* A93 (1990); Part 4, *67,* A125 (1990).

REVIEW PROBLEMS (Answers in Appendix 1)

3.1. From memory, draw a cutaway diagram of a superconducting magnet assembly (including the probe). Describe the purpose of each part of the assembly.

3.2. You are using a superconducting NMR spectrometer that has a proton operating frequency of 250 MHz and a computer with 64K of RAM to examine the ^{31}P spectrum of a compound. Based on prior experience, you know that all the frequency-domain ^{31}P signals (Δv_i) fall within a spectral width of 32,350 Hz, and you require a resolution of 1 Hz. You may assume that the various relaxation times of ^{31}P resemble those of ^1H.

(a)   What is the ^{31}P operating frequency of this instrument? (b) What values of the data acquisition parameters t_p, t_d, and t_{acq} would be appropriate? (c)   The frequency-domain spectrum from the initial FID exhibits a *S/N* of 2.4, but you require a value of at least 15 to see the detail you need. How many scans (repetitions) of the pulse sequence must be signal averaged to attain this? (d) What will be the total time required for data acquisition?

3.3.   A Gaussian lineshape (G_v) has the appearance of the curve in Figure 3.11, and conforms to the equation

$$G_v = a \exp\left[\frac{-(v - v_0)^2}{b^2}\right] \qquad (3.17)$$

where a is the maximum peak height at v_0 and v is any point along the frequency axis. Derive the relationship between b and $v_{1/2}$.

4

A LITTLE BIT OF SYMMETRY

4.1 SYMMETRY OPERATIONS AND DISTINGUISHABILITY

Before we can begin to predict the appearance of NMR spectra, we must be able to recognize when the nuclei (i.e., atoms) in a given structure will be distinguishable and when they will not. The test of distinguishability is based on symmetry relations among the nuclei, and it is these we will now explore.

Perhaps the best one-word synonym for symmetry is *balance*, but our use of the term requires a more detailed analysis. A **symmetry operation** is defined as some actual or imagined manipulation of an object that leaves it *completely indistinguishable* from the original object in *all* ways, including orientation. Doing *nothing* to an object leaves it indistinguishable, but we will not consider this trivial **identity operation** further. Symmetrical objects are said to possess **symmetry elements**, which are centers (points), axes, or planes that help describe symmetry operations. An object that possesses no symmetry elements is labeled *asymmetric*.

Imagine a perfect glass sphere. Assuming that we leave its center of mass unmoved, how might we manipulate it and still leave it indistinguishable? For one thing, we could rotate it around any imagined axis passing through its center, and the sphere would appear unchanged. Not only that, but we could also rotate it any number of degrees around such an axis, and it would still be unchanged. Such a symmetry element is called a C_∞ **rotational axis**, for reasons that will become clear shortly. Moreover, there is an infinite number of such axes passing through the center of the sphere.

Here is another manipulation we could perform. Suppose we were to pass an imaginary mirror directly through the center of the sphere. Certainly the mirror image of the left half would be indistinguishable from the actual right half, and vice versa. Thus, a sphere is said to have a **plane of symmetry**

(symbol σ) because one half of the object is the mirror image of the other half. In fact, the sphere has an infinite number of such mirror planes.

No other objects are as perfectly symmetrical as a sphere. Still, many objects, and most molecules, have at least some symmetry. Let's discuss a few examples.

Chloromethane, structure **4-1**, has a nearly tetrahedral structure, where the chlorine and hydrogen atoms form the corners of a pyramid (tetrahedron) with carbon at its center.

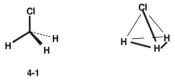

4-1

This molecule possesses several types of symmetry. Consider Figure 4.1. If you rotate the molecule around an imaginary axis passing through the carbon and chlorine atoms (colinear with the C–Cl bond), you'll notice that every 120° rotation brings each hydrogen to a position previously occupied by another hydrogen, leaving the structure as a whole indistinguishable from the original. These hydrogens are said to be *related by* this operation and are therefore *indistinguishable* from one another. Of course, after the third 120° rotation, we would return to where we started. And that is why this axis is called a C_3 **axis:** The object becomes indistinguishable from the original three times during one complete revolution (360°). To generalize, an object with a C_n **axis** becomes indistinguishable n times during one 360° rotation (once every 360°/n degrees).

Notice also that if we were to rotate **4-1** by any angle *other* than an integer multiple of 120°, the molecule would be *distinguishable* from the original. For example, rotation of the

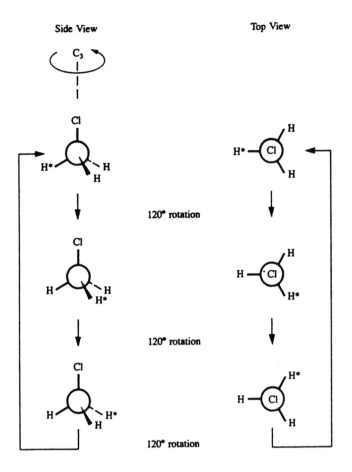

Side View   **Top View**

Figure 4.1. The C_3 axis of chloromethane. The circle represents the central carbon atom. One hydrogen is starred to follow its position during rotation. It is actually indistinguishable from either of the other hydrogens.

top structure in Figure 4.1 around its C_3 axis by 60° or 180° would give structures **4-1′** and **4-1″**, respectively. Do you see how these differ in orientation from the original structure, while all the structures in Figure 4.1 are indistinguishable?

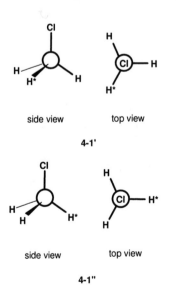

side view   top view

4-1′

side view   top view

4-1″

■ **EXAMPLE 4.1** A certain object has a C_6 axis. (a) How many times does it become indistinguishable from the original during one full rotation around the axis? (b) How many degrees of rotation bring it to an indistinguishable orientation?

☐ *Solution:* (a) From the definition of a C_n axis, this object becomes indistinguishable six times during one 360° rotation around the axis. (b) 360°/6 = 60°. Now you see why the rotational axes of a sphere are C_∞. The sphere becomes indistinguishable an infinite number of times during one complete rotation. ☐

Reflection in a mirror plane is also a symmetry operation. You may have realized that chloromethane also has three planes of symmetry, one for each C–H bond (and each containing the chlorine atom), as shown in Figure 4.2. Reflection in any of these planes relates one of the remaining (out-of-plane) hydrogens to the other, so mirror symmetry (like the C_3 axis) tells us that the three hydrogens are indistinguishable.

Here is the most important thing to remember. Atoms (i.e., nuclei) that are related by virtue of a symmetry operation are said to be *symmetry* (or *chemically) equivalent*. As such, they are indistinguishable in all respects, and they are **isochronous**; that is, they precess at *exactly* the same frequency. Thus, symmetry-equivalent nuclei *cannot* be distinguished by NMR. Returning to the case of chloromethane, the three hydrogens are symmetry equivalent by virtue of both the C_3 axis and the symmetry planes and therefore give rise to only one signal in the ^1H NMR spectrum of the compound.

Next, consider the three **isomers** (molecules with the same molecular formula but somehow a different structure) of dichloroethylene, structures **4-2**, **4-3**, and **4-4**:

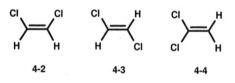

    4-2   **4-3**   **4-4**

Each of these structures is planar; that is, all six atoms in each molecule lie in the same plane, called the *molecular plane*. What symmetry relationships can you find in these structures? Figure 4.3 shows that, in all three cases, the pairs of hydrogens and chlorines are symmetry equivalent by virtue of C_2 axes and/or σ planes. Note also that for each of these three planar structures, the molecular plane itself constitutes a plane of symmetry. But this symmetry element does not relate any of the atoms to each other because all the atoms lie *in this plane*.

■ **EXAMPLE 4.2** How many ^1H, ^{13}C, and ^{35}Cl signals would you expect from each of the isomers in Figure 4.3?

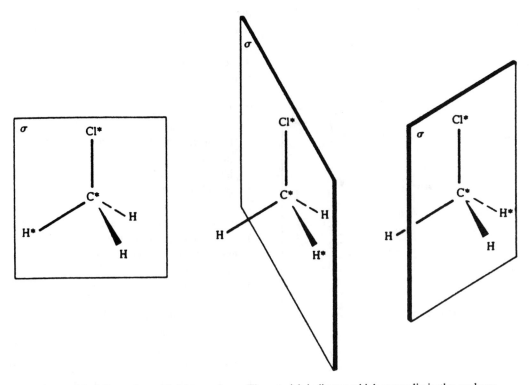

Figure 4.2. Mirror planes of chloromethane. The asterisk indicates which atoms lie in the σ plane; the remaining hydrogens are related by the σ plane.

☐ *Solution:* In each case the two hydrogens are symmetry equivalent, as are the two chlorines. Therefore, each isomer would exhibit a single ^1H signal and a single ^{35}Cl signal. In both **4-2** and **4-3** the two carbons are also equivalent, giving one ^{13}C signal in each case. However, in **4-4** the carbons are not related by any symmetry element, so this molecule would give rise to two ^{13}C signals.   ☐

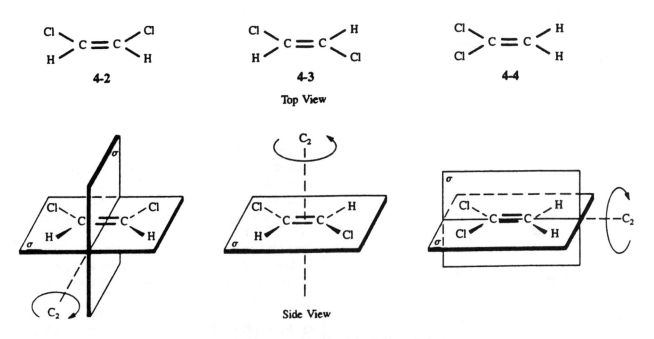

Figure 4.3. Symmetry elements of the dichloroethylenes.

■ **Example 4.3** (a) What symmetry elements can you discover for the planar molecule benzene, structure **4-5**? (b) How many ^{1}H and ^{13}C NMR signals would you predict for benzene?

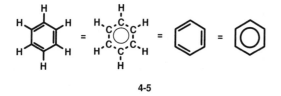

4-5

□ *Solution:* (a) All six hydrogens, as well as all six carbons, are symmetry equivalent by virtue of a C_6 axis perpendicular to the ring and passing through its center. Additionally, the molecule has six C_2 axes in the plane of the ring, six σ planes perpendicular to the ring, and one more σ in the plane of the ring. Some of these are shown in Figure 4.4. Can you find the rest? (b) Because of the equivalencies, benzene exhibits one ^{1}H signal and one ^{13}C signal. □

There is another type of symmetry element that, though less common than simple planes and axes, is nonetheless useful. If a molecule has a **center of symmetry** (symbol *i*, located at its center of mass), every atom in the structure has an indistinguishable companion atom located the same distance from the center, along a line through the center connecting the atoms. Though it lacks any planes or axes of symmetry, structure **4-6** has a center of symmetry. Do you see how each pair of companion atoms (carbons, hydrogens, chlorines, and bromines) are equivalent by virtue of the *i*?

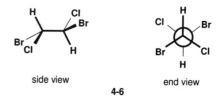

side view   end view
4-6

■ **EXAMPLE 4.4** Which other molecules (if any) described previously in this chapter possess a center of symmetry?

□ *Solution:* Structures **4-3** and **4-5** each possess a center of symmetry. □

4.2 CONFORMATIONS AND THEIR SYMMETRY

In Chapter 5 we will discuss the ^{1}H and ^{13}C NMR spectra of toluene (**3-1**, Section 3.4). Examine the two structures of toluene shown in Figure 4.5. These are referred to as **conformational structures** (or conformations, for short) because they are related by rotation around a single bond (in this case the ring-to-CH$_3$ bond), a process that is very facile. There is actually an infinite number of such conformations, differing only in the angle of rotation around this bond.

There is a significant difference in symmetry between these two conformations of toluene. In structure **B**, one of the methyl (CH$_3$) hydrogens (H*) is *in* the symmetry plane containing the benzene (or **phenyl**) ring, while in structure **A**, H* is in a symmetry plane *exactly perpendicular* to the ring. Careful inspection will reveal that the σ plane in structure **A** makes the two unstarred methyl hydrogens equivalent to each other, the two *ortho* ring hydrogens (H$_o$) and carbons equivalent to each other, and the two *meta* ring hydrogens (H$_m$) and carbons equivalent to each other. On this basis, we would

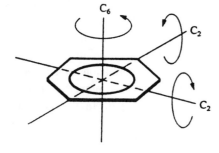

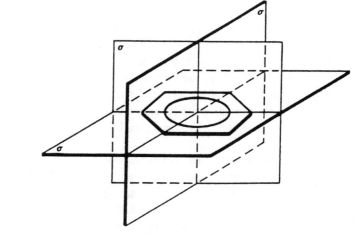

Figure 4.4. Some symmetry elements of benzene.

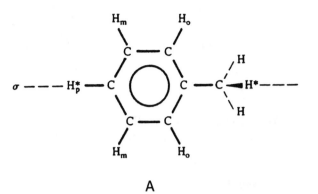

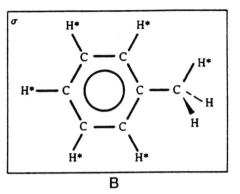

A

B

Top View

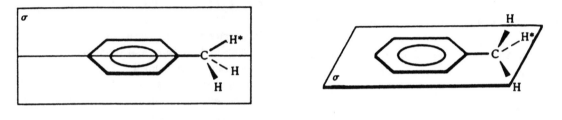

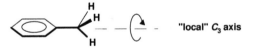

Side View

Figure 4.5. Symmetry elements of two conformations of toluene. The symbol H* indicates the hydrogen in the σ plane; *o*, *m*, and *p* indicate the ortho, meta, and para positions.

expect structure **A** to exhibit five carbon signals (the methyl carbon and the four sets of phenyl ring carbons) and five hydrogen signals (two types of methyl hydrogens and three sets of ring hydrogens). Note that toluene *lacks* the C_6 axis present in benzene (Example 4.3).

■ **EXAMPLE 4.5** (a) What symmetry element(s) can you find in structure **B**, Figure 4.5? (b) How many ¹H and ¹³C NMR signals do you predict for this conformation?

□ *Solution:* (a) There is a σ plane containing all six ring carbons, the five ring hydrogens, the methyl carbon, and one methyl hydrogen (H*). The two remaining methyl hydrogens are related by this plane and are thus equivalent. (b) Because of the location of H*, all seven carbons and all ring hydrogens are nonequivalent. Therefore, there will be six carbon signals and seven hydrogen signals. □

It will turn out in Chapter 5 that the actual ¹H and ¹³C NMR spectra of toluene match quite closely our expectations for structure **A**, with one important exception. We find that the three methyl hydrogens always prove to be equivalent. Why is this so? The answer lies in the rate at which the ring–methyl bond rotates. Because this process is extremely rapid on the NMR time scale (Section 1.4), the methyl group behaves as

if it had a *local* C_3 axis, rendering all three hydrogens symmetry equivalent *on the time average* (i.e., dynamically equivalent):

Thus, the methyl group can be treated as if it has threefold symmetry, without regard to the rest of the molecule. But remember, this is true only because of the high rate of rotation. If this rate of interconversion between the conformations were slow on the NMR time scale, we would indeed expect to see multiple ¹H signals from the methyl group hydrogens.

To summarize, different conformations of a molecule can exhibit different NMR spectroscopic properties. However, if the conformations interconvert rapidly (as they normally do), they generate a time-averaged spectrum that reflects the *most symmetrical conformation* of the molecule. In order to observe spectra of the separate conformations, it would be necessary to slow down the rotation of some of the single bonds in the molecule, and this can sometimes be done by cooling the sample to a very low temperature. There will be more about these dynamic processes in Chapter 10.

4.3 HOMOTOPIC, ENANTIOTOPIC, AND DIASTEREOTOPIC NUCLEI

Nuclei that are equivalent by virtue of any C_n axis ($n > 1$) are said to be *homotopic*. The hydrogens in **4-1** through **4-5** all fall into this category. A set of homotopic nuclei gives rise to only one NMR signal.

Consider now the structure of bromochloromethane, structure **4-7**. Careful inspection will reveal the presence of a σ plane containing the bromine, carbon, and chlorine atoms and bisecting the H–C–H angle:

4-7

Therefore, these two hydrogens are symmetry equivalent (and isochronous) since they are related by reflection in the mirror plane. But because they are not related by any axis of symmetry, they are not homotopic. They are instead referred to as *enantiotopic* hydrogens, ones related only by a symmetry plane.

If you have difficulty deciding whether two nuclei are related by a mirror plane, there is another test for whether or not two atoms are enantiotopically related; it is called the **isotope substitution test**. To make the test, mentally substitute first one, then the other, suspected atom with a different isotope, and compare the two resulting structures. If the two structures are **enantiomers** (i.e., non-superimposable mirror images, as are your left and right hands), the two suspected nuclei are said to be enantiotopically related. In the case of structure **4-7** the two substituted structures would be **4-7A** and **4-7B**:

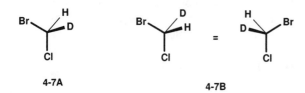

4-7A   **4-7B**

These two structures are, in fact, non-superimposable mirror images, though you may wish to verify this by constructing molecular models of each one. We can summarize by saying that all enantiotopic nuclei are symmetry equivalent (by virtue of a σ plane) but not all symmetry-equivalent nuclei are enantiotopic. To prove this, carry out the isotope substitution test on the equivalent nuclei in structures **4-1** through **4-5** and confirm that no two are enantiotopic.

Structures **4-8** and **4-9** are more challenging examples:

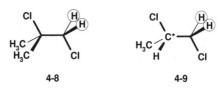

4-8   **4-9**

Are the two circled hydrogens in each structure equivalent or not? For structure **4-8** (in the conformation shown) there is a σ plane (can you find it?) that relates the two circled hydrogens. Application of the isotope substitution rule should also convince you that they are enantiotopic. But that same σ plane is *absent* in structure **4-9** because the carbon bearing the asterisk has four *different* substituents attached to it (H, Cl, CH$_3$, and CH$_2$Cl). In structure **4-8** the analogous carbon has only three different groups (the two methyls, like the two circled hydrogens, are enantiotopic and therefore equivalent). A tetrahedral atom with four different attached groups is called a **chiral** (or asymmetric) **center**. Applying the isotope substitution test to structure **4-9**, we generate structures **4-9A** and **4-9B**, which are not enantiomers, but rather are **diastereomers** (stereoisomers that are not enantiomers):

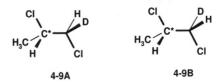

4-9A   **4-9B**

Therefore, the two circled hydrogens in **4-9** are said to be *diastereotopically related*, or simply *diastereotopic*. They are *not* equivalent. And this relationship persists even though rotation of the C*–C bond is fast on the NMR time scale. Here is the all-important bottom line: While NMR normally cannot distinguish between enantiotopic nuclei (but see Section 10.7.3), NMR *can* distinguish between diastereotopic nuclei (at least in principle).

■ **EXAMPLE 4.6**   (a) Indicate all homotopic nuclei, enantiotopic nuclei, and diastereotopic nuclei in structure **4-10**. You may assume rapid rotation of all bonds. (b) How many ^{1}H and ^{13}C NMR signals would you predict for the compound?

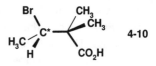

□ *Solution:* (a) First, notice that the carbon marked with an asterisk is a chiral center, so the molecule as a whole is asymmetric. In each methyl group the three hydrogens are equivalent by virtue of the rapid rotation of the C–C bonds (local C_3 axes). However, the two methyl groups attached to the same carbon are diastereotopic (and therefore distinguishable), and each is also distinguishable from the third methyl group. The remaining nuclei are all unique and therefore distinguishable.

(b) Each of the three methyl groups will exhibit its own ^1H and ^{13}C signal. In addition, the three remaining carbons and two remaining hydrogens will each give a separate signal, for a total of six carbon and five hydrogen signals. □

4.4 ACCIDENTAL EQUIVALENCE

In Chapter 3 we intimated that the true value of NMR comes from its ability to distinguish *identical* nuclei (e.g., ^1H) that have *nonidentical* molecular environments. And, as stated in the previous section, we can in many cases even distinguish two hydrogens attached to the same carbon if they are diastereotopic. However, it occasionally happens (especially in older low-field ^1H spectra) that two nuclei that are not symmetry equivalent in any way accidentally precess at exactly the same frequency and, hence, give rise to a single NMR signal. Such nuclei are said to exhibit *accidental equivalence*. The two classic examples of this are propyne (**4-11**) and cyclopentanone (**4-12**). Both compounds would be expected to exhibit two hydrogen signals, yet both (at 60 MHz) display only one!

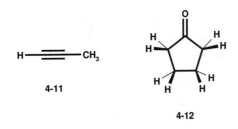

4-11   4-12

■ **EXAMPLE 4.7** (a) Verify that structure **4-12** would be expected to show only two hydrogen signals. (b) How many carbon signals are expected?

□ *Solution:* (a) The four hydrogens closer to the C=O form a symmetry-equivalent set by virtue of two σ planes and a C_2 axis. (Can you find these symmetry elements?) The same is true of the two carbons to which these hydrogens are attached. The four remaining hydrogens form a second set of equivalent hydrogens, and the carbons to which they are attached form a second set of equivalent carbons. Thus, there are only two hydrogen signals expected. (b) The two sets of carbons, plus the C=O carbon, give rise to three carbon signals. □

Now that we have an idea of what to expect, we can proceed to the next chapter and examine some actual NMR spectra.

CHAPTER SUMMARY

1. A symmetry operation is the manipulation (e.g., rotation or reflection) of an object in such a way that it is left indistinguishable in all respects from the original object. There are three principal types of symmetry elements: centers, axes, and planes.

2. An *n*-fold axis of symmetry (designated C_n) is an imaginary axis passing through the center of the object. Rotation around this axis by $360°/n$ leaves the object indistinguishable from the original.

3. A plane of symmetry (designated σ) is an imaginary (mirror) plane through the center of an object. The half of the object on one side of the plane is the mirror image of the half on the other side.

4. A molecule has a center of symmetry (symbol *i*) when every atom in the structure has an indistinguishable companion atom located the same distance from the center, along a line through the center connecting the atoms.

5. Conformations of a molecule are the instantaneous arrangements of the atoms in the molecule that interconvert by rotation around its single bonds.

6. Nuclei that are related by one or more symmetry elements are indistinguishable by NMR methods. They are said to be symmetry or chemically equivalent and precess at identical frequencies.

7. Nuclei that are related by an axis of symmetry are called homotopic.

8. Nuclei that are related only by a plane of symmetry are termed enantiotopic.

9. Nuclei that are not symmetry equivalent but nonetheless precess at the same frequency are described as accidentally equivalent.

ADDITIONAL RESOURCES

1. Mislow, K., *Introduction to Stereochemistry*, W. A. Benjamin, New York, 1966.

2. Orchin, M., and Jaffe, H. H., *Symmetry, Orbitals, and Spectra*, Wiley Interscience, New York, 1971.

REVIEW PROBLEMS (Answers in Appendix 1)

4.1. Are the hydrogens in toluene (**3-1**) homotopic or enantiotopic, assuming rapid rotation of the ring–methyl bond?

4.2. Describe the type and location of symmetry elements in each of the cyclopropane derivatives below. Indicate which nuclei are equivalent, and specify how many ^{1}H and ^{13}C NMR signals each structure will exhibit.

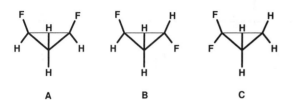

4.3. (a) What symmetry, if any, can you find in the conformation of 1,1,2-trifluoroethane shown below? How many ^{1}H NMR signals would you expect for this structure?

(b)   Redraw the molecule in its most symmetrical conformation. How many ^{1}H NMR signals would you now expect for this compound?

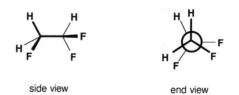

side view   end view

4.4. Are the hydrogens in structure **4-6** (page 51) enantiotopic?

5

THE ^{1}H AND ^{13}C NMR SPECTRA OF TOLUENE

5.1   THE ^{1}H NMR SPECTRUM OF TOLUENE AT 80 MHz

The ultimate value of any type of spectroscopy depends on our ability to interpret accurately the spectroscopic data we acquire. These data are usually plotted as a spectrum, which is nothing more than a graph of intensity versus frequency (also called *position*) for each signal. In this chapter we will examine and interpret the actual ^{1}H and ^{13}C NMR spectra of toluene.

From our discussion in Chapter 4 of its symmetry properties, we decided that toluene (structure **3-1**) should exhibit four ^{1}H signals: one for the three equivalent methyl hydrogens, one for the two equivalent ortho hydrogens (H_o), one for the two equivalent meta hydrogens (H_m), and one for the para hydrogen (H_p):

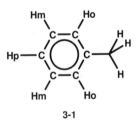

3-1

Figure 5.1 shows the actual ^{1}H frequency-domain NMR spectrum of toluene generated using pulse-mode technology on an instrument with a conventional electromagnet (field strength of 1.88 T) and an operating frequency of 80 MHz. The sample solution consisted of 25.5 mg of toluene and 0.40 mL of CDCl$_3$ solvent (containing 0.2% TMS, see below) in a 5-mm tube.

Not only does this spectrum show three signals instead of four, but the signal at "zero" is actually from another source

to be discussed later. Thus, toluene itself exhibits only two ^{1}H signals at 80 MHz, one at 188 Hz downfield (to the left) of zero and another at 573 Hz. The hertz scale is shown at the top of the spectrum. The ratio *S/N* (Section 3.3.3) exceeds 40, and the spinning sidebands are too small to be observed.

Recall from Section 3.2 that although it is virtually impossible to measure *absolute* frequencies with the high degree of precision and reproducibility necessary in NMR, it is possible to measure frequency *differences* very precisely. This is the reason that NMR signal (precession) frequencies are actually measured as the difference ($\Delta\nu_i$) between the precession frequency and the instrument's oscillator (operating) frequency: $\Delta\nu_i = \nu_i - \nu_0$.

Recognizing that there is a slight unavoidable drift in absolute frequencies but a constancy of frequency differences, how might we devise a reproducible scale for the position (frequency) axis of an NMR spectrum? A little thought should suggest the answer. We dissolve in each sample solution a small amount (<1% by volume) of a standard **internal reference compound** that gives rise to a sharp signal somewhat apart from the other signals of interest. Although the reference signal will drift along with the other signals from the sample, the difference in frequency between the reference signal and any other signal is always constant. We will *arbitrarily* assign the reference signal the frequency value zero and measure frequency differences in hertz downfield or upfield (to the right) of the reference signal. To differentiate the instrumentally measured signal frequency ($\Delta\nu_i$) from the signal position relative to the reference signal, we will use the symbol $\delta\nu_i$ (= $\Delta\nu_i - \Delta\nu_{ref}$) for the latter.

Because we can only examine one isotope at a time (Section 3.3), the reference signal must come from the *same isotope* as the nuclei of interest, e.g., a ^{1}H reference signal for ^{1}H spectra, a ^{13}C reference signal for ^{13}C spectra, and so on.

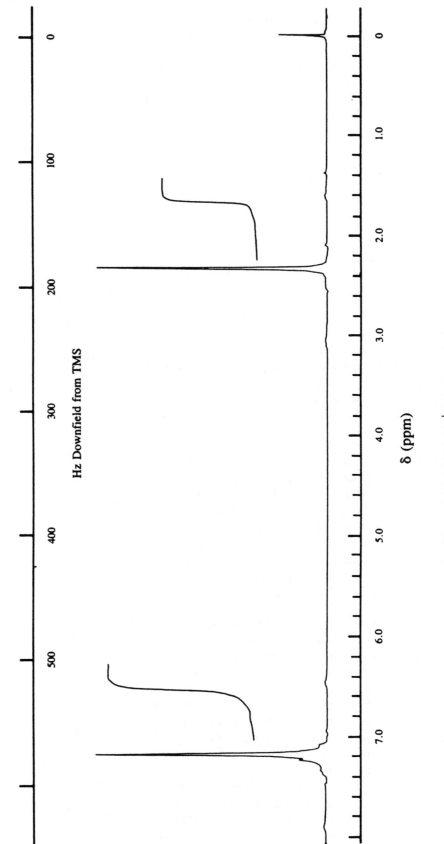

Figure 5.1. The 80-MHz ^{1}H spectrum of toluene.

The reference compound should be chemically inert, soluble in a variety of solvents, and readily removable if the sample is to be reclaimed. A perfect candidate in the case of ^{1}H, ^{13}C, and ^{29}Si spectroscopy is **tetramethylsilane** (TMS, structure **5-1**) in which all 12 hydrogens are symmetry equivalent (why?), as are the 4 carbons. The use of TMS as reference is so widespread that you may assume its signal defines zero frequency in all ^{1}H and ^{13}C spectra. TMS is responsible for the upfield zero marker signal in Figure 5.1. Unfortunately, TMS is not water soluble, so for samples requiring an aqueous (D_2O) medium, a water-soluble analog of TMS known as DSS (or SDDSS), compound **5-2**, is available. In this case the ^{1}H or ^{13}C signal of the $(CH_3)_3Si$ group serves as reference (see review problem 5.1):

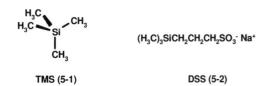

TMS (5-1)   **DSS (5-2)**

In some cases, notably ^{31}P NMR, the chosen reference compound is not inert toward most samples. One way around this problem is to put the reference compound in a small capillary tube, which itself is carefully centered within the sample tube. Such a reference compound is referred to as **external**, that is, not dissolved in the sample solution.

Reference compounds for other nuclei are more varied, and some of these are listed in Table 5.1. It is always good practice to verify what the reference compound is before attempting to interpret an NMR spectrum.

Perhaps this discussion of reference compounds may remind you of the lock substances mentioned in Section 3.2. Indeed, in some older cw spectrometers, the reference signal was used as the lock signal. There were two major problems with this, however. First, it required large amounts of the reference compound to be dissolved in the sample, and second, it was very difficult to observe signals located near the lock signal because of electronic interference between them. In modern spectrometers this problem is avoided by using a **heteronuclear lock**, that is, a lock substance containing none of the isotope that generates the signals of interest. Thus, a deuterium (^{2}H) lock is used in most cases.

Now that we have agreed on a position scale (hertz downfield or upfield from TMS, in the case of ^{1}H and ^{13}C spectra), we consider the scale for intensity. In Figure 5.1 the signals at 188 and 578 Hz seem to have about the same peak height, but signal height alone is not the best measure of intensity. Look at the three signals in Figure 5.2. Although each one has a different height, the actual intensities of the signals are equal. This is because signal intensity is best measured by the area under the curve, which is roughly equal to the peak height times halfwidth (Section 3.5.1).

TABLE 5.1   NMR Reference Compounds^a

Nucleus	Reference Compound
^{1}H	$(C\underline{H}_3)_4Si$ (TMS)^b
^{2}H (D)	$(CH_3)_3(CD\underline{H}_2)Si^c$
^{13}C	$(\underline{C}H_3)_4Si$ (TMS)^d
^{15}N	$(CH_3)_4N^+I^-$, NH_3, or CH_3NO_2
^{19}F	$CCl_3\underline{F}$
^{31}P	85% $H_3\underline{P}O_4$ (external)

^{a}The signal of the underlined atom(s) is used as reference.
^{b}DSS [$(\underline{H}_3C)_3SiCH_2CH_2CH_2SO_3^-Na^+$] is used for aqueous samples.
^{c}Natural abundance 0.2% in normal TMS.
^{d}DSS [$(H_3\underline{C})_3SiCH_2CH_2CH_2SO_3^-Na^+$] is used for aqueous samples.

■ **EXAMPLE 5.1**   Using a ruler, measure the peak height and halfwidth of each signal in Figure 5.2 and confirm that all three are equally intense.

☐ *Solution:* Measure the halfwidth of each peak and multiply this by the height of the peak. The exact numbers you calculate will depend on the scale of your ruler, but the three products should be essentially identical.   ☐

Look again at Figure 5.1. Can you measure halfwidths accurately enough to decide if the downfield signal is *really* more intense? In general, this is difficult to do, especially for sharp signals. But do not despair! All NMR spectrometers can electronically *integrate* (i.e., find the relative area under) any

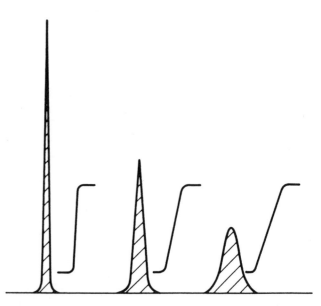

Figure 5.2. Three signal peaks of different height and halfwidth but equal area.

peak in the spectrum. This integral can be displayed numerically or shown by an S-shaped line like those beside each signal in Figures 5.1 and 5.2. The *vertical* displacement of the "integral" line is proportional to signal area and, hence, intensity.

The integrals in Figure 5.1 tell us that the ratio of intensity of the downfield signal to that of the upfield signal is 4.9 : 3.0. This is very significant, because (all other things being equal) the intensity of an NMR signal is directly proportional to the number of nuclei giving rise to that signal.

Since the closest integer ratio of the integrals for toluene is 5 : 3 and there are eight hydrogens in the molecule, we can infer that the downfield signal represents five hydrogens and the upfield signal three. But which hydrogens go with which signals?

Because toluene exhibits fewer ^1H signals than we expected on the basis of symmetry considerations (Chapter 4), there must be accidental equivalence (Section 4.4) of some of the nuclei. Which ones would you expect to have the most similar molecular environments? Judging from the structure of toluene and the integral ratios, it would be a reasonable guess (and is indeed correct!) to assign the downfield signal to the five accidentally equivalent hydrogens attached to the phenyl ring and the upfield signal to the three equivalent methyl hydrogens.

5.2   THE CHEMICAL SHIFT SCALE

A new problem now emerges. Even though we all agree to use TMS as our reference signal, the spectroscopic data, when expressed as δv (hertz downfield from TMS), will vary from one spectrometer to another if their operating frequency (and magnetic field) are different. This is again because precessional frequencies, and any differences between them, are directly proportional to field strength [Eq. (3.1)]. Is there a system we could all adopt that would give the same numerical position scale regardless of the spectrometer's field strength or operating frequency? Indeed, there is! We'll define a new quantity called the **chemical shift** of nucleus i (δ_i) by the equation

$$\delta_i = \frac{10^6 (\Delta v_i - \Delta v_{ref})}{v_0} = 10^6 \frac{\delta v_i}{v_0} = \frac{\delta v_i}{v_0'} \qquad (5.1)$$

where v_0 and v_0' are the operating frequencies in hertz and megahertz, respectively. The δ scale is actually dimensionless, but its "dimension" is often expressed as ppm (parts per million) as a consequence of the factor of 10^6 in Eq. (5.1).

■ **EXAMPLE 5.2**   (a) Calculate the chemical shifts of the two toluene signals using the 80-MHz data. (b) What would have been the chemical shift of each signal if a 250-MHz instrument had been used?

□ *Solution:* (a) For the upfield signal at 80-MHz

$$\delta_{upfield} = \frac{\delta v}{v_0'} = \frac{188 \text{ Hz}}{80 \text{ MHz}} = 2.35 \text{ ppm}$$

$$\delta_{downfield} = \frac{\delta v}{v_0'} = \frac{578 \text{ Hz}}{80 \text{ MHz}} = 7.23 \text{ ppm}$$

(b) Exactly the same! The δ scale is field and operating frequency independent.   □

■ **EXAMPLE 5.3**   The 250-MHz ^1H NMR spectrum of methyllithium (CH_3Li) shows a signal at 525 Hz upfield of TMS. What is its chemical shift?

□ *Solution:* A signal upfield of TMS has a negative δv_i (because $\delta v_{ref} > \delta v_i$), so its δ value will also have a negative value:

$$\delta (CH_3Li) = \frac{\delta v}{v_0'} = -\frac{525 \text{ Hz}}{250 \text{ MHz}} = -2.10 \text{ ppm}$$
□

As Example 5.3 shows, signals downfield from the reference signal have positive chemical shifts, and those upfield of the reference signal have negative chemical shifts. In the older literature there was some confusion about whether the upfield or downfield signals should carry positive chemical shifts, but the above convention has since been universally adopted.

Look back at Figure 5.1. In addition to the hertz scale, the δ scale is shown along the bottom axis. And here is something important to remember: Eq.(5.1) also requires that 1 δ unit equals 80 Hz at 80 MHz, 250 Hz at 250 MHz, and so on. For this reason, you can think of the units of v_0' as hertz per parts per million.

At one time an alternate ^1H chemical shift scale called the *tau (τ) scale* was used, defined by the relationship

$$\tau_i = 10 - \delta_i = 10 - \frac{\delta v_i}{v_0'} \qquad (5.2)$$

Although the τ scale is not used much any more, you may occasionally run across it in older literature. So beware!

■ **EXAMPLE 5.4**   Calculate the chemical shifts in τ units for the signals in Examples 5.2 and 5.3.

□ *Solution:* Substitute the δ values from Examples 5.2 and 5.3 into Eq. (5.2). For toluene

$$\tau_{upfield} = 10 - \delta_{upfield} = 10 - 2.35 = 7.65$$

$$\tau_{downfield} = 10 - \delta_{downfield} = 10 - 7.23 = 2.77$$

For methyllithium

$$\tau\,(CH_3Li) = 10 - \delta(CH_3Li) = 10 - (-2.10) = 12.10 \quad \square$$

5.3   THE 250- AND 400-MHZ ^{1}H NMR SPECTRA OF TOLUENE

By using a newer instrument equipped with a superconducting magnet and the same toluene sample as before, we can generate the 250-MHz ^{1}H spectrum of toluene shown in Figure 5.3. In this figure, only the δ scale is shown.

■ **EXAMPLE 5.5**   (a) What field strength is needed to generate a 250-MHz ^{1}H spectrum? (b) What is the chemical shift of the methyl signal in Figure 5.3? (c) What is the position of the methyl signal in hertz downfield of TMS ($\delta\nu$)?

(d) The tiny spinning side bands for the methyl signal are marked with asterisks. Estimate the spinning rate.

□ *Solution:* (a) 5.87 T (see Table 2.1). (b) Even before we look at Figure 5.3 we know the methyl signal should have a chemical shift of δ 2.35 (Example 5.2). Although we probably cannot interpolate the δ scale quite that precisely, notice that the exact δ value is written numerically at the top of the peak (2.34679, but usually reported only to three decimal places). Such a numerical presentation of chemical shifts is called **peak picking**. (c) Using a rearranged version of Eq. (5.1):

$$\delta\nu_{250\,MHz} = \delta\,(250\,Hz/ppm) = 2.347\,(250) = 586.75\,Hz$$

(d) At 250 MHz, each δ unit equals 250 Hz, so each tick on the position axis equals 25 Hz. The spacing between

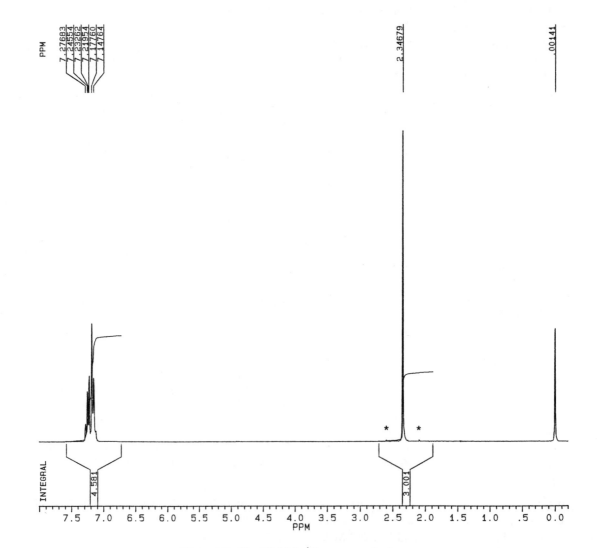

Figure 5.3. The 250-MHz ^{1}H spectrum of toluene.

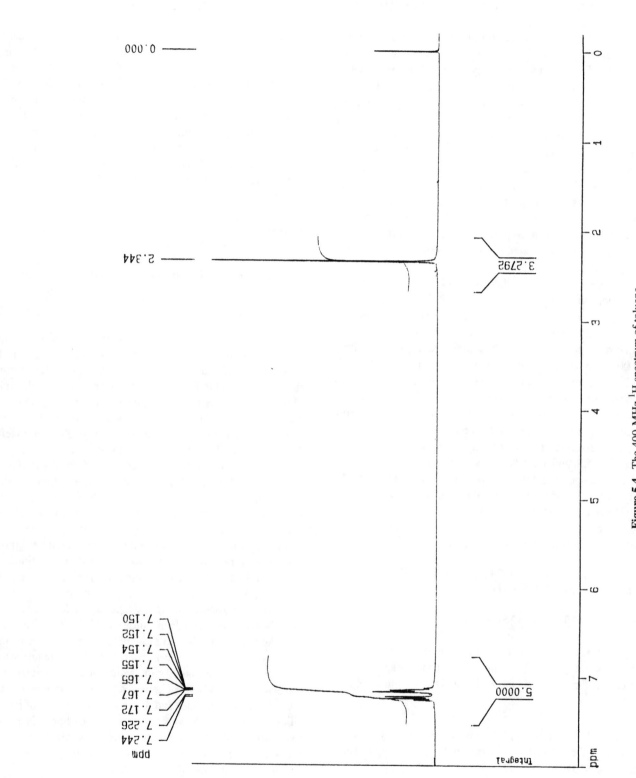

Figure 5.4. The 400-MHz ^{1}H spectrum of toluene.

each sideband and main peak is therefore approximately 65 Hz, and this equals the spin rate (see Example 3.5). □

The integral in this spectrum is shown both numerically and with the S-shaped curve. The numerical integration indicates a 4.58 : 3.00 ratio of phenyl ring hydrogens to methyl hydrogens, acceptably close to 5 : 3, the error due mainly to differences in relaxation times. But do you notice anything else different about this spectrum, compared to the 80-MHz spectrum? Certainly the signal-to-noise ratio is much improved, now exceeding 200. But even more important, the ring hydrogen signal is now split into six lines, rather than being a single peak as in Figure 5.1. This is because we are operating at such a high field strength that signals previously accidentally equivalent are now capable of being resolved. Two of these lines are separated by as little as 0.0129 ppm (3.2 Hz), and this is why the level of spectroscopic resolution [Eq. (3.8)] is so important.

Figure 5.4 shows the 400-MHz ^{1}H spectrum of toluene obtained on an instrument with a 9.39-T superconducting magnet using the same sample solution as before but this time in a 10-mm tube. It looks quite similar to the 250-MHz spectrum; the chemical shift of the methyl group is within 0.002 ppm (0.8 Hz) of the previous value, the integral ratio is 5.0 : 3.3, and *S/N* exceeds 100. Notice that the complex downfield signal has separated even further into two separate multiline signals. And the small signal at δ 1.46 is due to a trace of moisture (water) in the sample.

The lesson to be gained from comparisons between Figures 5.1, 5.3, and 5.4 is that for a given sample the higher the field (and operating frequency) of an NMR spectrometer, the higher is *S/N* and the better resolution it provides.

If you are wondering why there are more lines for the ring hydrogen signals than the three we expected on the basis of symmetry, you will find out in Chapter 8!

5.4   THE ^{13}C NMR SPECTRUM OF TOLUENE AT 20.1, 62.9, AND 100.6 MHZ

5.4.1   The 20-MHz ^{13}C Spectrum

Figure 5.5 is the ^{13}C NMR spectrum of toluene, obtained with the same sample and same instrument (same magnetic field strength) that provided the 80-MHz ^{1}H spectrum in Figure 5.1.

■   **EXAMPLE 5.6**   (a) What operating frequency was used to obtain the 20-MHz ^{13}C spectrum? (b) How many hertz correspond to each δ unit?

□   *Solution:* (a) The field strength that generated a ^{1}H frequency of 80-MHz is 1.88 T [from Section 5.1 or by solving Eq. (2.6) for B_0]. From this value, we calculate an actual operating frequency of 20.1 MHz. (b) At 20.1 MHz, each δ unit equals 20.1 Hz. □

Once again, TMS (its carbon signal) is used to define "zero" on the chemical shift scale. Furthermore, the three small signals near δ 77 ppm arise from the carbon atom in solvent ($\underline{C}$DCl$_3$), which also provides the deuterium for the lock signal. (The reason CDCl$_3$ exhibits three carbon signals, rather than just one, will be explained in Section 8.6.2.) Besides those peaks, there are five others exactly in accord with our expectations (Section 4.2). The exact positions of these signals, obtained by peak picking because visual interpolation is too imprecise, are δ 21.4, 125.6, 128.5, 129.2, and 137.6.

■   **EXAMPLE 5.7**   What is the position ($\delta\nu_i$) of each ^{13}C toluene signal in hertz downfield from TMS?

□   *Solution:* Multiply the chemical shift of each signal by 20.1 Hz, to give 430, 2525, 2583, 2597, and 2766 Hz, respectively. □

Because of the similarity in relative position of the signals in this spectrum to those in the previously discussed ^{1}H spectra, it seems reasonable (and is indeed correct!) to assign the upfield signal to the methyl carbon and the four relatively close signals to the four nonequivalent phenyl ring carbons. But wait! If the above assignments are correct, should the intensity ratios of these five signals not be (from right to left) 1 : 2 : 2 : 1 : 1? The answer is yes, and no.

It turns out there are several factors that control the *relative* intensity of NMR signals generated by the pulsed-mode technique. Besides the relative numbers of nuclei giving rise to the signal, the most important of these are relative T_1 relaxation times and the **nuclear Overhauser effect** (NOE). The dependence on spin-lattice relaxation time is far more significant with the pulsed-mode method than with the continuous-wave method. When using signal-averaging procedures, rapidly relaxing nuclei give stronger signals than slow-relaxing ones. This is because rapidly relaxing nuclei return more quickly to a Boltzmann distribution (Section 2.3) and are thus less likely to approach saturation. In the case of ^{1}H nuclei, the difference in relaxation times is fairly small, and therefore the relaxation factor is roughly the same for all ^{1}H nuclei in a molecule. With ^{13}C nuclei, however, there can be large differences in relaxation times and thus a significant span of intensities for equal numbers of nonequivalent carbons. One way to circumvent the relaxation problem is to provide long delay times (Section 3.4) between pulses, but because of time constraints, this is not always practical.

The nuclear Overhauser effect is the enhancement of intensity of an NMR signal generated by one nucleus when it is near another nonequivalent nucleus being simultaneously irradiated. We will discuss this effect in more detail later

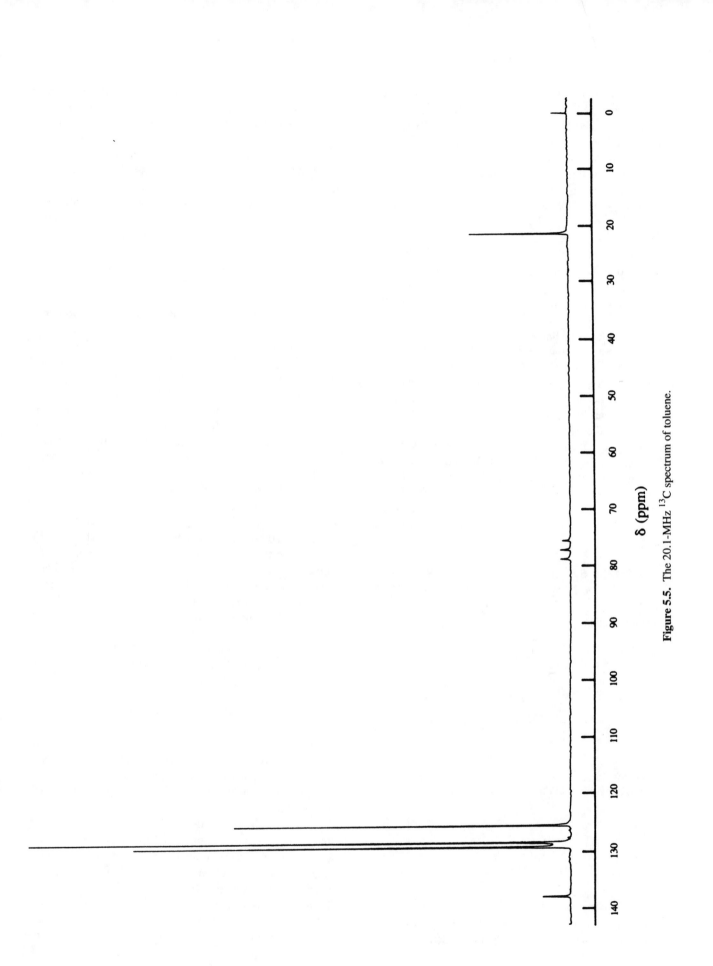

Figure 5.5. The 20.1-MHz ^{13}C spectrum of toluene.

(Chapter 12), but for now, here is its most significant result: Among carbons with otherwise similar molecular environments (e.g., tetrahedral), the more hydrogens directly attached to a given carbon, the more intense will be its signal. This fact helps us assign the phenyl ring carbon signals to the various positions in toluene. The weakest signal (at δ 137.6) must correspond to the carbon lacking hydrogens, the one bearing the methyl group. Each of the remaining ring carbons has a single hydrogen attached, but there are two equivalent ortho carbons, two equivalent meta carbons, and only one para carbon. The second weakest signal at 125.6 ppm can therefore be assigned to the para carbon. The remaining two signals are harder to assign, especially since their relative intensity depends on the exact parameters of the pulse sequence (see Table 5.2). We will learn more in Chapter 7 about assigning such signals, but it turns out that the δ 128.5 signal results from the meta carbons, while the δ 129.2 signal is due to the ortho carbons. Empirically, we notice that the closer a ring carbon is to the methyl group, the farther downfield its signal occurs.

Finally, notice that the signal for the methyl carbon, with three hydrogens directly attached, is slightly *less* intense than the signal for the para carbon (with one hydrogen attached). This is because the molecular environment of the methyl carbons involves sp^3 hybridization (four single bonds), while the phenyl ring carbons are all sp^2 hybridized (two single bonds and one aromatic double bond). These differences in bonding environment (and the associated changes in relaxation times) more than offset the Overhauser enhancement factor.

■   **EXAMPLE 5.8**   (a) How many signals do you expect to see in the ^{13}C NMR spectrum of cyclohexanone, structure **5-3**? (b) Based on NOE considerations, which signal should be the least intense?

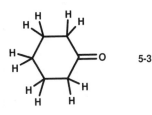   **5-3**

□   *Solution:* (a) Symmetry considerations (Chapter 4) lead us to conclude that there are four sets of carbons:

(b) Of the four signals, the carbonyl (C═O) signal will be weakest, for the carbonyl carbon lacks any hydrogens attached directly to it (see problem 6 in Self-Test I).   □

There is another very significant difference between the ^{1}H spectra we saw previously and this ^{13}C spectrum. Did you notice it? The ^{1}H signals of toluene differ in chemical shift by only about 5 ppm, which would correspond to 100 Hz at 20 MHz. The ^{13}C signals, on the other hand, occupy a span of nearly 120 ppm (2400 Hz)! In fact, the ^{1}H signals of most known compounds show up in a fairly narrow range of chemical shift, about δ −5–15 ppm (5000 Hz at 250 MHz), while ^{13}C signals span a range of 250 ppm (15715 Hz at 62.9 MHz). These spans, in fact, determine the spectral width discussed in Section 3.4. For this reason, the chance of accidental equivalence is far *smaller* in the case of ^{13}C spectra than with ^{1}H spectra.

5.4.2   The 62.9- and 110.6-MHz ^{13}C Spectra of Toluene

Before looking at Figures 5.6 and 5.7, try to predict the effect on the ^{13}C spectrum of toluene of increasing the field strength by factors of 3.125 and 5.0 using the same instruments and samples that provided the 250- and 400-MHz ^{1}H spectra.

Although the signal positions will spread out (when measured in hertz), the chemical shifts (δ) should remain unaffected. Relative intensities might vary a little because of different pulse parameters. But, by and large, the spectra should not change significantly, since there are no accidental equivalencies to resolve.

Now look at the 62.9-MHz ^{13}C spectrum in Figure 5.6. You are right! Except for the relatively larger signals for the solvent (δ 76.5–77.5), there are no significant changes! The chemical shifts of the toluene signals, now determined with a higher degree of both precision and accuracy, are all within a small fraction of a δ unit of the previous values. The relative intensities of the ring carbon signals have changed little, and we would have no difficulty in recognizing these to be ^{13}C spectra of toluene.

Figure 5.7 shows the 100.6-MHz ^{13}C spectrum. Again it is superimposable on the 62.9-MHz spectrum, except for one thing: See review problem 5.4.

5.5   DATA ACQUISITION PARAMETERS

As mentioned in Section 3.4, the success of the pulsed-mode NMR technique depends on the careful selection of data acquisition parameters. These data for the six spectra discussed in this chapter appear in Table 5.2. The same sample was used for all the spectra.

Notice the following trends. First, as expected, it takes many more scans to generate an adequate ^{13}C spectrum than a ^{1}H spectrum because of the lower sensitivity of the carbon nuclei (Table 2.1). On the other hand, the higher the field and operating frequency, the fewer scans needed to produce the desired *S/N*. Because the carbon spectral width is substantially larger than that of hydrogen at the same field strength, the dwell time is *shorter* for carbon, and since this fills the

TABLE 5.2 **Data Acquisition Parameters for NMR Spectra of Toluene**[a]

Figure	5.1	5.3	5.4	5.5	5.6	5.7
Nucleus	^1H	^1H	^1H	^{13}C	^{13}C	^{13}C
ν_0 (MHz)	80	250	400	20.1	62.9	110.6
SW (Hz)	1200	5000	4200	5000	21739	17857
Scans	8	8	4	2000	512	64
t_p (μs)	5.0	6.0	20.0	6.0	2.0	15.0
t_{acq} (s)	3.0	0.819	1.95	0.77	0.377	0.917
t_d (μs)	367	100.0	119.0	94	23.0	28.0
t_w (s)	3.0	3.0	1.0	15	3.0	1.00
S/N	>40	>200	>100	>100	>30	>140

[a]Refer to Figure 3.17.

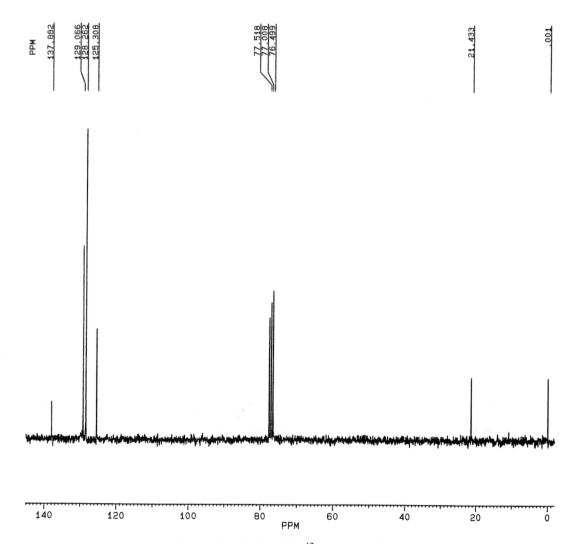

Figure 5.6. The 62.9-MHz ^{13}C spectrum of toluene.

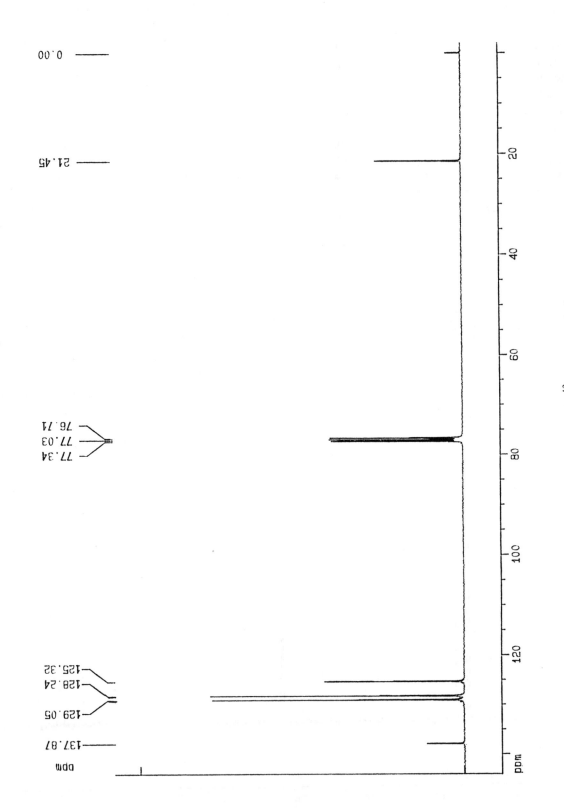

Figure 5.7. The 100.6-MHz ^{13}C spectrum of toluene.

microprocessor's memory faster, acquisition times are generally shorter for carbon than for hydrogen.

Before we leave the discussion of these ^{13}C spectra, there is one more qualifier to add. All three ^{13}C spectra discussed in this chapter (and all subsequent ones we will encounter, unless specifically noted to the contrary) involve simultaneous irradiation of the hydrogens as well as the carbons, a technique known as **^{1}H-spin decoupling**. (It is this simultaneous, or "double," irradiation that gives rise to the NOE mentioned in Section 5.4.1.) A ^{1}H-spin-decoupled ^{13}C spectrum is labeled $^{13}C\{^{1}H\}$. Spin decoupling will be described more fully in Chapter 12, but for now, suffice it to say that if we had not made use of this decoupling, these spectra would have looked far more complicated, and all intensity differences due to NOE would be absent.

So now we have seen and interpreted several actual ^{1}H and ^{13}C NMR spectra over a large span of frequency and field strength. We have made some empirical observations about factors that influence the position and intensity of the various signals. For example, there appears to be a rough correlation between the relative chemical shifts of ^{1}H nuclei and the relative chemical shifts of the carbons to which they are attached. In the next two chapters, our goal is to systematize a very large body of chemical shift data in order to be able to assign NMR signals to specific nuclei within any molecule and to predict where such signals might occur even before seeing the spectrum!

CHAPTER SUMMARY

1. Positions of NMR signals (Δv_i) are actually measured in hertz from the operating frequency ($\Delta v_i = v_i - v_0$). The signal position is then recast as δv_i, the frequency difference (in hertz) between Δv_i and the signal of a reference compound ($\delta v_i = \Delta v_i - \Delta v_{ref}$). The reference signal comes from a standard reference compound added to the sample. For ^{1}H and ^{13}C NMR, the reference compound is tetramethylsilane (TMS). Finally, the signal position is quoted in field- and operating-frequency-independent δ (ppm) units: $\delta_i = \delta v_i / v_0'$, where v_0' is the operating frequency in megahertz.

2. The intensity of an NMR signal is directly proportional to the area (rather than peak height) of the signal. This area is measured by electronically integrating the

signal. The intensity of ^{1}H NMR signals is directly proportional to the number of hydrogens giving rise to each signal. The intensity of ^{13}C signals is normally not directly proportional to the number of carbons giving rise to the signal. This is because different carbons usually have a greater range of relaxation times and because of NOE enhancements.

3. The higher the operating frequency (and field) of an NMR spectrometer, the better the signal-to-noise ratio and the resolution it provides in NMR spectra.

REVIEW PROBLEMS (Answers in Appendix 1)

5.1. How many ^{1}H and ^{13}C signals will DSS (structure **5-2**) exhibit? What will be the relative intensity (integral) of the proton signals?

5.2. How many ^{1}H signals will cyclohexanone (structure **5-3**, Example 5.8) exhibit, assuming that the ring is flat on the NMR time average? What will be the relative intensity (integral) of these signals?

5.3. Suppose that the ^{31}P spectrum acquired in review problem 3.2 exhibits two signals, one at 6396 Hz downfield from the reference compound and the other at 3937 Hz upfield from the reference compound. (a) What is the reference compound? (b) What is the chemical shift in ppm for each of the above signals?

5.4. (a) Deuteriochloroform ($CDCl_3$), the common NMR solvent, is usually 99.9% deuteriated. The ^{1}H NMR spectrum of this solvent shows a very weak signal at δ 7.24. To what do you attribute this signal? (b) You may have noticed that although the chemical shift of the middle line of the ^{13}C signal for deuteriochloroform was the same in Figures 5.5, 5.6, and 5.7, the two outer lines had different chemical shifts in each spectrum. What are the separations in hertz between the outer lines and the middle line in Figures 5.6 and 5.7? What is the significance of these differences?

5.5. The following questions pertain to the data in Table 5.2 for Figure 5.6. (a) How many δ units are covered by the sweep width? (b) How many data points were acquired in each scan for the spectrum? (c) What is the resolution in the spectrum?

6

CORRELATING PROTON CHEMICAL SHIFTS WITH MOLECULAR STRUCTURE

6.1 SHIELDING AND DESHIELDING

From our discussion in Chapter 5, we can begin to appreciate the ability of NMR spectroscopy to act as a sensitive probe of molecular structure. We saw that the ^1H NMR spectrum of toluene readily distinguishes the methyl hydrogens from the ring hydrogens (and even the individual ring hydrogens from one another at sufficiently high field) and how the ^{13}C spectrum distinguishes all the carbons from one another. In this chapter we will explore semiquantitative relationships between the chemical shift of a given ^1H nucleus and its molecular environment.

Let us return for a moment to the 80-MHz ^1H NMR spectrum of toluene (Figure 5.1). In it we saw three signals: TMS (reference) hydrogens at δ 0.00 ppm, methyl hydrogens at 2.35 ppm, and ring hydrogens at 7.23 ppm. Suppose you were asked to indicate which of these hydrogens precesses with the highest frequency at a given field strength. Recall from Section 3.3 that frequency increases from *right to left* in a typical spectrum. Therefore, the TMS hydrogens precess slowest, while the ring hydrogens precess fastest *at a given field strength*. Conversely, we could say that *at constant frequency* the TMS hydrogen signals occur at the highest field (i.e., the TMS hydrogens require the highest field to enter resonance), while the ring hydrogen signals occur at the lowest field. If we could obtain an NMR signal for a "bare" ^1H nucleus (a proton free of all electrons and other molecular entanglements), its signal would appear downfield (higher frequency) than even the toluene ring hydrogens. Why does a bare proton precess at such a high frequency, while ^1H nuclei within molecules (and surrounded by electrons) precess more slowly?

The answer lies in a very simple fact: The electron cloud surrounding each nucleus in a molecule serves to *shield* that

nucleus from the external magnetic field. Let us see how this happens. Figure 6.1 depicts a comparison between a bare proton and one shielded by an electron cloud. Recall from Section 3.1.1 that an electric current moving through a wire generates a magnetic field. In an analogous way, the external magnetic field (B_0 in Figure 6.1) causes each electron pair surrounding the nucleus to circulate through its orbital in such a way as to generate an *induced* magnetic field (B_i) *opposed* to the external field. As a result, while a bare proton experiences the full magnitude of the external field, the shielded nucleus experiences an effective field (B_{eff}) that is equal to the external field minus the induced field:

$$B_{eff} = B_0 - B_i \qquad (6.1)$$

Because the strength of the induced field is directly proportional to that of the external field, we can define a **shielding constant** σ (not to be confused with a σ bond) that is a function of the exact molecular (i.e., electronic) environment of the nucleus:

$$B_i = \sigma B_0 \qquad (6.2)$$

The value of σ (ca. 10^{-5} for ^1H, larger for heavier atoms with many electrons) is dimensionless and field independent. Many advanced texts on NMR tabulate signal positions in terms of σ rather than in terms of δ (chemical shift). See Example 6.4.

Using Eq. (6.2), Eq. (6.1) can be recast in the form

$$B_{eff} = (1 - \sigma)B_0 \qquad (6.3)$$

Substituting this result into Eq. (2.6) gives

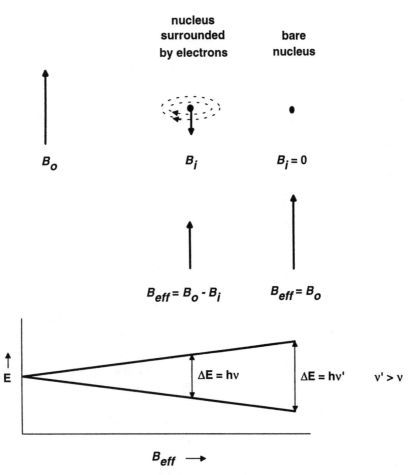

Figure 6.1. Effect of diamagnetic shielding. The dotted ellipses represent motion of electrons in their orbitals under the influence of **B**$_0$.

$$\nu_{\text{precession}} = \frac{\gamma(1 - \sigma)\, \mathbf{B}_0}{2\pi} \qquad (6.4)$$

Thus, the greater the shielding of the nucleus (the larger the value of σ), the lower will be its resonance frequency and the farther to the right it will appear in an NMR spectrum. Conversely, nuclei from which electron density has been withdrawn (resulting in a smaller σ) are said to be *deshielded* and appear toward the left of the spectrum (higher frequency).

From an earlier chemistry course you may recall the concept of **electronegativity**, the tendency of an atom *in a molecule* to attract bonding electrons toward itself. The electronegativity of an atom is a consequence of a high kernel charge (nuclear charge minus subvalence electrons) coupled with a small atomic radius. The more electronegative an atom, the lower the energy of its valence orbitals and the higher its ionization energy. Table 6.1 lists the Pauling electronegativity values for some of the common main-group elements. Note that fluorine is the most electronegative element, and electronegativity decreases as you move down or to the left within the periodic table (Appendix 2). It should come as no surprise that, in general, the more electronegative an atom or group of atoms, the greater its deshielding effect on neighboring atoms in the same molecule. The polarization (unequal sharing) of bonding electrons by virtue of differences in electronegativity among atoms is an example of **inductive effects**.

■ **EXAMPLE 6.1** (a) Arrange the following atoms from most electronegative to least: H, Li, C, Si. (b) Arrange the following underlined hydrogens from most deshielded to least: $\underline{H}_3C–H$, $\underline{H}_3C–Li$, $\underline{H}_3C–C$, $\underline{H}_3C–Si$.

TABLE 6.1   Pauling Electronegativities

H (2.1)					
Li (1.0)	Be (1.5)	B (2.0)	C (2.5)	N (3.0)	O (3.5)	F (4.0)
Na (0.9)	Mg (1.2)	Al (1.5)	Si (1.8)	P (2.1)	S (2.5)	Cl (3.0)
						Br (2.8)
						I (2.5)

☐ *Solution:* (a) C (2.5) > H (2.1) > Si (1.8) > Li (1.0). (b) $\underline{H}_3C–C > \underline{H}_3C–H > \underline{H}_3C–Si > \underline{H}_3C–Li.$ ☐

The type of shielding and deshielding we have been discussing so far is a consequence of the **isotropic** (spherically symmetric) distribution of paired (Section 2.1.2) electrons and is referred to as **diamagnetic shielding**. In the case of heavier atoms there can also be a nonsymmetrical (anisotropic) distribution of valence electrons, and circulation of these electrons in a magnetic field gives rise to an induced field aligned *with* the external field. This is called **paramagnetic shielding**. Thus, the net σ value for an atom includes both terms, $\sigma_{diamagnetic} - \sigma_{paramagnetic}$, and is proportional to the electron density around the nucleus.

■ **EXAMPLE 6.2** Does paramagnetic shielding cause a nucleus to precess at higher or lower frequency?

☐ *Solution:* Referring to Figure 6.1, we note that if $\mathbf{B}_i$ is aligned with $\mathbf{B}_0$, then $\mathbf{B}_{eff}$ will equal $\mathbf{B}_0 + \mathbf{B}_i$. This *larger* net field will cause the nucleus to precess at a higher frequency. Thus, paramagnetic "shielding," in a sense, deshields nearby nuclei, that is, causes them to precess at a higher frequency. ☐

In the case of hydrogen (which has its electrons in spherically symmetric orbitals), diamagnetic shielding is the predominant term. But for other nuclei (e.g., carbon, fluorine, and phosphorus) paramagnetic shielding (which has the opposite effect) becomes increasingly important. It is possible, by means of molecular orbital calculations, to estimate the electron distribution around each atom in a molecule and thereby to make quantitative estimates of shielding at each nucleus. However, we will adopt a semiempirical approach, trying to discern from large amounts of chemical shift data the likely position of signals for nuclei in a variety of molecular environments.

6.2 CHEMICAL SHIFTS OF HYDROGENS ATTACHED TO TETRAHEDRAL CARBON

6.2.1 Methyl Hydrogens

A carbon atom that is singly (i.e., σ) bonded to four other atoms is called **aliphatic**, **saturated**, or **tetrahedral** because the four attached atoms lie approximately at the corners of a tetrahedron (Section 4.1). Such a carbon is nominally sp^3 hybridized. The simplest molecule containing a tetrahedral carbon is methane, CH_4. If one of the hydrogens in methane is substituted by a different atom or group (X), the resulting molecule, CH_3X, is still tetrahedral at carbon and consists of a methyl group attached to X. Sequential substitution of two more hydrogens leads first to XCH_2Y (a methylene group with two substituents) and then to $XCH(Y)Z$ (a methine group with three substituents), both still tetrahedral at carbon.

So far we have discussed the ^1H chemical shifts of the methyl groups in toluene (δ 2.35 ppm), TMS (0.00 ppm), and CH_3Li (−2.10 ppm; Example 5.3) as well as the ring hydrogens of toluene (ca. 7.23 ppm). In Figure 6.2, the relative positions of these signals are depicted graphically. Of these, the CH_3Li has the most shielded hydrogens and the most upfield chemical shift, owing to the low electronegativity and resulting electron-donating effect of the lithium atom (Example 6.1). The methyl hydrogens in TMS are the next most shielded, due to the nearby electron-rich silicon atom. The methyl group in toluene is somewhat less shielded (or more deshielded), and the ring hydrogens are the most deshielded of all, for reasons we will explain in Section 6.5. One way to begin to organize such data, at least for the methyl groups, is to regard each of these compounds as a derivative (CH_3X) of methane whose ^1H signal occurs at δ 0.23 ppm. Next, we can define a **substituent parameter** ($\Delta\delta_X$) for substituent X by the equation

$$\Delta\delta_X = \delta(CH_3X) - \delta(CH_4) = \delta(CH_3X) - 0.23 \quad (6.5a)$$

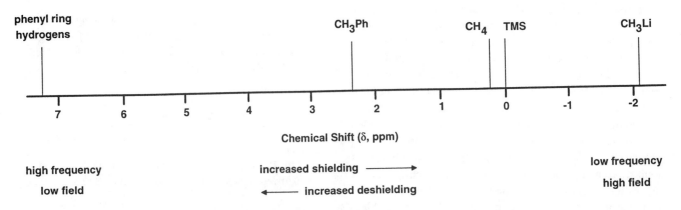

Figure 6.2. Partial ^1H chemical shift range.

■ **EXAMPLE 6.3** Calculate $\Delta\delta_X$ for the following X groups: Li, Si(CH$_3$)$_3$, H, and phenyl (C$_6$H$_5$).

☐ *Solution:* Using Eq. (6.5a),

$$\Delta\delta_{Li} = \delta(CH_3Li) - \delta(CH_4) = -2.10 - 0.23 = -2.33$$

In like manner, the $\Delta\delta_X$ values for Si(CH$_3$)$_3$, H, and phenyl are -0.23, 0.00, and 2.12, respectively, the same order we saw in Example 6.1. ☐

Notice from Example 6.3 that a positive value of $\Delta\delta_X$ corresponds to a downfield shift in signal position (resulting from deshielding of the nucleus by the substituent) and a negative value corresponds to an upfield shift (shielding of the nucleus by the substituent), relative to the effect of X = H.

■ **EXAMPLE 6.4** Derive the relationship between the shielding constant σ [Eqs. (6.2)–(6.4)] and the substituent parameter $\Delta\delta_X$. [*Hint:* You will need to remember Eq. (5.1).]

☐ *Solution:* From Eq. (6.5a)

$$\Delta\delta_X = \delta(CH_3X) - \delta(CH_4)$$

Using Eq. (5.1), we can recast this equation as

$$\Delta\delta_X = \frac{[\Delta\nu(CH_3X) - \Delta\nu(TMS)]}{\nu_0'} - \frac{\Delta\nu(CH_4) - \Delta\nu(TMS)}{\nu_0'}$$

$$= \frac{\Delta\nu(CH_3X) - \Delta\nu(CH_4)]}{\nu_0'}$$

Applying Eq. (6.4) gives us

$$\Delta\delta_X = \frac{[(1 - \sigma_X) - (1 - \sigma_H)]\gamma \mathbf{B}_0}{2\pi\nu_0'} = \frac{[(\sigma_H - \sigma_X)]\gamma \mathbf{B}_0}{2\pi\nu_0'}$$

Thus, a substituent (X) that is more strongly shielding than H ($\sigma_X > \sigma_H$) has a negative value of $\Delta\delta_X$ (corresponding to an upfield shift), while a deshielding substituent ($\sigma_X < \sigma_H$) has a positive $\Delta\delta_X$ (corresponding to a downfield shift). ☐

When ^{1}H NMR spectroscopy was in its infancy, J. W. Shoolery began analyzing the spectra of literally thousands of compounds in an attempt to extract average values of $\Delta\delta_X$ for a wide variety of common substituents in a number of different molecular environments. Over the years these values have been refined by him and others, and the concept has been extended to ^{13}C chemical shifts as well as to those of many other common nuclei (Chapter 7). With these values in hand,

it is possible to predict the chemical shift of a given hydrogen by using a rearranged version of Eq. (6.5a):

$$\delta_{calc} = \text{base value} + \sum (\Delta\delta_X) \qquad (6.5b)$$

where the base value is the chemical shift of the appropriate unsubstituted molecule (e.g., δ 0.23 for methane) and $\sum (\Delta\delta)$ is the sum of $\Delta\delta_X$ values for all contributing substituents.

Table 6.2 lists some of the most commonly encountered substituent groups and their average substituent parameters when attached to tetrahedral carbon. The values labeled $\Delta\delta_{\alpha-X}$ are to be used when X is directly attached to the methyl group, while the values labeled $\Delta\delta_{\beta-X}$ are used when a methylene group separated the methyl from the X (CH$_3$CH$_2$X). When using one or more $\Delta\delta_{\beta-X}$ values in a calculation, be sure to add 0.62 for the deshielding effect of the intervening CH$_2$ group.

Notice in Table 6.2 that the deshielding ability of substituents increases as you go down the table. Note also that the $\Delta\delta_{\alpha-X}$ value phenyl (2.00) is slightly different from the value we calculated in Example 6.3 (2.12). This is because the latter value comes from a specific compound (toluene) while the former value represents an average value from many phenyl-containing molecules.

■ **EXAMPLE 6.5** Using the data in Table 6.2 together with Eq. (6.5b), estimate the chemical shift of the methyl hydrogens in the following compounds:

☐ *Solution:* (a) We will use the base value of δ 0.23 (CH$_4$) and the $\Delta\delta$ value for X = I (1.94) from Table 6.2.

$$\delta(CH_3X) = 0.23 + \Delta\delta_I = 0.23 + 1.94 = 2.17 \text{ ppm}$$

The actual value from the spectrum of CH$_3$I is 2.15 ppm.[2]

(b) By symmetry (Chapter 4), both methyl groups in this molecule are equivalent. The $\Delta\delta_{\alpha-X}$ for C(=O)R is 1.87, leading to a predicted chemical shift of δ 2.10 (1.87 + 0.23) for both methyl groups. The observed value is δ 2.17.[2]

(c) The two methyl groups in this structure are *not* equivalent; one is attached to an oxygen atom, the other to a carbonyl carbon atom. The chemical shift of the former is calculated from the $\Delta\delta_{\alpha-X}$ value for the –OC(=O)R group (3.40) while the latter requires the value for the –C(=O)OR group (1.77):

TABLE 6.2 ^{1}H Substituent Parameters ($\Delta\delta_X$, ppm) for Substituents on Tetrahedral Carbons^a

Group X^b	$\Delta\delta_{\alpha-X}$	$\Delta\delta_{\beta-X}{}^c$	Group X^b	$\Delta\delta_{\alpha-X}$	$\Delta\delta_{\beta-X}{}^c$
–R	0.62	0.01	–SPh	2.27	
–CF$_3$	1.2		–S(=O)$_{1,2}$R	2.37	
–CH=C(R/H)$_2$	1.37	0.15	–Br	2.47	0.95
–C≡C(R/H)	1.50	0.35	–SC≡N	2.47	
–C(=O)OR	1.77	0.33	–N=CR$_2$	2.67	
–C(=O)N(R/H)$_2$	1.77	0.25	–N$^+$(R/H)$_3$	2.72	0.55
–C(=O)OH	1.87	0.33	–NHC(=O)R	2.72	0.25
–S(R/H)	1.87	0.43	–SO$_3$(R/H)	2.77	
–C(=O)R	1.87	0.20	–Cl	2.80	0.70
–C≡N	1.92	0.43	–O(R/H)	2.97	0.35
–I	1.94	0.90	–P$^+$Cl$_3$	3.07	
–C(=O)H	1.97	0.25	–N=C=S	3.17	
–NR$_2$	2.00	0.20	–OC(=O)(R/H)	3.40	0.45
–Ph	2.00	0.33	–OSO$_2$R	3.47	
–PR$_2$,–P(=O)R$_2$	2.00		–OPh	3.60	0.45
–C(=O)Ph	2.17	0.33	–OC(=O)Ph	3.60	0.80
–SSR	2.17		–NO$_2$	3.82	0.75
–NH$_2$	2.27		–F	4.00	0.70

^{a}Compiled from data in ref. 1. All numerical data in ppm. When calculating the chemical shift of a methylene group (X–CH$_2$–Y), decrease by 10% the value calculated from Eq. (6.6a) [see Eq. (6.6b)].
^{b}R represents any alkyl group; R/H represents either alkyl or hydrogen; Ph represents phenyl.
^{c}When using these values to calculate the methyl chemical shift of CH$_3$CH$_2$X, be sure to add 0.62 ppm for the effect of the CH$_2$ group.

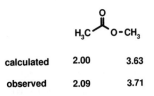

calculated	2.00	3.63
observed	2.09	3.71

■ **EXAMPLE 6.6**   Estimate the chemical shift of the methyl hydrogens in CH$_3$CH$_2$NO$_2$.

□ **Solution:** In this case the NO$_2$ group is β to the CH$_3$ (separated by a CH$_2$), so we will use a base value of δ 0.23, the $\Delta\delta_{\beta-X}$ for NO$_2$ (0.75), and add 0.62 for the effect of the CH$_2$ group, leading to a predicted chemical shift of δ 1.60.   □

■ **EXAMPLE 6.7**   A certain compound has molecular formula C$_2$H$_3$N but its structure is unknown. The compound exhibits a single peak in its ^{1}H NMR spectrum at δ 2.05.^{3} Propose a structure for the molecule.

□ **Solution:** Because there is only one peak, all three of the hydrogens are equivalent, probably occurring as a methyl group. This would leave one carbon and one nitrogen as "X" to account for the observed chemical shift. Notice in

Table 6.2 that a –C≡N group has a $\Delta\delta_{\alpha-X}$ value of 1.92, so a methyl attached to a CN group should occur near δ 1.92 + 0.23 = 2.15. Indeed, CH$_3$CN is the unknown compound.   □

In most cases the use of the substituent parameters in Table 6.2 gives a reasonable estimate (± 0.2 ppm) of the expected chemical shift of a given methyl hydrogen. Once you have tentatively identified the compound from its spectrum, you can in many cases look up the actual spectrum of the assigned structure in references (such as those listed at the end of this chapter) to confirm the identification.

6.2.2   Acyclic Methylene and Methine Hydrogens

Suppose a methylene group were connected to *two* of the substituent groups from Table 6.2. Could you predict the chemical shift of the methylene hydrogens? If the two substituents (X and Y) exert their (de)shielding effects independently, then perhaps the chemical shift of the methylene group could be calculated by simply adding the substituent parameters of both substituents to the chemical shift of methane:

$$\delta(XCH_2Y) = 0.23 + \Delta\delta_{\alpha-X} + \Delta\delta_{\alpha-Y} \qquad (6.6a)$$

2.17
2.47
4.64
0.23

Let us see how well the **additivity principle** implied in Eq. (6.6a) really works.

■ **EXAMPLE 6.8** Predict the chemical shift of the methylene hydrogens in (a) $Cl-CH_2-Cl$, (b) $Br-CH_2-C(=O)Ph$, (c) $Ph-CH_2-C(=O)CH_3$.

☐ *Solution:* (a) By substituting the substituent parameter for Cl (2.80) into Eq. (6.6a) for *both* X and Y, we calculate a chemical shift of δ 5.83. The observed ¹H chemical shift of this compound is δ 5.30.[2] (b) In this case the relevant $\Delta\delta_{\alpha-X}$ values are 2.47 for Br and 2.17 for C(=O)Ph, leading to a predicted chemical shift of δ 4.87. The observed value is δ 4.44.[3] (c) Remember, we are after the methylene chemical shift, so we will use the $\Delta\delta_{\alpha-X}$ values for Ph (2.00) and C(=O)R (1.87), giving a predicted chemical shift of δ 4.10. The actual value is δ 3.67.[3] ☐

We note from the above example that the additivity principle works reasonably well when X = R, but it consistently overestimates the chemical shift by ca. 10%. Applying this 10% correction, Eq. (6.6b) yields quite good results when X and Y ≠ R:

$$\delta(XCH_2Y) = 0.90\,(0.23 + \Delta\delta_{\alpha-X} + \Delta\delta_{\alpha-Y}) \quad (6.6b)$$

You might wonder if such an approach can be extended to **methine** groups. Unfortunately, in such cases the three substituent groups tend to interfere significantly with each other so that their (de)shielding effects are *not* simply additive.

■ **EXAMPLE 6.9** Predict the chemical shift of the methine hydrogen in $H-C(OCH_2CH_3)_3$.

☐ *Solution:* The $\Delta\delta_{\alpha-X}$ value for the OR group is 2.97. Adding three of these to δ 0.23 gives a predicted chemical shift of δ 9.14. The value actually observed is δ 5.16,[3] so the additivity principle [Eq. (6.6a)] greatly overestimates the chemical shift of methine hydrogens. ☐

6.2.3 Methylene and Methine Groups That Are Part of Rings

Many chemical compounds contain rings formed by the cyclic connection of three or more atoms. Toluene (Chapters 3–5), for example, possesses a ring composed of six carbons. Let us see if we can use the approach developed so far to estimate the chemical shifts of some cyclic compounds whose rings are composed solely of tetrahedral carbons.

■ **EXAMPLE 6.10** Predict the chemical shift of the hydrogens in the cycloalkanes below. For the purposes of this problem you may assume that the rings are planar, though actually they are not:

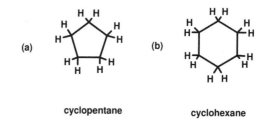

(a) cyclopentane (b) cyclohexane

☐ *Solution:* (a) By symmetry (a C_5 axis), all five of the methylene groups are equivalent. We can estimate their chemical shift by treating them as $R-CH_2-R$ groups, where $\Delta\delta_{\alpha-R}$ is 0.62. Using Eq. (6.6a), we calculate a chemical shift of δ 1.47; the observed value is δ 1.51.[3] (b) Using the same reasoning, we predict that all six equivalent methylenes in cyclohexane should appear at the same chemical shift, δ 1.47. In actuality, the signal occurs at δ 1.43.[2] ☐

Example 6.10 tempts us to predict that all simple cyclic structures composed of methylenes should exhibit ¹H NMR signals at about the same position. Although this is true for rings with five or more methylenes, the smaller rings show significant variations (see Table 6.3). The reason for the anomalous behavior of the small rings (n = 3, 4) will be described in Section 6.4. But more important for the present purpose is that we can use the chemical shifts in Table 6.3, rather than the value δ 0.23 (methane), as base values when estimating the chemical shifts of methylene and methine hydrogens in cyclic compounds.

■ **EXAMPLE 6.11** Estimate the chemical shift of the underlined hydrogen in chlorocyclohexane:

TABLE 6.3 ¹H Chemical Shifts of Unsubstituted Cycloalkanes[a]

Name	Ring Size	δ (ppm)
Cyclopropane	3	0.22
Cyclobutane	4	1.96
Cyclopentane	5	1.51
Cyclohexane	6	1.43
Cycloheptane	7	1.53
Cyclooctane	8	1.57
Cyclodecane	10	1.51

[a]Data from refs. 1 and 3.

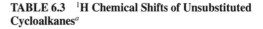

□ *Solution:* Using a base value of δ 1.51 for a CH_2 group in cyclopentane (Table 6.3) and the $\Delta\delta_{\alpha-Cl}$ value of 2.80 (Table 6.2), we predict a chemical shift of δ 4.31. The observed value is δ 4.37.[3]   □

By comparing the $\Delta\delta_\alpha$ and $\Delta\delta_\beta$ for substituents in Table 6.2, you have probably noted that the (de)shielding effect of a substituent normally decreases as the number of intervening bonds increases. The chlorine is separated from the underlined hydrogen in Example 6.11 by *two* bonds; this is called a **geminal** relationship, and here the effect is greatest ($\Delta\delta_{\alpha-Cl}$ = 2.80). The hydrogens (H*) *three* bonds away from the chlorine (called a **vicinal** relationship) are deshielded only slightly ($\Delta\delta_{\beta-Cl}$ = 0.70), while the deshielding effect on those hydrogens (H′) separated by four bonds is negligible.

Before we proceed any further, there is something else that might be troubling you. Why do we need all these correlations and tables if we can simply look up the actual spectra of so many compounds? There are two answers to this question. First, in trying to identify an unknown compound, we need to start somewhere, and acquiring its NMR spectrum is the best place. With only that information, we can usually (if it is not too complicated a molecule) make a decent guess about its structure, then go to the literature for confirmation. But what if the spectrum of our compound has never been reported before? In that case our main evidence for its structure may be how well its NMR spectrum matches expectations based on the above correlations.

6.2.4 Index of Unsaturation and the Nitrogen Rule

In the course of this book we are going to be solving many problems like the one in Example 6.7. Here is a little "trick" to make the process a little easier. Whenever you encounter a molecular formula $C_cH_hN_nO_oX_x$ (where X represents halogen: F, Cl, Br, or I), begin by calculating its **index of unsaturation (IOU)** from the equation

$$IOU = \tfrac{1}{2}\left[2c + 2 + n - (h + x)\right] \qquad (6.7)$$

This equation derives from the normal valences of (number of bonds to) atoms of carbon, nitrogen, hydrogen, and halogen (4, 3, 1, and 1, respectively) in molecules where all valences are satisfied (divalent atoms such as oxygen and sulfur do not appear in the equation). The IOU tells us how many rings plus π bonds there must be in any legitimate isomer with that molecular formula. Recall that a double bond is usually pictured as one cylindrically symmetric σ bond and one π bond between two parallel *p* orbitals. A triple bonds consists of one σ bond and two π bonds. The IOU, therefore, is very useful in quickly eliminating from consideration any structures that do not possess the correct number of rings plus π bonds. Thus, the IOU of C_2H_3N (Example 6.7) is [2(2) + 2 +

1 – 3]/2 = 2, so the correct structure must have either two rings or one ring and one π bond (i.e., one double bond) or two π bonds (two double bonds or one triple bond). The $CH_3C\equiv N$ molecule fits in the last category.

The "nitrogen rule" applies to molecules made up of carbon, hydrogen, nitrogen, oxygen, sulfur, and halogen. When the number of nitrogens in the molecular formula is *even*, the number of hydrogens plus halogens, as well as the nominal molecular mass, must both be even. When the number of nitrogens is *odd*, the number of hydrogens plus halogens, as well as the nominal molecular mass, must both be odd (zero is considered even in this context). This rule can be useful in making guesses about an unknown molecular formula, when the number of hydrogens plus halogens is known (e.g., from NMR data).

■ **EXAMPLE 6.12** Deduce the structure of the compound whose 250-MHz ^{1}H NMR spectrum is reproduced in Figure 6.3, given only that its molecular formula is $C_6H_{12}O$.

□ *Solution:* Begin by calculating the IOU: [2(6) + 2 – 12]/2 = 1. The molecule must therefore possess one ring or one π (that is, double) bond. We note immediately that there are 12 hydrogens, and therefore an even number (zero) of nitrogens. These 12 are divided into 9 equivalent hydrogens at δ 1.15 and 3 other equivalent ones at δ 2.14. Whenever you encounter a 9-proton signal near δ 1.0, it is probably due to 3 equivalent methyl groups, such as those in a *tertiary* butyl group [$(CH_3)_3C-$]. Note how all 3 of the methyls in such a group are equivalent, just as the 3 hydrogens in a methyl group are equivalent (Section 4.2). The δ 2.14 signal fits well for a methyl group connected directly to a C(=O)R group (see Table 6.2 and Example 6.5). The correct structure, therefore, is pinacolone (3,3-dimethyl-2-butanone):

□

6.3 VINYL AND FORMYL HYDROGEN CHEMICAL SHIFTS

Two carbons connected by a double (i.e., σ + π) bond are called **vinyl** or **olefinic** carbons. Each vinyl carbon is nominally sp^2 hybridized and attached to three atoms (one of which is the other vinyl carbon). These three atoms describe an approximately equilateral triangle, which is why vinyl carbons are also described as **trigonal**. Hydrogens attached directly to vinyl carbons are called vinyl (or olefinic) hydrogens.

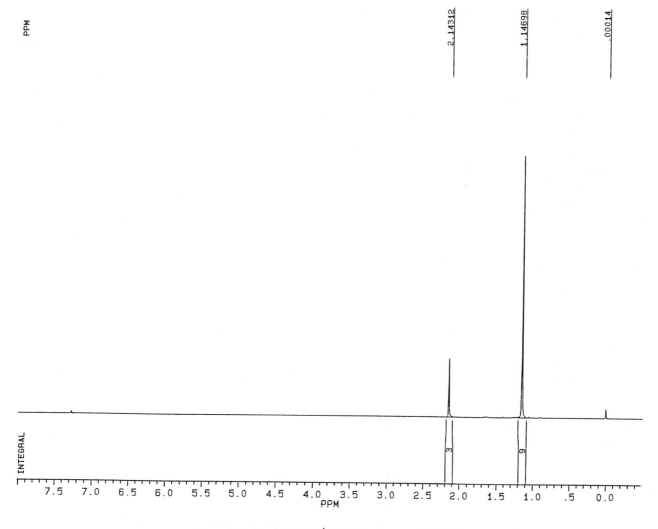

PPM

2.14312

1.14698

.00014

INTEGRAL

3

9

7.5 7.0 6.5 6.0 5.5 5.0 4.5 4.0 3.5 3.0 2.5 2.0 1.5 .5 0.0
PPM

Figure 6.3. The 250-MHz ^{1}H NMR spectrum of $C_6H_{12}O$.

■ **EXAMPLE 6.13** (a) Circle the vinyl hydrogens in structure **6-1**. (b) What is the IOU of **6-1**?

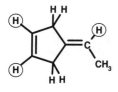

6-1

□ *Solution:* (a)

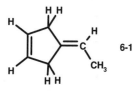

(b) Hopefully you did not go to the trouble of first generating the molecular formula, then using Eq. (6.7). Just look at the structure. There are two π bonds and one ring; IOU = 3.   □

The sp^2 orbitals with which a vinyl carbon forms σ bonds are somewhat more electronegative than the sp^3 orbitals that tetrahedral carbons use. This is because an sp^2 hybrid has greater s orbital character than an sp^3 hybrid (33.3 vs. 25%), and since s orbitals are lower in energy than the corresponding p orbitals used to create the hybrids, sp^2 hybrids are lower energy than sp^3 hybrids. The lower the energy of a valence orbital, the more it tends to attract bonding electrons toward itself. Vinyl hydrogens, therefore, have less electron density around them than do hydrogens attached to tetrahedral carbons. As a direct result (see Section 6.1), vinyl hydrogens are deshielded, and their signals appear downfield of hydrogens attached to tetrahedral carbons. For example, the simplest alkene, ethylene (**6-2**), exhibits one ^{1}H signal (all four hydrogens are equivalent, right?) at δ 5.28:

6-2

Vinyl hydrogens typically appear in the δ 4.5–7.0 region of the ^{1}H spectrum. We can establish a list of substituent parameters for vinyl substituents (Table 6.4) similar to the one we had for methyl substituents (Table 6.2). Notice that the magnitude of a substituent's (de)shielding effect is strongly dependent on its location relative to the vinyl hydrogen of interest. In our calculations of vinyl hydrogen chemical shifts, we will use δ 5.28 (the chemical shift of ethylene) as our base value in Eq. (6.5b).

■ **EXAMPLE 6.14** Predict the chemical shift of each of the three vinyl hydrogens in styrene (**6-3**):

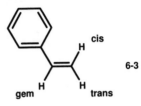

6-3

TABLE 6.4   Substituent Parameters ($\Delta\delta_X$, ppm) for Vinyl Hydrogen Chemical Shifts[a]

–X	$\Delta\delta_{gem}$	$\Delta\delta_{cis}$	$\Delta\delta_{trans}$
–C≡N	0.23	0.78	0.58
–R (alkyl)	0.44	–0.26	–0.29
–C≡CR	0.50	0.35	0.10
–CH$_2$SR	0.53	–0.15	–0.15
–CH$_2$NR$_2$	0.66	–0.05	–0.23
–CH$_2$OR	0.67	–0.02	–0.07
–CH$_2$I	0.67	–0.02	–0.07
–NR$_2$	0.69 (2.30)	–1.19 (–0.73)	–1.31 (–0.81)
–Cycloalkenyl[b]	0.71	–0.33	–0.30
–CH$_2$Cl	0.72	0.12	0.07
–CH$_2$Br	0.72	0.12	0.07
–C(=O)OR	0.84 (0.68)	1.15 (1.02)	0.56 (0.33)
–CH=CH$_2$	0.98 (1.26)	–0.04 (0.08)	–0.21 (–0.01)
–C(=O)OH	1.00 (0.69)	1.35 (0.97)	0.74 (0.39)
–Cl	1.00	0.19	0.03
–SR	1.00	–0.24	–0.04
–C(=O)H	1.03	0.97	1.21
–Br	1.04	0.40	0.55
–C(=O)R	1.10 (1.06)	1.13 (1.01)	0.81 (0.95)
–C(=O)Cl	1.10	1.41	0.99
–OR	1.18 (1.14)	–1.06 (–0.65)	–1.28 (–1.05)
–Ph	1.35	0.37	–0.10
–C(=O)NR$_2$	1.37	0.93	0.35
–SO$_2$R	1.58	1.15	0.95
–OC(=O)R	2.09	–0.40	–0.67

[a]Data (in ppm) from Pascual, C., Meier, J., and Simon, W., *Helv. Chim. Acta, 49*, 164 (1966), as quoted in ref. 1. Recall that a negative value of $\Delta\delta$ corresponds to an upfield shift. The data in parentheses are to be used if either the X group or the C=C is further conjugated.

[b]The double bond is endocyclic to a ring.

□ *Solution:* For each hydrogen, add the appropriate value of Δδ for a Ph group to the base value of δ 5.28:

$$\delta_{gem} = 5.28 + 1.35 = 6.63 \text{ (observed: } \delta\ 6.66^2)$$

$$\delta_{cis} = 5.28 + 0.37 = 5.65 \text{ (observed: } \delta\ 5.64^2)$$

$$\delta_{trans} = 5.28 + (-0.10) = 5.18 \text{ (observed: } \delta\ 5.18^2)$$

The close agreement of these numbers is no accident. Styrene was the model compound used to generate the Δδ values for a phenyl substituent! Now, let us try a more challenging example.   □

■ **EXAMPLE 6.15** Deduce the structure of a compound whose molecular formula is $C_5H_8O_2$ and that exhibits ^{1}H NMR signals at δ 6.13 (1H), 5.59 (1H), 3.79 (3H), and 1.98 (3H); the number in parentheses is the relative integration of each signal.

□ *Solution:* This problem requires us to use everything we have learned so far in this chapter. First, the IOU is [2(5) + 2 − 8]/2 = 2; two rings plus π bonds. From the chemical shifts and integrals it looks as though this molecule possesses two nonequivalent vinyl hydrogens and two nonequivalent methyl groups. From Table 6.2, the methyl group at δ 3.79 could be connected to a –OC(=O)R group (0.23 + 3.40 = 3.66), suggesting the partial H_3C–O–C(=O)–R, where R must consist of the remaining three carbons and five hydrogens. Furthermore, the methyl group at δ 1.98 must be attached to a vinyl carbon. Here are several possible structures:

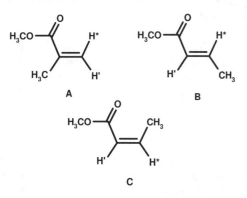

How should we decide among the three structures? Right! We'll use the data in Table 6.4 to calculate the expected chemical shifts of the two vinyl hydrogens in each structure. In **A**, H* is cis to the CO_2CH_3 group and trans to the CH_3 group, so we predict a chemical shift of δ 6.14 (5.28 + 1.15 − 0.29). Similarly, H′ should occur near δ 5.59 (5.28 + 0.56 − 0.26). For structure **B** the calculated chemicals shifts are H* δ 6.87 (5.28 + 1.15 + 0.44) and H′ δ 5.86 (5.28 + 0.84 − 0.26). For structure **C** the calculated

chemicals shifts are H* δ 6.28 (5.28 + 0.56 + 0.44) and H′ δ 5.83 (5.28 + 0.84 − 0.29). Clearly structure **A** fits the observed data best, and this result supports our extending the additivity principle to vinyl hydrogens.   □

There is another type of hydrogen that, at first glance, appears quite similar to a vinyl hydrogen. It is called a **formyl** (or aldehydic) hydrogen and is directly bonded to a carbonyl (i.e., C=O) carbon:

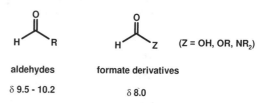

The carbonyl group has a strong deshielding effect (see Tables 6.2 and 6.4) by virtue of the highly electronegative oxygen. For this reason, formyl hydrogens are significantly more deshielded than even vinyl hydrogens. They are found in a rather remote region of the ^{1}H spectrum, around δ 8–10.5, where little else appears.

6.4 MAGNETIC ANISOTROPY

Knowing that a vinyl hydrogen is quite a bit more deshielded than typical methyl or methylene hydrogens, what would you predict about the properties of a hydrogen (called an **acetylenic** hydrogen) directly attached to a triply bonded carbon (H–C≡C–)? We might predict such a hydrogen to be even more deshielded than a vinyl hydrogen, because the *sp* orbital carbon uses to form the σ bond with the hydrogen is more electronegative than an sp^2 orbital (Why?). In fact, acetylenic hydrogens normally appear in the region around δ 2.0–3.0, significantly upfield of their vinyl cousins. Why is this so?

The explanation for this apparent anomaly involves the special way in which the "cylinder" of electrons in the triple bond behaves when immersed in a magnetic field. Recall (Section 6.1) that when electrons are subjected to a strong external magnetic field, they circulate in such a way as to induce a smaller magnetic field that opposes the external field. An external magnetic field causes the electrons in the triple bond to circulate around the internuclear (bond) axis, thereby inducing a field along the internuclear axis in opposition to the external field. This is depicted in Figure 6.4, where the lines of magnetic flux are shown as dotted lines. Note that in the cone-shaped regions around the internuclear axis the induced field, being opposed to the external field, will shield nuclei situated there. A hydrogen attached to a triply bonded carbon lies along this axis and therefore experiences a substantial upfield shift in the position of its NMR signal.

Notice also that in the regions outside the cones (i.e., perpendicular to the triple-bond axis) the induced field is

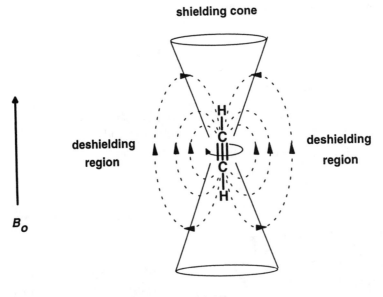

Figure 6.4 Anisotropic-induced magnetic field (dotted lines) in the proximity of a triple bond.

aligned *with* the external field. Because the direction and magnitude of the induced field vary as a function of position, we say the induced field is **anisotropic**. Similar anisotropic induced fields in the regions around double bonds and strained σ bonds in small rings account for the anomalous chemical shifts of nearby nuclei (e.g., cyclopropane; Table 6.3).

■ **EXAMPLE 6.16** In cyclooctyne would you expect the H* hydrogens to be shielded or deshielded by the induced field of the triple bond? Explain.

□ **Solution:** The induced field in the region of the H* hydrogens is aligned with the external field (see Figure 6.4), so the effective field (Section 6.1) experienced by the H* hydrogens is greater than the external field alone. This causes the signal to move downfield (to higher frequency), the result of apparent deshielding. However, this deshielding effect would be smaller than the shielding effect on acetylenic hydrogens because the H* hydrogens are farther away from the triple bond.   □

It should be noted that molecules like acetylene in Figure 6.4 are not, as the figure might suggest, statically aligned

parallel to the external field. Instead, they are constantly tumbling around in solution millions of times per second (Section 2.5). However, during the times when they are approximately parallel to the field, the anisotropic-induced field will be operative. When the molecules are approximately perpendicular to the field, there will be little or no anisotropic-induced field. The overall field surrounding the triple bond will be the time average of all possible orientations.

6.5 AROMATIC HYDROGEN CHEMICAL SHIFT CORRELATIONS

Recall from Chapter 5 that in toluene the five hydrogens attached to the phenyl ring appear at *ca.* δ 7.23, somewhat more deshielded than typical vinyl hydrogens. These hydrogens belong to a special subclass of vinyl hydrogens called aromatic hydrogens. You probably remember from an earlier organic chemistry course that molecules possessing a planar ring composed of sp^2 hybridized atoms and containing $4n + 2\pi$ electrons (e.g., 2, 6, 10, 14) have very special chemical and physical properties. Compounds with such rings are referred to as **aromatic**, referring to the unique resonance characteristics of such a cyclic network of alternating double and single bonds.

The fact that aromatic hydrogens are deshielded compared to their vinyl cousins is a direct consequence of the magnetic anisotropy of the aromatic π system. Below is a depiction of the toluene molecule showing how the three π bonds (six π electrons) are conjugated, that is, all the p orbitals (one on each sp^2 ring carbon) are parallel. This makes all C–C bonds in the ring equivalent, halfway between single and double bonds in character. This cyclic array of six π electrons can be viewed as forming "donuts" of electron density above and below the ring:

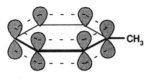

(This is the reason that structures of aromatic compounds are often shown with circles inside the ring; see Section 3.4 and Examples 4.3 and 4.5.) When immersed in an external magnetic field, these π electrons begin to circulate just as the electrons in a triple bond did in Figure 6.4. This circulation, called a **ring current** in the context of aromatic molecules, generates an analogous induced field whose lines of magnetic flux are shown by the dashed lines in Figure 6.5. Look carefully at the *direction* of the induced field. Above and below the center of the ring, the induced field is *opposed* to the external field, giving rise to a shielding effect on nuclei located in that region. However, in the donut-shaped region *outside* the periphery of the ring where aromatic hydrogens are located, the induced field is aligned *with* the external field, causing deshielding of nuclei in that region. Thus, aromatic hydrogens experience a deshielding effect due to the ring current of the aromatic π electrons.

Aromatic hydrogens typically occur in the δ 6.5–8.0 chemical shift range. And just as for vinyl hydrogens, correlation tables of substituent parameters have been developed for aromatic substituents. In this case we will use the chemical shift of benzene (δ 7.27) as our base value in the calculations. Table 6.5 lists many common aromatic substituent parameters.

■ **EXAMPLE 6.17** Predict the chemical shift of the aromatic hydrogens in the compound whose structure is shown below:

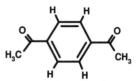

□ *Solution:* First, we recognize from symmetry considerations that all four aromatic hydrogens are equivalent whenever there are two equivalent functional groups para to each other (see review problem 6.2). Further, each of these hydrogens is both ortho *and* meta to a C(=O)R group. Therefore, using the data in Table 6.5,

$$\delta = 7.27 + 0.64 + 0.09 = 8.00$$

The value actually observed is δ 8.08.[3]   □

■ **EXAMPLE 6.18** Identify the compound C_8H_6 that exhibits signals at δ 7.40 (5H) and 3.09 (1H) in its ^1H NMR spectrum.

□ *Solution:* The IOU is [2(8) + 2 − 6]/2 = 6. Clearly this compound must have a monosubstituted benzene ring,

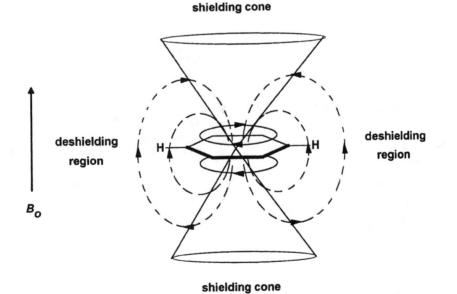

Figure 6.5. Anisotropic-induced magnetic field (dotted lines) in the proximity of an aromatic ring. The ellipses above and below the ring represent the ring current of π electrons.

TABLE 6.5   Aromatic Substituent Parameters ($\Delta\delta_X$, ppm)[a]

X	H_{ortho}	H_{meta}	H_{para}	X	H_{ortho}	H_{meta}	H_{para}
–CH₃	–0.17	–0.09	–0.18	–I	0.40	–0.26	–0.03
–CH₂CH₃	–0.15	–0.06	–0.18	–OH	–0.50	–0.14	–0.4
–CH(CH₃)₂	–0.14	–0.09	–0.18	–OR	–0.27	–0.08	–0.27
–C(CH₃)₃	0.01	–0.10	–0.24	–OC(=O)R	–0.22	0	0
–CH=CH₂	0	0	0	–OSO₂Ar	–0.26	–0.05	0
–C≡CH	0.20	0	0	–C(=O)H	0.58	0.21	0.27
–Ph	0.18	0	0.08	–C(=O)R	0.64	0.09	0.3
–CF₃	0.25	0.25	0.25	–C(=O)OH	0.8	0.14	0.20
–CH₂Cl	0	0.01	0	–C(=O)OR	0.74	0.07	0.20
–CHCl₂	0.10	0.06	0.10	–C(=O)Cl	0.83	0.16	0.3
–CCl₃	0.8	0.2	0.2	–C≡N	0.27	0.11	0.3
–CH₂OH	–0.10	–0.10	–0.10	–NH₂	–0.75	–0.24	–0.63
–CH₂OR	0	0	0	–NR₂	–0.60	–0.10	–0.62
–CH₂NH₂	0.0	0.0	0.0	–NHC(=O)R	0.23		
–SR	–0.03	0	0	–N⁺H₃	0.63	0.25	0.25
–F	–0.30	–0.02	–0.22	–NO₂	0.95	0.17	0.33
–Cl	0.02	–0.06	–0.04	–N=C=O	–0.20	–0.20	–0.20
–Br	0.22	–0.13	–0.03				

[a]Data abstracted from refs. 1 and 4.

judging from the *five* equivalent aromatic hydrogens. Since these five hydrogens cannot be symmetry equivalent, the lone substituent must have nearly the same effect on ortho, meta, and para hydrogens. The only reasonable structure is phenylacetylene:

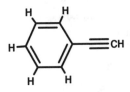

Yet, is the δ 3.09 value a reasonable chemical shift for the acetylenic hydrogen? In Section 6.4 we saw that such hydrogens usually fall in the region δ 2–3 ppm. The observed signal in this case is at the deshielded end of this range, probably because the acetylenic hydrogen lies in the deshielded region of the aromatic ring.   □

■ **EXAMPLE 6.19**   Identify the compound with molecular formula $C_8H_6O_2$ whose ¹H NMR spectrum exhibits signals at δ 8.06 (4H) and 10.15 (2H).

□ *Solution:* The IOU is 6. The ratio of hydrogens and the signal at δ 8.09 suggest an aromatic ring with two equivalent deshielding substituents para to each other. But what kind of hydrogens occur at δ 10? Look back at Section 6.3. Aldehydic hydrogens! The spectrum belongs to the dialdehyde below:

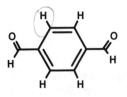

The predicted chemical shift of these aromatic hydrogens is δ 7.27 + 0.58 + 0.21 = 8.06. Note the close similarity of its spectrum to the one described in Example 6.17. □

■ **EXAMPLE 6.20** Explain why the two equivalent methyl groups in structure **6-4** appear at δ –4.25,[5] far upfield of TMS:

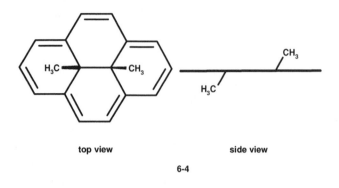

top view   side view

6-4

□ *Solution:* The alternating single and double bonds around the periphery of the 14-membered ring give rise to an aromatic 14 π electron cloud, with its associated ring current. The methyl groups lie directly above and below the plane of the ring, in the strongly shielding region of the induced field. □

Not all cyclic molecules with alternating single and double bonds are aromatic, only those with 4n + 2π electrons. Furthermore, not all aromatic rings are six membered (**benzenoid aromatics**); some are five membered with a π system that includes an unshared pair of electrons donated by a **heteroatom** (an atom other than carbon). For example, the structures **6-5** through **6-9** all exhibit aromatic properties, and hydrogens attached to the rings fall in the aromatic region. The chemical shift of each set of equivalent hydrogens is given in parentheses. Compounds such as **6-5**, **6-6**, and **6-7** are called **het-**

eroaromatic compounds because their rings contain atoms other than carbon. Compounds **6-8** and **6-9** are examples of **polycyclic aromatic hydrocarbons** (PAH):

6.6 HYDROGEN ATTACHED TO ELEMENTS OTHER THAN CARBON

With more than 100 elements besides carbon in the periodic table (Appendix 2), you might fear that the number of ^{1}H chemical shift correlations is endless. However, except for a few specialized applications, the most important heteroatoms to which hydrogen finds itself bonded are oxygen and nitrogen. But before we discuss these two specific cases, here is a useful generalization: As the electronegativity (Table 6.1) of X increases, both the acidity and chemical shift of a hydrogen bonded directly to X increase.

6.6.1 Hydrogens Attached to Oxygen

A hydrogen attached to an oxygen (O–H) constitutes a **hydroxyl group**. Hydroxyl groups appear in several classes of organic molecules, including **alcohols** (where the carbon bearing the O–H is tetrahedral), **enols** (where the O–H group is directly bonded to a vinyl carbon), **phenols** (where the O–H group is bonded directly to an aromatic ring), and **carboxylic acids** (where the O–H group is directly bonded to a carbonyl carbon):

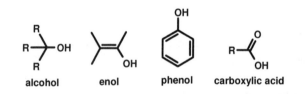

alcohol   enol   phenol   carboxylic acid

The order of acidity of these four classes of compounds is alcohols < enols < phenols < carboxylic acids. Therefore, it

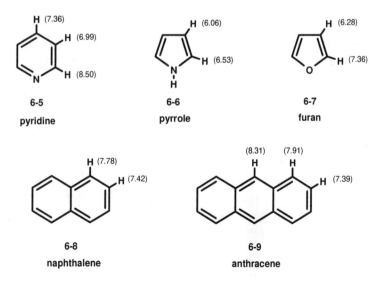

6-5
pyridine

6-6
pyrrole

6-7
furan

6-8
naphthalene

6-9
anthracene

should not surprise us that the chemical shifts (in $CDCl_3$ solvent) follow pretty much the same order: alcohols (δ 1–4) < phenols (δ 4.0–7.5) < enols (δ 6–7 for enols of cyclic α-diketones, δ 14.5–16.5 for enols of β-dicarbonyls) < carboxylic acids (δ 10–14). In fact, carboxylic acid hydrogens and β-dicarbonyl enol hydrogens are among the most deshielded of all hydrogens, thereby defining the high-frequency (low-field) limit of the ^1H spectral width.

However, hydroxyl proton NMR signals have several additional characteristics of which you should be aware. To describe these, we need to know a little about an important property of acidic hydrogens. Unlike most hydrogens attached to carbon, those attached to oxygen and nitrogen are subject to **hydrogen bonding** and **exchange**. Hydrogen bonding is a special type of charge dipole–dipole interaction involving a relatively acidic, partially positively charged hydrogen still bonded to X ($X–H^{\delta+}$) but also attracted to another electronegative atom, $Y^{\delta-}$ ($\delta+$ and $\delta-$ represent partial charges). This is usually written X–H--Y or X--H–Y. Thus, two hydrogen-bonded alcohol molecules might look something like this:

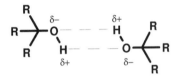

Under certain conditions (such as the presence of a small amount of acid or base catalyst), the hydrogens can actually be traded (i.e., exchanged) between the two hydroxyl groups. Such exchange processes can be very rapid, occurring at rates comparable to the NMR time scale (Section 1.4). The results of hydrogen bonding and exchange are that hydroxyl proton signals are often broader (larger halfwidth, Section 3.5.1) than C–H proton signals, and their chemical shifts are quite dependent on temperature, concentration, and the nature of the solvent (Section 3.5.2) because all three of these variables affect the rate of hydrogen exchange and the strength of the hydrogen bonds. The upshot of all this is that hydroxyl proton chemical shifts are quite variable, and it is difficult to be as accurate in our predictions of their chemical shifts as we were with hydrogens attached to carbon.

Finally, it is important to recognize that hydrogen bonding can be either *inter*molecular (X–H and Y are part of separate molecules) or *intra*molecular (X–H and Y are part of the same molecule).

■ **EXAMPLE 6.21** Examine the 250-MHz ^1H NMR spectrum (Figure 6.6) of 2,4-dimethyl-2,4-pentanediol, **6-10**:

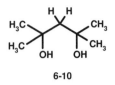

(a) Give the chemical shift and integration of each signal. (b) Assign each signal to specific hydrogens in the molecule, and compare the observed chemicals shifts with predicted values. (c) Account for the halfwidth of the signal at δ 4.29.

□ *Solution:* (a) δ 1.32 (12H), 1.72 (2H), and 4.29 (2H). (b) The 12H signal must be due to the four equivalent methyl groups, whose predicted chemical shift is δ 1.22 [0.23 + 0.62 (α-R) + 2 (0.01, β-R) + 0.35 (β-OH)]. The methylene hydrogens are predicted to appear at δ 1.99 {0.9[0.23 + 2(0.62) + 4(0.01) + 2(0.35)]}, and this corresponds to the signal observed at δ 1.72. The alcoholic OH signal should appear in the range δ 1–4, consistent with the signal observed at δ 4.29. (c) This signal, being due to the OH groups, is broadened by hydrogen bonding and exchange. □

■ **EXAMPLE 6.22** A dilute solution of CH_3OH in $CDCl_3$ exhibits its O–H signal at δ 1.43 ppm. When a slight excess of D_2O (deuterated water) is added to the sample, the hydroxyl signal disappears and is replaced by a signal at δ 4.75.[1] Explain.

□ *Solution:* Remember that deuterium (^2H, Table 2.1) does not show up in a ^1H spectrum. But *chemically* deuterium acts just like hydrogen. So there is exchange between the OH hydrogens of the alcohol and the deuteria of D_2O, thereby generating CH_3OD (which exhibits only a methyl signal) and H–O–D. The latter gives rise to the δ 4.75 signal. This **deuterium exchange** technique is a useful way of confirming the presence of a readily exchangeable hydrogen. □

6.6.2 Hydrogens Attached to Nitrogen

Hydrogens directly bonded to nitrogen occurs in several classes of compounds, as shown in structures **6-11** through **6-15**. Compare these structures with those of alcohols, enols, phenols, and acids:

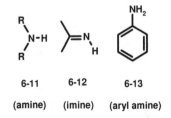

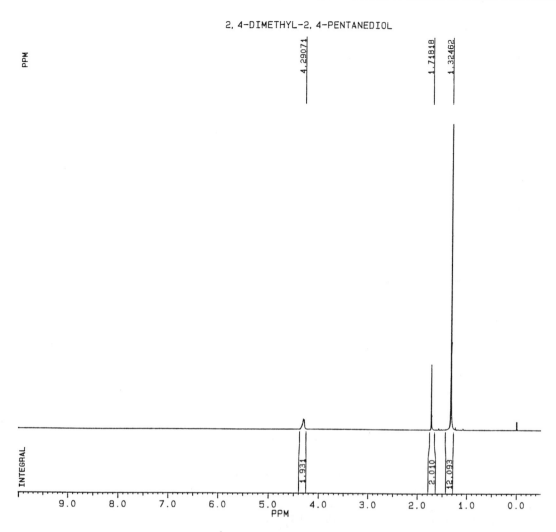

Figure 6.6. The 250-MHz ^{1}H NMR spectrum of 2,4-dimethyl-2,4-pentanediol.

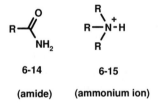

6-14 6-15

(amide) (ammonium ion)

Because nitrogen is somewhat less electronegative than oxygen (why?), N–H hydrogen signals tend to be less deshielded and appear at higher field (lower frequency) than their O–H counterparts. The normal chemical shift ranges are amines (δ 0.5–3.0) < aryl amine (δ 3–5) < amides (δ 4–7) < ammonium salts (δ 6.0–8.5) < imines (δ 5–11), again paralleling the acidity of these hydrogens. As in the case of O–H hydrogens, N–H signals are also affected by hydrogen bonding and exchange and are therefore also dependent on temperature, concentration, and solvent. And there is one further complication. Nitrogen (^{14}N) has a nuclear spin of 1 and is

therefore a quadrupolar nucleus (Section 2.1). For this reason, nitrogen nuclei undergo rapid spin–lattice relaxation (Section 2.5), and this facilitates similar fast relaxation of the attached hydrogens. In addition, as we will see in Chapter 8, nitrogen exhibits "spin coupling" with the attached hydrogens. These factors combine to make the NMR signals of protons attached to nitrogen exceptionally broad, so broad, in fact, that they are sometimes difficult to differentiate from baseline noise (Section 3.3.3).

■ **EXAMPLE 6.23** Identify the compound C_2H_5NO whose 250-MHz ^{1}H spectrum is shown in Figure 6.7. Also account for the small signal at δ 7.27.

□ *Solution:* The IOU in this case is 1. The three-hydrogen signal at δ 2.01 is exactly where we would expect a methyl group bonded directly to an amide carbonyl (0.23 + 1.77; Table 6.2). Moreover, the two remaining hydrogens give

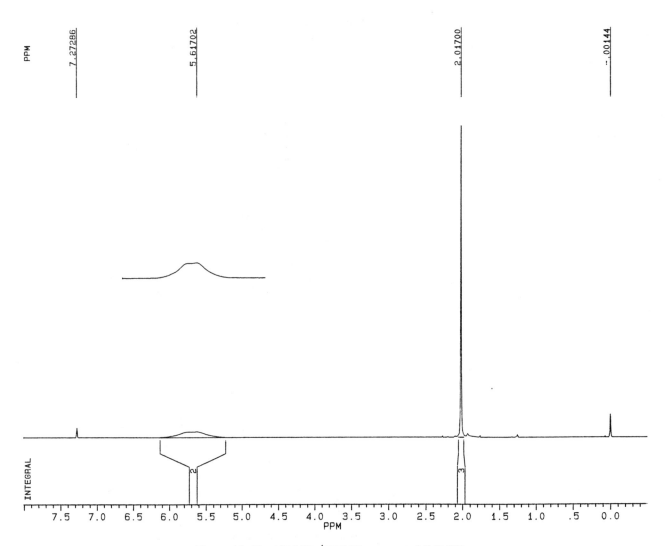

Figure 6.7. The 250-MHz ^{1}H NMR spectrum of C_2H_5NO.

rise to an exceptionally broad signal centered at δ 5.6, in the middle of the amide hydrogen region. The large halfwidth is due mainly to the quadrupolar effects of nitrogen as well as to hydrogen bonding and exchange effects. The compound is acetamide, and the small signal at δ 7.27 is due to the small amount of $CHCl_3$ in the $CDCl_3$ solvent [see review problem 5.4(a)]:

We will learn in Chapter 10 that there is one additional complication in the NMR spectra of amides: the two amide hydrogens are not equivalent because rotation of the C–N bond is often slow on the NMR time scale!   □

In this chapter we have developed a familiarity with the factors that influence the NMR chemical shift of hydrogens,

principally their molecular environments. Figure 6.8 reviews graphically what we have learned. We have also begun to use our knowledge to assign structures to unknown compounds. Let us finish this chapter with one final example to see how well you have mastered the material.

■ **EXAMPLE 6.24**   Identify the compound $C_2H_8N_2$ whose 250-MHz ^{1}H NMR spectrum is shown in Figure 6.9. Assign both signals to specific hydrogens in your structure and show that their chemical shifts are within expected limits.

□ *Solution:* The fact that there are only two signals, each resulting from four equivalent hydrogens, suggests a high degree of symmetry in the structure. The signal at δ 1.33, while equal in intensity to the one at δ 2.75, is nonetheless very broad. This may be due to nitrogen quadrupolar effects, suggesting the signal is due to hydrogens directly

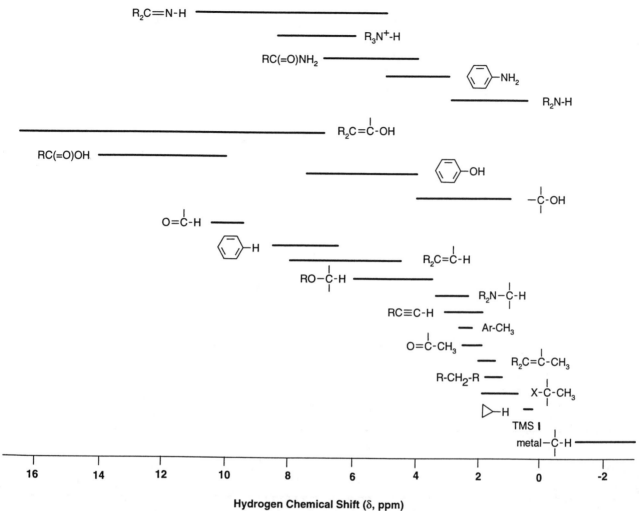

Figure 6.8. Complete ^{1}H chemical shift range.

bonded to nitrogen. Also, the δ 1.38 chemical shift is squarely in the range for amine N–H. Moreover, the δ 2.75 signal is correct for the equivalent methylene hydrogens in a structure of the type $R_2N–CH_2–CH_2–NR_2$ [δ = 0.9(0.23 + 2.20 + 0.62 + 0.20) = 2.93]. Putting all of this together in one structure leads to $H_2N–CH_2–CH_2–NH_2$. □

CHAPTER SUMMARY

1. Although all nuclei of a given isotope have exactly the same magnetogyric ratio, they may precess at slightly different frequencies (and hence give rise to separate NMR signals) because of differences in their molecular (electronic) environments. In general, the greater the electron density around a nucleus, the more it is shielded from the effects of the external magnetic field.

The more shielded a nucleus is, the lower will be the frequency of its NMR signal.

2. Hydrogen nuclei attached to carbon are divided among the following groups: aliphatic (methyl, methylene, methine), vinyl (olefinic), acetylenic, aromatic, and formyl.

3. The chemical shift of a hydrogen attached to carbon can be predicted quite accurately from a version of Eq. (6.5b), with the appropriate base value and substituent parameter(s) from Tables 6.2–6.5.

4. Certain types of functional groups, most notably carbon–carbon triple bonds and aromatic rings, give rise to pronounced anisotropic-induced magnetic fields. As a result, acetylenic hydrogens are unusually shielded, while aromatic hydrogens are unusually deshielded.

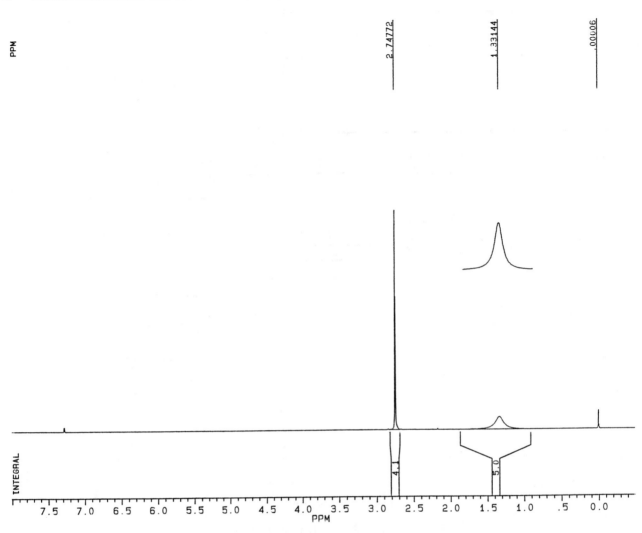

Figure 6.9. The 250-MHz ^{1}H NMR spectrum of $C_2H_8N_2$.

5. Hydrogens attached to oxygen or nitrogen give NMR signals whose positions are highly dependent on concentration, temperature, and solvent. This is because such hydrogens undergo hydrogen bonding and exchange. Often, such signals are quite broad, especially when subject to the quadrupolar effects of nitrogen.

6. The IOU of a molecular formula [Eq. (6.7)] is the sum of rings plus π bonds in the structure. The nitrogen rule says that the parity (odd or even) of the number of nitrogens in the molecular formula must match the parity of the nominal molecular mass and the number of hydrogens plus halogens.

REFERENCES

[1]Silverstein, R. M., Bassler, G. C., and. Morrill, T. C., *Spectrometric Identification of Organic Compounds*, 5th ed., Wiley, New York, 1991.

[2]Bhacca, S., Hollis, D. P., Johnson, L. F., and Pier, E. A., *High Resolution NMR Spectra Catalog*, Varian Associates, Palo Alto, CA, 1963.

[3]Pouchert, C. J., *The Aldrich Library of NMR Spectra*, 2nd ed., Aldrich Chemical Company, Milwaukee, 1983.

[4]Gunther, H., *NMR Spectroscopy, An Introduction*, Wiley, New York, 1973.

[5]Boekelheide, V., and Phillips, J. B., *J. Am. Chem. Soc. 89*, 1695 (1967).

ADDITIONAL RESOURCES

If you need to review the relevant basics of organic chemistry, you might take a look at: Macomber, R. S., *Organic Chemistry*, Vols. I and II, University Science Books, Sausalito, CA, 1996.

Other ^{1}H NMR spectral compilations, in addition to the references cited above, include:

1. *The Sadtler Standard NMR Spectra*, Sadtler Research Laboratories, Philadelphia, 1972.

2. *Handbook of Proton NMR Spectra and Data*, Asahi Research Center, Academic, Orlando, 1985.

REVIEW PROBLEMS (Answers in Appendix 1)

6.1. (a) Differentiate the quantities $\Delta\nu$, $\delta\nu$, and $\Delta\delta$. (b) What is the $\Delta\delta$ value for hydrogen?

6.2. How many ^{1}H NMR signals would you predict for each structure below when X = Y? When X ≠ Y?

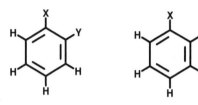

ortho   meta   para

6.3. Explain why the circled CH_2 hydrogens appear at δ 0.8 while the **benzylic** CH_2 groups (those directly bonded to the aromatic ring) have a chemical shift of δ 2.6.

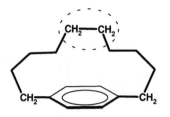

6.4. Identify each compound below from its molecular formula and ^{1}H NMR spectrum. Assign all signals to specific hydrogens in your structure, and show that the observed chemical shifts are in accord with your structure: (a) $C_5H_{10}O_2$; δ 1.26 (9H), 12.05 (1H, broad); (b) $C_2H_2Cl_2$; δ 6.40; (c) $C_{12}H_{18}$; δ 2.20; (d) $C_{12}H_{18}$; δ 1.08 (6H), 1.60 (12H); (e) $C_{15}H_{24}O$; δ 1.47 (18H), 2.29 (3H), 4.99 (1H, broad), 7.00 (2H).

7

CHEMICAL SHIFT CORRELATIONS FOR ^{13}C AND OTHER ELEMENTS

7.1 ^{13}C CHEMICAL SHIFTS REVISITED

In Chapter 6 we developed an understanding of the relationships between the molecular environment of a hydrogen atom and the chemical shift of its nucleus. Now let us see if the same approach of base values plus substituent parameters [Eq. (6.5b)] will allow us to predict ^{13}C chemical shifts. Recall from Sections 2.1 and 2.2 that ^{13}C (like ^{1}H) has a nuclear spin of $\frac{1}{2}$ but undergoes resonance at a much lower frequency than ^{1}H because of the lower magnetogyric ratio of ^{13}C. Also, the low natural abundance and low relative sensitivity of ^{13}C (Table 2.1) require that the signal-averaged pulsed-mode technique (Section 3.4) be used for data collection. Remember that TMS (its carbon signal now) still defines the zero point of our chemical shift scale and that ^{13}C chemical shifts span a range of about 250 ppm. And finally, unless otherwise noted, all ^{13}C spectra discussed in this chapter are proton decoupled (Section 5.5).

7.2 TETRAHEDRAL (sp^3 HYBRIDIZED) CARBONS

Take a moment to review the ^{13}C spectra of toluene in Section 5.4. How would you go about predicting the ^{13}C chemical shift of the methyl group in toluene? The logical base value is the ^{13}C chemical shift of methane (which, it turns out, is δ −2.3, Table 7.1), to which we add the ^{13}C substituent parameter of a phenyl ring connected directly to the methyl carbon ($\Delta\delta = 23$, Table 7.2). The predicted valued, therefore, is δ 20.7, in excellent agreement with the observed value of δ 21.4. Table 7.1 lists ^{13}C chemical shifts for a number of common alkanes and cycloalkanes, to be used as base values. Table 7.2 gives the substituent parameters for many common substituent groups as a function of their proximity to the carbon of interest (α = one-bond separation, β = two-bond separation, γ = three-bond separation). Notice once again that the

(de)shielding effect of a substituent tends to decrease as the number of intervening bonds increases. In fact, most groups exert a modest shielding effect (shown by the negative value of $\Delta\delta$) on the γ carbon.

■ **EXAMPLE 7.1** Why does butane (Table 7.1) exhibit only two ^{13}C signals rather than four?

□ *Solution:* By symmetry, the two methyl groups are equivalent (δ 13.4), as are the two methylenes (δ 25.2). □

TABLE 7.1 ^{13}C Chemical Shifts for Common Alkanes and Cycloalkanes[a]

Name	Structure	C1[b]	C2	C3
Methane	CH_4	−2.3		
Ethane	CH_3CH_3	5.7		
Propane	$CH_3CH_2CH_3$	15.8	16.3	
Butane	$CH_3CH_2CH_2CH_3$	13.4	25.2	
Pentane	$CH_3CH_2CH_2CH_2CH_3$	13.9	22.8	34.7
Hexane	$CH_3CH_2CH_2CH_2CH_2CH_3$	14.1	23.1	32.2
Cyclopropane	Cyclo-$(CH_2)_3$	−3.5		
Cyclobutane	Cyclo-$(CH_2)_4$	22.4		
Cyclopentane	Cyclo-$(CH_2)_5$	25.6		
Cyclohexane	Cyclo-$(CH_2)_6$	26.9		
Cycloheptane	Cyclo-$(CH_2)_7$	28.4		
Cyclooctane	Cyclo-$(CH_2)_8$	26.9		

[a]Data from ref. 1.
[b]Numbering starts at a terminal carbon.

TABLE 7.2 ^{13}C Substituent Parameters ($\Delta\delta$, ppm)^a

X	Terminal X $X-C_\alpha-C_\beta-C_\gamma$			Internal X $C_\gamma-C_\beta-C_\alpha(X)-C_\beta-C_\gamma$		
	α	β	γ	α	β	γ
–F	68	9	–4	63	6	–1
–NO$_2$	63	4	—	57	4	
–OR	58	8	–4	51	5	–4
–OC(=O)R	51	6	–3	45	5	–3
–OH	48	10	–5	41	8	–5
–NR$_2$	42	6	–3			
–NHR	37	8	–4	31	6	–4
–C(=O)Cl	33	—	—	28	2	
–Cl	31	11	–4	32	10	–4
–C(=O)H		31	–2	0	—	–2
–C(=O)R	30	1	–2	24	1	–2
–NH$_2$	29	11	–5	24	10	–5
–N$^+$H$_3$	26	8	–5	24	6	–5
–C(=O)O$^-$	25	5	–2	20	3	–2
–Phenyl	23	9	–2	17	7	–2
–C(=O)NH$_2$	22	—	–0.5	2.5	—	–0.5
–C(=O)OH	21	3	–2	16	2	–2
–CH=CH$_2$	20	6	–0.5			
–C(=O)OR	20	3	–2	17	2	–2
–Br	20	11	–3	25	10	–3
–SR	20	7	–3			
–SH	11	12	–4	11	11	–4
–CH$_3$	9	10	–2	6	8	–2
–C≡C–H	4.5	5.5	–3.5			
–C≡N	4	3	–3	1	3	–3
–I	–6	11	–1	4	12	–1

^{a}Data from Wehrli, F.W., and Wirthlin, T., *Interpretation of Carbon-13 NMR Spectra,* 2nd Ed., Heyden, London,1983, as quoted in ref. 1.

■ **EXAMPLE 7.2** Predict the chemical shift of each carbon in cyclopentanol (the C–H hydrogens are not shown in the structure below):

□ *Solution:* Using as a base value the chemical shift of cyclopentane (δ 25.6, Table 7.1) and the internal substituent parameters for the –OH group, we calculate:

$$\delta_\alpha = 25.6 + 41 = 66.6$$

$$\delta_\beta = 25.6 + 8 = 33.6$$

$$\delta_\gamma = 25.6 + (-5) = 20.6$$

These are in reasonable agreement with the observed values of δ 73.3, 35.0, and 23.4.[1] □

■ **EXAMPLE 7.3** The ^{13}C spectrum of 2-chlorobutane (Figure 7.1) exhibits signals at δ 11.01, 24.85, 33.33, and 60.38. Assign each signal to a specific carbon in the structure by correlating the observed chemical shifts with the predicted values:

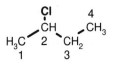

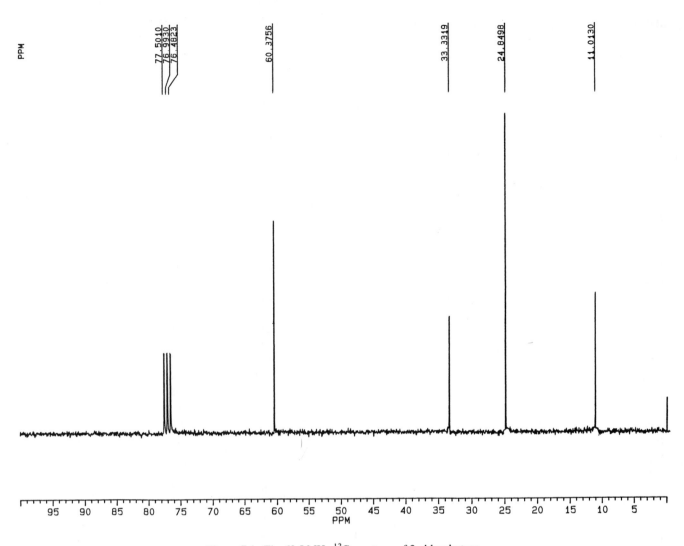

Figure 7.1. The 62.5 MHz ^{13}C spectrum of 2-chlorobutane.

☐ *Solution:* For carbon 1 we use the C1 base value for butane (δ 13.4 ppm, Table 7.1) and add the Δδ value for an internal β chlorine (10 ppm, Table 7.2). Thus,

$$\delta_1 = 13.4 + 10 = 23.4 \text{ ppm}$$

Similarly,

$$\delta_2 = 25.2 + 32 = 57.2 \text{ ppm}$$

$$\delta_3 = 25.2 + 10 = 35.2 \text{ ppm}$$

$$\delta_4 = 13.4 + (-4) = 9.4 \text{ ppm}$$

Clearly, δ_1 corresponds to the peak at δ 24.85, δ_2 to the peak at 60.38, δ_3 to the peak at 33.33, and δ_4 to the peak at 11.01. Note that ^{13}C chemical shifts can be ascertained

only to ±1 ppm when determined visually from a plotted spectrum. These more precise values come from peak picking (the number plotted above each signal) or a computer listing that accompanies the spectrum.   ☐

Did you notice the relative intensities of the lines in Figure 7.1? From Section 5.4 we know that ^{13}C signal intensities are not necessarily proportional to the numbers of carbons giving rise to the signal (as they are in the case of hydrogen signals) because of differences in relaxation times and NOE enhancements. Nonetheless, as is usually the case, the relative intensities of the four tetrahedral carbon signals in the spectrum of 2-chlorobutane vary according to the number of hydrogens directly attached to each carbon, that is, methyl > methylene > methine.

■ **EXAMPLE 7.4**   Now let us try to identify an unknown. Suggest a structure for C_5H_3N, whose ^{13}C spectrum [2] exhibits signals at δ 10.4, 30.4, and 54.4. Assign each signal to a carbon in your structure; then calculate the expected chemical shift of each carbon in your structure.

☐ *Solution:* The IOU for this molecular formula is zero; therefore, it is saturated, with no rings or multiple bonds. The fact that there are only three signals for the five carbons indicates that the structure has some symmetry. The only substituent groups with a single nitrogen that give a signal in the δ 50–60 region are **primary** (1°), **secondary** (2°), or **tertiary** (3°) amines (structures **7-1** through **7-3**):

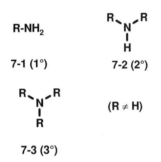

R-NH₂

7-1 (1°)

7-2 (2°)

7-3 (3°)

(R ≠ H)

The signal at δ 54.4 represents either one carbon or two or three *equivalent* carbons attached directly to nitrogen. A little thought should convince you that there is no way to distribute *five* carbons among two or three equivalent R groups to yield a secondary or tertiary amine unless there were *two* different signals in the δ 50–60 range. Therefore, we must be dealing with a primary amine with just *one* carbon attached to nitrogen. The four remaining carbons must be divided into two sets of two to account for the two remaining signals. The only structure that fulfills these criteria is 3-aminopentane:

To confirm our assignment, let us calculate the predicted chemical shifts using the base values for pentane (Table 7.1) and the Δδ values for an NH₂ substituent (Table 7.2):

$$\delta_1 = 13.9 + (-5) = 8.9$$

$$\delta_2 = 22.8 + 10 = 32.8$$

$$\delta_3 = 34.7 + 24 = 58.7$$

Once again, the agreement with the observed values is close enough to give us confidence in our structural assignment. ☐

7.3 HETEROCYCLIC STRUCTURES

A cyclic molecule in which one or more of the ring atoms is a heteroatom is referred to as a **heterocyclic molecule.**

■ **EXAMPLE 7.5** Predict the chemical shift of the carbons in the heterocyclic molecule below, known as dioxane:

☐ *Solution:* First, we recognize by symmetry that all four carbons are equivalent. We can think of the molecule as a derivative of ethane:

Thus, each carbon is α to one and β to another (equivalent) O–R group. Using the data in Tables 7.1 and 7.2, we calculate

$$\delta = \delta(\text{ethane}) + \Delta\delta\,(\text{OR},\,\alpha) + \Delta\delta\,(\text{OR},\,\beta)$$

$$= 5.7 + 58 + 8 = 71.7$$

The observed value is 66.5.[1] ☐

When predicting chemical shifts for substituted heterocycles, you will find it preferable to use the chemical shifts of the parent heterocycle as base values. The ¹³C chemical shifts for a few typical examples are shown in structures **7-4** through **7-9**.[1,3,4] Consult the references listed at the end of this chapter for additional examples:

tetrahydrofuran (7-4)   pyrrolidine (7-5)   thiacyclopentane (7-6)

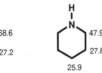

tetrahydropyran (7-7)   piperidine (7-8)   thiacyclohexane (7-9)

■ **EXAMPLE 7.6** Predict the chemical shift of each carbon in the structure below:

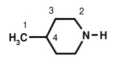

Solution: Using the base values in structure **7-8** and the substituent parameters for an internal CH_3 group (Table 7.2), we predict the following for the ring carbons:

$$\delta_2 = \delta(\text{piperidine, C2}) + \Delta\delta\,(CH_3, \gamma) = 47.9 + (-2) = 45.9$$

$$\delta_3 = \delta(\text{piperidine, C3}) + \Delta\delta(CH_3, \beta) = 27.8 + 8 = 35.8$$

$$\delta_4 = \delta(\text{piperidine, C4}) + \Delta\delta\,(CH_3, \alpha) = 25.9 + 6 = 31.9$$

We can calculate the chemical shift of the methyl carbon by viewing the molecule as a substituted derivative of ethane (Table 7.2):

$$\underset{1}{H_3C}-\underset{\underset{R}{|}}{\overset{\overset{R}{|}}{CH}}$$

$$\delta_1 = \delta(\text{ethane}) + 2[\Delta\delta(CH_3, \beta)] = 5.7 + 2(10) = 25.7$$

These turn out to be very close to the observed values of δ 46.8, 35.7, 31.3, and 22.5, respectively.[3] □

7.4   TRIGONAL CARBONS

You remember that the *aromatic* carbons of toluene (Section 5.4) appear far downfield of the methyl carbon. Just as with vinyl and aromatic hydrogens (Sections 6.3 and 6.5), sp^2 hybridized vinyl and aromatic carbons are deshielded in comparison to their sp^3 hybridized tetrahedral counterparts. For this reason, their signals usually fall in the region of δ 100–165.

7.4.1   Vinyl Carbons

We will treat molecules with vinyl carbons as derivatives of ethylene ($H_2C{=}CH_2$, Section 6.3), whose ^{13}C signal appears at δ 123.3.[1] To this base value, we will apply $\Delta\delta$ corrections (Table 7.3) for substituents according to their location with respect to the carbon of interest.

■ **EXAMPLE 7.7**   Predict the chemical shift of the two vinyl carbons in *trans*-2-pentene:

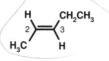

□ **Solution:** Carbon 2 has alkyl carbon substituents at positions α, a, and b. Therefore, $\delta_2 = \delta(C_2H_4) + \Delta\delta(R, \alpha) +$

TABLE 7.3   Vinyl Carbon Substituent Parameters ($\Delta\delta$, ppm)^{a,b}

X	α	β	γ	a	b	c
–R	10.6	7.2	–1.5	–7.9	–1.8	–1.5
–OR	29	2	—	–39	–1	
–OC(=O)R	18	—	—	–27		
–C(=O)R	15	—	—	6		
–Phenyl	12	—	—	–11		
–C(=O)OR	6	—	—	7		
–C(=O)OH	4	—	—	9		
–OH	—	6	—	—	–1	
–Cl	3	–1	—	–6	2	
–Br	–8	0	—	–1	2	
–C≡N	–16	—	—	15		
–I	–38	—	—	7		

^{a}Data from refs. 1 and 4. R = alkyl.
^{b}When a group is in the b (or β) position, X_a (or X_α) is assumed to be carbon; when a group is in the c (or γ) position, both X_a and X_b (or X_α and X_β) are assumed to be carbon.

$$\Delta\delta(R, a) + \Delta\delta(R, b) = 123.3 + 10.6 + (-7.9) + (-1.8) = 124.2$$

Similarly,

$$\delta_3 = \delta(C_2H_4) + \Delta\delta(R, \alpha) + \Delta\delta(R, a) + \Delta\delta(R, \beta)$$

$$= 123.3 + 10.6 + (-1.8) + 7.2 = 133.2$$

The observed values are δ 123.7 and 133.3, respectively.   □

Table 7.3 includes substituent parameters for several heteroatom-containing substituents. Notice that there is a very large difference between the $\Delta\delta$ value for a group located at the α position and the value for the same group at the *a* position. This is because the π electrons in the carbon–carbon double bond interact directly through **resonance** with many of these substituents, causing substantial shielding when at the α position and deshielding when at the *a* position, or vice versa.

For example, an atom X with an unshared pair of electrons (e.g., X = OR, NR₂, S, or halogen) can donate the pair to (i.e., share it with) the π system, as shown in the resonance structures below. This interaction increases the electron density at carbon 2 (thereby shielding it), while decreasing the electron density on carbon 1 and X (deshielding them):

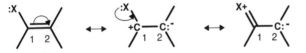

Conversely, if atom X has an empty *p*-type valence orbital (e.g., X = C=O, C≡N, NO₂), it can withdraw a pair of electrons from the π bond, significantly deshielding carbon 2 while shielding X:

Thus, the net effect of any such **polar** substituent is the sum of its resonance effects and its inductive (electronegativity) effects.

■ **EXAMPLE 7.8** Predict the chemical shifts for the vinyl carbons in the molecule below; then draw a resonance structure of the molecule that shows why one vinyl carbon is exceptionally shielded while the other is deshielded:

□ *Solution:* Using the data in Table 7.3,

$$\delta_1 = \delta(\text{ethylene}) + \Delta\delta(OR, \alpha) = 123.3 + 29 = 152.3$$

$$\delta_2 = \delta(\text{ethylene}) + \Delta\delta(OR, a) = 123.3 + (-39) = 84.3$$

Recall that a divalent oxygen has two unshared electron pairs, either of which can be donated to the π system:

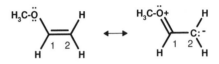

The observed signals occur at δ 153.2 and 84.2, respectively.[1] Notice that carbon 1 is deshielded by the electronegative oxygen directly connected to it, while carbon 2 is shielded by resonance between the π bond and the unshared pair on oxygen.

Because of such resonance interactions, the best policy to follow when you are predicting chemical shifts of vinyl carbons with polar substituents directly connected is to find a model compound as similar in structure as possible to serve as the source of base values.   □

In organometallic complexes of alkenes, where a transition metal uses an empty *d* orbital to coordinate to the π bond of the alkene, the vinyl carbons move upfield to the range of δ 7–110, varying widely and depending on the exact structure of the complex.[5] The reasons for this upfield shift include both shielding by the electron-rich electropositive metal, and a change in hybridization of the vinyl carbons toward less *s* character (i.e., more *sp*³ like).

7.4.2 Aromatic Carbons

In the case of aromatic carbons, we will use as our base value the ¹³C chemical shift of benzene (Example 4.3), δ 128.5. Notice that aromatic carbons appear slightly downfield of vinyl carbons, just as aromatic hydrogens appear downfield of vinyl hydrogens (Sections 6.3 and 6.5). Table 7.4 lists substituent parameters for a number of aromatic substituents as a function of proximity to a given carbon. Observe how, for many of these groups, their effect alternates back and forth from deshielding to shielding as the number of intervening bonds increases. This is a consequence of the changing balance between competing inductive and resonance effects of the substituent as a function of location.

■ **EXAMPLE 7.9** Predict the chemical shifts of each aromatic carbon in *para*-nitroaniline:

TABLE 7.4 Aromatic Substituent Parameters ($\Delta\delta$, ppm)[a]

X	α	o (Ortho)	m (Meta)	p (Para)
–F	35.1	–14.3	0.9	–4.5
–OCH$_3$	31.4	–14.4	1.0	–7.7
–OPh	29.0	–9.4	1.6	–5.3
–OH	26.6	–12.7	1.6	–7.3
–OC(=O)CH$_3$	22.4	–7.1	–0.4	–3.2
–N(CH$_3$)$_2$	22.4	–15.7	0.8	–11.8
–C(CH$_3$)$_3$	22.2	–3.4	–0.4	–3.1
–CH(CH$_3$)$_2$	20.1	–2.0	0	–2.5
–NO$_2$	19.6	–5.3	0.9	6.0
–NH$_2$	19.2	–12.4	1.3	–9.5
–CH$_2$CH$_3$	15.6	–0.5	0.0	–2.6
–S(O)$_2$NH$_2$	15.3	–2.9	0.4	3.3
–Si(CH$_3$)$_3$	13.4	4.4	–1.1	–1.1
–CH$_2$OH	13.3	–0.8	–0.6	–0.4
–phenyl	12.1	–1.8	–0.1	–1.6
–NHC(=O)CH$_3$	11.1	–9.9	0.2	–5.6
–SCH$_3$	10.2	–1.8	0.4	–3.6
–CH$_3$	9.3	0.7	–0.1	–2.9
–CH=CH$_2$	9.1	–2.4	0.2	–0.5
–C(=O)Ph	9.1	1.5	–0.2	3.8
–C(=O)H	8.2	1.2	0.6	5.8
–C(=O)CH$_3$	7.8	–0.4	–0.4	2.8
–CH$_2$OC(=O)CH$_3$	7.7	0	0	0
–Cl	6.4	0.2	1.0	–2.0
–N=C=O	5.7	–3.6	1.2	–2.8
–C(=O)Cl	4.6	2.9	0.6	7.0
–C(=O)OH	2.9	1.3	0.4	4.3
–CF$_3$	2.6	–3.1	0.4	3.4
–SH	2.3	0.6	0.2	–3.3
–C(=O)OCH$_3$	2.0	1.2	–0.1	4.8
–Br	–5.4	3.4	2.2	–1.0
–C(=O)CF$_3$	–5.6	1.8	0.7	6.7
–C≡CH	–5.8	6.9	0.1	0.4
–C≡N	–16.0	3.6	0.6	4.3
–I	–32.2	9.9	2.6	–7.3

[a]Data abstracted from ref. 1.

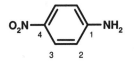

□ *Solution:* By symmetry, the aromatic carbons are divided into four sets in the ratio of 1 : 2 : 2 : 1. Carbon 1 is α to an NH_2 group, and para to an NO_2 group. Therefore,

$$\delta_1 = \delta(\text{benzene}) + \Delta\delta(NH_2, \alpha) + \Delta\delta(NO_2, p)$$

$$= 128.5 + 19.2 + 6.0 = 153.7$$

Likewise,

$$\delta_2 = \delta(\text{benzene}) + \Delta\delta(NH_2, o) + \Delta\delta(NO_2, m)$$

$$= 128.5 + (-12.4) + 0.9 = 117.0$$

$$\delta_3 = \delta(\text{benzene}) + \Delta\delta(NH_2, m) + \Delta\delta(NO_2, o)$$

$$= 128.5 + 1.3 + (-5.3) = 124.5$$

$$\delta_4 = \delta(\text{benzene}) + \Delta\delta(NH_2, p) + \Delta\delta(NO_2, \alpha)$$

$$= 128.5 + (-9.5) + 19.6 = 138.6$$

These predictions agree quite nicely with the observed values of δ 155.1, 112.8, 126.3, and 136.9, respectively. Notice how carbons 1 and 4 are deshielded (the former quite strongly) while carbons 2 and 3 are shielded, relative to benzene. □

■ **EXAMPLE 7.10** Which of the three isomers of xylene gives a ^{13}C spectrum with signals at δ 137.7, 130.0, 128.2, 126.2, and 21.3? To confirm your choice, predict the chemical shift of each carbon in your structure:

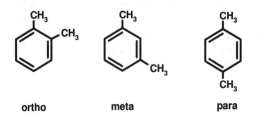

ortho   meta   para

□ *Solution:* If you took more than 1 minute to pick out the correct answer, you have forgotten what you learned about symmetry in Chapter 4. In each of these three structures the two methyl groups are equivalent. But the number of aromatic carbon signals would vary: The ortho isomer would show three, the meta four, and the para only two. Since the first four ^{13}C signals are squarely in the aromatic

region, only the meta isomer fits the observed spectrum. To confirm our identification, let us predict chemical shifts:

$$\delta_1 = \delta(\text{benzene}) + \Delta\delta(CH_3, \alpha) + \Delta\delta(CH_3, m)$$

$$= 128.5 + 9.3 + (-0.1) = 137.7$$

$$\delta_2 = \delta(\text{benzene}) + 2[\Delta\delta(CH_3, o)]$$

$$= 128.5 + 2(0.7) = 129.9$$

$$\delta_3 = \delta(\text{benzene}) + \Delta\delta(CH_3, o) + \Delta\delta(CH_3, p)$$

$$= 128.5 + 0.7 + (-2.9) = 126.3$$

$$\delta_4 = \delta(\text{benzene}) + 2[\Delta\delta(CH_3, m)]$$

$$= 128.5 + 2(-0.1) = 128.3$$

The observed chemical shift of the methyl groups (δ 21.3) could have been predicted from the data for toluene in Section 7.2. □

As we saw with metal complexed vinyl carbons, organometallic complexes of aromatic compounds exhibit ^{13}C chemical shifts upfield of the uncomplexed compound. For example, metal-coordinated benzene carbons are found in the range δ 74–111, depending on the exact structure of the complex. Similarly, metal-coordinated cyclopentadienide (C_5H_5-, abbreviated Cp^-) appears in the range δ 75–123.

7.4.3 Heteroaromatic Compounds

As was true for the hydrogens attached to heteroaromatic rings and polycyclic aromatic hydrocarbons (Section 6.5), the carbons of such rings also appear in the aromatic region. Some examples are given below,[1] and these can be used as base values when calculating chemical shifts for substituted derivatives of these compounds:

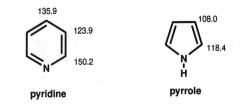

pyridine   pyrrole

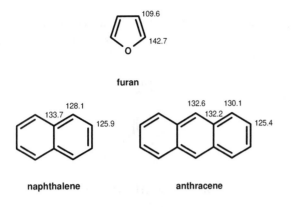

furan

naphthalene   anthracene

7.5   TRIPLY BONDED CARBONS

Recall (Section 6.4) that hydrogens attached to triply bonded carbons (*acetylenic* hydrogens) are unusually shielded because of the magnetic field anisotropy in the region of the triple bond. Exactly the same is true of acetylenic carbons themselves. Triply bonded (*sp*-hybridized) carbons usually appear in the δ 70–90 region, significantly upfield of typical vinyl and aromatic carbons. This is a region of the spectrum where relatively few other types of carbons are found. Acetylenic carbons are classified as either *terminal* (if they have a hydrogen directly attached) or *internal* (if they have a carbon attached). Signals for internal acetylenic carbons are usually downfield of and less intense than the signals for terminal ones. This is because of the deshielding effect of the carbon substituent (Table 7.2) and the absence of the intensity-increasing NOE (Section 5.4.1) due to the attached hydrogen.

■ **EXAMPLE 7.11**   Propose a structure for the compound C_8H_6 whose ^{13}C spectrum exhibits signals at δ 77.4, 83.8 (the latter being only 20% as intense as the former), 122.4, 128.3, 128.7, and 132.2. Assign all signals by calculating expected chemical shifts.

☐ *Solution:* The IOU is 6. We note immediately that there are four signals in the aromatic region and two in the acetylenic region. Thus, a likely candidate would be a monosubstituted aromatic ring with one acetylenic substituent:

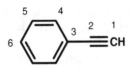

Carbons 1 and 2 can be assigned on the basis of relative position and intensity (remember the NOE: fewer attached hydrogens, weaker signal) to the signals at δ 77.4 and 83.8, respectively. To assign the aromatic signals, let us predict where each should appear, with observed values:

$$\delta_3 = \delta(\text{benzene}) + \Delta\delta\ (-C \equiv CH, \alpha)$$

$$= 128.5 + (-5.8) = 122.7$$

$$\delta_4 = \delta(\text{benzene}) + \Delta\delta(-C \equiv CH, o)$$

$$= 128.5 + 6.9 = 135.4$$

$$\delta_5 = \delta(\text{benzene}) + \Delta\delta(-C \equiv CH, m)$$

$$= 128.5 + 0.1 = 128.6$$

$$\delta_6 = \delta(\text{benzene}) + \Delta\delta(-C \equiv CH, p)$$

$$= 128.5 + 0.4 = 128.9$$

The signals for carbons 5 (of which there are two) and 6 (of which there is one) are so close that assignments based on calculated chemical shifts alone are not likely to be 100% reliable. But notice that relative signal intensity is also consistent with the above assignment.

Look back at Example 6.18. Notice once again how the relative chemical shifts of the hydrogens parallel the relative chemical shifts of the carbons to which they are bonded.   ☐

Another important triply bonded carbon is the one in a **cyano** group (–C≡N). Because of the higher electronegativity (and hence deshielding effect) of the nitrogen, cyano carbons occur downfield of acetylenic carbons, usually around δ 115–120. And remember: With no hydrogens attached, they usually give fairly weak signals.

Finally, in the quite unusual –C≡P group, the carbon is deshielded to δ 185, while metal carbyne complexes (M≡C–R) are even further deshielded into the region δ 230–365. This indicates not only less shielding by the magnetic anisotropy of the CP triple bond (suggesting that there is less efficient circulation of the C≡P triple bond electrons around the internuclear axis) but also paramagnetic deshielding by the phosphorus or metal atom.

7.6   CARBONYL CARBONS

Among the most deshielded carbon atoms are those that are doubly bonded to oxygen (C=O). Because of the high electronegativity of oxygen, carbonyl carbons generally appear in the δ 165–220 range. The carbonyl group occurs in many types of organic and organometallic compounds, and its ^{13}C chemical shift varies accordingly, as shown in Table 7.5. Because the regions characteristic of each functional group tend to overlap, one must usually consider the chemical shifts

TABLE 7.5 ^{13}C Chemical Shift Ranges of Carbonyl Compounds

Compound class	Structure	δ (ppm)
Ketone	R–C(=O)–R	195–220
Aldehyde	R–C(=O)–H	190–200
Carboxylic acid	R–C(=O)–OH	170–185
Carboxylate ester	R–C(=O)–OR	165–175
Anhydride	R–C(=O)–O–C(=O)–R	165–175
Amide	R–C(=O)–NR$_2$	160–170
Acid halide	R–C(=O)–X(X = Cl, Br, I)	160–170
Metal-coordinated CO	M–C=O	150–285
Carbon monoxide	CO	183.4[a]
Carbon dioxide	CO$_2$	124.8[b]
Ketene	R$_2$C=C=O	See Table 7.6

[a]CO terminally coordinated to a metal in an organometallic complex (M–C=O) is usually deshielded and appears in the range δ 180–250 (ref. 5).

[b]CO$_2$-saturated CDCl$_3$.

of the other carbons in order to suggest a unique structure for an unknown compound.

In addition to their characteristic downfield positions, carbonyl carbons can also be recognized by their relatively low intensity. Except in the case of a formyl carbonyl [one attached directly to a hydrogen, –C(=O)H], there are no hydrogens attached to carbonyl carbons, so their signal cannot benefit from NOE enhancement. Carbonyl carbons also tend to have longer T_1 relaxation times than many other carbons, another cause of lower intensities in a signal-averaged pulsed-mode spectrum (Section 5.4.1).

■ **EXAMPLE 7.12**   Which of the isomeric structures below has a ^{13}C spectrum consisting of signals at δ 190.6, 135.2, 132.3, 130.8, and 129.4? Assign all signals by correlating them with predicted chemical shifts:

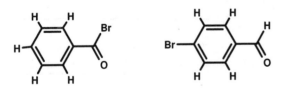

□ *Solution:* It is easy to pick out the carbonyl signal; it is the one at δ 190.6. From Table 7.5 we can see that this signal is within the normal range for aldehydic carbons but too far downfield for an acid halide carbonyl. So, we pick the structure on the right:

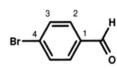

$δ_1 = δ(benzene) + Δδ[–C(=O)H, α] + Δδ(Br, p)$

$= 128.5 + 8.2 + (–1.0) = 135.7$

$δ_2 = δ(benzene) + Δδ[–C(=O)H, o] + Δδ(Br, m)$

$= 128.5 + 1.2 + 2.2 = 131.9$

$δ_3 = δ(benzene) + Δδ[–C(=O)H, m] + Δδ(Br, o)$

$= 128.5 + 0.6 + 3.4 = 132.5$

$δ_4 = δ(benzene) + Δδ[–C(=O)H, p] + Δδ(Br, α)$

$= 128.5 + 5.8 + (–5.4) = 128.9$ □

7.7   MISCELLANEOUS UNSATURATED CARBONS

There is a host of less common functional groups that involve unsaturated (i.e., multiply bonded) carbon and appear in the same downfield region of the ^{13}C spectrum as carbonyls, for example, the *central* (sp hybridized) carbon of an **allene** linkage (C=C=C). Such carbons give very weak signals

TABLE 7.6 ^{13}C Chemical Shifts of Other Functional Groups

Functional Group	Structure	δ(ppm)
Oxime	$R_2C{=}N{-}OH$	145–170
Isocyanate	$R{-}N{=}C{=}O$	110–135
Isothiocyanate	$R{-}N{=}C{=}S$	120–140
Metallo-carbene	$M{=}CR_2$	180–400
Carbenium ions	R_3C^+	212–320
Acylium ions	$R{-}C^+{=}O$	145–155
Carbon disulfide	$S{=}C{=}S$	192.3
Ketenes[a]	$Z_2C_\beta{=}C_\alpha{=}O$	Z = H, alkyl, aryl: α, 194–206; β, 2.5–48
Ketenes[a]	$Z_2C_\beta{=}C_\alpha{=}O$	Z = heteroatom: α, 161–183; β, −20–125

[a]See ref. 6.

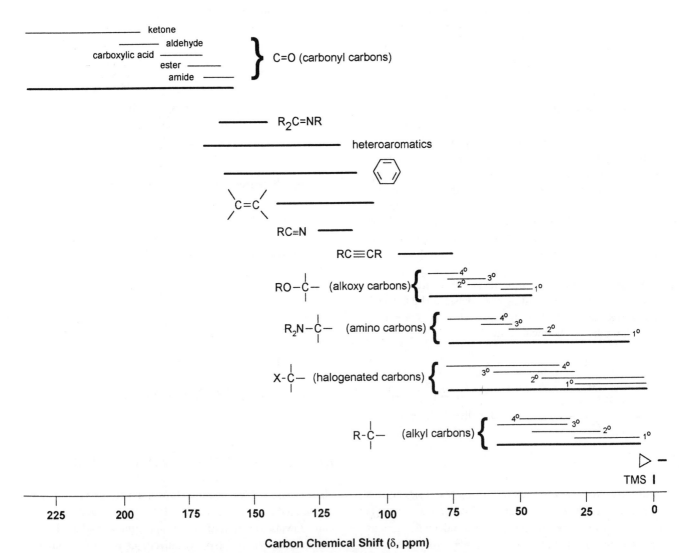

Figure 7.2. Pictorial representation of the ^{13}C chemical shift ranges for various classes of carbons.

(why?) in the δ 200–215 region. The two outer carbons, by contrast, are more shielded than typical vinyl carbons (Section 7.4) and appear in the δ 75–95 range. Table 7.6 lists some additional examples.

7.8 SUMMARY OF ^{13}C CHEMICAL SHIFTS

The range of ^{13}C chemical shifts is shown graphically in Figure 7.2. This is a useful place to start when trying to identify an unknown compound from its ^{13}C spectrum. But remember that these ranges represent only generalizations and that certain combinations of substituents will cause a carbon signal to show up outside the confines of its "normal" region.

Notice the similarity of Figure 7.2 with the corresponding one for hydrogen chemical shifts (Figure 6.8). This parallel behavior is because *in general* (there *are* exceptions) a carbon and the hydrogens attached directly to it both experience similar shielding and deshielding effects of neighboring substituents. To demonstrate the validity of this assertion, Figure 7.3 shows a plot of ^{1}H chemical shifts for 335 different hydrogens versus the ^{13}C chemical shifts of the carbons to which they are directly bonded.[7] The linear relationship, expressed by Eqs. (7.1a) and (7.1b), exhibit a correlation coefficient of 0.995 over a wide variety of molecular environments:

$$\delta_H = 0.0479\delta_C + 0.472 \qquad (7.1a)$$

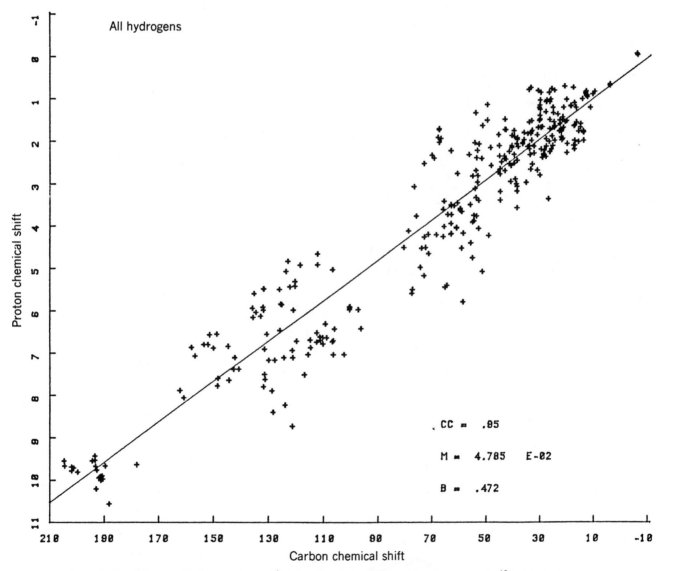

Figure 7.3. Graphical comparison of ^{1}H chemical shifts of C–H hydrogens versus the ^{13}C chemical shift of the attached carbon. [From "Proton–Carbon Chemical Shift Correlations," by R. S. Macomber, *Journal of Chemical Education*, *68*, 284–285 (1991). Reprinted with permission.]

$$\delta_C = 20.9\delta_H - 9.85 \qquad (7.1b)$$

Thus, armed with only the ^{1}H spectrum of a molecule, one can estimate the chemical shift of the attached carbons (or vice versa) with a fair level of confidence.

■ **EXAMPLE 7.13** Suggest a structure for C_8H_5NO, whose ^{13}C spectrum exhibits signals at δ 112.9, 129.6, 130.3, 133.4, 136.9, and 167.8. Assign all signals by calculating the expected chemical shift of each carbon in your suggested structure.

☐ *Solution:* The IOU for this compound is 7. There are six signals in this spectrum, indicative of some symmetry equivalences in the molecule. The four signals at δ 129.6, 130.3, 133.4, and 136.9 fall squarely in the aromatic carbon region, indicating either a monosubstituted benzene ring or one with two dissimilar groups para to each other. The signal at δ 167.8 is likely to be some sort of carbonyl carbon. But what kind of nitrogen-containing group would appear at δ 112.9? A cyano group! Now, see if you can put all this information into some tentative isomeric structures:

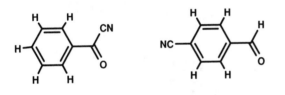

The structure on the left has a functional group we have not encountered previously, one with a cyano group attached directly to a carbonyl. It is somewhat like a ketone, whose carbonyls usually appear around δ 195–220, not δ 168. Similarly, the one on the right is an aldehyde, so its carbonyl carbon should appear around δ 190. For some reason our carbonyl signal must be unusually shielded. Perhaps the triple bond of the CN group can give rise to anisotropic shielding just like a carbon–carbon triple bond. Let us see how close the aromatic carbons in the left-hand structure come to our expectations, by treating the –C(=O)CN group as the closest analog to be found in Table 7.4, a –C(=O)CH$_3$ group:

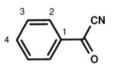

$$\delta_1 = \delta(benzene) + \Delta\delta[-C(=O)CH_3, \alpha]$$

$$= 128.5 + 7.8 = 136.3$$

$$\delta_2 = \delta(benzene) + \Delta\delta[-C(=O)CH_3, o]$$

$$= 128.5 + (-0.4) = 128.1$$

$$\delta_3 = \delta(benzene) + \Delta\delta[-C(=O)CH_3, m]$$

TABLE 7.7 Representative ^{19}F Chemical Shifts^a

Compound	δ, ppm^b	Compound	δ, ppm^b
ClF	−441.5	CF$_3$H	−78.5
CH$_3$F	−275.4	CF$_2$ClH	−71.8
HF	−214.4	CF$_4$	−63.4
trans-FHC=CHF	−186	CF$_3$CN	−56.5
SiF$_4$	−167.6	CF$_3$Cl	−28.8
Cis-FHC=CHF	−165	CF$_3$Br	−18.2
CH$_2$F$_2$	−143.5	CF$_2$Cl$_2$	−7.0
F$_2$C=CF$_2$	−135.1	CF$_3$I	−4.0
BF$_3$	−131.6	CFCl$_3$	0
CH$_3$CHF$_2$	−110.8	ClO$_2$F	332.1
Ph–F	−106.3	F$_2$	428.4
O=PF$_3$	−92.6	XeF$_4$	446.6
C$_2$F$_6$	−87.4	FNO	485.6
H$_2$C=CF$_2$	−83.4	XeF$_6$	552.6
CFCl$_2$H	−80.7	FOOF	871.6

^{a}From ref. 8.
^{b}All chemical shifts referenced to CFCl$_3$.

$$= 128.5 + (-0.4) = 128.1$$

$$\delta_4 = \delta(\text{benzene}) + \Delta\delta[-C(=O)CH_3, p]$$

$$= 128.5 + 2.8 = 131.3$$

Although the signal observed at δ 136.9 is assignable to C1, the remaining signals are difficult, if not impossible, to assign unambiguously among C2, C3, and C4. In Chapter 13 we will find ways to make more definite assignments. Just to set your mind at ease, the cyano–ketone structure above *is* the correct answer. This example demonstrates that in many real-life situations we have to use all the data at our disposal and still make some educated judgments. That, after all, is what makes life interesting! □

7.9 CHEMICAL SHIFTS OF OTHER ELEMENTS

As stated in Chapters 2 and 5, the applications of NMR extend well beyond just hydrogen and carbon, although these two are arguably among the most commonly studied. Other nuclei subjected to *routine* NMR examination include ^2H, ^{14}N, ^{19}F, and ^{31}P; scores of others are studied somewhat less routinely. For several of these nuclei, semiempirical chemical shift correlations of the type we have seen for ^1H and ^{13}C have been generated. In addition, there are several computer software packages now available that use these types of correlations to estimate the hydrogen and carbon chemical shifts for virtually any structure input to the program.

Table 7.7 lists the ^{19}F chemical shifts for a variety of fluorine-containing compounds; notice that these values span a range of 1300 ppm! Figures 7.4 and 7.5 show, respectively, the chemical shift ranges for ^{31}P (data from ref. 9) and ^{14}N

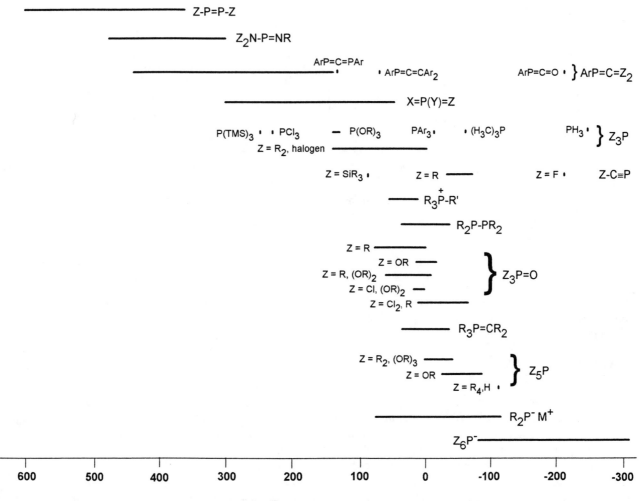

Figure 7.4. Pictorial representation of the ^{31}P chemical shift ranges for various classes of phosphorus atoms. Chemical shifts are referred to 85% H_3PO_4.

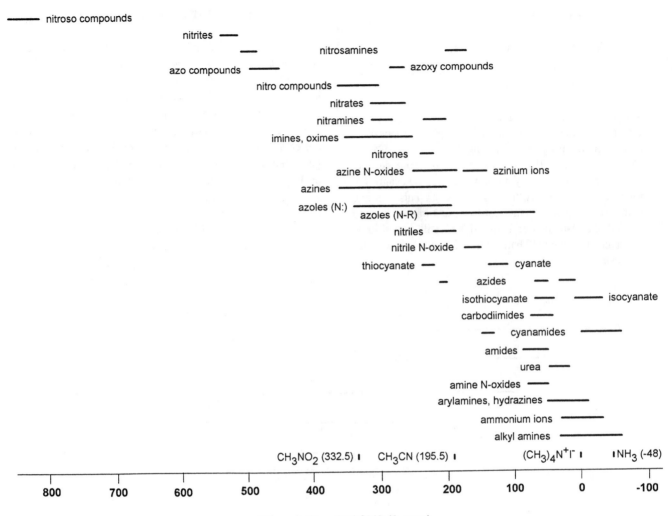

Figure 7.5. Pictorial representation of the ^{14}N chemical shift ranges for venous classes of nitrogen atoms. Chemical shifts are referred to $(CH_3)_4N^+I^-$.

(data from ref. 10); both span ranges of more than 900 ppm. Compare these chemicals shift ranges with those for ^{1}H (20 ppm) and ^{13}C (less than 400 ppm).

CHAPTER SUMMARY

1. Carbon atoms in molecules are classified, as are hydrogens, on the basis of the structural unit of which they are a part: tetrahedral (methyl, methylene, methine, or quaternary), trigonal (e.g., vinyl, aromatic, or carbonyl), and triply bonded (e.g., acetylenic).

2. Because of the low natural abundance and low sensitivity of ^{13}C, the acquisition of carbon NMR spectra requires the use of signal-averaged pulsed-mode meth-ods. Further simplification is afforded by applying ^{1}H decoupling.

3. The chemical shift of each carbon in a molecule can be predicted accurately by the same type of correlations used for hydrogens, namely, an appropriate base value plus substituent parameters [Eq. (6.5b)].

4. There is a consistent parallel between the relative order of hydrogen chemical shifts and the order of chemical shifts of the carbons to which they're attached.

REFERENCES

[1]Silverstein, R. M., Bassler, G. C., and Morrill, T. C., *Spectrometric Identification of Organic Compounds*, 5th ed., Wiley, New York, 1991.

[2]Simons, W. W., Ed., *The Sadtler Guide to Carbon-13 NMR Spectra*, Sadtler Research Laboratories, Division of Bio-Rad Laboratories, Philadelphia, 1983.

[3]Johnson, L. F., and Jankowski, W. C., *Carbon-13 NMR Spectra,* Krieger, Huntington, NY, 1978.

[4]Cooper, J. W., *Spectroscopic Techniques for Organic Chemists,* Wiley, New York, 1980.

[5]Kegley, S. E., and Pinhas, A. R., *Problems and Solutions in Organometallic Chemistry*, University Science Books, Mill Valley, CA, 1986.

[6]Tidwell, T. T., *Ketenes*, Wiley, New York, 1995.

[7]Macomber, R. S., *J. Chem. Educ.*, *68*, 284 (1991).

[8]Mason, J., Ed., *Multinuclear NMR*, Plenum, New York, 1987.

[9]Verkade, J. G., and Quin, L. D., Ed., *Phosphorus-31 NMR Spectroscopy in Stereochemical Analysis. Organic Compounds and Metal Complexes*, VCH Publishers, New York, 1987.

[10]Witanowski, M., and Webb, G. A., Ed., *Nitrogen NMR*, Plenum, New York, 1973.

REVIEW PROBLEMS (Answers in Appendix 1)

7.1. Why are aromatic carbons only slightly more deshielded than vinyl carbons, whereas aromatic hydrogens are considerably more deshielded than vinyl hydrogens.

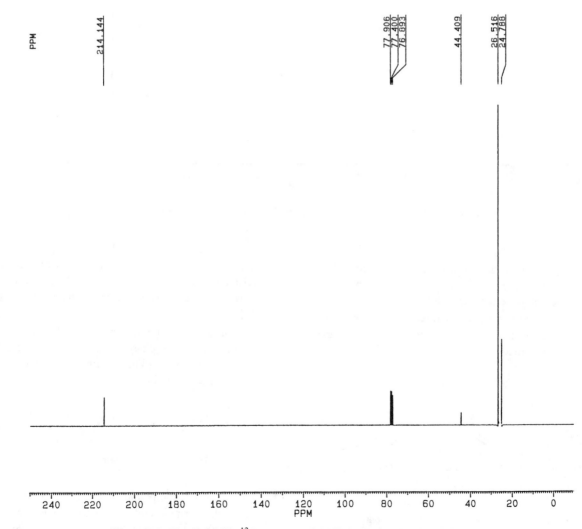

Figure 7.6. The 62.5-MHz ^{13}C spectrum of $C_6H_{12}O$ [Review Problem 7.6(a)].

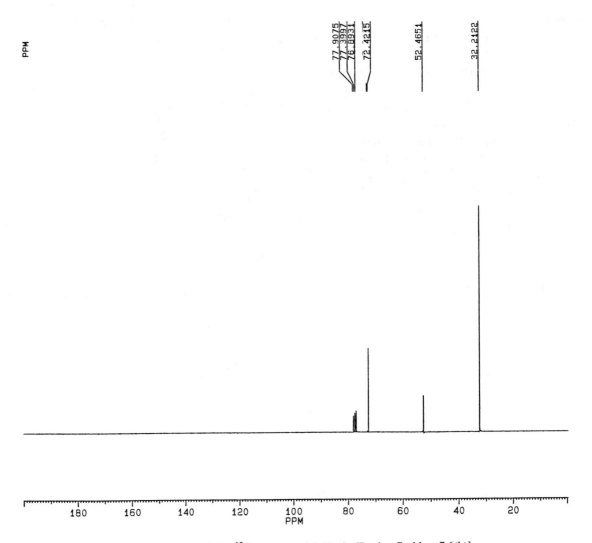

Figure 7.7. The 62.5-MHz ^{13}C spectrum of $C_7H_{16}O_2$ [Review Problem 7.6(b)].

7.2. Draw one or more resonance structures of pyrrole (Section 7.4.3) that show why the carbon chemical shifts have the relative order they exhibit.

7.3. How many carbon signals will each of the structures below exhibit if X ≠ Y? Repeat your analysis for the situation where X = Y.

7.4. What would be the ^{14}N chemical shift of acetonitrile (H$_3$CCN) if, instead of (CH$_3$)$_4$N$^+$I$^-$, nitromethane (CH$_3$NO$_2$) were used as internal standard? (See Figure 7.5.)

7.5. One of the carbenium ions below exhibits a ^{13}C signal at δ 320.6 for the charged carbon, while the other's occurs at δ 250.3. Which chemical shift goes with which structure and why?

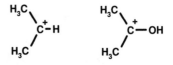

7.6. Identify each compound below based solely on its molecular formula and ^{13}C spectroscopic data. Then assign each ^{13}C signal by calculating the expected

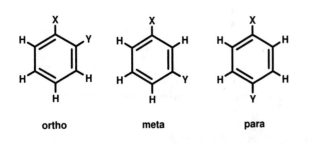

ortho   meta   para

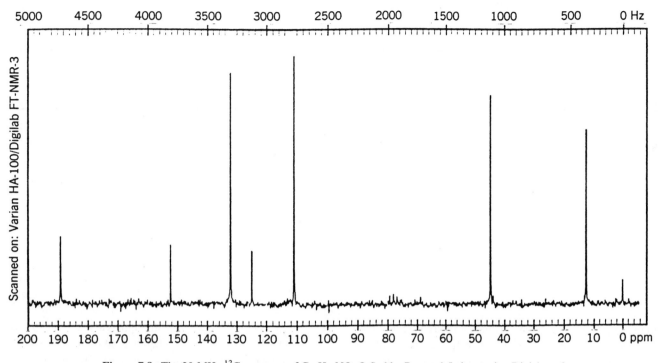

Figure 7.8. The 20-MHz ^{13}C spectrum of $C_{11}H_{15}NO$. © Sadtler Research Laboratories, Division of Bio-Rad Laboratories, Inc. (1983).

chemical shift: (a) $C_6H_{12}O$; its ^{13}C spectrum appears in Figure 7.6 (page 103); (b) $C_7H_{16}O_2$; its ^{13}C spectrum appears in Figure 7.7; (c) C_9H_{10}; δ 19.6, 19.7, 30.6, 85.2, 92.0; and (d) $C_{10}H_8O_4Cr$; δ 55.6, 78.3, 85.6, 95.2, 143.4, 233.3. (*Hint*: The chromium is coordinated to three equivalent carbon monoxide molecules and to the six π electrons of an aromatic ring.)

7.7. What compound gives rise to the three signals at δ 76.9, 77.4, and 77.9 in Figures 7.1, 7.7, and 7.8?

7.8. Identify each compound below based on its molecular formula and NMR spectroscopic data. Be prepared to support your structure by assigning all signals and

demonstrating that their chemical shifts are within appropriate ranges: (a) C_5H_9P; ^1H δ 1.15; ^{13}C δ 31.3, 36.4, 184.8; ^{31}P δ –69.2; (b) C_2H_3N; ^1H δ 1.95; ^{13}C δ 1.7, 117.9; ^{14}N δ 195.5; (c) C_6H_4ClF; ^{13}C δ 116.7, 121.2, 125.0, 128.3, 130.9, 158.5; ^{19}F δ –109.

7.9. Suggest a structure for $C_{11}H_{15}NO$ whose ^{13}C {^1H} spectrum (shown in Figure 7.8) exhibits signals at δ 12.5, 44.5, 110.7, 124.9, 132.0, 152.2, and 189.2 . Be prepared to support your structure by assigning all signals and demonstrating that their chemical shifts are within appropriate ranges.

SELF-TEST I

At this point it is time to demonstrate that you can really apply the things you have learned so far from this book by solving some problems. Refer back as necessary to the previous chapters to get values for physical constants, $\Delta\delta$ values, and so on. If you can solve these problems without looking at the answers in Appendix 1, you are ready to proceed to the next topics. If not, perhaps you should review the relevant sections of the first seven chapters.

1. What is the significance of the "500" in the model number of the Bruker DMX 500 NMR spectrometer shown in Figure 3.5a?

2. Consider the 250-MHz ^{1}H NMR spectrum of $C_6H_{12}O_2$ shown in Figure ST1.1. (a) In this spectrum, 0.0–8.0 ppm on the δ scale corresponds to how many hertz? In what part of the electromagnetic spectrum does 250-MHz radiation appear? What is the energy of a photon of this frequency? (b) At what magnetic field strength was this spectrum obtained? How many spin states are available to a ^{1}H nucleus in such a field? How fast do ^{1}H nuclei precess when immersed in a field of this magnitude? (c) What is the source of the signals at δ 0.00 and 7.27? (d) What is the chemical shift and relative intensity (expressed as the number of hydrogens represented) of the other signals in the spectrum? (e) What is the IOU of $C_6H_{12}O_2$? Suggest a structure consistent with the spectrum. (f) Assign all signals to hydrogens in your structure and show that the observed chemical shifts match expected values.

3. Consider the ^{13}C NMR spectrum of $C_9H_{10}O_2$ shown in Figure ST1.2 (a) A ^{13}C nucleus possesses how many protons? Neutrons? (b) At what nominal frequency was this spectrum obtained? (c) What is the nuclear spin of a ^{13}C nucleus and what is the importance of this number? (d) To what are the signals at δ 0.0 and 75–80 due? (e) What is the chemical shift of each signal [other than those mentioned in part (d)]? What is the separation in hertz between the two most downfield signals? (f) Suggest a structure consistent with the spectrum. (g) Assign the signals in part (e) to carbons in your structure and show that the observed chemical shifts match expected values.

4. The ^{1}H NMR spectrum for a certain compound with molecular weight 116 shows two sharp signals, one at δ 1.41 and the other at δ 1.90, with intensity ratio 3 : 1, respectively. The ^{13}C NMR spectrum of the same compound is shown in Figure ST1.3.

   (a) Without being given the molecular formula, suggest a structure consistent with both spectra.

   (b) Assign each NMR signal to appropriate nuclei in your structure and correlate observed chemical shifts with expected values.

5. For each set of spectroscopic data below, suggest a structure consistent with the data. Assign each NMR signal to the appropriate nucleus.

	Molecular Formula	Nucleus	Signal Position (Relative Intensity)
(a)	C_7H_8O	^{1}H	δ 2.43 (1H); 4.58 (2H); 7.28 (m, 5H)^a
(b)	$C_3H_5Cl_3$	^{1}H	δ 2.20 (3H); 4.02 (2H)^a
(c)	$C_{14}H_{12}O$	^{1}H	δ 3.88 (2H); 7.38 (m, 10H)^a
(d)	$C_{14}H_{12}O$	^{1}H	δ 4.37 (2H); 7.20 (m, 10H)^a
(e)	$C_{10}H_{20}$	^{1}H	δ 0.98 (18H); 5.37 (2H)
(f)	$C_5H_{12}O$	^{1}H	δ 0.93 (9H); 1.74 (1H, broad); 3.34 (2H)
(g)	C_7H_7Cl	^{1}H	δ 4.50 (2H); 7.23 (m, 5H)
(h)	$C_6H_4Cl_2$	^{1}H	δ 7.20

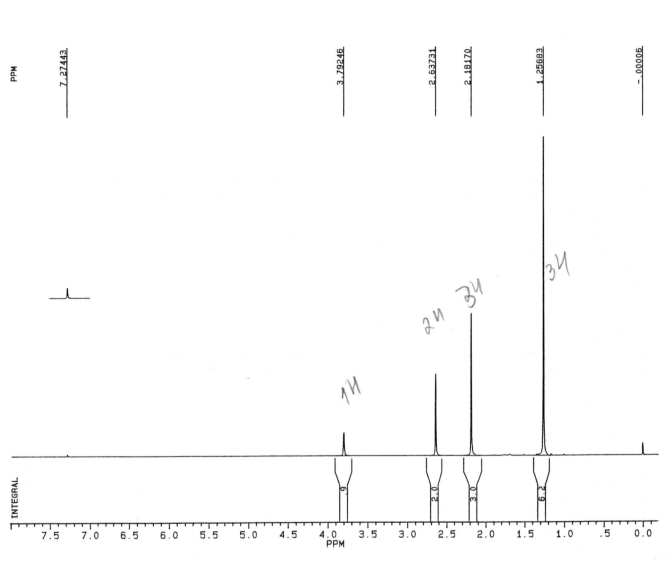

Figure ST1.1. The 250-MHz ^{1}H spectrum of $C_6H_{12}O_2$, problem 2.

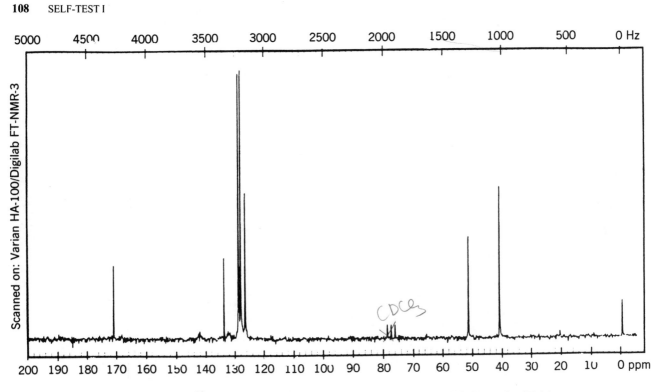

Figure ST1.2. The ^{13}C spectrum of $C_9H_{10}O_2$, problem 3. © Sadtler Research Laboratories, Division of Bio-Rad Laboratories, Inc. (1983).

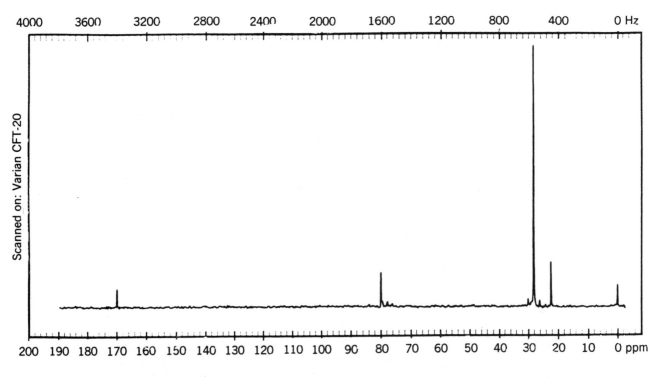

Figure ST1.3. The ^{13}C spectrum of the compound in Problem 4. © Sadtler Research Laboratories. Division of Bio-Rad Laboratories, Inc. (1983).

	Molecular Formula	Nucleus	Signal Position (Relative Intensity)
(i)	$C_{18}H_{30}$	^1H	δ 1.19 (18H); 1.21 (12H)
(j)	$C_{16}H_{16}$	^1H	δ 3.09 (8H); 6.52 (8H)
(k)	$C_{10}H_{14}O$	^{13}C	δ 31.5 (86); 33.7 (10); 114.9 (55); 125.9 (36); 141.6 (7); 154.7 (7)[b]
(l)	C_5H_7N	^{13}C	δ 35.6 (18); 108.3 (88);121.6 (58)[b]
(m)	$C_7H_{12}O_2$	^{13}C	δ 25.5 (60); 25.9 (70); 29.0 (66); 43.1 (38); 183.0 (25)[b]
(n)	$C_3H_9O_3P$	^1H	δ 3.34
		^{13}C	δ 61.9
		^{31}P	δ 137.7
(o)	C_3H_9N	^1H	δ 2.18
		^{13}C	δ 45.7
		^{14}N	δ −30
(p)	$C_4H_9O_2N$	^1H	δ 1.59
		^{14}N	δ 522

[a] From *High Resolution NMR Spectra Catalog*, Varian Instruments, Palo Alto, CA, 1963. The symbol *m* indicates a multiplet, several close lines that are not completely resolved.

[b] From *The Sadtler Guide to Carbon-13 NMR Spectra*, Sadtler Research Laboratories, Division of Bio-Rad Laboratories, Philadelphia, 1983.

6. Explain why the compound below exhibits only two ^1H signals δ 2.99 and 9.28.

7. Explain why the indicated hydrogens in phenanthrene appear at δ 8.93, while in the ethynyl derivative on the right **H** appears at δ 10.64:

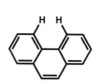

phenanthrene

8

FIRST-ORDER (WEAK) SPIN–SPIN COUPLING

8.1 UNEXPECTED LINES IN AN NMR SPECTRUM

Alas, things are not as simple as they may have appeared. From the discussions in Chapters 6 and 7 you might be under the impression that typical ^{1}H NMR spectra exhibit just one sharp signal line for each ^{1}H nucleus (or each set of equivalent ^{1}H nuclei) and that the same thing is true for ^{13}C spectra as well as for spectra of any other isotope. Actually, this is not usually the case. Instead, the individual signals expected on the basis of the molecule's symmetry are themselves often *split* into symmetrical patterns (**multiplets**) consisting of two or more lines. While these extra lines *do* make a spectrum appear more complex, they also offer valuable structural information that complements the chemical shift data. This chapter explains the source of these extra lines and shows how useful they can be for confirming the structures of molecules.

Take a moment to look back at the 250- and 400-MHz ^{1}H spectra of toluene in Figures 5.3 and 5.4. Do you recall that there were more lines (at least six) for the aromatic hydrogens than the three we predicted from symmetry considerations? And do you recall from our discussions in Chapters 5 and 7 that all of the ^{13}C spectra involved a technique called proton spin decoupling? These phenomena are related.

8.2 THE ^{1}H SPECTRUM OF DIETHYL ETHER

Before looking at Figure 8.1, try to predict the appearance of the ^{1}H spectrum of diethyl ether, CH_3CH_2–O–CH_2CH_3. By using symmetry and the data in Table 6.2, we would predict two signals: one near δ 1.2 for the six equivalent methyl hydrogens and one near δ 3.4 for the four equivalent methylene hydrogens. Now look at the actual 250-MHz ^{1}H spectrum in Figure 8.1. There *are* two signals, a three-line pattern centered at δ 1.21 and a four-line pattern centered at δ 3.48. The ratio of the total areas of these two multiplets is 6

: 4, respectively. Let us examine these signals a little more closely.

The three-line signal for the methyl hydrogens is a **triplet.** The **multiplicity** of a signal is the number of lines in the signal, so a triplet has multiplicity 3. The three lines that constitute the triplet have an intensity ratio of 1 : 2 : 1, and the spacings between neighboring lines are equal. From the exact position of each line (expressed in hertz at the top of the spectrum), the magnitude of these spacings (both 6.9 Hz in this case) can be determined. The chemical shift of a multiplet is measured at its center, which in the case of a triplet corresponds to the middle line.

The four-line pattern for the methylene hydrogens is a **quartet** and has multiplicity 4. Its chemical shift is midway between the second and third lines. Again, the three spacings between neighboring lines are all equal to the spacings in the triplet (6.9 Hz). However, in a quartet the relative intensity of the lines is 1 : 3 : 3 : 1. These intensity ratios, and the fact that the spacings in both multiplets are equal, are no accident.

■ **EXAMPLE 8.1** How many lines constitute each of the following multiplets: **doublet, quintet, sextet, septet**?

□ *Solution:* Two, five, six, and seven, respectively. *Note:* A signal with only one line is a **singlet.**

■ **EXAMPLE 8.2** (a) What is the separation in hertz between the triplet and the quartet in Figure 8.1? (b) What would be the separation if the spectrum were run at 400 MHz?

□ *Solution:* (a) You can either measure the center-to-center spacing using the hertz scale at the top of the spectrum (869.83 − 301.97 = 567.86 Hz) or calculate it from the difference in chemical shifts. From Eq. (5.1)

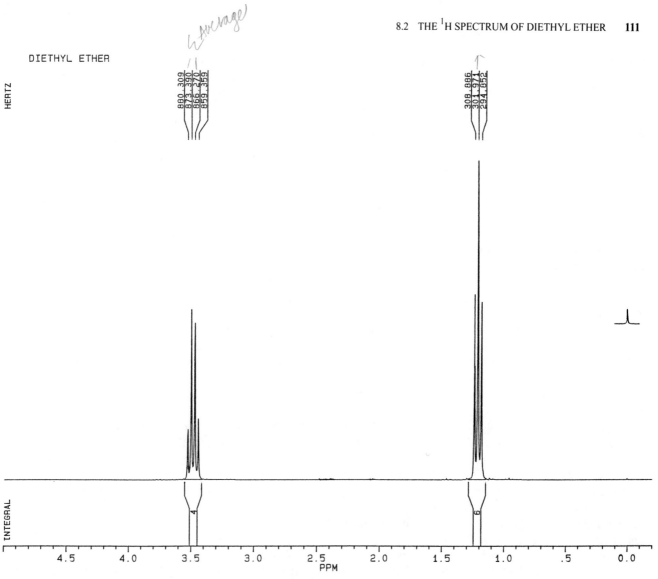

DIETHYL ETHER

HERTZ

880.309
873.390
866.270
859.359

308.886
301.971
294.852

INTEGRAL

4.5   4.0   3.5   3.0   2.5   2.0   1.5   1.0   .5   0.0

PPM

Figure 8.1. The 250-MHz ^{1}H spectrum of diethyl ether.

$$\Delta \nu = (\delta_1 - \delta_2)\nu_0'$$

$$= (3.48 - 1.21 \text{ ppm}) (250 \text{ Hz/ppm})$$

$$= 568 \text{ Hz}$$

(b) Use a proportion of the type described in Section 5.2:

$$\frac{\Delta \nu_1}{(\nu_0')_1} = \frac{\Delta \nu_2}{(\nu_0')_2}$$

$$\Delta \nu_2 = \frac{\Delta \nu_1 (\nu_0')_2}{(\nu_0')_1} = 568 \text{ Hz} \left(\frac{400}{250}\right) = 909 \text{ Hz}$$

Of course, the separation expressed in δ (ppm) would be unchanged, 2.27 ppm.   □

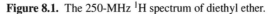

In Example 8.2, you found that the separation *between* the multiplets would increase from 568 to 909 Hz if the spectrometer's operating frequency were increased from 250 to 400 MHz. That is nothing new. But here is the surprise. Even at 400 MHz, the spacings between the lines within each multiplet of diethyl ether is still 6.9 Hz! This spacing between the lines of a multiplet is called the (spin–spin) **coupling constant** and is given the symbol *J*. Do not forget: While multiplet positions (and their differences) in hertz vary directly with the operating frequency and field strength of the instrument, coupling constants (in hertz) are *independent* of these instrumental parameters.

When reporting NMR data in **condensed format,** you list the chemical shift in δ (ppm) of each multiplet, followed in parentheses by the type of multiplet (often abbreviated), coupling constant (in hertz), and, in the case of ^{1}H spectra, relative signal area (intensity). By the way, the magnitude of *J* must usually exceed 1 Hz for the individual lines within a multiplet to be resolved. Values of J less than 1 Hz normally render the signal a somewhat broadened singlet.

■ **EXAMPLE 8.3**   Describe in condensed format the data from the spectrum in Figure 8.1.

☐ *Solution:*   δ 1.21 (triplet, J = 6.9 Hz, 6H), 3.48 (quartet, J = 6.9 Hz, 4H); or δ 1.21 (t, J = 6.9, 6H), 3.48 (q, J = 6.9, 4H).   ☐

8.3   HOMONUCLEAR ^{1}H COUPLING: THE SIMPLIFIED PICTURE

I hope that by now your interest has been piqued. Why do equivalent hydrogens give rise to singlets in some cases and multiplets in others? To understand the phenomenon of spin coupling, recall how the magnitude of the effective magnetic field experienced by a nucleus determines its precessional frequency and thereby its chemical shift. If that is not second nature to you by now, perhaps you should review Section 6.1 before proceeding further.

Let us examine the two equivalent hydrogen nuclei in one of the methylene groups of diethyl ether. Each of these hydrogens has two possible spin states (orientations, Section 2.1.3): *up* (aligned *with* the $\mathbf{B}_0$ external field, $m = \frac{1}{2}$) or *down* (*opposed* to the field, $m = -\frac{1}{2}$). These two spin states are essentially equally populated (Section 2.3), and nuclei in each state precess at the same frequency (Section 2.2). Since n equivalent nuclei can exist in any one of $(2I + 1)^n$ spin *combinations*, the two hydrogens *together* can adopt four possible spin combinations: both up, both down, or one of each, as shown in Figure 8.2. We can label each resulting spin *state* by its M value, where M, the total magnetization, is the sum of the individual m values:

$$M = m_1 + m_2 + m_3 + \cdots \tag{8.1}$$

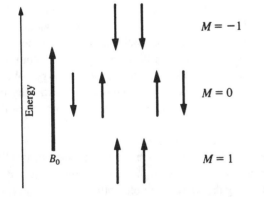

Figure 8.2. Pictorial representation of the spin states of two hydrogens (or other $I = \frac{1}{2}$) nuclei.

The number of different spin *states* (M values) that n equivalent nuclei can adopt is $2nI + 1$. Notice that there is only one way to have both up ($M = \frac{1}{2} + \frac{1}{2} = 1$) and one way for both to be down ($M = -\frac{1}{2} - \frac{1}{2} = -1$) but two combinations where one is up and one is down ($M = \frac{1}{2} - \frac{1}{2} = 0$). Thus, two equivalent hydrogens together give rise to three spin states ($M = 1, 0, -1$) with population ratio of 1 : 2 : 1. Does this ratio ring a bell?

For a moment, consider the methyl hydrogens. Just three bonds away are the two methylene hydrogens, which can be in any one of the three possible spin states. Suppose the methylene hydrogens happen to be in the (slightly) more stable $M = 1$ state. Since their spins are aligned with the external field, is it not reasonable to expect that the effective field the methyl hydrogens experience is *slightly increased* by the "both up" methylene hydrogens? If so, the methyl hydrogens will precess at a somewhat greater frequency, and their signal will be correspondingly shifted downfield. Alternatively, suppose the two methylene hydrogens were in the $M = -1$ state. Their spins, being opposed to the external field, would slightly diminish the effective field experienced by the methyl hydrogens, causing them to precess more slowly and their signal to move upfield.

■ **EXAMPLE 8.4**   If the methylene hydrogens were in the $M = 0$ state, what effect would they have on the precession rate of the methyl hydrogens?

☐ *Solution:*   None! When $M = 0$ the two methylene hydrogens are in opposite spin states, so their effects on the effective magnetic field cancel each other.   ☐

Consider the implications of this. In a collection of diethyl ether molecules, the methyl hydrogens are next to methylene hydrogens that can be in any one of three spin states with relative probability 1 : 2 : 1. As a result, the methyl signal will be split into three lines with intensity ratio 1 : 2 : 1, as depicted in Figure 8.3. This phenomenon is therefore termed **spin–spin coupling** (or **splitting**). It is further described as **homonuclear** coupling because the coupling takes place between nuclei of the same isotope (here, ^{1}H).

■ **EXAMPLE 8.5**   Describe what happens to the methylene signal as a consequence of the three neighboring (separated by three bonds) methyl hydrogens. (a) How many spin states can three equivalent hydrogens adopt? Draw all possible combinations and group them by M value. (b) What multiplicity do you predict for the methylene signal? What will be the intensity ratio of the lines? (c) Draw a diagram similar to Figure 8.3 for the methylene signal.

☐ *Solution:* (a) The three methyl hydrogens together can adopt eight (2^3) possible spin combinations distributed among four [$2(3)(\frac{1}{2}) + 1$] spin states. Of these, one has all three up ($M = \frac{3}{2}$), three have two up and one down ($M = \frac{1}{2}$),

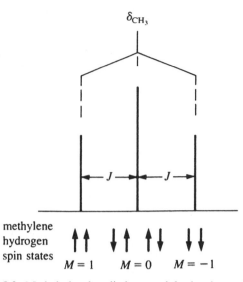

Figure 8.3. Methyl signal, split into a triplet by the methylene hydrogens.

three have one up and two down ($M = -\frac{1}{2}$), and one has all three down ($M = -\frac{3}{2}$); see Figure 8.4. (b) Quartet (four lines) in the ratio 1 : 3 : 3 : 1. (c) See Figure 8.4. □

Remember this: If nucleus **A** is coupled to nucleus **B**, then nucleus **B** is also coupled to nucleus **A**, and the coupling

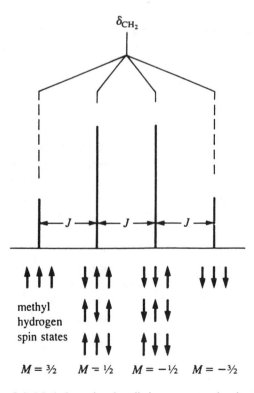

Figure 8.4. Methylene signal, split into a quartet by the methyl hydrogens.

constant (J) will have the *same magnitude* in the multiplets of *both* **A** and **B**.

8.4 THE SPIN–SPIN COUPLING CHECKLIST

By now it should be apparent to you that the multiplicity of a given signal is related to the number of *neighboring* nuclei, *not* to the number of nuclei in the set giving rise to the signal. Let us review what we know so far.

1. For two nuclei to engage in spin coupling, *both* must have nonzero nuclear spin ($I \neq 0$, Section 2.1). If the neighboring nucleus had $I = 0$, it could adopt only one orientation in a magnetic field and therefore could not split the signal of the other nucleus.

2. To produce the multiplet characteristic of spin coupling, the coupling nuclei must be *nonequivalent.* We have already decided (from symmetry considerations) that all four methylene hydrogens are equivalent, as are the six methyl hydrogens. Why is it that the three hydrogens within each methyl group do not couple with each other? The answer is *they do!* But, for reasons we will discuss in the next chapter, we normally cannot see the effects of this coupling because the three methyl hydrogens are all equivalent to each other.

3. Finally, you might wonder why the methyl hydrogens are coupled only to the two *nearest* (three bonds away) methylene hydrogens and not to the other two located five bonds away. Here is the reason. The interaction between nuclear spins takes place predominantly through the intervening bonds (electrons). In the next chapter we will discuss the factors that influence the magnitude of J, but for now we can generalize by saying that, normally, coupling is observed only when the nuclei are separated by no more than three bonds.

To summarize, here is our spin-coupling checklist:

1. Are both (sets of) nuclei magnetic ($I \neq 0$)?
2. Are they nonequivalent?
3. Are they separated by three or fewer bonds?

■ **EXAMPLE 8.6** Which hydrogens in structure **8-1** are expected to exhibit coupling with each other?

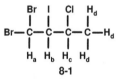

8-1

□ *Solution:* By symmetry (or the lack of it!), we see that H_a, H_b, and H_c are all nonequivalent, but the three methyl hydrogens (H_d) are equivalent. We would therefore expect four signals. Recalling that all ^1H nuclei have $I = \frac{1}{2}$ and counting the bonds separating them, we can complete our checklist:

Interaction	Separation (bonds)	Coupling Observable?
H_a–H_b	3	Yes
H_a–H_c	4	No
H_a–H_d	5	No
H_b–H_c	3	Yes
H_b–H_d	4	No
H_c–H_d	3	Yes
H_d–H_d	2	No (equivalent!)

Thus, the homonuclear couplings we will observe are H_a–H_b, H_b–H_c, and H_c–H_d. ☐

8.5 THE $n + 1$ RULE

Let us attack the general problem. How do we predict the number of lines and their relative intensity in a given multiplet once we know the number of neighboring nuclei? We know that if there are two equivalent neighboring hydrogens, a triplet results (Figure 8.3), while if there are three equivalent neighboring hydrogens, a quartet results (Figure 8.4). A little inductive reasoning should convince you that if a nucleus has n equivalent neighboring hydrogens, its signal will be split into $n + 1$ lines. This is because n neighboring hydrogens (or other nuclei with $I = \frac{1}{2}$) can generate only $n + 1$ different spin states (M values) from 2^n spin combinations.

■ **EXAMPLE 8.7** (a) Diagram all possible spin combinations of four equivalent hydrogens. Arrange them into spin states by their value of M. (b) If another hydrogen were coupled to these four hydrogens, how many lines would comprise its multiplet and what would be their relative intensities?

☐ *Solution:* (a) There are 16 (2^4) combinations distributed among 5 (4 + 1) spin states:

M						Multiplicity
−2			↓↓↓↓			1
−1	↓↓↓↑	↓↓↑↓	↓↑↓↓	↑↓↓↓		4
0	↑↑↓↓	↑↓↑↓	↓↑↑↓	↓↑↓↑	↓↓↑↑ ↑↓↓↑	6
1		↑↑↑↓	↑↑↓↑	↑↓↑↑	↓↑↑↑	4
2			↑↑↑↑			1

(b) Five lines (a quintet) with intensity ratio 1 : 4 : 6 : 4 : 1. ☐

Review the intensity ratios we have encountered so far: triplet, 1 : 2 : 1; quartet, 1 : 3 : 3 : 1; and quintet, 1 : 4 : 6 : 4 : 1. Do these ratios sound familiar? Actually, they are the coefficients of the **binomial distribution.** There is a cute

TABLE 8.1 Pascal's Triangle (Coefficients of the Binomial Distribution)

n	$n+1$	Intensity Ratio	Multiplicity	Multiplet
0	1	1	1	Singlet
1	2	1 1	2	Doublet
2	3	1 2 1	3	Triplet
3	4	1 3 3 1	4	Quartet
4	5	1 4 6 4 1	5	Quintet
5	6	1 5 10 10 5 1	6	Sextet

mnemonic device called **Pascal's triangle** that will enable you to generate these ratios in short order. A portion of it is shown in Table 8.1 (the triangle can be extended downward as far as you wish to go). Note how each number in the triangle is the sum of the two numbers diagonally above it. For example, each 10 in the last line is the sum of the 6 and 4 above it.

■ **EXAMPLE 8.8** What is the ratio of intensities in a septet?

☐ *Solution:* Generate the next line of Pascal's triangle to get 1 : 6 : 15 : 20 : 15 : 6 : 1. ☐

We should reemphasize at this point that these ratios apply only to nuclei with $I = \frac{1}{2}$. Coupling to nuclei with $I > \frac{1}{2}$ will be covered in the next section.

Let us return to our discussion of the ^1H spectrum of structure **8.1** in Example 8.6. The signal for H_a will be split into a doublet by its one neighbor, H_b. Similarly, the methyl (H_d) signal will be split into a doublet by H_c. But what about H_b? Its signal will be split by *both* H_a and H_c, but how? The answer depends on the *relative magnitudes* of $^3J_{ab}$ versus $^3J_{bc}$ (where the superscript 3 indicates the number of intervening bonds). Suppose for the moment that $^3J_{ab} > ^3J_{bc}$. The signal for H_b would first be split into a doublet by H_a (with line separation $^3J_{ab}$), and each of the resulting lines would be further split into a doublet by H_c (with line separation $^3J_{bc}$). The resulting pattern, a **doublet of doublets**, is shown in Figure 8.5. Note that the intensity ratio of the lines is 1 : 1 : 1 : 1. If $^3J_{bc}$ were larger than $^3J_{ab}$, the pattern would look the same, except that the two coupling constants would be interchanged.

■ **EXAMPLE 8.9** How would you describe the appearance of Figure 8.5 if the two coupling constants ($^3J_{ab}$ and $^3J_{bc}$) were accidentally equal?

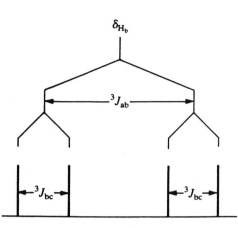

Figure 8.5. Signal for H_b of structure **8-1**, split into a doublet of doublets with $J_{ab} > J_{bc}$.

□ *Solution:* The two middle lines would be superimposed to give a triplet with line intensities 1 : 2 : 1 (see Figure 8.6).   □

You can infer from Example 8.9 that a triplet is really a doublet of doublets with both coupling constants are equal. Similarly, a quartet can be seen to be a doublet of doublets of doublets, with all three coupling constants equal.

But what about the signal for H_c? It will be split by both H_b as well as the three methyl hydrogens. Suppose that $^3J_{bc} > ^3J_{cd}$. Each line of the doublet due to coupling with H_b will be split further into a quartet by the three methyl hydrogens, as in Figure 8.7. This multiplet is called a doublet of quartets, and the line intensity ratio is 1 : 3 : 3 : 1 : 1 : 3 : 3 : 1.

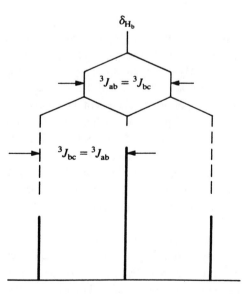

Figure 8.6. Signal for H_b of structure **8-1**, collapsed to a triplet because $J_{ab} = J_{bc}$.

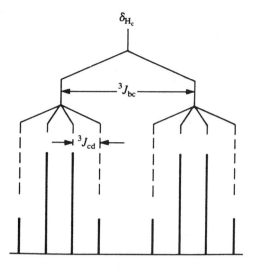

Figure 8.7. Signal for H_c of structure **8-1**, split into a doublet of quartets with $J_{bc} > J_{cd}$.

■ **EXAMPLE 8.10** How would Figure 8.7 appear if (a) $^3J_{bc}$ were less than $^3J_{cd}$ or (b) $^3J_{bc} = ^3J_{cd}$?

□ *Solution:* (a) A quartet of doublets with intensity ratio 1 : 1 : 3 : 3 : 3 : 3 : 1 : 1, as shown in Figure 8.8a. (b) A quintet with intensity ratio 1 : 4 : 6 : 4 : 1, as shown in Figure 8.8b.   □

We can generalize as follows:

1. If a nucleus is coupled to *n* other $I = \frac{1}{2}$ nuclei and all coupling constants are equal, the signal for the nucleus will be split into *n* + 1 lines whose intensity ratio can be predicted from Pascal's triangle.

2. If a nucleus is coupled to several sets of $I = \frac{1}{2}$ nuclei (n_a of nucleus a, n_b of nucleus b, etc.), the total multiplicity (*L*) of the signal (assuming no overlap of any lines) is given by the product

$$L = (n_a + 1)(n_b + 1) + \cdots$$

$$= \prod (n_i + 1) \quad (I = \tfrac{1}{2} \text{ nuclei only}) \tag{8.2}$$

where the $\prod$ indicates a product of *i* terms. Thus, in the case of H_c in structure **8.1** (with $n_b = 1$ and $n_d = 3$) there will be $(1+1)(3+1) = 8$ lines (barring accidental equivalence and overlap). The intensity ratio of the lines will depend on the relative magnitudes of the two coupling constants (compare Figures 8.7, 8.8a, and 8.8b).

■ **EXAMPLE 8.11** (a) Predict in detail the appearance of the 250-MHz ^{1}H spectrum of isobutyl alcohol (**8-2**), given the

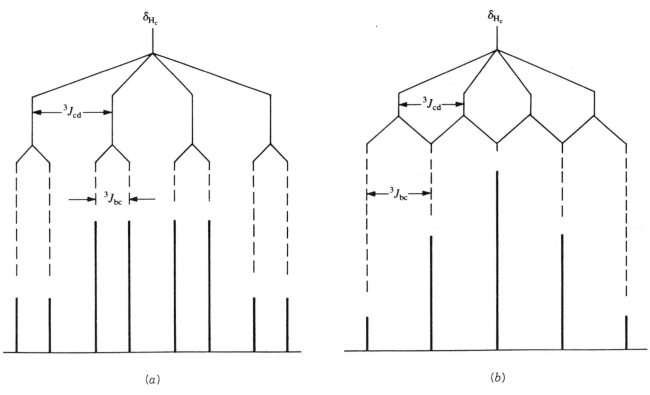

Figure 8.8. (*a*) Signal for H_c of structure **8-1**, split into a quartet of doublets with $J_{bc} < J_{cd}$; (*b*) the same signal collapsed to a quintet with $J_{bc} = J_{cd}$.

spectroscopic parameters below. Draw the spectrum on graph paper with the *x* axis scaled to 0.01-ppm divisions:

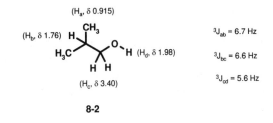

(H$_a$, δ 0.915)

(H$_b$, δ 1.76)

(H$_d$, δ 1.98)

(H$_c$, δ 3.40)

$^3J_{ab}$ = 6.7 Hz

$^3J_{bc}$ = 6.6 Hz

$^3J_{cd}$ = 5.6 Hz

8-2

(b) Describe the resulting spectrum in condensed format.

☐ *Solution:* (a) Let us begin by transforming the *J* values into ppm downfield from TMS by dividing each by 250 Hz/ppm (Section 5.1). Thus, the $^3J_{ab}$ is 0.027 ppm, $^3J_{bc}$ is 0.026 ppm, and $^3J_{cd}$ is 0.022 ppm. Note also that all three coupling constants are within about 1 Hz. (As we will see in Chapter 9, 6.5 Hz is a typical value for $^3J_{HH}$ in a freely rotating system.)

The six equivalent methyl hydrogens (H$_a$) are coupled only to H$_b$ (why?). The H$_a$ signal therefore appears as a doublet with a separation between the lines of 0.027 ppm

(6.7 Hz, $^3J_{ab}$). Draw two equally intense lines (1 : 1 ratio, right?), one at δ 0.915 – (0.027/2) = 0.90; the other at δ 0.915 + (0.027/2) = 0.93. Remember that the total area of this doublet is six hydrogens, so each of these lines represents three hydrogens.

Let us shift our attention to H$_d$. It is coupled to the two equivalent H$_c$'s with a *J* value ($^3J_{cd}$) of 5.6 Hz (0.022 ppm). Therefore, the H$_d$ signal will appear as a 1 : 2 : 1 triplet, with lines at δ 1.96, 1.98, and 2.00. The total area (height) of this triplet is $\frac{1}{6}$ of the total height of the doublet, so each outer line of the triplet is $\frac{1}{12}$ as tall as the lines in the H$_a$ doublet. [It should be noted that because OH hydrogens undergo rapid exchange if there is even a trace of acidic or basic impurities present (Section 6.6.1), they do not exhibit the effects of spin coupling unless the sample has been rigorously purified to prevent exchange.]

Since $^3J_{bc}$ and $^3J_{cd}$ are within 1 Hz of the same value, H$_c$ will think it is essentially coupled to two equivalent hydrogens with an average *J* value of 6 Hz (0.024 ppm), so its signal will be split into a triplet. Draw three lines, each twice as intense as the H$_d$ signal (since there are two H$_c$'s), at δ 3.38, 3.40, and 3.42.

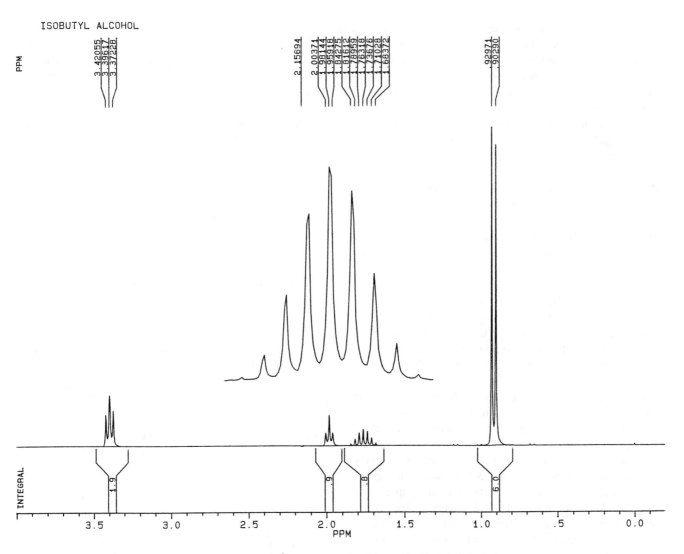

ISOBUTYL ALCOHOL

Figure 8.9. The 250-MHz ^{1}H spectrum of highly purified isobutyl alcohol.

Last but not least is H_b. Since it is coupled to two H_c's and six H_a's, will it appear as a triplet of septets (21 lines) or a septet of triplets? *Answer*: Neither! Because both coupling constants ($^3J_{ab}$ and $^3J_{bc}$) are essentially equal (6.6 Hz or 0.026 ppm), H_b will think it is coupled to eight equivalent nuclei. Therefore, its signal will be split into a nine-line (8 + 1) multiplet (nonet) with intensity ratio 1 : 8 : 28 : 56 : 70 : 56 : 28 : 8 : 1. Furthermore, the total area of this multiplet will equal one hydrogen. Try to draw this on your diagram!

The actual spectrum of this compound is shown in Figure 8.9. How does it compare with your drawing? Notice how the nonet has to be amplified to show all the lines.

(b) δ 0.915 (J = 6.7, 6H), 1.76 (J = 6.6, 1H), 1.98 (J = 5.6, 1H), 3.40 (J = 6.0, 2H). □

8.6 HETERONUCLEAR SPIN–SPIN COUPLING

Spin-coupling interactions are not limited to hydrogens. Any magnetic ($I \neq 0$) nuclei within three bonds can split a signal. Let us divide our discussion into heteronuclei with $I = \frac{1}{2}$, and then all others.

8.6.1 Heteronuclear Coupling to $I = \frac{1}{2}$ Nuclei

From Chapter 2 we know that the common $I = \frac{1}{2}$ nuclei are ^{1}H, ^{13}C, ^{19}F, and ^{31}P. Consider the molecule H_2CF_2 (difluoromethane), and assume that the carbon is ^{12}C ($I = 0$):

The two hydrogens are equivalent, as are the two fluorines. Since we can only examine the NMR spectrum of one isotope at a time, can you predict the appearance of the ^1H spectrum of this compound? Because there are two neighboring flourines (i.e., within three bonds), the hydrogen signal is split into a triplet (intensity ratio 1 : 2 : 1), and the spacing between the lines will be $^2J_{HF}$. This is an example of two-bond **heteronuclear coupling** because it occurs between different isotopes or different elements.

■ **EXAMPLE 8.12** Predict the appearance of the ^{19}F spectrum of difluoromethane.

☐ *Solution:* The ^{19}F signal is also split into a triplet (intensity ratio 1 : 2 : 1) by the two equivalent hydrogens, with the same line spacing ($^2J_{HF}$) as in the ^1H spectrum.   ☐

But then, why have we not seen couplings between ^1H and ^{13}C in all the ^1H spectra we have examined so far? The answer is that the natural abundance of ^{13}C is only 1% of all carbon (Table 2.1). So only 1% of the hydrogens bonded to carbon are attached to a ^{13}C nucleus, and this small fraction of coupled hydrogens can normally be neglected. Nonetheless, in some cases you *can* see the effect of ^{13}C–H coupling in ^1H spectra. Look at the highly amplified singlet in the ^1H spectrum of TMS (Figure 8.10) or any other sharp ^1H singlet for that matter. The two small satellite peaks (sidebands) flanking the singlet (each integrating to 0.5% of the main singlet) are *not* spinning sidebands (Section 3.3.3) because changing the sample spinning rate does not affect their position. In fact, the satellites are due to one-bond ^{13}C–H coupling. One percent of the methyl groups in a collection of TMS molecules possesses a ^{13}C atom. This atom splits the methyl hydrogen signal (the one we use as our reference signal) into a doublet,

with line spacing $^1J_{CH}$. Of course, you will see these satellites only if you greatly increase the signal amplification (gain). But they *are* there, and the magnitude of $^1J_{CH}$ is approximately 125 Hz, independent of sample spin rate or instrument operating frequency .

■ **EXAMPLE 8.13** From your knowledge of heteronuclear coupling, predict which ^{13}C–^1H couplings should be observed in the ^{13}C spectrum of 2-chlorobutane:

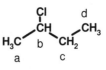

☐ *Solution:* Let us prepare a coupling checklist (the *J* values are from Chapter 9):

Coupling	Bonds	Observable?	*J*, Hz
C_a–H_a	1	Yes	125
C_a–H_b	2	Yes	5–6
C_a–H_c	3	Yes	3–5
C_a–H_d	4	No	
C_b–H_a	2	Yes	5–6
C_b–H_b	1	Yes	125
C_b–H_c	2	Yes	5–6
C_b–H_d	3	Yes	3–5
C_c–H_a	3	Yes	3–5
C_c–H_b	2	Yes	5–6
C_c–H_c	1	Yes	125
C_c–H_d	2	Yes	5–6
C_d–H_a	4	No	
C_d–H_b	3	Yes	3–5
C_d–H_c	2	Yes	5–6
C_d–H_d	1	Yes	125

☐

Figure 8.11 shows the actual *undecoupled* 62.5-MHz ^{13}C spectrum of 2-chlorobutane. The signal for C_d (δ 11.0) is split into a quartet (by H_d, *J* = 125 Hz) of barely discernable triplets (by H_c, *J* = 6 Hz); coupling to H_b is too small to observe. The remaining signals clearly exhibit multiplets due to one-bond H/C coupling (C_a, δ 24.9, q; C_c, δ 33.3, t; C_b, δ 60.4, d); the two- and three-bond couplings in these three signals are too small to be resolved. Compare this spectrum with the one in Figure 7.1. Why is the latter spectrum so simple, showing only a singlet for each carbon? If you look carefully in Chapters 5 and 7, you will see the proviso that all ^{13}C spectra were obtained using a technique called **proton spin decoupling.**

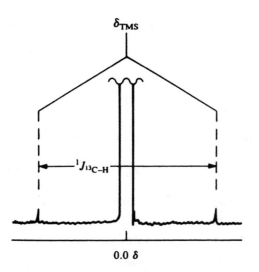

Figure 8.10. The ^{13}C satellite peaks flanking a greatly amplified TMS singlet.

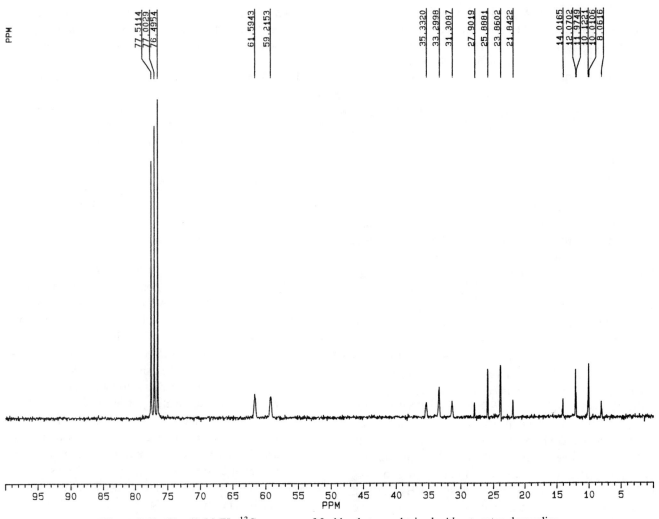

Figure 8.11. The 62.5-MHz ^{13}C spectrum of 2-chlorobutane, obtained *without* proton decoupling. Compare with Figure 7.1.

We will discuss this technique again in Chapter 12, but for now we can say that spin *decoupling* allows us to electronically "erase" the effect of H/C coupling, greatly simplifying the appearance of the spectrum. Can this same technique be used to decouple all *homonuclear* coupling ^{1}H spectra? No, not exactly. But in Chapter 13 we will see an indirect way to accomplish this feat!

■ **EXAMPLE 8.14** Predict the relative position and multiplicity of each signal in the ^{1}H spectrum of the molecule represented by structure **8-3**. You need only consider one-, two-, and three-bond couplings. *Hint*: (1) A P=O (**phosphoryl**) group behaves somewhat like a carbonyl (C=O) group and (2) the methylene hydrogens (H$_c$) are rendered equivalent by the rapid intra- and intermolecular exchange of H$_d$ between the two exocyclic oxygens:

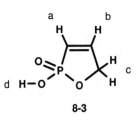

8-3

□ *Solution:* First, remember that ^{31}P is magnetic ($I = \frac{1}{2}$). There will be four ^{1}H signals (*a, b, c,* and *d*) with intensity ratio 1 : 1 : 2 : 1. The signal with the farthest downfield shift is that of H$_d$ because of its similarity to a carboxylic acid hydrogen (near δ 12; Section 6.6.1). Although H$_d$ is only two bonds away from the phosphorus, coupling to P is not observed because of rapid exchange of the OH hydrogens (Example 8.11). The signal for vinyl hydrogen H$_a$ will occur near δ 6.0 [Table 6.4, using the data for X =

C(=O)OR and CH$_2$OR] and will be split into a doublet (by phosphorus) of doublets (by H$_b$), but it is not coupled to H$_d$ (why?). The signal for H$_c$ will appear near δ 4.7 (Table 6.2) and is split into a doublet (by phosphorus) of doublets (by H$_b$). The signal for H$_b$ (near δ 6.5; Table 6.4) will be split into a doublet (by phosphorus) of doublets (by H$_a$) of triplets (by the two Hc's), for a total of 12 lines! We will discuss this spectrum again in Section 9.10. ☐

8.6.2 Heteronuclear Coupling Involving Nuclei with $I > \frac{1}{2}$

In Section 5.4 (and Self-Test I, problem 3d) we saw that CDCl$_3$, a very common NMR solvent, gives rise to three lines in its ^{13}C spectrum. Why is this so? Recall that D represents ^{2}H (deuterium), an isotope of hydrogen with $I = 1$, which can therefore adopt three $(2nI + 1 = 2 \times 1 \times 1 + 1 = 3)$ spin states $(M = m = -1, 0, 1)$ with essentially equal populations in a magnetic field (Section 2.2). Thus, the carbon signal of CDCl$_3$ will be split into three equally intense lines, a 1 : 1 : 1 triplet, with line spacing $^1J_{CD}$.

Next, consider the molecule D$_2$CH$_2$. Each of the two D nuclei can adopt three orientations. Therefore, two D nuclei can adopt nine $[(2I + 1)^2]$ spin *combinations* corresponding to five $(2nI + 1)$ spin *states* $(M = 2, 1, 0, -1, -2)$. These are shown in Figure 8.12. Notice that the 1 : 2 : 3 : 2 : 1 population ratio of these states (which is the same as the relative intensity of the multiplet lines coupled to them) is *different* from that predicted by Pascal's triangle, which only applies to $I = \frac{1}{2}$ nuclei. At any rate, the ^{1}H spectrum of D$_2$CH$_2$ shows a five-line pattern, with intensity ratio 1 : 2 : 3 : 2 : 1 and line spacing $^2J_{HD}$.

EXAMPLE 8.15 Predict the appearance of the deuterium (^{2}H) spectrum of D$_2$CH$_2$.

☐ *Solution:* The (only) deuterium signal will be split into a triplet (1 : 2 : 1) by the two equivalent hydrogens. The line spacing $^2J_{HD}$ will be the same as in the ^{1}H spectrum.

EXAMPLE 8.16 Fully deuterated acetone (acetone–D$_6$) is often used as a proton-free NMR solvent. However, it usually contains a small amount of acetone–D$_5$. (a) Predict the appearance of the ^{1}H spectrum of acetone–D$_5$. (b) Predict the appearance of the ^{13}C spectrum of acetone–D$_6$:

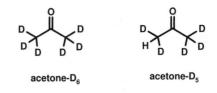

acetone-D$_6$ acetone-D$_5$

☐ *Solution:* (a) There are two deuterium nuclei two bonds away and three others four bonds away from the hydrogen. Only the former ones will couple with the hydrogen. Therefore, the ^{1}H spectrum will exhibit a five-line pattern centered at δ 2.05, with intensity ratio 1 : 2 : 3 : 2 : 1, and line spacing $^2J_{HD}$. (b) The carbonyl carbon signal appears at δ 206 (Table 7.5) and is split by the six equivalent deuterium nuclei (two bonds away) into $2 \times 6 \times 1 + 1 = 13$ lines, with line spacing $^2J_{CD}$. However, this coupling constant is much smaller than 1 Hz, so the individual lines are not resolved, and the signal appears as a slightly broadened singlet. The two equivalent methyl carbons will appear at δ 20.8 (Tables 7.1 and 7.2). The signal is split by a one-bond coupling to the nearest three deuteriums and by a much smaller (< 1 Hz) three-bond coupling to the three more remote deuteriums. The latter coupling is too small to observe. Thus, the total multiplicity of this signal is $2 \times 3 \times 1 + 1 = 7$ lines, with line spacing $^1J_{CD}$. (See problem 4a in Self-Test II.) ☐

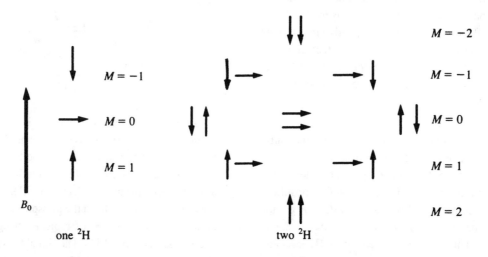

Figure 8.12. Possible spin states of one and two ^{2}H (or other $I = 1$) nuclei.

We can now modify Eq. (8.2) into a more general form. Since the number of different spin states (different M values) available to n nuclei with nuclear spin I is $2nI + 1$, the total multiplicity L is given by

$$L = (2n_a I_a + 1)(2n_b I_b + 1) + \cdots$$

$$= \prod (2n_i I_i + 1) \quad (I \geq \tfrac{1}{2}) \tag{8.3}$$

assuming there are no overlaps of lines. Notice how Eq. (8.3) reduces to Eq. (8.2) if we substitute a value of $\tfrac{1}{2}$ for I.

■ **EXAMPLE 8.17** Predict the chemical shift and multiplicity of the ^1H and ^{19}F NMR signals and the multiplicity of the D (^2H) signal for $CHDF_2$. Assume that the carbon is ^{12}C ($I = 0$) and that the deuterium can be treated as a hydrogen in chemical shift estimations.

☐ *Solution:* Let us summarize the properties of each magnetic nucleus:

Nucleus	n_i	I_i	$2n_i I_i + 1$
H	1	$\tfrac{1}{2}$	2
D	1	1	3
F	2	$\tfrac{1}{2}$	3

The ^1H chemical shift should be δ 7.4 (Table 6.2), and the signal will be split into a total of $3 \times 3 = 9$ [Eq. (8.3)] lines, a 1 : 2 : 1 triplet due to coupling with the two fluorines (with line spacing $^2J_{HF}$), with each of these three lines being further split into a 1 : 1 : 1 triplet by the deuterium (line spacing $^2J_{HD}$).

The ^{19}F signal appears at δ –143.5 (Table 7.7) and will be split into $2 \times 3 = 6$ lines, a 1 : 1 doublet due to the hydrogen (with line spacing $^2J_{HF}$), with each line further split into a 1 : 1 : 1 triplet by the deuterium (line spacing $^2J_{DF}$).

Table 8.2 ^1H and ^{13}C Data for Common NMR Solvents[a]

Compound	Structure	^1H δ[b]	^{13}C δ	^1H δ (H_2O)[c]
Acetic acid–D_4	H_3CCO_2H	2.03 (5), 11.59	20.0 (7), 178.4	11.59[d]
Acetone–D_6	$D_3CC(=O)CD_3$	2.05 (5)	20.8 (7), 206.2	2.85
Acetonitrile–D_3	D_3CCN	1.94 (5)	1.24 (7), 118.1	2.16
Benzene–D_6	Example 4.3[e]	7.15	128.0 (3)	0.5
Carbon tetrachloride	CCl_4		97	
Chloroform–D	$CDCl_3$	7.24	77.0 (3)	1.54
Cyclohexane–D_{12}	Example 6.10[e]	1.38	26.4 (5)	
Diethyl ether–D_{10}	C_2D_5–O–C_2D_5	1.07 (m), 3.34 (m)	13.4 (7), 64.3 (5)	
Dimethyl formamide–D_7	$DC(=O)N(CD_3)_2$	2.74 (5), 2.91 (5), 8.02	30.1 (7), 35.2 (7), 162.7 (3)	3.48
Dimethyl sulfoxide–D_6	$D_3CS(=O)CD_3$	2.50 (5)	39.4 (7)	3.32
Dioxane–D_8	Example 7.5[e]	3.53 (m)	66.5 (5)	2.43
Ethanol–D_6	D_3CCD_2OD	1.1 (br), 3.55 (br), 5.26	17.2 (7), 56.8 (5)	5.26[d]
Hexamethyl phos-phoramide–D_{18}	$O=P[N(CD_3)_2]_3$	2.53 (2 × 5)	35.8 (7 × 2)	
Methanol–D_4	D_3COD	3.30 (5), 4.84	59.05 (7)	4.84[d]
Methylene chloride–D_2	CD_2Cl_2	5.31 (3)	53.8 (5)	1.52
Nitromethane–D_3	D_3CNO_2	4.33 (5)	62.8 (7)	
Pyridine–D_5	Section 6.5[e]	7.20, 7.57, 8.73	123.4 (3), 135.4 (3), 149.8 (3)	4.96
Tetrahydrofuran–D_8	Section 7.3[e]	1.72 (br), 3.57 (br)	25.3 (5), 67.4 (5)	2.23
Toluene–D_8	Section 3.4[e]	2.08 (5), 6.97 (m), 7.01 (m), 7.09 (m)	20.4 (7), 125.1 (3), 149.8 (3) 128.9 (3), 137.5	4.96
Trifluoroacetic acid–D	F_3CCO_2D	11.50	121 (4), 164	11.50[d]
Water–D_2	D_2O	4.82	—	4.82[d]

[a] Number in parentheses is multiplicity, if > 1; br = broad, m = unresolved multiplet.

[b] Residual monoprotonated compound (one D replaced by H).

[c] Signal for adventitious water.

[d] Exchange with the OD of solvent.

[e] Refer to this section for the structure of the undeuteriated compound.

The ^2H signal appears as $2 \times 3 = 6$ lines, a 1 : 1 doublet due to coupling with the hydrogen (with line spacing $^2J_{DH}$), with each line further split into a 1 : 1 : 1 triplet by the deuterium (line spacing $^2J_{DF}$).   □

Now that we know when to expect homo- and heteronuclear couplings, you might want to look at the NMR data for the most common solvents used in NMR, shown in Table 8.2. See if you can account for the multiplicity of each signal.

8.7   REVIEW EXAMPLES

Let us look at two more examples that review the concepts we have seen in the preceding sections.

■ **EXAMPLE 8.18**   Consider the 250-MHz ^1H spectrum of $C_{11}H_{12}O_2$, which is shown in Figure 8.13. (a) From the spectrum determine the value (in hertz) of each H/H coupling

constant except those in the aromatic multiplets. (b) Suggest a structure for the molecule, given that its ^{13}C spectrum (not shown) exhibits, among others, a signal at δ 170. Be sure your structure accounts for all the observed multiplicities in the ^1H spectrum.

□ *Solution:* (a) First, we notice a triplet centered at δ 1.34, a quartet at δ 4.27, two doublets (δ 6.44 and 7.69), and two complex aromatic multiplets at δ 7.38 and 7.52. The coupling constants in each of the first four multiplets, obtained by multiplying the line spacing (in ppm) by 250 Hz/ppm, are 7.2, 7.1, 16.1, and 16.0 Hz, respectively. This shows that the triplet and quartet are coupled to each other (as in a CH_3CH_2 group), as are the two doublets (as in an H_a–C–C–H_b group or an H_a–C–H_b group). In accord with these partial assignments, the integral (intensity) of each multiplet indicates 3H, 2H, 1H, and 1H, respectively, with 5H in the aromatic multiplets.

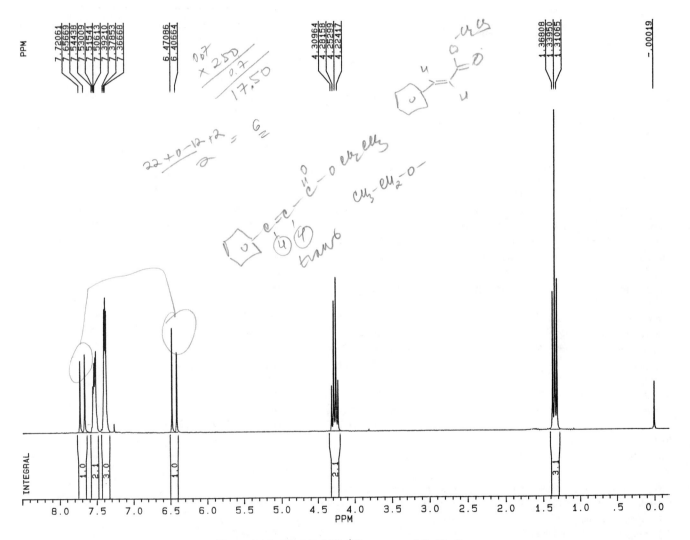

Figure 8.13. The 250-MHz ^1H spectrum of $C_{11}H_{12}O_2$.

(b) From the similarity of the triplet and quartet to those in Figure 8.1, we can assign them to an ethoxy (OCH_2CH_3) group, probably part of an ester group judging from the ^{13}C signal at δ 170. The five aromatic hydrogens suggest a monosubstituted aromatic ring. From their quite different chemical shifts, the two doublets are most likely due to vinyl hydrogens at opposite ends of a polarized double bond. We might propose two structures, the ethyl esters of *cis-* or *trans-*cinnamic acid (**A** and **B**), respectively:

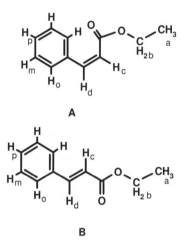

The predicted chemical shifts of vinyl hydrogens H_c and H_d (Table 6.3) are δ 6.02 and 7.19, respectively, for **A**, and δ 6.49 and 7.78, respectively, for **B**. Structure **B** is a much better fit. As we will see in Chapter 9, the magnitude of $^3J_{cd}$ is also very informative with regard to selecting the correct stereoisomer. The coupling constant $^3J_{cd}$ for trans alkenes is uniformly larger (16–18 Hz) than for cis alkenes (10–12 Hz), so the observed value of 16 Hz further supports the trans isomer.

You already know that the five aromatic hydrogens are divided into three sets (two ortho, two meta, and one para; Section 5.1) that occur at nearly the same chemical shift. If each signal were completely resolved from the other two, the ortho hydrogen signals would be split into a doublet by the meta hydrogen three bonds away. The meta hydrogens are coupled to both the ortho and para hydrogens three bonds away, so the signal should be split into a doublet of doublets (or a triplet, if the two J values are comparable). The para hydrogen will be coupled into a triplet by the two equivalent meta hydrogens. Chemical shift predictions (Table 6.5, with X = CO_2R) suggest that the downfield 2H multiplet is the ortho hydrogen signal, while the upfield 3H signal (a triplet) is due to the accidentally equivalent meta and para hydrogens. The reason the ortho signal is not simply a doublet is due to additional long-range (more than three-bond) coupling, to be dis-

cussed in Chapter 9. It is difficult to measure all these coupling constants directly from this spectrum, though there are alternative techniques for doing so, to be discussed later. □

■ **EXAMPLE 8.19** The ^1H of an unknown compound with molecular formula $C_6H_4FNO_2$ consists of signals at δ 7.19 [doublet of doublet (dd), J = 9 and 8 Hz, 2H], 8.20 (dd, J = 9 and 5 Hz, 2H). The proton-decoupled 20-MHz ^{13}C spectrum for the compound is shown in Figure 8.14 and consists of lines at δ 116.0, 117.2, 126.3, 126.9, 144.9, 160.3, and 173.1. (a) Suggest a structure for the compound. Account for the multiplicity of each signal in both spectra. *Note:* you need not worry about coupling with nitrogen; the J value is too small. (b) Determine the magnitude of any couplings in the ^{13}C spectrum.

□ *Solution:* (a) Remember: With fluorine ($I = \frac{1}{2}$) present there are going to be extra coupling interactions! The two doublets of doublets in the ^1H spectrum are in the aromatic region, with two of the four hydrogens in each. The ^{13}C spectrum has seven lines for six carbons; the extra line(s) must be due to coupling(s) between carbon and fluorine. The four lines from δ 116 to 127 suggest a monosubstituted or para-disubstituted aromatic ring. Yet, if there are only six carbons in the structure, all of them must be part of the aromatic ring. This requires that the signals in the region δ 145–173 also represent aromatic carbons, ones that are highly deshielded by substituent groups. A quick review of Table 7.4 shows that two of the most deshielding groups are F and NO_2. On the basis of the analysis so far we can suggest three isomeric structures:

The table below shows the predicted number of signals (before couplings are considered) for each structure:

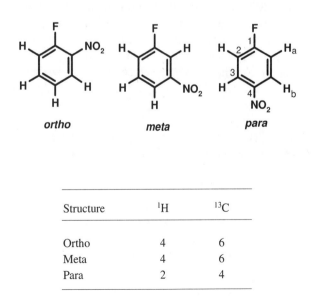

Structure	^1H	^{13}C
Ortho	4	6
Meta	4	6
Para	2	4

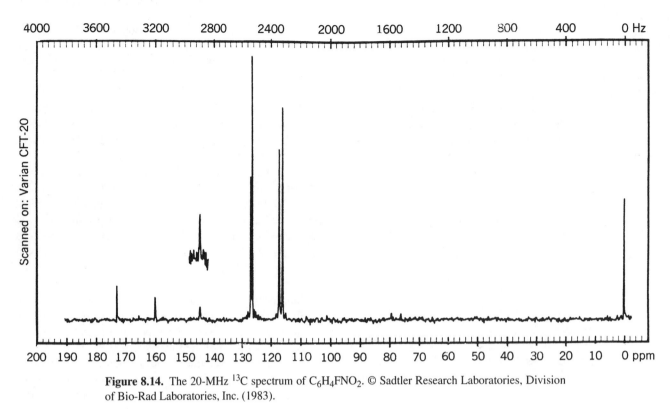

Figure 8.14. The 20-MHz ^{13}C spectrum of $C_6H_4FNO_2$. © Sadtler Research Laboratories, Division of Bio-Rad Laboratories, Inc. (1983).

When you include both homonuclear and heteronuclear couplings, you can quickly see how the ortho and meta structures would give much more complicated ^{1}H and ^{13}C spectra than those actually observed.

Let us focus our attention on the para isomer. The signal for H_a should occur (Table 6.5) at δ 7.14 and should be split into a doublet (by F) of doublets (by H_b). The reason that H_a is split by only the H_b ortho to it (and not the H_b para to it) is because of the number of intervening bonds. Similarly, the H_b signal is predicted to appear at δ 8.20, split into a doublet by H_a. The fact that the actual signal appears as a doublet of doublets shows that there is also a long-range (four-bond) coupling between H_b and F (see Chapter 9).

Using the data in Table 7.4, we can compute the predicted chemical shifts of each carbon in the para isomer:

$$\delta_1 = 128.5 + 35.1 + 6.0 = 169.6$$

$$\delta_2 = 128.5 - 14.3 + 0.9 = 115.1$$

$$\delta_3 = 128.5 + 0.9 - 5.3 = 124.1$$

$$\delta_4 = 128.5 - 4.5 + 19.6 = 143.6$$

From these calculated values we must conclude that the two signals near δ 116 are actually the two lines of a doublet for C_2, centered at δ 116.6. The same is true of the two lines centered at δ 126.6 for C_3. The weak singlet at δ 144.9 must correspond to C_4, not coupled to F. This leaves the last two lines, which must constitute a doublet centered at δ 166.7 for C_1.

(b) The ^{13}C–F coupling constants can be estimated by computing the line spacings in ppm, then multiplying by 20 Hz/ppm. The values $^1J(C_1–F) = 256$ Hz, $^2J(C_2–F) = 24$ Hz, and $^3J(C_3–F) = 12$ Hz. Note how the magnitude of the coupling constant decreases as the number of intervening bonds increases. More on this in Chapter 9.

In like manner, the H–H and H–F coupling constants can be determined as follows. Remembering that $^3J_{ab}$ must equal $^3J_{ba}$, we look in the two multiplets for the coupling constant that is common to both. That value, 9 Hz, must be the three-bond H_a–H_b coupling constant. The remaining spacings are the H–F coupling constants. From the lower field multiplet we extract a value of $^4J(H_b–F) = 5$ Hz; from the upfield one, $^3J(H_a–F) = 8$ Hz.   □

In the next chapter we will discuss the various factors that determine the magnitude of typical homo- and heteronuclear coupling constants.

CHAPTER SUMMARY

1. If two nuclei in a molecule meet the criteria of the spin-coupling checklist (below), the NMR signal for each nucleus will be split into multiple lines. **a.** Are both nuclei magnetic ($I \neq 0$)? **b.** Are they nonequivalent? **c.** Are they separated by no more than three bonds?

2. The maximum multiplicity (number of lines, L) in a given NMR signal is determined by the number (n) of neighboring coupled nuclei according the Eq. (8.3), $L = \prod (2n_i I_i + 1)$.

3. In the case of coupling to n equivalent hydrogens (or other $I = \frac{1}{2}$ nuclei), Eq. (8.3) reduces to $L = n + 1$.

4. The NMR signals consisting of multiple lines are characterized by the number of lines: doublet (two), triplet (three), quartet (four), and so on. A signal with just one line is a singlet. A signal with an unspecified number of lines is a multiplet.

5. The relative intensity of lines within a multiplet can be predicted from consideration of the various possible spin combinations among the spin states of the neighboring nuclei. In the case of coupling to n equivalent hydrogens (or other $I = \frac{1}{2}$ nuclei), the relative intensities of the $n + 1$ lines are given by Pascal's triangle.

6. The separation between neighboring lines of a multiplet is called the coupling constant J and is measured

in hertz. The value of J is dependent on the type of and structural relationship between the nuclei; it is not affected by the field strength or operating frequency of the spectrometer. Here, $^3J_{ab}$ indicates a three-bond coupling between nuclei a and b. Note also that $J_{ab} = J_{ba}$.

REVIEW PROBLEMS (Answers in Appendix 1)

8.1. (a) Predict the multiplicity of the ^{13}C signal of CD_3OD, assuming that only one-bond C–D coupling is observable. (b) Draw a diagram similar to Figure 8.12 showing all possible spin orientations for three deuterium nuclei grouped by M values. From this, give the relative intensity of the lines predicted in part (a).

8.2. Predict the multiplicity of each ^1H signal for *cis*-1,2-dimethylcyclopropane:

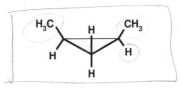

8.3. (a) Figure 8.15 shows the 60-MHz ^1H spectrum of $C_4H_6BrF_3O$. Determine the structure of the molecule and assign each signal. (b) Report the NMR data in condensed format.

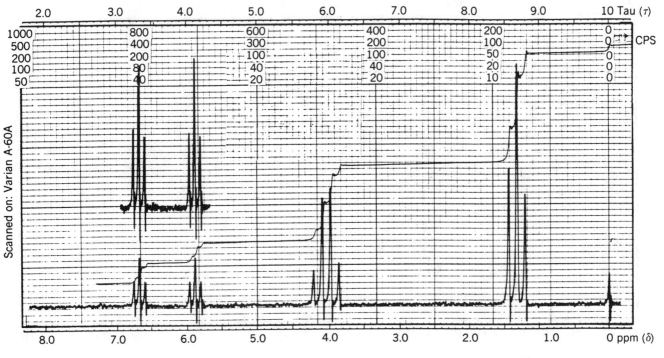

Figure 8.15. The 60-MHz ^1H spectrum of $C_4H_6BrF_3O$. © Sadtler Research Laboratories, Division of Bio-Rad Laboratories, Inc. (1970).

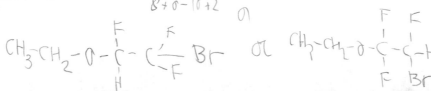

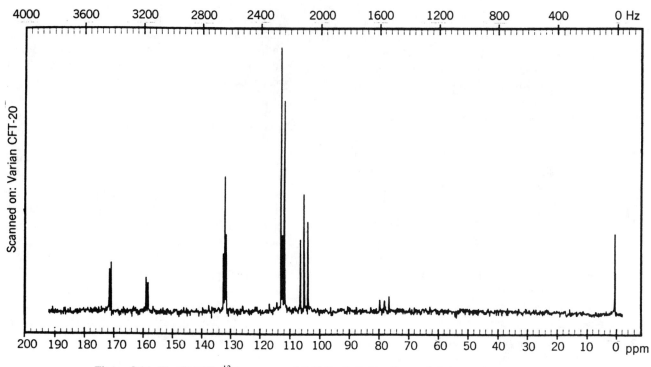

Figure 8.16. The 20-MHz ^{13}C spectrum of $C_6H_4F_2$. © Sadtler Research Laboratories, Division of Bio-Rad Laboratories, Inc. (1983).

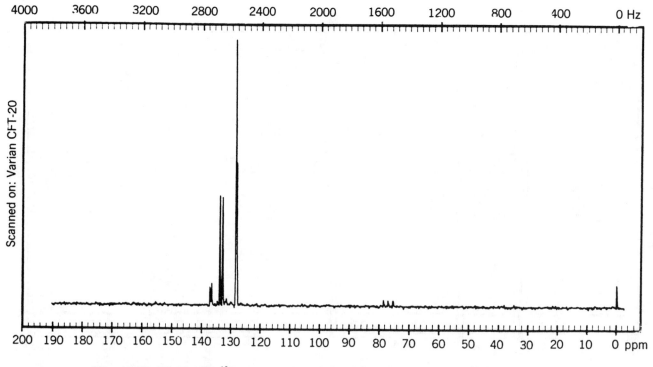

Figure 8.17. The 20-MHz ^{13}C spectrum of $C_{18}H_{15}P$. © Sadtler Research Laboratories, Division of Bio-Rad Laboratories, Inc. (1983).

8.4. (a) Figure 8.16 shows the 20-MHz ^{13}C spectrum of $C_6H_4F_2$, with lines at δ 103.0, 104.3, 105.6, 110.9, 112.1, 130.5, 131.0, 131.5, 158.4, 159.0, 170.8, and 172.4. Determine the structure of the molecule and assign each signal. (b) Determine the value of each C–F coupling constant.

8.5. (a) Figure 8.17 shows the 20-MHz ^{13}C spectrum of $C_{18}H_{15}P$, with lines at δ 128.4, 128.8 (two superimposed lines), 133.1, 134.1, 136.9, and 137.5. The ^{31}P spectrum of the compound exhibits one signal at δ −8. Determine the structure of the molecule and assign each signal. (b) Determine the value of each C–P coupling constant.

8.6. Peter Stang and co-workers at the University of Utah have synthesized a family of alkynyliodonium triflate salts with the generic structure

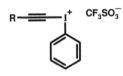

In Figures 8.18*a* and *b* are shown the 300-MHz ^{1}H and 75-MHz ^{13}C spectra for one such compound (**A**) with molecular formula $C_{11}H_8F_3IO_5S$. In Figures 8.19*a* and *b* are shown the ^{1}H and ^{13}C spectra for another member of the class (**B**) with molecular formula $C_{10}H_7ClF_3IO_3S$. (a) To obtain the NMR spectra, one of the above compounds was dissolved in CDCl$_3$. What solvent was used to dissolve the other compound? (b) Both ^{13}C spectra exhibit a quartet centered at approximately δ 120. What species is giving rise to this signal? (c) Each of the ^{1}H spectra exhibit three signals in the range δ 7.5–8.2, with integrals of 2H, 1H, and 2H. What is the multiplicity of each signal and why? (d) Suggest structures for **A** and **B**. Assign the remaining signals in both spectra and account for any anomalous chemical shifts.

8.7. Figures 8.20*a* and *b* show, respectively, the proton-decoupled 121.4-MHz ^{31}P and partial 300-MHz ^{1}H spectra (also contributed by Stang) of the novel

organometallic complex shown below:

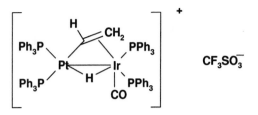

Note that platinum is composed of six different isotopes but all are nonmagnetic ($I = 0$) except ^{195}Pt ($I = \frac{1}{2}$), which has a natural abundance of 33.8%. (Iridium, though composed of two $I = \frac{3}{2}$ isotopes, does not give rise to observable coupling.) (a) As best you can, assign all the signals in the ^{31}P spectrum. Account for the multiplicity of each signal, and estimate the magnitude of each platinum–phosphorus coupling constant. (b) Figure 8.17*b* shows the 300-MHz ^{1}H signal for the bridging hydrido hydrogen, a complex multiplet centered at δ −11.55 that is highly shielded by the two nearby metal atoms. Account for the multiplicity of the signal, and estimate the magnitude of all phosphorus and platinum coupling constants to this hydrogen.

8.8. Consider the structure of the bicyclic diol below:

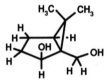

(a) How many ^{1}H signals do you expect for this structure and what should be the multiplicity of each signal? (b) The 300-MHz ^{1}H spectrum (contributed by John Bender and Soren Giese of Fred West's group at the University of Utah) of a highly purified sample of the diol is shown in Figure 8.21. Noting the chemical shift and multiplicity of each signal, assign as many as you can to the hydrogens in the structure.

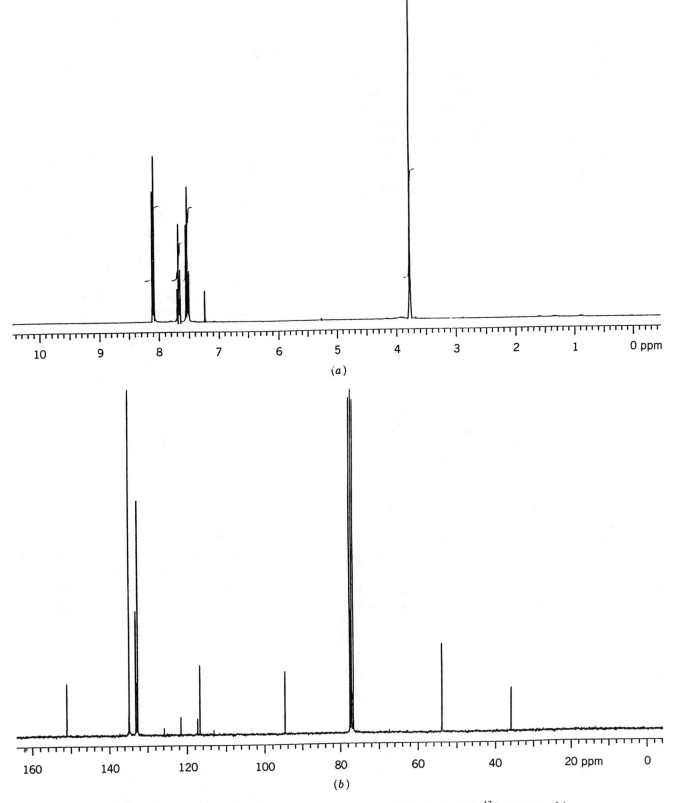

Figure 8.18 (*a*) The 300-MHz ^{1}H spectrum of **A**, problem 8.6. (*b*) The 75-MHz ^{13}C spectrum of **A**, problem 8.6.

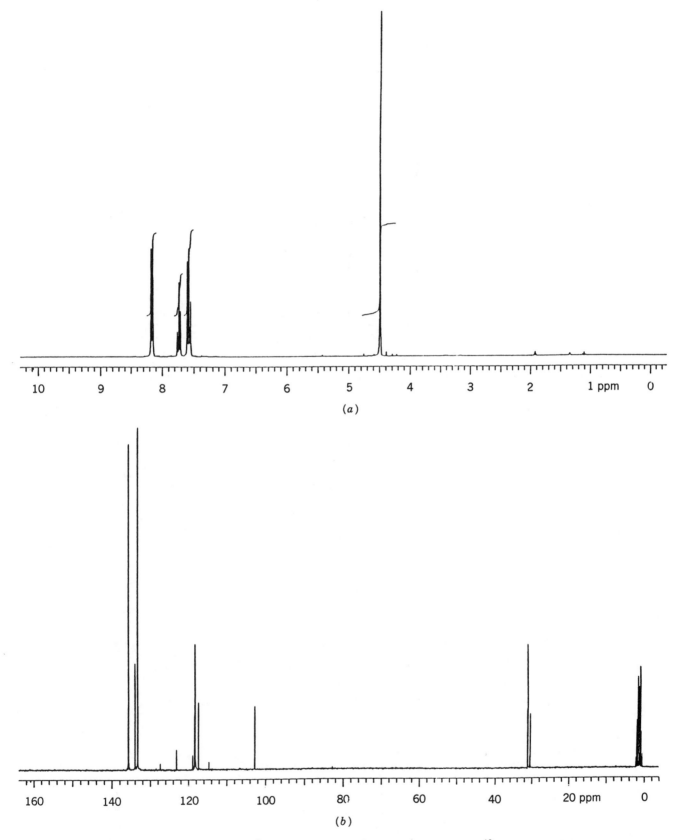

Figure 8.19 (*a*) The 300-MHz ^{1}H spectrum of **B**, problem 8.6. (*b*) The 75-MHz ^{13}C spectrum of **B**, problem 8.6.

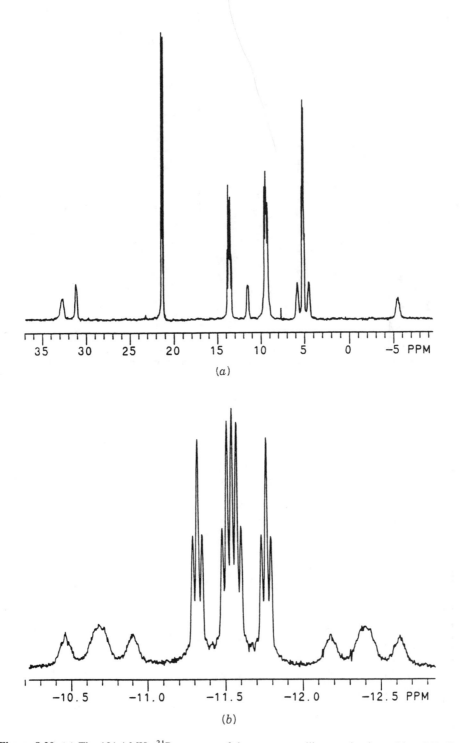

Figure 8.20. (*a*) The 121.4-MHz ^{31}P spectrum of the organometallic complex in problem 8.7. (*b*) Partial 300-MHz ^{1}H spectrum of the organometallic complex in problem 8.7.

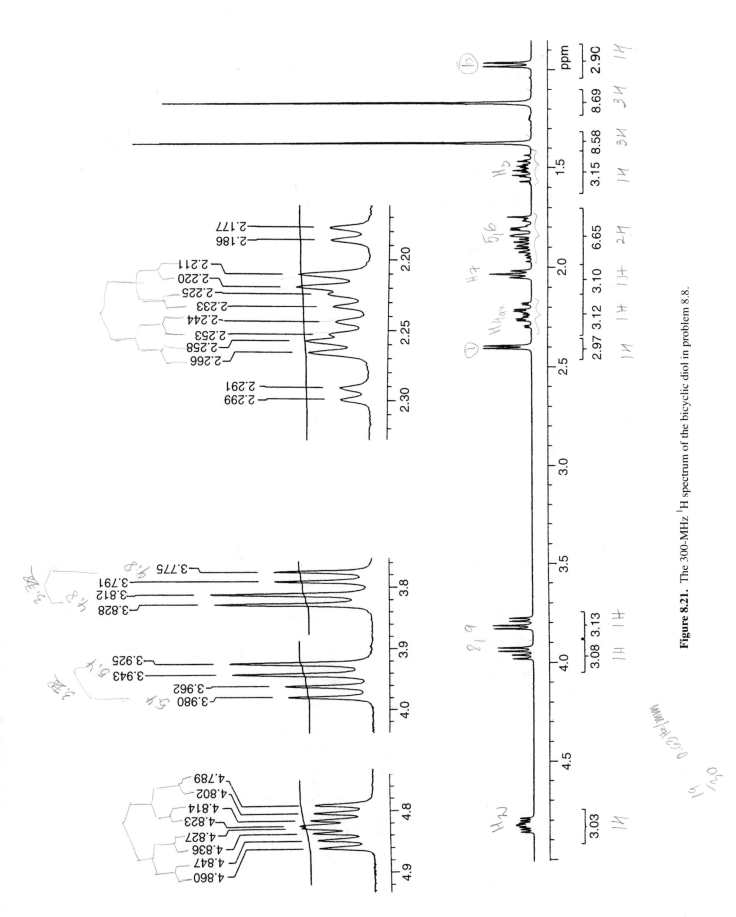

Figure 8.21. The 300-MHz ^{1}H spectrum of the bicyclic diol in problem 8.8.

9

FACTORS THAT INFLUENCE THE SIGN AND MAGNITUDE OF *J*: SECOND-ORDER (STRONG) COUPLING EFFECTS

9.1 NUCLEAR SPIN ENERGY DIAGRAMS AND THE SIGN OF *J*

Most numbers have three parts: a magnitude, a sign (positive or negative), and units (dimensions). In Chapter 8 we found that coupling constants have widely varying magnitude, and they have units (hertz). Now we are going to discover that they also have signs.

Consider the situation where we have two nonequivalent hydrogens in a molecule, H_a (with chemical shift δ_a) and H_b (with chemical shift δ_b). For the sake of this discussion, we will assume that $\delta_a > \delta_b$ and that the two nuclei are *not* spin coupled to each other. Each of the ^1H nuclei can exist in either of two spin orientations (or spin states), and transitions between these spin states give rise to their NMR signals (Section 2.3). Figure 9.1 shows the relative energies of the two states for H_a, the two states for H_b, as well as the four spin states (numbered from the bottom up) that result from a combination of H_a and H_b. Recall that, since $\delta_a > \delta_b$, H_a has the higher precessional frequency (ν_a) and the larger energy gap ($h\nu_a$) between its two spin states.

If asked how many transitions are possible among the four spin states, you might be tempted to list all of the following: $1{\rightarrow}2$, $1{\rightarrow}3$, $1{\rightarrow}4$, $2{\rightarrow}3$, $2{\rightarrow}4$, and $3{\rightarrow}4$. But there is a selection rule that controls the probability (and hence intensity) of each transition: An *allowed* transition involves only the flip of one nuclear spin; all transitions involving more than one flip are forbidden. From the list of transitions we can therefore delete $1{\rightarrow}4$ and $2{\rightarrow}3$, because these involve the simultaneous flip of both spins. Transitions $1{\rightarrow}2$ and $3{\rightarrow}4$ result from the flip of only the H_b spin and are thus allowed and responsible for the H_b signal. Likewise, transitions $1{\rightarrow}3$ and $2{\rightarrow}4$ result from the flip of only the H_a spin and give rise to the H_a signal. Further, notice (from symmetry) that transi-

tions $1{\rightarrow}2$ and $3{\rightarrow}4$ are *degenerate* (involve an identical energy gap $h\nu_b$), as are transitions $1{\rightarrow}3$ and $2{\rightarrow}4$ (energy gap $h\nu_a$).

In Figure 9.2*b* are the same four (uncoupled) spin states we saw in Figure 9.1. In this diagram, however, we have numbered the allowed *transitions* rather than the spin states themselves. So, transition $1{\rightarrow}2$ has become transition 1, $1{\rightarrow}3$ becomes transition 2, $2{\rightarrow}4$ becomes 3, and $3{\rightarrow}4$ becomes 4. The two-singlet spectrum that would result from these transitions is shown below the spin state diagram. Note how transitions 2 and 3 (which are degenerate) define δ_a, while transitions 1 and 4 define δ_b.

Now, let us complicate the picture by including a coupling interaction between H_a and H_b. This interaction will have one of two effects. Either the parallel spin states (1 and 4) will be raised in energy (by an amount Δ) and the antiparallel states (2 and 3) lowered by the same amount or vice versa. This is easier to show in a diagram. Figure 9.2*a* shows the spin state energies after the *parallel* states are raised by an amount Δ and the *antiparallel* ones are lowered by Δ. In Figure 9.2*c* it is the parallel states that are *lowered* (by Δ) and the antiparallel ones that are *raised* (by Δ). Inspection of the gaps between the energy levels in the two coupled cases reveals that transitions 1 and 4 are no longer degenerate, nor are transitions 2 and 3, as they were in the uncoupled case. In Figure 9.2*a*, transitions 3 and 4 have increased in energy gap (and hence frequency) by 2Δ while transitions 1 and 2 have decreased by 2Δ. Therefore, the lines resulting from transitions 1 and 2 move slightly upfield, while the lines corresponding to transitions 3 and 4 move an equal amount downfield. The resulting spectrum (shown below the spin state diagram) consists of two doublets, one centered at δ_a, the other at δ_b, with coupling constant *J* separating the two lines of each doublet.

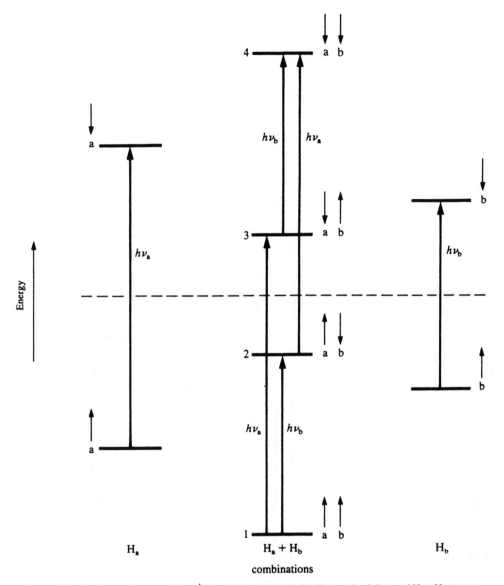

Figure 9.1. Nuclear spin states of two ^{1}H nuclei. H_a (on the left), H_b (on the right), and $H_a + H_b$ (center, with $J = 0$).

■ **EXAMPLE 9.1** What is the relationship between Δ (expressed in *hertz*) and J?

☐ *Solution:* Because the separation between lines 2 and 3 in the coupled case is J *hertz*, each line must have moved $J/2$ *hertz* away from δ_a. Line 2 involves a transition between spin state 1 (which was raised by Δ) to spin state 3 (which was lowered by Δ). Therefore, $J/2$ must equal 2Δ, or $J = 4\Delta$.   ☐

Next, compare Figure 9.2c with 9.2b. Because this time the antiparallel states are raised (by Δ) and the parallel ones lowered, the positions of lines 2 and 3 have been reversed, as

have the positions of lines 1 and 4. But the resulting spectrum itself is indistinguishable from the one in Figure 9.2a.

If the (a) and (c) (coupled) spectra are identical (except for the ordering of lines), what difference, if any, is there between the two cases? The answer is the *sign* of the coupling constant. A *positive* value ($J > 0$) implies that antiparallel spin states are lowered and parallel ones raised, while a *negative* value ($J < 0$) implies the converse. Thus, while a coupling constant possesses a sign as well as a magnitude, the sign (whether positive or negative) normally has *no* effect on the appearance of the spectrum. But the *relationship* between the signs of the coupling constants in a molecule (i.e., whether the signs are like or unlike) *can* sometimes have an effect when **second-or-**

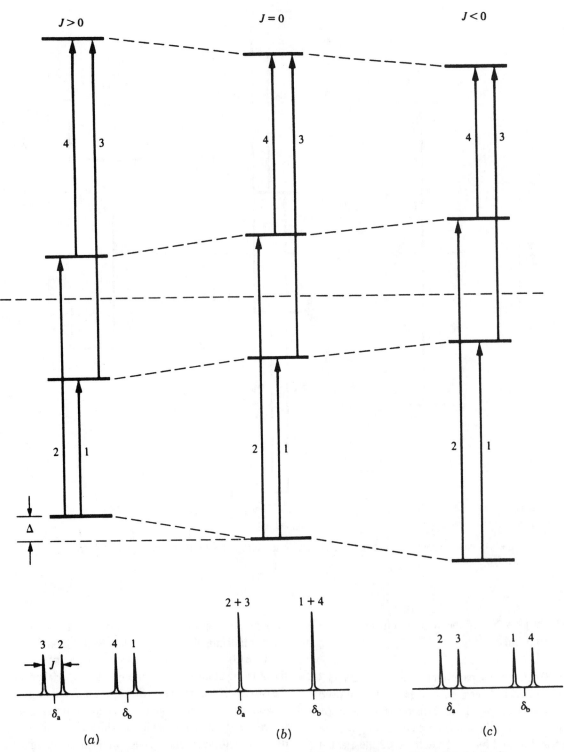

Figure 9.2. Effects of spin coupling on the $H_a + H_b$ spin state diagram in Figure 9.1: (*a*) $J > 0$; (*b*) $J = 0$; (*c*) $J < 0$.

der coupling (Section 9.9) is involved. Furthermore, when measuring *J* values from a spectrum, we can get only the *magnitude* of *J*, not its sign.

9.2 FACTORS THAT INFLUENCE *J*: PRELIMINARY CONSIDERATIONS

9.2.1 Factors Influencing the Sign of *J*

From Chapter 8 we know that coupling constants have a magnitude that is somehow related to the number of intervening bonds. Furthermore, in the previous section we learned that each coupling constant has a sign that indicates whether antiparallel spin states are lowered as a result of coupling (*J* > 0) or parallel states are lowered (*J* < 0). The sign of *J*, it turns out, is also affected by the number of intervening bonds.

Theoretical chemists have a relatively complete understanding of the factors that control the sign and magnitude of *J*. In fact, they have derived some impressive equations that describe the spin-coupling interaction. We will try our best to steer clear of the high-powered math and take a more qualitative approach.

Recall from Section 8.4 that the information about a nucleus's spin state is communicated to other nearby nuclei predominantly through the intervening bonding electrons. Thus, it comes as no surprise that, in general, the magnitude of *J* tends to decrease as the number of intervening bonds increases. A chemical bond consists of a pair of electrons occupying a **molecular orbital,** a region of space around two (or more) nuclei. Electrons, like protons, have magnetic spin ($s = \pm\frac{1}{2}$, Section 2.1.2), and in order for two electrons to occupy the same orbital, their spins must be *paired* (antiparallel, one with $s = +\frac{1}{2}$, the other with $s = -\frac{1}{2}$). The simplest case of coupling between two hydrogen nuclei is the one-bond coupling in the hydrogen (H_2) molecule depicted in Figure 9.3. Suppose that the preferred (more stable, lower energy) orientation of a nuclear spin is *opposite* to the spin of the nearest electron. Since the electron spins are paired, the two nuclear spins would also prefer to be paired (antiparallel). And

this is exactly the definition of a positive coupling constant. Thus, the coupling constant between two nuclei is positive if the spins of the coupled nuclei prefer to be paired (antiparallel).

■ **EXAMPLE 9.2** Using the spin-pairing model above, predict the sign of the two-bond coupling constant ($^2J_{HH}$) in H–^{13}C–H.

□ *Solution:* Building an extra bond onto the more stable sequence in Figure 9.3 gives a preferred parallel orientation of the two ^{1}H nuclei (Figure 9.4). Therefore, we predict that $^2J_{HH}$ should be negative. □

Although this model is somewhat naive, it does help us rationalize why one- and three-bond H–H, C–H, and C–C coupling constants are usually positive, while two-bond coupling constants are usually negative.

9.2.2 Factors Influencing the Magnitude of *J*

So much for the sign of *J*. In general, the magnitude (absolute value) of a coupling constant between nuclei a and b ($|J_{ab}|$) is determined by the generalized equation.

$$|J_{ab}| \propto \gamma_a \gamma_b f_a f_b \, F(\text{angle}) \tag{9.1}$$

where γ is the magnetogyric ratio (Section 2.2), *f* is the fraction of *s*-orbital character in the atomic orbital used to create the molecular orbital, *F*(angle) is a functional dependence on the angle between interacting molecular orbitals (in the case of multibond couplings), and ∝ indicates "proportional to." Thus, for example, all other things being equal, the magnitude

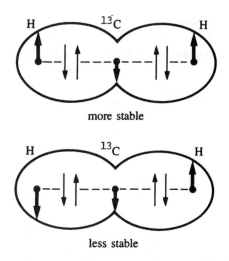

Figure 9.4. Nuclear and electronic spin orientations in an H–^{13}C–H unit. Boldface arrows represent nuclear spins, while regular arrows represent the electronic spins. The more stable arrangement is the one with ^{1}H nuclear spins parallel (unpaired).

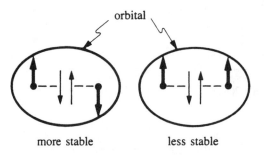

Figure 9.3. Nuclear and electronic spin orientations in the H_2 molecule. Boldface arrows represent nuclear spins, while regular arrows represent the electronic spins. The more stable arrangement is the one with nuclear spins paired (opposite).

of *J* is directly proportional to the product of the magnetogyric ratios of the two coupled nuclei. [Note, however, that since a few elements have a negative γ value (Table 2.1), coupling between nuclei with γ values of opposite sign will reverse some of the generalizations described below.]

■ **EXAMPLE 9.3**   (a) In the hydrogen molecule $^1J_{HH}$ is + 280 Hz. What is the value (sign and magnitude) of $^1J_{HD}$ in H–D? (b) The 21.4-MHz ^{13}C spectrum of DCCl₃ (Self Test 1, problem 3d) consists of signals at δ 75.5, 77.0, and 78.5. Predict the sign and value of $^1J_{CH}$ for HCCl₃.

☐ **Solution:**   (a) The values of γ (Table 2.1) are 267.512 × 10⁶ rad T⁻¹ s⁻¹ for ¹H and 41.0648 × 10⁶ rad T⁻¹ s⁻¹ for D (²H). Since the magnitude of *J* is directly proportional to the product of these values and all other variable are the same, we can set up a proportion:

$$\frac{^1J_{HD}}{^1J_{HH}} = \frac{\gamma_H\gamma_D}{\gamma_H\gamma_H} = \frac{\gamma_D}{\gamma_H}$$

$$^1J_{HD} = {^1J_{HH}}\left(\frac{\gamma_D}{\gamma_H}\right) = (280 \text{ Hz})\left(\frac{267.512}{41.0648}\right) = 43 \text{ Hz}$$

And because this is a one-bond coupling constant, it is positive.

(b) These three lines are due to one-bond C–D coupling (Section 8.6.2). To find the value of the corresponding C–H coupling constant, first find the value of $^1J_{CD}$ by applying Eq. (5.1):

$$^1J_{CD} = (\Delta\delta)(\nu_0')(78.5 - 77.0 \text{ ppm}) (21.4 \text{ Hz / ppm})$$

$$= 32.1 \text{ Hz}$$

Then use the same type of relationship as in part (a):

$$^1J_{CH} = {^1J_{CD}}\left(\frac{\gamma_H}{\gamma_D}\right) = 32.1 \text{ Hz} \left(\frac{267.512}{41.0648}\right) = 209 \text{ Hz} \qquad \Box$$

9.3   ONE-BOND COUPLING CONSTANTS

In review problem 8.7 we saw two examples of one-bond phosphorus–platinum coupling constant of 2380 and 4640 Hz. These are exceptionally large *J* values. The one-bond coupling constants most commonly encountered in NMR are $^1J_{CH}$, $^1J_{CD}$, $^1J_{CF}$, $^1J_{CP}$, and $^1J_{PH}$. In fact, we saw examples of some of these in Section 8.6, Example 8.19, and problems 8.1, 8.4, 8.5, and 8.6. These *J* values are generally positive and involve nuclei with positive γ values.

Though the sign of all these one-bond coupling constants is normally positive, the magnitude depends on the nature of the orbital connecting the two nuclei. For example, the σ molecular orbital that comprises a carbon–hydrogen bond results from the overlap of two **atomic orbitals,** one centered on the carbon and one on the hydrogen. (An atomic orbital is an orbital centered on a single nucleus.) All hydrogens use the same type of orbital (a 1*s* orbital) to make their bonds. Carbon, on the other hand, uses a variety of different **hybrid orbitals** (combinations of *s* and *p* atomic orbitals) to construct its σ bonds. The specific type of carbon hybrid depends on the shape of the molecule. Table 9.1 lists the hybrid carbon orbitals that occur in σ bonds of three common situations (single-, double-, and triple-bonded carbon), along with the corresponding value of $^1J_{CH}$. Note that an sp^n hybrid is one where the ratio of *s* character to *p* character (the relative contributions of *s* and *p* orbitals to the hybrid) is 1 : *n*, so the fraction of *s* character (f_s) equals $(n + 1)^{-1}$.

From the data in Table 9.1, we can see that $^1J_{CH}$ is directly proportional to the fraction of *s* character in the carbon hybrid:

TABLE 9.1   Effect of Carbon Hybridization on $^1J_{CH}{}^a$

	Tetrahedral	Trigonal Planar	Linear
Structure (Shape at Carbon)			
C–C–H Angle	109.45°	120°	180°
Carbon hybrids	sp^3	sp^2	sp
f_s	0.25	0.333	0.50
f_p	0.75	0.667	0.50
$^1J_{CH}$, Hz	125	156	249

^{a}Note that $f_s + f_p = 1$.

$$^1J_{CH} = (500 \text{ Hz}) (f_s) \qquad (9.2)$$

The reason for this effect is that s orbitals interact more directly (i.e., have more "contact") with the nucleus than do p orbitals. So, the greater an orbital's s character, the more nuclear spin information it communicates between neighboring nuclei.

■ **EXAMPLE 9.4** What type of carbon hybrid is involved in each of the eight equivalent C–H bonds (only one is explicitly shown) of the **cubane**, the structure below? *Hint*: The value of $^1J_{CH}$ is 160 Hz.

☐ *Solution:* Solving Eq. (9.2) for f_s, we find

$$f_s = \frac{^1J_{CH}}{500 \text{ Hz}} = \frac{160 \text{ Hz}}{500 \text{ Hz}} = 0.32$$

Since, for any s/p hybrid, $f_s + f_p = 1$,

$$f_p = 1 - f_s = 1 - 0.32 = 0.68$$

$$n = f_p / f_s = 0.68 / 0.32 = 2.13$$

Therefore, the hybrid orbital is $sp^{2.13}$.   ☐

Many other homonuclear and heteronuclear one-bond coupling constants are known; ranges of J values for some of the most common ones are listed in Table 9.2.

■ **EXAMPLE 9.5** From the value of $^1J(^{13}C≡^{13}C)$ listed in Table 9.2, devise a relationship to predict the values of $^1J(^{13}C=^{13}C)$ and $^1J(^{13}C-^{13}C)$.

☐ *Solution:* Because $^1J_{CC}$ is sensitive to the fraction of s character (f_s) in *both* carbon orbitals of the σ bond, a relationship similar to Eq. (9.2) should prove valid for carbon–carbon coupling. The σ component of a triple bond is formed in part by the overlap of two sp orbitals, that of a double bond involves overlap of two sp^2 orbitals, and a single bond results from overlap of two sp^3 orbitals. Thus, the desired relationship should include the product of both f values:

$$^1J_{C≡C} = (\text{const})(f_s)^2 \qquad (9.3)$$

To evaluate the constant, we enter the data for the triple bond:

$$^1J_{C≡C} = 170 \text{ Hz} = (\text{const})(0.50)^2$$

$$\text{const} = 170 \text{ Hz}/(0.50)^2 = 680 \text{ Hz}$$

Using this constant for $^1J_{C-C}$ and $^1J_{C=C}$ gives

$$^1J_{C-C} = (680 \text{ Hz})(0.25)^2 = 42.5 \text{ Hz}$$

$$^1J_{C=C} = (680 \text{ Hz})(0.33)^2 = 75 \text{ Hz}$$

These are in quite good agreement with the observed values in Table 9.2.   ☐

TABLE 9.2   Some Representative One-Bond Coupling Constants[a]

Type	^1J (Hz)	Type	^1J (Hz)
$^1H-^1H$	280	$^{13}C-^{19}F$	−165 to −350[c]
$^1H-^{13}C$	110–270	$^{13}C(sp^3)-^{31}P$	48–56
$^1H-^{31}P$	140–1115	$^{13}C(sp^2)-P^{31}$	P73–159
$^1H-^{31}P=O$	500–700	$^{13}C=^{31}P$	50–95
$^{13}C-^{13}C$	35	$^{13}C≡^{31}P$	150–200
$^{13}C=^{13}C$	70	$^{15}N-^{15}N$	14
$^{13}C≡^{13}C$	170	$^{19}F-^{31}P=O$	1000
$^{13}C-^{15}N$	−4–18[b]	$^{31}P-^{31}P$	−100 to −500[c]
$^{13}C≡^{15}N$	−17[b]		

[a]Most data quoted from ref. 1.

[b]Negative sign due to negative γ value for ^{15}N.

[c]Negative sign due to the effect of the unshared pair on the heteroatom.

9.4 TWO-BOND (GEMINAL) COUPLING CONSTANTS

The most important two-bond coupling constants are H–H, H–F, and H–P. In general, two-bond (**geminal**) coupling constants have a smaller magnitude than one-bond coupling constants and often (but not always) have negative signs (Section 9.2.1). Geminal coupling constants are sensitive not only to magnetogyric ratios and *s*-character effects but also to the angle between the two bonding molecular orbitals (i.e., the bond angle).

As Table 9.3 shows, two-bond hydrogen–hydrogen coupling constants (as in H–C–H) usually fall in the range –9 to –15 Hz. In the cyclic molecules the value of $^2J_{HH}$ becomes less negative (i.e., more positive) as the ring gets smaller (and hence, the H–C–H angle gets larger). This is because the fraction of *s* character (*f*) in the carbon orbitals increases with the angle between them.

Recall that two nuclei attached to the same atom can be nonequivalent [as required to produce observable coupling between them (Section 8.4)] if they are diastereotopic (Section 4.3). Even if the two nuclei are equivalent, there is a way to measure the coupling constant between them. For example, if two hydrogens are equivalent, one can (in principle, at least)

substitute a deuterium for one of them, measure $^2J_{HD}$, then use an equation like the one in Example 9.3(a) to calculate $^2J_{HH}$. Nonetheless, it is true that geminal nuclei often are equivalent, so that coupling *between* them is not observed in the spectrum.

9.5 THREE-BOND (VICINAL) COUPLING CONSTANTS

Of all the types of coupling, **vicinal** (three-bond) coupling constants (which are normally positive) can tell us the most about the three-dimensional arrangement of the atoms within a molecule. To understand why this is so, we will be making reference to **dihedral angles**, the angle between two planes. If you open this book and lay it on your desk, the angle between the facing pages is 180°; if you close the book, the angle between the facing pages is 0° (Figure 9.5). In an analogous way, vicinal bonds describe a dihedral angle, as also shown in Figure 9.5. The angle of interest is the one between the H_a–C–C plane and the C–C–H_b plane. In actual molecular structures, this dihedral angle can vary continuously (and rapidly) from 0° to 360° by rotation around the C–C single bond. Certain conformations (Section 4.2), however, are more stable (and more prevalent) than others, and each one has its own specific dihedral angle. For example,

TABLE 9.3 Some Representative Two-Bond Coupling Constants[a]

Structure	$^2J_{HH}$ (Hz)	Structure	^2J (Hz)
$H_2C\langle^H_H$	–12.4	$^{13}C(sp^3)-C(sp^3)-H$	–5–6
$(CH_2)_n$ C$\langle^H_H$ n = 5	–13	$^{13}C(sp^2)-C(sp^2)-H$	–2.4–27
n = 4	–10.5	$^{13}C(sp)-C(sp)-H$	49–61
n = 3	–9	$^{19}F-C(sp^3)-H$	44–81[b]
n = 2	–4.3		
$\rangle{=}\langle^H_H$	+ 2.5	$^{19}F-C(sp^2)-H$	70–80[b]
		$^{31}P-C(sp^3)-H$	7–14
$H_2Si\langle^H_H$	+ 2.8	$^{31}P-C(sp^2)-H$	33
$H_2Sn\langle^H_H$	+ 15.3	$^{13}C-C-^{31}P{=}O$	10–15
$^{19}F-C(sp^3)-^{19}F$	155–225	$^{13}C{=}C-^{31}P{=}O$	7–40
		$C{=}^{13}C{=}C-^{31}P{=}O$	0
$^{19}F-C(sp^2)-^{19}F$	28–87	$^{31}P(V)-C(sp^3)-F$	6–190

[a] Data from ref. 1–4.

[b] Absolute values.

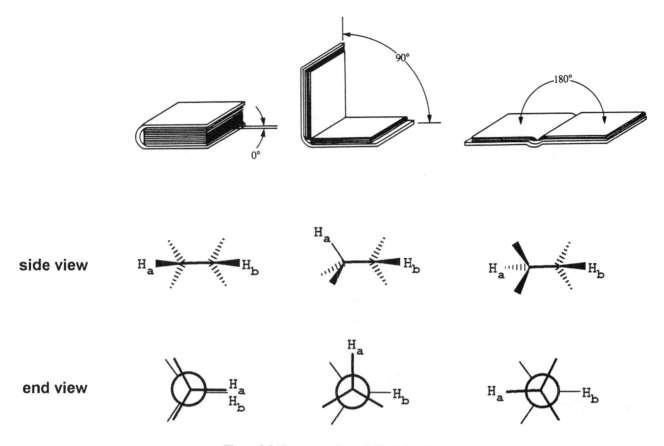

side view

end view

Figure 9.5. Representations of dihedral angles.

dihedral angles of 180° and 60°, called **anti** and **staggered**, respectively, are more stable than 0° or 120° (**eclipsed**) because there is less electronic and steric repulsion with the former angles.

The important fact is that the magnitude of $^3J_{ab}$ (e.g., between H_a and H_b in Figure 9.5) is (in addition to its dependence on γ values and s-character effects) a function of the dihedral angle (θ) between the nuclei. The **Karplus equation** (named for the discoverer of the relationship) for vicinal H–H coupling has the form

$$^3J_{HH} = (7 - \cos\theta + 5\cos 2\theta) \text{ Hz} \qquad (9.4)$$

Figure 9.6 is a graph of this equation. Notice how J reaches its maximum values at 0° ($J = 11$ Hz) and 180° (13 Hz) and its minimum value at 90° (2 Hz). This is because the **stereoelectronic** interaction between the two vicinal molecular orbitals (bonds) is at its maximum when the orbitals are parallel (dihedral angles of 0° or 180°) and decreases to nearly zero when the orbitals are perpendicular (a dihedral angle of 90°). This is quite analogous to the requirement for p orbitals to be parallel to form a π bond.

■ **EXAMPLE 9.6**   Cyclohexane and its derivatives usually exist mainly in a nonplanar conformation called a **chair** form, with the attached substituents occupying sites that are either **axial** (a) or **equatorial** (e):

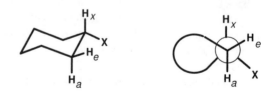

Predict the values of ^3J for H_x–H_e and H_x–H_a and ^2J for H_a–H_e.

□ *Solution:* From the structural diagrams it is clear that the H_x–H_e dihedral angle is 60°, while the H_x–H_a angle is 180°. Using Eq. (9.4) or Figure 9.6.

$$^3J_{xe} = 7 - \cos 60° + 5\cos 120° = 4 \text{ Hz}$$

$$^3J_{xa} = 7 - \cos 180° + 5\cos 360° = 13 \text{ Hz}$$

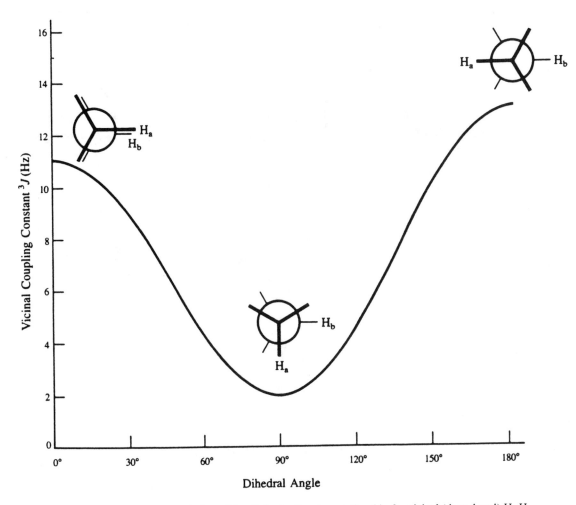

Figure 9.6. Graphical depiction of Eq. (9.4); the Karplus relationship for vicinal (three-bond) H–H coupling.

To get the value of $^2J_{ae}$ (a geminal coupling constant), refer back to Table 9.3; the value of $^2J_{HH}$ for cyclohexane is –13 Hz. *Note*: Cyclohexane is another example of a molecule in which two hydrogens attached to the same carbon (H_e and H_a) are nonequivalent.   □

As mentioned above, most single bonds that are not part of a ring rotate constantly and rapidly (on the NMR time scale), interconverting all possible conformations and dihedral angles. Since each conformation has a different value of each vicinal coupling constant, the *observed* values of ^3J represent a weighted average over all conformations. Such coupling constants can also show a dependence on temperature, because the ratio of the various conformations (of different energy) can vary with temperature [the Boltzmann distribution; Eq. (2.8)].

■ **EXAMPLE 9.7** Suppose a molecule were limited to just two conformations, **A** (where $^3J_{HH}$ = 13 Hz) and **B** ($^3J_{HH}$

= 4 Hz). Furthermore, suppose there is rapid interconversion between these two conformations and that the observed value of $^3J_{HH}$ is 7 Hz (a typical value in freely rotating systems). Calculate the average fraction of the population in each conformation.

□ ***Solution:*** Let f_A equal the fraction of conformation **A**. Therefore, $f_B = 1 - f_A$. Since the observed coupling constant is the weighted average over both conformations,

$$^3J_{obs} = 7 \text{ Hz} = f_A\,(13 \text{ Hz}) + f_B\,(4 \text{ Hz})$$

$$= f_A\,(13 \text{ Hz}) + (1 - f_A)\,(4 \text{ Hz})$$

Thus, $f_A = 0.33$ and $f_B = 0.67$.   □

Vicinal coupling across double bonds ($\sigma + \pi$) shows a similar variation with stereochemistry. But the presence of the double bond has two important consequences. First, the double bond (owing to the presence of the π-bond component) does not allow rotation and therefore limits vicinal relationships to either 0° (cis) or 180° (trans):

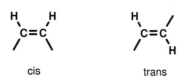

Second, because of the extra intervening (π) bond, there are more electrons communicating spin information between the nuclei, and $^3J_{HH}$ is about 20% greater than in the case of all single bonds. Thus, cis $^3J_{HH}$ values across a double bond are typically 10–12 Hz (8 Hz for ortho hydrogens on an aromatic ring), while the corresponding trans values are 16–18 Hz.

■ **EXAMPLE 9.8** Refer back to the solution of Example 8.18. From the observed value of $^3J_{cd}$ (16 Hz), which stereoisomer (**A** or **B**) better fits the data?

□ *Solution:* The H–C=C–H dihedral angle is 0° (cis) in **A** and 180° (trans) in **B**. The 16-Hz coupling constant is further evidence for structure **B**.   □

It may seem somewhat anomalous that the value of $^3J_{HH}$ in the linear molecule acetylene (H–C≡C–H) is only 9 Hz, because both the presence of the triple bond ($\sigma + 2\,\pi$) and s-character effects might have led us to predict a larger value. The problem here is that the two C–H molecular orbitals are collinear (and face opposite directions), so there is no dihedral angle.

So far in this section we have dealt only with homonuclear (HH) vicinal coupling. But it should come as no surprise that

TABLE 9.4   Typical Heteronuclear Vicinal Coupling Constants[a]

Structure	^3J(Hz)	Structure	^3J(Hz)
^{13}C–C–C–H	3–5	P–C–C–H	12–14
F–C–C–H	0–30	P, H (alkene)	30–55
F, H (alkene)	2–20	P, H (alkene)	50
F, (alkene) H	30–50	P–O–C–H	5–15
F–C–C–F	–3 to –20	P–X–Y–^{13}C	–5–30
F, F (alkene)	20–58	P, ^{13}C (alkene)	7–11
F, F (alkene)	–95 to –120	P, ^{13}C (alkene)	18–26
(fluorobenzene)	X = H: 6–10 X = F: 20	P–C–C–P	24–44

[a]Abstracted mainly from refs. 1 and 3.

the same type of Karplus angular dependence is exhibited by heteronuclear vicinal coupling constants as well. Table 9.4 lists some representative examples.

9.6 LONG-RANGE COUPLING CONSTANTS

As mentioned in Section 8.4, coupling interactions are normally observed only when the nuclei are separated by no more than three bonds because only then does the magnitude of *J*

exceed 1 Hz. (A multiple bond counts as only one bond separation.) Yet, there are special cases (e.g., Example 8.19) where coupling is observed over four, five, or even more bonds. Such **long-range coupling** can be anticipated under the following two circumstances:

1. There are one or more multiple bonds between the coupling nuclei.

TABLE 9.5   Long-Range Coupling Mediated by Multiple Bonds[a]

Structure		Number of Bonds (n)	^nJ(Hz)
(benzene: H, H_m, H_p)	H-H_m	4	2
	H-H_p	5	0.5
(fluorobenzene: F, H_m, H_p)	F-H_m	4	7
	F-H_p	5	2
(alkene H / C—H)		4	−2
H≡C—H		4	−2
(alkene H / H)		4	6
(H—C / C—H alkene)		5	1–2
(H / C—H alkene)		5	3
(diene H / H)		5	±1
H—C≡≡C—H		7	1
(P / C—H alkene)		5	12
(P / ^{13}C alkene)		4	5

[a]Taken mainly from refs. 1 and 4.

2. The molecule is rigid and the two nuclei are part of a **W geometrical relationship**.

Electron pairs involved in the π component of multiple (i.e., double or triple) bonds behave somewhat differently from those in single (σ) bonds (Section 7.4.1). Not only are there differences in geometry, hybridization, and *s* character (Section 9.3), but π-bond electrons interact with (*delocalize into*) neighboring σ molecular orbitals in the molecule (**hyperconjugation**). Because of this delocalization, π-bond electrons can communicate nuclear spin information over distances further than three bonds. Several examples of this type of long-range coupling are listed in Table 9.5.

The other situation where long-range coupling can be expected is when the orbitals connecting the two coupling nuclei (X and Y) are forced by a rigid molecular architecture to adopt a *W* relationship:

In such a structure the nuclear spin information is communicated between nuclei X and Y by overlap of the "tails" of the C–X and C–Y σ molecular orbitals. This overlap "short circuits" or sidesteps the other two intervening C–C bonds, making this type of four-bond interaction resemble a three-bond coupling. But, again, the molecular structure must constrain these bonds into this W relationship for long-range coupling of this sort to be significant. For example, compare the value of the four-bond H_e–$H_{e'}$ (W relationship) coupling

constant in cyclohexane (Table 9.6) with the negligibly small values of the four-bond H_e–$H_{a'}$ and H_a–$H_{a'}$ (not W) coupling constants.

The bottom line is that observable long-range coupling (over more than three bonds) *is* possible, but only under limited circumstances. And even then, it usually involves relatively small coupling constants.

9.7 MAGNETIC EQUIVALENCE

In Chapter 4 we encountered the concept of equivalence. If two (or more) nuclei are related by virtue of an axis, center, or plane of symmetry, they are said to be **symmetry** (or **chemically**) **equivalent**. Furthermore, chemically equivalent nuclei precess at *exactly* the same frequency and hence give rise to one NMR signal (coupling notwithstanding).

There is another kind of equivalence we need to introduce at this point. Let us examine the structure discussed in Example 8.19, *p*-fluoronitrobenzene:

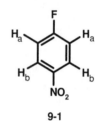

9-1

The two H_a's are chemically equivalent, being related by both a plane of symmetry and a C_2 axis; so are the two H_b's. Yet, as we found in Example 8.19, a given H_a couples only with

TABLE 9-6 Long-Range Coupling Mediated by W-Type Overlap

Structure	Coupling	^4J(Hz)
(chair cyclohexane with H_e, $H_{e'}$, H_a, $H_{a'}$)	ee'	1.8 (W)
	$ea'(=ae')$	−0.4 (not W)
	aa'	−0.9 (not W)
(norbornane with H, H)	exo–exo	1
(norbornene with H, H)		1
(bicyclo structure with H, H)		7

the H$_b$ ortho to it (three bonds away) and not to the other H$_b$ (five bonds away). Thus, as far as either H$_a$ is concerned, the two H$_b$'s are not equivalent, even though we know they are equivalent by symmetry. How can we resolve this paradox?

We describe this situation by saying that the two H$_b$'s are chemically equivalent (i.e., they are symmetry equivalent and occur at the same chemical shift), but they are not **magnetically equivalent.** For two (or more) nuclei to be magnetically equivalent, they must not only be chemically equivalent but also be equally coupled to any other nucleus. Thus, all magnetically equivalent nuclei are also chemically equivalent, but not all chemically equivalent nuclei are magnetically equivalent. We will use the same letter with a prime to indicate when one nucleus is chemically but not magnetically equivalent to another. So, we can relabel structure **9-1** as follows:

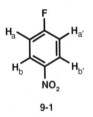

9-1

■ **EXAMPLE 9.9**   Use letter subscripts in the following structures to indicate which nuclei are chemically equivalent (by symmetry). Then use primes to indicate which are chemically equivalent nuclei but not magnetically equivalent. Consider only H–H and H–F couplings.

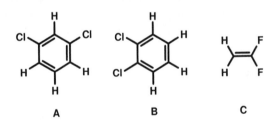

A   B   C

□ *Solution:*

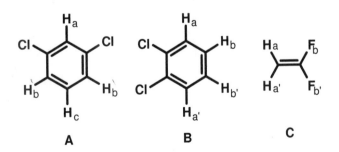

A   B   C

In **A**, the H$_b$'s are magnetically equivalent (no prime) because their couplings to H$_c$ are equal (ca. 8 Hz) and their couplings to H$_a$ (<1 Hz) are equal. In **B**, the H$_a$'s are not

magnetically equivalent, nor are the H$_b$'s, because ortho (three-bond) coupling between H$_a$ and H$_b$ is different from the meta (four-bond) coupling between H$_a$ and H$_{b'}$ (or H$_b$ and H$_{a'}$). And in **C**, the hydrogens are not magnetically equivalent, nor are the fluorines, because $^3J_{trans}$ is greater than $^3J_{cis}$. □

The situation becomes more complicated when we consider molecules with multiple conformations. For example, look carefully at the structure of 1,1,1-trifluoroethane:

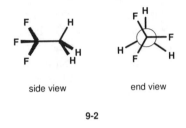

side view   end view

9-2

If this molecule were "frozen" in the conformation shown, the hydrogens would not be magnetically equivalent because their couplings to the three fluorines would involve three different dihedral angles and hence three different values of $^3J_{HF}$. Thus, we would label them H, H', H'', F, F', and F''. However, just as we saw with the methyl group of toluene (Section 4.2), there is rapid rotation around the C–C single bond, which averages all the H–F couplings (see Example 9.7). Thus, all three hydrogens are rendered magnetically equivalent on the time average, as are the three fluorines. But if this rotation were slowed or stopped, the effects of their magnetic nonequivalence would reemerge.

9.8   POPLE SPIN SYSTEM NOTATION

We are now ready to introduce a shorthand method developed by Pople for labeling spin-coupled systems. These labels will help us recognize some of the more common coupled multiplet patterns. Here is how it works. Assign a capital letter to each magnetic nucleus involved in the coupled system. Select letters whose alphabetic relationship reflects the relative chemical shifts of the various nuclei. Thus, for two nuclei of very different chemical shift (or different isotopes), pick A and X; for nuclei that have very similar (but not identical) chemical shifts, use A and B. For three nuclei of moderately different chemical shifts, use A, M, and X. (Just how much of a difference constitutes "very different" or "very similar" is discussed in the next section.) If two (or more) nuclei are chemically equivalent, give them the same letter. Finally, if two (or more) chemically equivalent nuclei are not also magnetically equivalent, distinguish them by adding a prime (or multiple primes).

Refer back to the structure of 1,1,1-trifluoroethane (**9-2**). Assuming rapid bond rotation, we would designate it as an

A_3X_3 spin system, where A_3 represents the three magnetically equivalent hydrogens and X_3 the three magnetically equivalent fluorines. Of course, if bond rotation were stopped, the system would become an AA′A″XX′X″ system, because the magnetic equivalence would have been lost.

■ **EXAMPLE 9.10**   Label the H/F spin system (neglecting carbon coupling) for structures **A**, **B**, and **C** in Example 9.9.

☐ *Solution:* **A**: AB$_2$C (all chemical shifts are similar); **B**: AA′BB′; **C**: AA′XX′.   ☐

■ **EXAMPLE 9.11**   Label the H/F spin system (neglecting carbon coupling) in *p*-fluoronitrobenzene (**9-1**).

☐ *Solution:* The completely correct answer is AA′BB′X. But since para (five-bond) couplings (A–B′ and A′–B) are essentially zero, we can regard the spin system as two equivalent (and superimposed) *ABX* systems.   ☐

But why go to the trouble of labeling spin systems? The answer is that spin systems with identical labels give spectra that are very similar in appearance. We can learn to recognize these recurring patterns and realize immediately what aspects of molecular structure generates them. For example, an ethoxy group (CH$_3$–CH$_2$–O–) is an example of an A_3X_2 spin system and always gives the familiar triplet (for three A nuclei split by two X nuclei) and quartet (for two X nuclei split by three A nuclei), as we saw in Section 8.1 and Example 8.18. See also Sections 9.11 and 9.12.

■ **EXAMPLE 9.12**   What type of spectra would you expect from the following spin systems: (a) AX; (b) AA′XX′; (c) AM$_2$X? (Assume all letters represent nuclei of the same isotope.)

☐ *Solution:* (a) Two doublets, one for A (split by X) and one for X (split by A). (b) The A signal will be split into a doublet of doublets by X and X′; the X signal will also be split into a doublet of doublets by A and A′.

(c) The A signal will be split into a triplet (by the two M nuclei) of doublets (by X); the M signal will be split into a doublet (by A) of doublets (by X), while the X signal will be split into a triplet (by the two M nuclei) of doublets (by A).   ☐

9.9 SLANTING MULTIPLETS AND SECOND-ORDER (STRONG COUPLING) EFFECTS

All of the coupling and multiplets we described in Chapter 8 result from what is called **first order** (or **weak**) coupling. Now we will introduce a complication.

Take a moment to compare the 250-MHz ^{1}H spectrum of ethyl cinnamate (Figure 8.13) with the 60-MHz spectrum of the same compound, shown in Figure 9.7. What differences do you see?

We expect the quartet lines in each spectrum to exhibit (from Pascal's triangle) an intensity ratio of 1 : 3 : 3 : 1, while the triplet lines should have a ratio of 1 : 2 : 1. Although the observed ratios match these expectations in Figure 8.13, there is an unmistakable trend in Figure 9.7. The higher field lines of the quartet are slightly more intense than the corresponding lower field lines, while the lowest field line of the triplet is slightly more intense than the highest field line. Another way of saying this is that the *inner* lines of both multiplets (those lines closer to the other coupled multiplet) are more intense than the *outer* lines. This asymmetry of the multiplets, sometimes referred to as **slanting**, canting, or leaning of the multiplets, is the result of perturbations in spin state populations caused by **second-order effects.** Such effects can either help or hinder the interpretation of complex spectra.

Here is one way that multiplet slanting can be helpful. Notice how, in Figure 9.7, the inner lines of the two vinyl hydrogen doublets are larger than the outer lines (though the inner line of the downfield doublet is partially obscured by the aromatic hydrogen signals). The two doublets are slanted *toward* each other, helping us to know where to look for the other multiplet(s) that are coupled to the multiplet of interest.

Why is the asymmetry very slight in the higher field spectrum and much more pronounced in the lower field spectrum? To answer that question, we need to examine the quantity $\Delta\,\delta v/J$, where $\Delta\,\delta v$ is the *difference* in chemical shift (measured in hertz; Section 5.1) between the two multiplets and *J* (in hertz) is the absolute value (magnitude without sign) of the coupling constant they share.

■ **EXAMPLE 9.13**   Referring again to Figures 8.13 and 9.7, calculate $\Delta\,\delta v/J$ for (a) the pair of doublets and (b) the quartet/triplet.

☐ *Solution:* (a) From the solution to Example 8.18, we know that the two doublets are centered at δ 6.44 and 7.69 ppm and that they share a ^3J value of 16 Hz. The spectrum was run at 250 MHz; therefore,

$$\Delta\,\delta v = (7.69 - 6.44)\ \text{ppm}\ (250\ \text{Hz/ppm}) = 312.5\ \text{Hz}$$

$$\frac{\Delta\,\delta v}{J} = 312.5\ \text{Hz}/16\ \text{Hz} = 19.5$$

In the 60-MHz spectrum $\Delta\,\delta v/J = 4.7$. (b) Again using the data from Example 8.18,

$$\Delta\,\delta v = (4.27 - ,.34)\ \text{ppm}\ (250\ \text{Hz/ppm}) = 732.5\ \text{Hz}$$

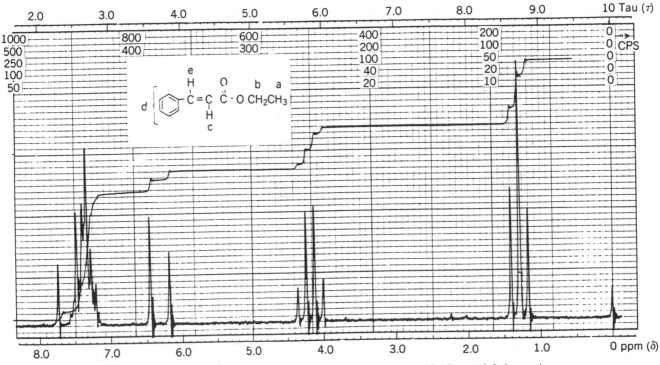

Figure 9.7. The 60-MHz ^{1}H spectrum of *trans*-ethyl cinnamate. © Sadtler Research Laboratories, Division of Bio-Rad Laboratories, Inc. (1966).

$$\frac{\Delta\,\delta\nu}{J} = 732.5\ \text{Hz}/7.1\ \text{Hz} = 10$$

In the 60-MHz spectrum $\Delta\,\delta\nu/J = 2.5$ □

From the above example, you can infer that the extent to which the intensity ratios depart from first-order (Pascal's triangle) expectations is a function of $\Delta\,\delta\nu/J$. If this ratio is large ($\Delta\,\delta\nu/J > 10$), we describe the spin system as **weakly coupled**, and the resulting multiplets will exhibit essentially first-order intensity ratios (as do the triplets and quartets in Figures 8.1 and 8.13). But as $\Delta\,\delta\nu/J$ decreases, second-order effects (multiplet slanting and even the appearance of "extra" lines) become increasingly apparent. Such a spin system is said to be **strongly coupled**.

In Chapter 5 we discussed the effect of increasing the operating frequency (and field strength) of an NMR spectrometer. You can now appreciate one of the major advantages of high-field instruments. From Sections 2.2 and 5.2 we know that $\Delta\,\delta\nu$ (in hertz) increases *linearly* with field strength, while (from Section 8.2) J remains unaffected. Thus, the ratio $\Delta\,\delta\nu/J$ also increases linearly with field strength, and coupled spectra become more nearly first order in appearance as the field strength increases.

For a more dramatic example of this effect, compare the 60-MHz (1.41-T) ^{1}H spectrum of I–CH$_2$–CH$_2$–CO$_2$H (Figure 9.8) with its (partial) 300-MHz (7.04-T) spectrum (Figure

9.9). Notice how the signals for the two coupled CH$_2$ groups (an A_2B_2 system) are transformed from a complex (but symmetrical) multiplet at 60 MHz (where $\Delta\,\delta\nu/J = 2.1$) into a pair of well-resolved triplets (as expected from first-order considerations for an A_2M_2 system) at 300 MHz ($\Delta\,\delta\nu/J = 10.5$). From Figure 9.8 you can also see that in strongly coupled systems (that is, where $\Delta\,\delta\nu/J$ is small), not only are the multiplets slanted, but extra lines also appear in the multiplets. We will talk more about this in the next three sections, but for now remember this: Strongly coupled systems lead to more complicated spectra than first-order rules (Chapter 8) predict.

■ **EXAMPLE 9.14** Explain why multiplets due to heteronuclear coupling exhibit no significant departure from first-order predictions, even when older low-field instruments were used.

□ *Solution:* Although heteronuclear coupling constants can be quite large (Table 9.2), the differences in chemical shift are huge. For example, consider the F–C–H system at field strength of 1.41 T (where the ^{1}H precessional frequency is 60 MHz and that of ^{19}F is 56.5 MHz). Even though $^2J_{HF}$ is 45 Hz, $\delta\nu$ at this field strength is 3.5 MHz (3,500,000 Hz), and the resulting $\Delta\,\delta\nu/J$ ratio is 78,000! First-order AX system behavior can be predicted with confidence.   □

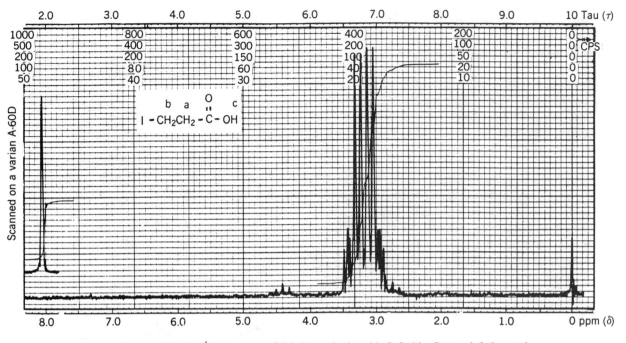

Figure 9.8. The 60-MHz ^{1}H spectrum of 3-iodopropionic acid. © Sadtler Research Laboratories. Division of Bio-Rad Laboratories, Inc. (1973).

■ **EXAMPLE 9.15** Consider the molecular structure of compound **9-3**:

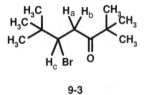

9-3

The 60-MHz ^{1}H spectrum of this compound shows, in addition to the two 9-hydrogen singlets for the $(CH_3)_3C-$ groups, a set of signals for H_a, H_b, and H_c. Their chemical shifts ($\delta\nu$, in hertz downfield from TMS) and coupling constants (in hertz) are $\delta\nu_a$, 163; $\delta\nu_b$, 195; $\delta\nu_c$, 263; $^2J_{ab}$, -18; $^3J_{ac}$, 2; and $^3J_{bc}$, 10. Here are the questions: (a) Why are H_a and H_b nonequivalent? (b) Calculate $\Delta \delta\nu/J$ for each coupling interaction. What type of spin system do these three nuclei constitute? (c) Using graph paper, draw the predicted first-order spectrum for the three hydrogens. (d) By successively "turning on" the slanting due to a–b, a–c, then b–c coupling, predict the final appearance of the spectrum, including all appropriate multiplet slanting. (e) What do the magnitudes of $^3J_{ac}$ and $^3J_{bc}$ indicate about the molecule's preferred conformation? (f) What would this spectrum look like at 250 MHz?

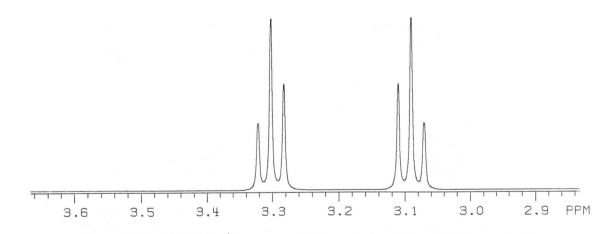

Figure 9.9. Partial 300-MHz ^{1}H spectrum of 3-iodopropionic acid (simulated), showing the two triplets characteristic of an A_2M_2 spectrum.

□ **Solution:** (a) Here, H_a and H_b are rendered diastereotopic (and hence nonequivalent) by the neighboring asymmetric center (the carbon bearing H_c; Section 4.3).

(b) $(\Delta\ \delta v/J)_{ab} = (195 - 163)$ Hz/18 Hz = 1.8 (strongly coupled); $(\Delta\ \delta v/J)_{ac} = (263 - 163)$ Hz/2 Hz = 50 (weakly coupled); and $(\Delta\ \delta v/J)_{bc} = (263 - 195)$ Hz/10 Hz = 6.8 (moderately coupled). Thus, we would designate this system either ABM or ABX.

(c) Under first-order conditions each of the hydrogens would give rise to a doublet of doublets. The position of each of the 12 equally intense lines can be predicted from the values of δv and J, as in Example 8.11. (The term $|J|$ indicates the absolute value of J; the first four lines are due to H_c, the next four to H_b, and the last four to H_a.)

Line	Position (Hz)
1	$\delta v_c + (|J_{bc}| + |J_{ac}|)/2 = 269$
2	$\delta v_c + (|J_{bc}| - |J_{ac}|)/2 = 267$
3	$\delta v_c - (|J_{bc}| - |J_{ac}|)/2 = 259$
4	$\delta v_c - (|J_{bc}| + |J_{ac}|)/2 = 257$
5	$\delta v_b + (|J_{ab}| + |J_{bc}|)/2 = 209$
6	$\delta v_b + (J_{ab}| - |J_{bc}|)/2 = 199$
7	$\delta v_b - (|J_{ab}| - |J_{bc}|)/2 = 191$
8	$\delta v_b - (|J_{ab}| + |J_{bc}|)/2 = 181$
9	$\delta v_a + (|J_{ab}| + |J_{ac}|)/2 = 173$

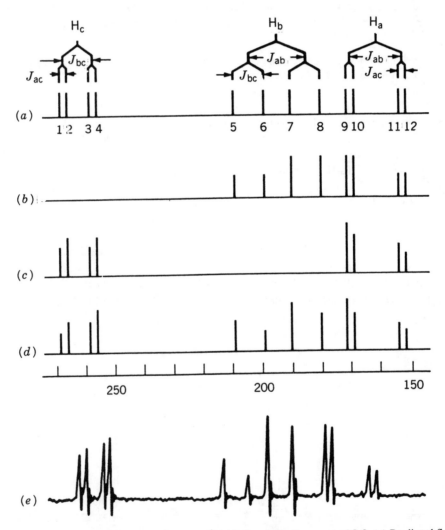

Figure 9.10. Predicted and observed 60-MHz ^{1}H ABC spectra of compound **9-3**. (*a*) Predicted first order spectrum; (*b*)–(*d*) multiplet slanting due to second-order effects; (*e*) observed spectrum [From "Prediction of the Appearance of Non-First-Order Proton NMR Spectra," by R. S. Macomber, *Journal of Chemical Education*, *60*, 525 (1983). Reprinted by permission.

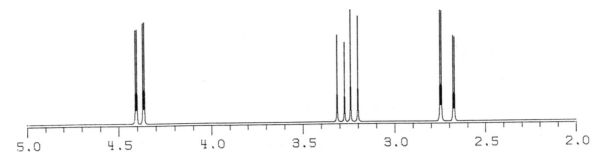

Figure 9.11. The 250-MHz ^{1}H ABC spectrum (simulated) of compound **9-3**.

Line	Position (Hz)
10	$\delta v_a + (|J_{ab}| - |J_{ac}|)/2 = 171$
11	$\delta v_a - (|J_{ab}| - |J_{ac}|)/2 = 155$
12	$\delta v_a - (|J_{ab}| + |J_{ac}|)/2 = 153$

These lines are shown in Figure 9.10*a*.

(d) Figure 9.10*b* shows the slanting effects due to strong H_a–H_b coupling: The inner lines of the two coupled multiplets (lines 7–10) increase in intensity at the expense of the outer lines (5, 6, 11, and 12). Figure 9.10*c* adds slight slanting due to the weak H_a–H_c coupling: Inner lines 2, 4, 9, and 11 increase at the expense of outer lines 1, 3, 10, and 12. Finally, Figure 9.10*d* adds the effects of moderate H_b–H_c coupling: Inner lines 3–5 and 7 increase; outer lines 1, 2, 6, and 8 decrease. Note how well this qualitative approach predicts the relative intensities observed in the actual spectrum; see Figure 9.10*e*.

(e) Comparing the two ^3J values with the Karplus relation [Figure 9.6 or Eq. (9.4)], we note that $^3J_{ac}$ (2 Hz) corresponds to a dihedral angle of 90°, while $^3J_{bc}$ (10 Hz) could correspond to either 20° or 150°. By redrawing the structure in end view, we can see that only the 150° angle is possible:

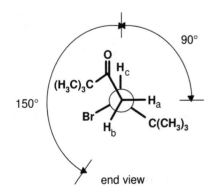

end view

(f) At 250 MHz the $\Delta\,\delta v/J$ values become $(\Delta\,\delta v/J)_{ab} = 7.5$ (moderately coupled), $(\Delta\,\delta v/J)_{ac} = 208$ (weakly coupled), and $(\Delta\,\delta v/J)_{bc} = 28$ (weakly coupled). The resulting spectrum of this *AMX* system, three doublets of doublets, is shown in Figure 9.11. ☐

Having completed the above example, you can appreciate just how complicated a spectrum can become, even with relatively few coupled nuclei. What recourse do we have when a spectrum is so complex that first-order analysis is virtually impossible?

9.10 CALCULATED SPECTRA

Quantum mechanics again comes to our rescue! If we know the values of the chemical shifts and coupling constants for a given structure, it is possible to calculate the exact position and intensity of every line in each of its NMR spectra. This requires the simultaneous solution of all the quantum-mechanical wave equations that describe the spin systems. For all but the simplest systems, such calculations require enough number crunching that they must be performed by a computer. One program for this purpose is LAOCOON,[5] which requires estimates of all δ and *J* values. Advanced Chemistry Development (Toronto, Canada) now markets software for calculating ^{1}H and ^{13}C spectra given only the *structure* of a molecule.

But suppose you have a very complicated spectrum and only a crude idea of the chemical shifts and coupling constants. (Certainly, after reading Chapters 6–9, you should have at least a *crude* idea!) No problem. You simply input your estimates along with the observed position of each line in the actual spectrum. The program iteratively matches calculated versus observed line positions, then varies the input data, and keeps repeating the sequence until a best (least-squares) fit is found. The resulting calculated values of all chemical shifts and coupling constants can be listed and the corresponding spectrum plotted.

The results of such calculations are impressive, often giving calculated spectra that are superimposable on the observed spectra. For example, Figure 9.12 shows the calculated 60-

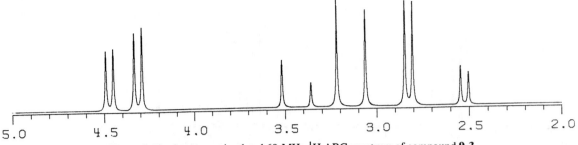

Figure 9.12. Computer-simulated 60-MHz ^{1}H ABC spectrum of compound **9-3**.

MHz ^{1}H spectrum of compound **9-3** in Example 9.15. Figures 9.13*a* and *b* show, respectively, the computer-generated 60- and 250-MHz ^{1}H spectra of structure **8-3** (Example 8.14), resulting from these input data:

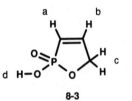

8-3

H	δν (Hz)	J values (Hz)
a	368	P–H$_a$, 33; H$_a$–H$_b$, 8.5
b	399	P–H$_b$, 49; H$_b$–H$_c$, 0
c	294	P–H$_c$, 5.5; H$_b$–H$_c$, 2
d	720 (not shown)	

(All couplings to H$_d$ are prevented by exchange processes.) The chemical shift of the phosphorus (which would not be

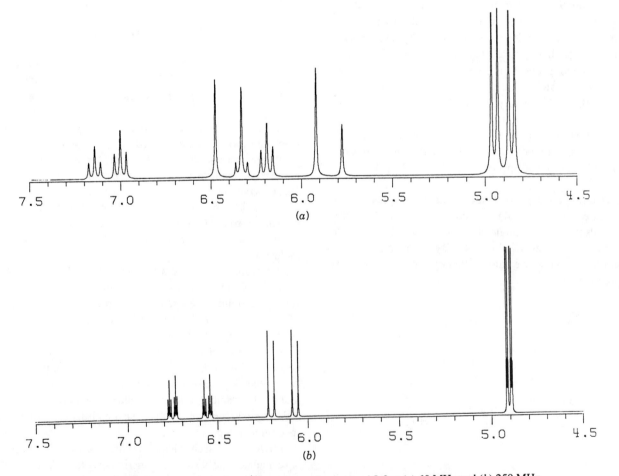

Figure 9.13. Computer-simulated ^{1}H spectrum of compound **8-3** at (*a*) 60 MHz and (*b*) 250 MHz.

part of the ^{1}H spectrum) can be set to any off-scale value. However, you must enter some value for $\delta\nu_P$ (e.g., 2000 Hz); otherwise the program will not include P–H couplings in its calculations. Notice once again that the 250-MHz spectrum shows all the signals as first-order multiplets (an A_2MX system), while in the 60-MHz spectrum many of the multiplets overlap and are unrecognizable (an A_2BC system).

9.11 THE AX → AB → A₂ CONTINUUM

A detailed examination of the calculated spectra for a host of common Pople spin systems has been published.[6] Still, it is instructive to examine certain aspects of the simplest coupled systems, those of two and three spins.

Let us start by considering a molecule with two coupled nuclei (A and B) of the same isotope (e.g., ^{1}H). There are three independent variables that describe the system completely: the chemical shifts (δ or $\delta\nu$) of A and B and their homonuclear coupling constant J. The exact appearance of the NMR spectrum for this system, that is, the position and intensity of each line, can be calculated from the values of these three variables (and the operating frequency of the instrument if δ values are used). The general solution for the two-spin system is a four-line spectrum, with each line having the position and intensity listed below:

Line	$\delta\nu$ (Hz downfield of TMS)	Relative Intensity
1	$\nu_{av} + C + (J/2)$	$1 - (J/2C)$
2	$\nu_{av} + C - (J/2)$	$1 + (J/2C)$
3	$\nu_{av} - C + (J/2)$	$1 + (J/2C)$
4	$\nu_{av} - C - (J/2)$	$1 - (J/2C)$

where $\nu_{av} = (\delta\nu_A + \delta\nu_B)/2$; $\Delta\delta\nu = \delta\nu_A - \delta\nu_B$; and $C = (1/2)[(\Delta\delta\nu)^2 + J^2]^{1/2}$

A graphical representation of the general solution for the AB system is shown in Figure 9.14b. The most significant features to remember are (1) the AB spectrum is symmetrical around its midpoint (ν_{av}), with the inner lines larger and the outer lines smaller; (2) the doublets are centered not at δ_A and δ_B but rather at $\nu_{av} \pm C$. The latter feature is what makes the measurement of exact chemical shifts difficult in spectra that show second-order effects.

■ **EXAMPLE 9.16** Consider two extremes of the two-spin system. Predict the appearance of the spectrum (a) when $\Delta\delta\nu \gg J$ and (b) when $\Delta\delta\nu = 0$.

□ *Solution:* (a) Because $\Delta\delta\nu \gg J$, $\Delta\delta\nu/J$ is very large. This is the weakly coupled limit, an example of an *AX* spectrum. In this case $C \approx (1/2)[(\Delta\delta\nu)^2]^{1/2} = \Delta\delta\nu/2$, and $J/2C = J/\Delta\delta\nu \approx 0$. Substituting these values for C and

$J/2C$ into the general solution gives a spectrum described by the following:

Line	$\delta\nu$ (Hz downfield of TMS)	Relative Intensity
1	$\nu_{av} + \Delta\delta\nu/2 + (J/2)$	1
2	$\nu_{av} + \Delta\delta\nu/2 - (J/2)$	1
3	$\nu_{av} - \Delta\delta\nu/2 + (J/2)$	1
4	$\nu_{av} - \Delta\delta\nu/2 - (J/2)$	1

This result, shown in Figure 9.14a, is exactly the two doublets we would have expected from the first-order ($n + 1$) rule.

(b) If $\Delta\delta\nu = 0$, nuclei A and B have the same chemical shift and so are chemically equivalent, making this an A_2 spin system. In this case, $C = (1/2)[J^2]^{1/2} = J/2$, and $J/2C = 1$. Notice how these values affect the position and especially the intensity of each line:

Line	$\delta\nu$ (Hz downfield of TMS)	Relative Intensity
1	$\nu_{av} + J$	0
2	ν_{av}	1
3	ν_{av}	1
4	$\nu_{av} - J$	0

This result, shown in Figure 9.14c, has lines 2 and 3 superimposed at ν_{av} (= $\delta\nu_A = \delta\nu_B$), while outer lines 1 and 4 have zero intensity. Now you see why coupling between equivalent nuclei, though it does occur, is not observed in the spectrum (Section 8.4). □

■ **EXAMPLE 9.17** Suppose there were an AB spectrum in which the spacing between lines 2 and 3 was accidentally equal to J. (a) What would be the position and intensity of each line? (b) What would be the value of $\Delta\delta\nu/J$? (c) How would this spectrum compare with a first-order quartet?

□ *Solution:* (a) From either Figure 9.14b or the general equations for the two-spin system, we can see that the separation between lines 2 and 3 is $2C - J$. But in this case we have arbitrarily set this separation equal to J. Therefore, $2C - J = J$, or $C = J$. Using this relation, we can calculate the position and intensity of each line:

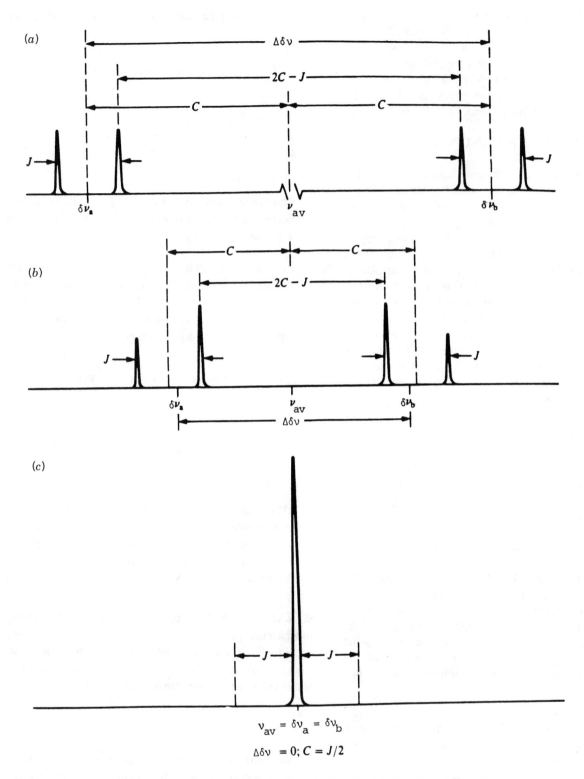

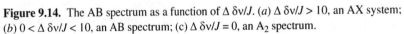

Figure 9.14. The AB spectrum as a function of $\Delta \delta \nu / J$. (*a*) $\Delta \delta \nu / J > 10$, an AX system; (*b*) $0 < \Delta \delta \nu / J < 10$, an AB spectrum; (c) $\Delta \delta \nu / J = 0$, an A_2 spectrum.

Line	δν (Hz downfield of TMS)	Relative Intensity
1	$\nu_{av} + 3J/2$	$\frac{1}{2}$
2	$\nu_{av} + J/2$	$\frac{3}{2}$
3	$\nu_{av} - J/2$	$\frac{3}{2}$
4	$\nu_{av} - 3J/2$	$\frac{1}{2}$

(b) From part (a), we know that $C = J$. Therefore,

$$J = C = \tfrac{1}{2}[(\Delta\,\delta\nu)^2 + J^2]^{1/2}$$

$$4J^2 = [(\Delta\,\delta\nu)^2 + J^2]$$

$$\frac{\Delta\,\delta\nu}{J} = \sqrt{3}$$

(c) The result in (a), shown in Figure 9.15, is identical to a first-order quartet with the intensity ratios predicted by Pascal's triangle. Are you wondering how to tell the difference between a quartet and an **AB quartet**? There are two ways. If it is a "real" quartet, there are three other equivalent $I = \frac{1}{2}$ nuclei somewhere in the molecule responsible for the coupling; try to find them. If they do not appear in the spectrum at hand (they may be heteroatoms such as F, P, etc.), try regenerating the spectrum at higher field strength. A real quartet will be unaffected, whereas in an *AB* quartet the spacing (in hertz) between lines 2 and 3 will increase while the spacing (in hertz) between lines 1 and 2 and between lines 3 and 4 (J) will remain constant (Section 9.9).   ☐

■ **EXAMPLE 9.18**   The values of how many parameters are needed to calculate the appearance of an *AA′BB′* spectrum?

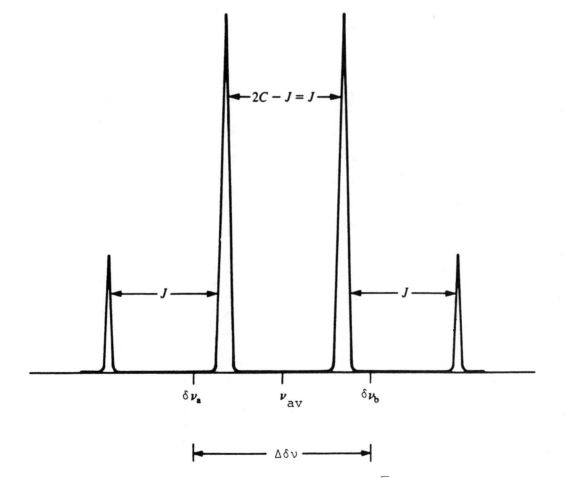

Figure 9.15. An AB quartet with $\Delta\,\delta\nu/J = \sqrt{3}$.

□ *Solution:* Two chemical shifts, for A (=A′) and B (=B′), and four coupling constants, J_{A-B}, $J_{A'-B'}$, $J_{A'-B}$, and $J_{A-B'}$. Often, fewer coupling constants are needed, either because $J_{A-B} = J_{A'-B'}$ and $J_{A'-B} = J_{A-B'}$ or because $J_{A'-B} = J_{A-B'} = 0$.   □

9.12   MORE ABOUT THE ABX SYSTEM: DECEPTIVE SIMPLICITY AND VIRTUAL COUPLING

Before we leave the topic of second-order effects, we revisit the three-spin system previously introduced in Example 9.15. There we established that if the chemical shifts of the three nuclei in a structure such as **9-3** are sufficiently different (i.e., an AMX system), the spectrum will consist of 12 lines: 3 (first-order) doublets of doublets, each one centered at the appropriate chemical shift and exhibiting line spacings equal to the appropriate coupling constants.

Let us now consider the consequences of changing some of the relationships between the nuclei. Suppose we were to make two of the nuclei magnetically equivalent to give an A_2X system. The resulting spectrum is determined by just three parameters: the chemical shifts of nuclei A and X and the coupling constant between them (J_{AX}); any coupling between the two equivalent A nuclei would not affect the appearance

of the spectrum (Section 8.4). The spectrum would exhibit just two signals, a doublet (integral = 2) for the A nuclei and a triplet (integral = 1) for the X nucleus, both with line spacings equal to J_{AX}.

Suppose we now make the two A nuclei magnetically nonequivalent but still chemically equivalent, giving an AA′X system. We still have just two chemical shifts, but there are now two coupling constants, J_{AX} and $J_{A'X}$. We might therefore expect the X signal to be split into a doublet of doublets, but computer simulation shows that the X signal is still a triplet. More importantly, the line spacings in this triplet (the apparent J value) are neither J_{AX} nor $J_{A'X}$, but in fact the *average* of the two. If you encountered such a spectrum without knowing the actual structure of the compound, you would likely be misled into thinking it represented a simple A_2X system. Such a spectrum is described as **deceptively simple**, and this averaging of coupling constants is another manifestation of second-order effects.

Now let the two A nuclei become very slightly nonequivalent, to give an ABX system. We will assume A and B are strongly coupled to each other, but only A is coupled to X. That is, J_{BX} is zero. Our first-order expectation is that the X signal should now be a doublet, with line spacing J_{AX}. But once again the computer simulation shows otherwise. The X signal is still a doublet of doublets, with one larger "apparent"

TABLE 9.7   Simulated Spectra for Two ABX Spectra Differing Only in the Sign of One *J* Value[a]

J_{AX} = +3 Hz			J_{AX} = −3 Hz		
Position (Hz)	Intensity	Apparent J	Position (Hz)	Intensity	Apparent J
286.1	0.33		286.5	0.38	
292.5	0.25		292.0	0.18	
		15.0, 6.4			15.0, 5.5
301.1	1.63		301.5	1.58	
307.5	1.79		307.0	1.85	
308.8	1.64		310.0	1.83	
312.4	1.78		311.5	1.60	
		15.0, 3.6			15.0, 1.5
323.8	0.34		325.0	0.16	
327.4	0.24		326.5	0.41	
			578.5	0.03	
595.0	1.03		596.5	1.00	
598.7	1.00		598.0	1.01	
		6.4, 3.6			5.5, 1.5
601.4	0.99		602.0	0.99	
605.0	0.97		603.5	0.95	
			621.6	0.02	

[a] $\delta_A = 1.00$, $\delta_B = 1.05$, $\delta_C = 2.00$, $J_{AB} = 15$ Hz, $J_{BX} = 7$ Hz.

coupling constant (just slightly smaller than J_{AX}), and a smaller apparent coupling constant, seemingly indicative of weak coupling between B and X. The sum of these two apparent coupling constants equals the true magnitude of J_{AX}. Actually, the extra splitting, called **virtual coupling**, is not due to any direct interaction between nuclei B and X, but rather to the second order effect that nucleus B has on the spin states of nucleus A, which lifts the degeneracy of lines that were previously superimposed.

It was stated in Section 9.1 that although the absolute sign of a coupling constant does not affect the appearance of an NMR multiplet, the *relative* signs of two coupling constants that are part of the same spin system *can* have an observable effect. In Table 9.7*a* and *b* are listed the position and intensity of each line in two closely related 300-MHz ABX spectra, differing only in the sign of the smallest of the three coupling constants. The input data are $\delta_A = 1.00$, $\delta_B = 1.05$, $\delta_C = 2.00$, $J_{AB} = 15$ Hz, $J_{BX} = 7$ Hz, and $J_{AX} = \pm 3$ Hz. Take a moment to compare the two sets of line positions, intensities, and apparent J values.

Notice that when J_{AX} and J_{BX} have the *same* sign, there are twelve lines but the apparent J values differ significantly from the "true" (i.e., input) values. When J_{AX} and J_{BX} have *opposite* signs, all positions, intensities, and apparent J values change, and there are even some new (but very weak) lines shown in boldface. The message is to be aware of these potential pitfalls whenever you are dealing with signals where the ratio $\Delta \delta\nu/J$ is small.

Now that we have spent two chapters describing spin–spin coupling, we will find in Chapter 12 ways to get rid of the effects of coupling!

CHAPTER SUMMARY

1. Nuclear spin coupling constants (J values) are either positive or negative. If the value of J is positive, the antiparallel arrangement of nuclear spins of the coupled nuclei is lower in energy than the parallel arrangement; if the value of J is negative, the parallel arrangement is lower in energy than the antiparallel arrangement.

2. In general, the signs of HH, CH, and CC coupling constants are a function of the number of intervening bonds: positive if the number of bonds is odd, negative if the number of bonds is even. In the case of coupling to atoms with unshared pairs, the sign of J is often reversed from expectations based on number of intervening bonds.

3. Normally, the sign of a coupling constant has no effect on the appearance of the NMR spectrum.

4. The magnitude of coupling constants is dependent on many factors [Eq. (9.1)]:

    a. the number of intervening bonds (J normally decreases as the number of bonds increases);

    b. the product of the magnetogyric ratios of the coupled nuclei;

    c. the fraction of s character (f) of the hybrid orbitals connecting the nuclei;

    d. for geminal (two-bond) coupling constants, the internuclear angle: J increases (becomes more positive) as the angle increases;

    e. for vicinal (three-bond) coupling constants, the dihedral angle [Eq. (9-4)]; and

    f. for long-range coupling constants (over more than three bonds), the number of intervening multiple bonds and the geometric relationship between the orbitals involved.

5. Two or more nuclei are magnetically equivalent if they are chemically equivalent (i.e., possess the same chemical shift) and are equally coupled to any other nucleus.

6. Spin systems (collections of interacting nuclei) are often labeled (Pople notation) by assigning a letter from the alphabet to each set of magnetically equivalent nuclei. Nuclei that are close (but not identical) in chemical shift are given letters that are close in the alphabet (e.g., AB). Two nuclei that are chemically equivalent but not magnetically equivalent are assigned the same letter but one letter is primed (e.g., AA′).

7. As the ratio of $\Delta \delta\nu$ (the difference in chemical shift between two coupled nuclei) to J decreases, the relative intensities of the lines in a multiplet deviate further from first-order (e.g., Pascal triangle) ratios. Inner lines (those facing the coupled multiplet) increase in intensity, while outer lines lose intensity. This slanting of the multiplets is one type of second-order effect. At very small values of $\Delta \delta\nu/J$, not only may extra lines appear in the multiplets but also apparent line positions and spacings may not equate with true chemical shifts and coupling constants (e.g., deceptive simplicity and virtual coupling).

8. Computer software exists that can calculate the exact position and intensity of each line in an NMR spectrum for which all δ and J values are known (or can be estimated). Such programs can also extract exact δ and J values from an observed spectrum by iterative data fitting.

REFERENCES

[1]Becker, E. D., *High Resolution NMR*, 2nd ed., Academic, New York, 1980.

[2]Lazlo, P., and Stang, P. J., *Organic Spectroscopy*, Harper & Row, New York, 1971.

[3]J. G. and Quin, L. D., Ed., *Phosphorus-31 NMR Spectroscopy in Stereochemical Analysis*. Organic Compounds and Metal Complexes, Verkade, VCH Publishers, Deerfield Beach, FL, 1987.

[4]Silverstein, R. M., Bassler, G. C., and Morrill, T. C., *Spectrometric Identification of Organic Compounds*, 5th ed., Wiley, New York, 1991.

[5]S. Castellano and A. A. Bothner-By, available from the Quantum Chemistry Program Exchange at Indiana University, Bloomington, IN 47401.

[6]Wiberg, K. B., and Nist, B. J., *Identification of NMR Spectra*, Benjamin, New York, 1962.

REVIEW PROBLEMS (Answers in Appendix 1)

9.1   Use Pople notation to identify the types of spin systems that give rise to the spectra shown in Figures 8.20*a* and *b* (review problem 8.7).

9.2   Use Pople notation to identify the type of hydrogen/phosphorus spin system in the structure below:

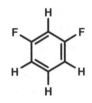

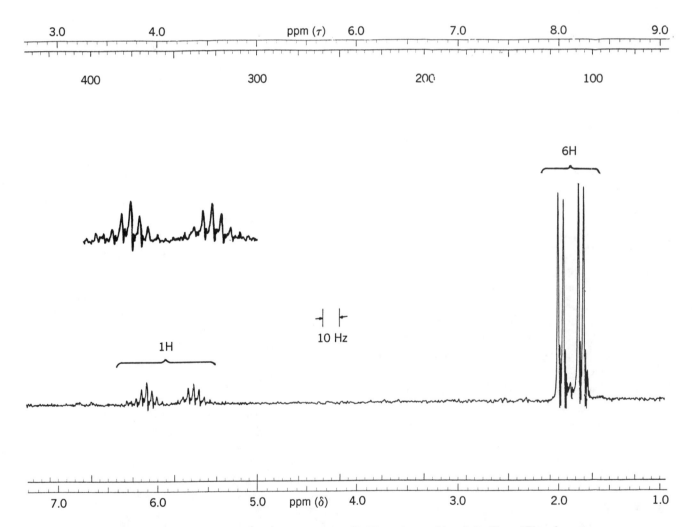

Figure 9.16. The 60-MHz ^{1}H spectrum of $C_5H_7Cl_2OP$ (review problem 9.4). [From "Phosphorus Coupling in ^{13}C and ^{1}H NMR," by R. S. Macomber, *Journal of Chemical Education*, *56*, 109 (1979). Reprinted by permission.

9.3 Referring to review problem 8.8, explain why the cyclo-propyl hydrogen (H_c) appears as a doublet rather than a triplet or doublet of doublets.

9.4 (a) From its 60-MHz ^{1}H spectrum (Figure 9.16), identify the compound whose molecular formula is $C_5H_7Cl_2OP$. *Hint*: The ^{31}P chemical shift is consistent with a dichlorophosphonyl ($O{=}PCl_2$) group. (b) Account for the chemical shift and multiplicity of each signal, and report the value of each coupling constant. (c) Predict the multiplicity of the ^{31}P signal.

9.5 (a) From the 20-MHz ^{13}C spectroscopic data below, identify the compound whose molecular formula is $C_6H_{11}O_3P$. *Hint*: The ^{31}P chemical shift is consistent with a phosphonate [$O{=}P(OR)_2$] group:

Line	δ (ppm)	Intensity	Line	δ (ppm)	Intensity
1	26.6	70	6	85.5	9
2	27.4	72	7	110.5	24
3	52.7	12	8	118.5	20
4	52.9	14	9	156.4	40
5	85.0	7	10	157.2	30

(b) Account for the chemical shift and multiplicity of each signal, and report the value of each coupling constant. (c) Predict the multiplicity of the ^{31}P signal.

10

THE STUDY OF DYNAMIC PROCESSES BY NMR

10.1 REVERSIBLE AND IRREVERSIBLE DYNAMIC PROCESSES

In chemistry the term **dynamic** implies *changing with time*. Dynamic physical and chemical processes are of two basic types, irreversible and reversible.

Irreversible processes are normally encountered in the context of chemical reactions. When the **free energy** ($G°$) of the product(s) of a reaction is sufficiently lower than the free energy of the reactant(s), the final (equilibrium) reaction mixture will comprise essentially all product(s) and no reactant(s). For example, a free-energy difference as little as 5 kcal/mol (ca. 21 kJ/mol) means the equilibrium mixture will consist of >99.97% products at 25°C [Eq. (2.8)]. In such cases, we can regard the reaction as essentially irreversible, that is, going in only one direction, or "going to completion."

If the *rate* of an irreversible reaction is slow enough that there is no significant change in the composition of the reaction mixture during the time it takes to acquire a spectrum, the reaction can be monitored by NMR, collecting spectra at regular intervals. It is essential that the temperature of the reaction mixture be held constant (±0.1°C) to generate precise kinetic data. Subsequent Fourier transformation followed by integration of the appropriate signals allows us to follow either the disappearance of reactant(s) or the appearance of product(s) as a function of time. Such composition-versus-time data can then be fit to appropriate rate laws to determine rate constants, half-lives, and activation parameters for reaction. Because collection of suitable FID data requires a minute or so for a typical ^1H spectrum, the *half-life* of an irreversible reaction should be at least on the order of several minutes to be monitored by NMR techniques.

When studying irreversible reactions by NMR, it is often useful to include an **internal standard**, an inert substance added to the reaction mixture at the beginning of the reaction, that generates a signal whose intensity is constant during the reaction. The intensity of a reactant (or product) signal can then be expressed as a *ratio* of that signal's integral to the integral of the internal standard signal. This ratio technique assures that any changes in signal intensity are due to decreases (or increases) in the amount of reactant (or product) and not to some inadvertent experimental variable (e.g., a change in sample volume or the instability of a reactant or product). It is even possible to use the reference signal (e.g., TMS) as an internal standard, provided that the sample tube is well sealed.

In Chapter 11 we will discuss a special class of irreversible reactions that involve radical-pair intermediates. Such reactions give rise to some very odd-looking NMR spectra.

A **reversible** process is one in which a molecule (or set of molecules) changes back and forth between two (or more) different structures (e.g., different conformations, different stereoisomers, even different structural isomers), forming an equilibrium mixture of both (or all) the structures. Recall that although the concentration of each component of an equilibrium mixture does not change with time, equilibrium is nonetheless dynamic because interconversion between the components continues at rates that preserve the composition of the mixture.

Virtually all of the molecular structures we have encountered in this book have dynamic aspects to their structures in the sense that there is rapid interconversion between all possible conformations (Section 4.2) of each molecule. As we shall see in the next few sections, many types of both physical and chemical processes occur reversibly, at rates that range from much slower than the NMR time scale (Section 1.4) to much faster. These processes are often ideally suited to study by NMR methods.

10.2 REVERSIBLE INTRAMOLECULAR PROCESSES INVOLVING ROTATION AROUND BONDS

Molecules are constantly undergoing many types of motion. One type is **translation**, motion in which the molecule as a whole (including its center of mass) changes position. In addition, there is **rotation** of the entire molecule around its center of mass and **vibration** (alternate stretching and compression) of each bond. There is also **internal rotation** (or **torsion**), rotation around individual single bonds that converts one conformation into another. (Double bonds cannot rotate without breaking the π bond, a prohibitively costly process energetically.) These types of motion are considered *physical* processes because they do not involve making or breaking chemical bonds or a change in the sequence (connectivity) of atoms in the molecule. For the most part, these processes are much faster than the NMR time scale, so an NMR spectrum normally exhibits signals with chemical shifts and coupling constants that are averaged over all possible conformations. We previously discussed the manifestations of this averaging in Section 4.2 when we examined the hydrogens of a methyl group attached to a benzene ring.

However, there are many molecules where rotation around a particular single bond is restricted, that is, prevented from rotating freely. If rotation is much slower than the NMR time scale, the molecule is essentially "locked" in one conformation. One example of such a compound is *N,N*-dimethylformamide (DMF, **10-1**), whose NMR data are given in Table 8.1:

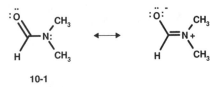

10-1

Because of resonance (Section 7.4) between the unshared electron pair on nitrogen and the C=O bond, there is some "double-bond character" to the formally single C–N bond. At room temperature there is not enough average thermal energy available to overcome the double-bond character. For this reason, rotation around the bond is restricted, and the two methyl groups are rendered nonequivalent: one is cis to the oxygen, the other is trans. At this **slow-exchange limit** each methyl group exhibits its own signal (^{1}H δ 2.74 and 2.91, ^{13}C δ 30.1 and 35.2).

As the temperature of a collection of molecules increases, the molecules acquire more thermal energy, and the rates of physical processes such as translation, rotation, and torsion increase. At sufficiently high temperature, the rate of rotation of the C–N bond in DMF becomes significant, bringing about **reversible exchange** of the two methyl groups: The cis methyl exchanges position with the trans:

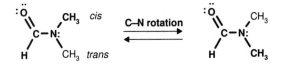

By the time DMF reaches 130°C, rotation becomes fast on the NMR time scale (the **fast-exchange limit**) and the two methyl signals are "averaged" to give one sharp signal midway between the two slow-exchange signals (^{1}H δ 2.825, ^{13}C δ 32.65).[1]

10.3 SIMPLE TWO-SITE INTRAMOLECULAR EXCHANGE

Just as we can calculate the appearance of an NMR spectrum if all the chemical shifts and coupling constants are known (Section 9.10), we can accurately predict the appearance of a spectrum of a molecule undergoing dynamic reversible exchange if we know the following:

1. the chemical shifts and coupling constants for each of the exchanging structures at the slow-exchange limit and, from this, the difference in resonance frequency ($\Delta \nu_{0} = \nu_{A} - \nu_{B}$) of the exchanging signals;

2. the populations (relative amounts) of the exchanging structures (or sites); and

3. the **rate constants** (k) for interconversion of the various structures.

For exchange between two *equally* populated sites (as with the methyl groups in DMF; Section 10.2), the **lineshape equation** [Eq. (10.1)] that describes NMR signal intensity [$I(\nu)$] at each point along the frequency (ν) axis[2] is:

$$I(\nu) = \frac{Ck(\Delta \nu_{0})^{2}}{k^{2}(\nu_{av} - \nu)^{2} + 4\pi^{2}(\nu_{A} - \nu)^{2}(\nu_{B} - \nu)^{2}} \tag{10.1}$$

where $\nu_{av} = (\nu_{A} + \nu_{B})/2$ and the scaling constant C has units of hertz. Equation (10.1) neglects the contribution of effective transverse relaxation (T_{2}^{*}; Section 3.5.1) to the linewidth of each signal: $\nu_{1/2}^{0} = 1/T_{2}^{*}$.[3] The individual rate constants k for exchange from one site to another are inversely related to the **lifetime** (τ) of each site (the average time that a specific nucleus occupies that site):

$$\tau_{A} = (k_{A\to B})^{-1} \qquad \tau_{B} = (k_{b\to A})^{-1} \qquad \tau = (k)^{-1}$$

$$\text{where} \quad k = k_{A\to B} + k_{B\to A} \tag{10.2}$$

If the two sites are equally populated, both individual rate constants are equal, as are both lifetimes, and the total lifetime

(τ) is one-half of τ_A (or τ_B). Notice also that *slow* processes are ones with *small* rate constants and long lifetimes.

At this point we can be more explicit about what we mean by "NMR time scale." We will define the dimensionless **exchange ratio** (R) as the total rate constant (k, in reciprocal seconds) divided by the difference in frequency between two exchanging signals (Δv_0, in hertz):

$$R = \frac{k}{\Delta v_0} \qquad (10.3)$$

A *large* value of R ($k \gg \Delta v_0$) indicates *fast* exchange, while a *small* value ($k \ll \Delta v_0$) indicates *slow* exchange.

Let us investigate what happens to the NMR spectrum of DMF as we increase the exchange ratio from the slow-exchange limit to the fast-exchange limit.

1. When R is less than ~0.1, the system is essentially "frozen" at the slow-exchange limit, and Eq. (10.1) predicts two sharp signals, one at v_A, the other at v_B (Figure 10.1, $R = 0.10$). The halfwidth of these signals ($v^0_{1/2}$ is the inverse of the effective spin–spin relaxation time.

2. As the value of R increases (by increasing the temperature to increase k), the signals broaden and move closer together (Figure 10.1; $R = 2.66$). The separation (Δv) between these broadened signals during slow exchange is given by

$$\Delta v = \Delta v_0 \sqrt{1 - \frac{R^2}{2\pi^2}} = \sqrt{\Delta_0^2 - \frac{k^2}{2\pi^2}} \qquad (10.4)$$

■ **EXAMPLE 10.1**   (a) Rearrange Eq. (10.4) to express the slow-exchange rate constant k as a function of Δv and Δv_0. (b) What is the value of Δv_0 for the 250-MHz ^{1}H spectrum of DMF? (c) At a certain temperature the 250-MHz ^{1}H spectrum of DMF exhibits two broad signals with a separation of 34.0 Hz. Calculate the values of k and R at this temperature.

□ *Solution:*   (a)

$$\Delta v = \Delta v_0 \sqrt{1 - \frac{k^2}{2\pi^2 \Delta v_0^2}} \qquad (10.4')$$

$$\frac{k^2}{2\pi^2 \Delta v_0^2} = 1 - \left(\frac{\Delta v}{\Delta v_0}\right)^2 = \frac{\Delta v_0^2 - \Delta v^2}{\Delta v_0^2}$$

$$k^2 = 2\pi^2 (\Delta v_0^2 - \Delta v^2)$$

$$k = \pi\sqrt{2(\Delta v_0^2 - \Delta v^2)} \qquad (10.5)$$

(b) At 250 MHz, $\Delta v_0 = (2.91 - 2.74$ ppm$)$ (250 Hz/ppm) $= 42.5$ Hz. Using Eq. (10.5)

$$k = \pi\sqrt{2[(42.5)^2 - (34.0)^2]} = 113 \text{ Hz} = 113 \text{ s}^{-1}$$

From Eq. (10.3)

$$R = \frac{k}{\Delta v_0} = \frac{113}{42.5} = 2.66 \qquad \square$$

3. At a higher value of R, the two peaks **coalesce** into a single peak centered at v_{av} (Figure 10.1, $R = 4.44$). However, this peak is sometimes so broad that it can be difficult to distinguish from background noise.

■ **EXAMPLE 10.2** (a) In terms of Δv_0, calculate the values of k and R at coalescence. (b) At what value of k will the methyl signals in the 250-MHz ^{1}H spectrum of DMF coalesce?

□ **Solution:**(a) By the definition of coalescence, Δv must equal zero. Substituting this value into Eq. (10.5), we find that k_c (the value of k at coalescence) is given by

$$k_c = \pi[2(\Delta v_0^2)]^{1/2} = 4.44 \Delta v_0 \qquad (10.6)$$

and therefore

$$R_c = \frac{k_c}{\Delta v_0} = \frac{4.44 \Delta v_0}{\Delta v_0} = 4.44$$

(b) Using Eq. (10.6)

$$k_c = 4.44 \ (42.5 \text{ Hz}) = 189 \text{ s}^{-1} \qquad \square$$

4. As R increases further, the single peak at v_{av} sharpens (Figure 10.1, $R = 8.90$). The halfwidth ($v_{1/2}$) at moderately fast exchange is

$$v_{1/2} = v^0_{1/2} + \frac{(\pi \Delta v_0^2)}{k} \qquad (10.7)$$

where $v^0_{1/2}$ is the line width due to spin–spin relaxation (usually taken to be equal to the halfwidth of the signal at the fast-exchange limit) and $\pi \Delta v_0^2 / k$ is the additional broadening due to exchange.

■ **EXAMPLE 10.3**   (a) Solve Eq. (10.7) for the k at moderately fast exchange. (b) Calculate the values of k and R for the methyl exchange of DMF if $v_{1/2} = 16$ Hz and $v^0_{1/2} = 1.0$ Hz.

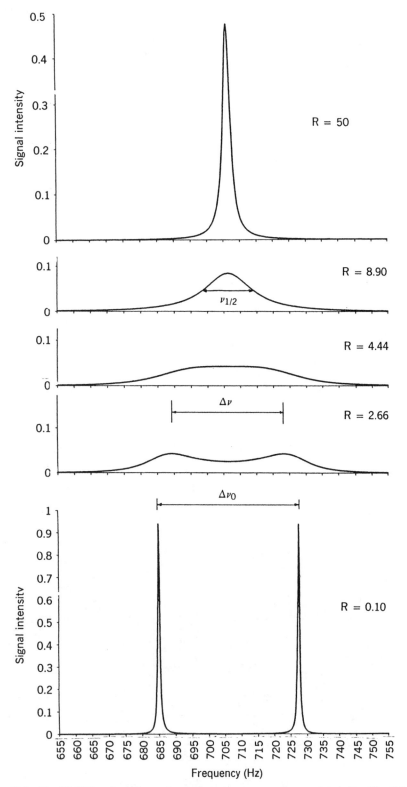

Figure 10.1. The NMR lineshape for exchange between two equally populated sites [Eq. (10.1)] as a function of exchange ratio R. The following conditions are shown: the slow-exchange limit ($R = 0.10$), moderately slow exchange ($R = 2.66$), coalescence ($R = 4.44$), moderately fast exchange ($R = 8.90$), and approaching the fast-exchange limit ($R = 50$).

☐ *Solution:* (a)

$$k = \frac{\pi \Delta \nu_0^2}{\nu_{1/2} - \nu_{1/2}^0} \tag{10.8}$$

(b) Applying Eq. (10.8)

$$k = \frac{\pi (42.5)^2}{16 - 1.0} = 378 \ s^{-1}$$

$$R = \frac{378}{42.5} = 8.90$$

☐

5. Finally, when R is greater than ~50, the system is essentially at the fast-exchange limit, and the spectrum consists of one sharp signal at ν_{av} (Figure 10.1, $R = 50$). At higher values of R the halfwidth of this signal approaches $\nu_{1/2}^0$.

The curves in Figure 10.1 were generated from Eq. (10.1) using the above values of k and the following data: $\nu_A = 727.5$ Hz and $\nu_B = 685$ Hz, $\nu_{av} = 706.3$ Hz, $\Delta \nu_0 = 42.5$ Hz, and $C = 1$ Hz.

Whenever an NMR spectrum exhibits reversible changes as the sample temperature is varied, we should suspect that dynamic processes are at work. In a full investigation of a

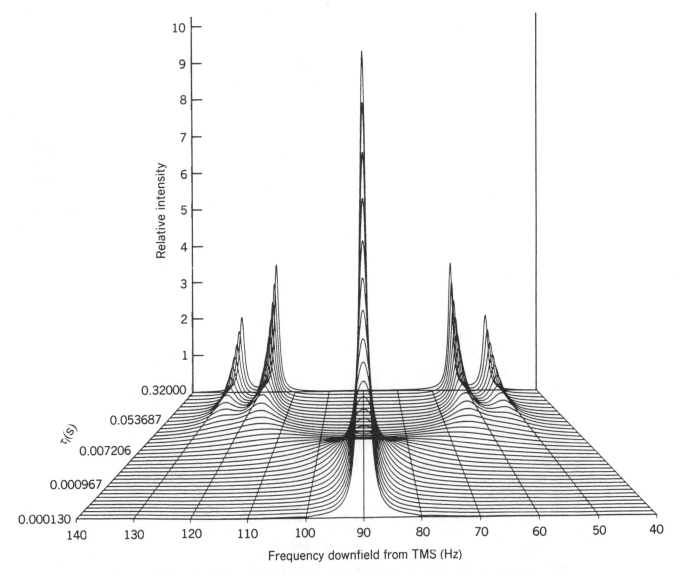

Figure 10.2. Stacked plot rendering of computer-simulated spectra for an AB ⇔ A′B′ exchanging system. [From "A Simple Three Dimensional Perspective Plotting Program," by R. S. Macomber, *Journal of Chemical Education*, *53*, 279 (1976). Reprinted by permission.]

dynamic process by NMR, the first step is to determine the values of v_A and v_B and the fractions of the molecular population in each structure or site at the slow exchange limit. Next, spectra are recorded at various higher temperatures, including at coalescence and into the fast-exchange region. The value of $v_{1/2}^0$ is measured at the fast-exchange limit. Finally, the sample is cooled back to the slow-exchange limit to insure that no irreversible processes have occurred at the higher temperatures.

The rate constant corresponding to each spectrum can be calculated using equations similar to Eq. (10.1)–(10.8). For systems involving more complex exchange processes than simple two-site exchange, such calculations require computer simulation, and programs for this purpose (such as DNMR3)[4] are available. The practice of matching an actual spectrum with a computer-generated spectrum is called **complete line-shape analysis**.

Figure 10.2 is a computer-simulated 60-MHz ^{1}H spectrum of a coupled AB spin system (Section 9.11) undergoing exchange. An example of such a system is the "ring flipping" (a series of connected rotations of the C–C bonds) of cyclohexane derivative **10-2** from one chair conformation to the other. Notice how this process exchanges the sites (environments) of the two indicated hydrogens from **axial** (*a*) to **equatorial** (*e*), though the two structures **i** and **ii** are otherwise equivalent:

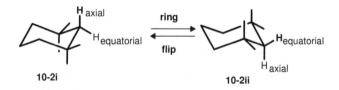

10-2i   **10-2ii**

Values used in creating Figure 10.2 were $v_a = 120$ Hz, $v_e = 60$ Hz, $^2J_{ae} = 10$ Hz, and $v_{1/2}^0 = 2$ Hz. The spectra are plotted as a function of τ_i, the lifetime of structure **10-2i** ($\tau_i = 1/k_{i \to ii}$).

■ **EXAMPLE 10.4** From the data used to generate Figure 10.2, calculate Δv_0 and τ_i at coalescence.

□ *Solution*

$$\Delta v_0 = v_a - v_e = (120 - 60 \text{ Hz}) = 60 \text{ Hz}$$

Next, substitute the value of the total rate constant k_c at coalescence [4.44 Δv_0, from Eq. (10.6)] into Eq. (10.2):

$$\tau_i = 2\tau = \frac{2}{k} = \frac{2}{4.44 \, \Delta v_0} = \frac{1}{(4.44)(60 \text{ Hz})} = 0.0075 \text{ s}$$

Note in Figure 10.2 that at this value of τ_i the spectrum consists of one very broad signal.   □

■ **EXAMPLE 10.5** Suppose you are using a 400-MHz instrument to investigate the dynamic behavior of an exchange process involving the two methyl groups in DMF. (a) At what value of k will the ^{1}H signals coalesce? Will this occur at a higher or lower temperature than coalescence at 250 MHz? (b) At what value of k will the ^{13}C signals coalesce? Will this occur at a higher or lower temperature than coalescence of the ^{1}H signals?

□ *Solution:* (a) In going from 250 to 400 MHz, Δv_0 for the ^{1}H signals will increase from 42.5 to 68 Hz. This means that that k_c increases from 189 to 302 s^{-1}. This larger rate constant will require a higher temperature. (b) A 400-MHz instrument will function at 100 MHz for ^{13}C. The chemical shift difference for the carbon signals (35.2 – 30.1 = 5.1 ppm) thus corresponds to a Δv_0 of 510 Hz. Coalescence of the carbon signals therefore occurs at a k value of 4.44 (510 Hz) = 2260 s^{-1}, requiring a much higher temperature than coalescence of the ^{1}H signals at 400 MHz.   □

10.4 REVERSIBLE INTRAMOLECULAR CHEMICAL PROCESSES

Many molecules undergo reversible *chemical* changes, that is, reorganization of bonds and changes in the sequence of atoms. Such chemical changes can be either intramolecular or intermolecular. Molecules that undergo reversible *intra*molecular chemical changes are called **fluxional**, and two (or more) structures that are interconverted by this type of reorganization are called **valence isomers**. As with dynamic physical processes, such dynamic chemical processes are also particularly amenable to study by NMR.

One classic example of valence isomerization is bullvalene[5]:

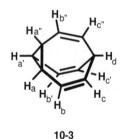

10-3

This unique structure has a C_3 axis (Section 4.1), and its hydrogens (as well as its carbons) constitute an AA′A″B-B′B″CC′C″D spin system. At low temperature (< –85°C) nothing extraordinary occurs, and the 100-MHz ^{1}H spectrum shows four complex multiplets centered at δ 2.07 (H_a), 2.13 (H_d), 5.62 (H_b), and 5.70 (H_c). But as the temperature of the sample is raised, these multiplets broaden and move together. At 15°C the spectrum coalesces to one extremely broad unsymmetrical signal ($v_{1/2}$ = 150 Hz). And at 120°C there is but *one* sharp singlet!

The way in which all 10 hydrogens (or carbons) in bullvalene become equivalent involves a series of **degenerate** (i.e., leading to an equivalent structure) isomerizations known as **Cope rearrangements** (or **[3.3] sigmatropic shifts**). One such rearrangement is shown below (the dot is there to keep track of C_d):

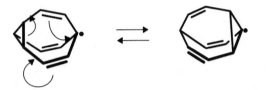

Note how this particular bond reorganization of two π bonds and one cyclopropane bond interconverts three pairs of hydrogens (and the attached carbons): H_a with H_c, $H_{a''}$ with $H_{c''}$, and $H_{a'}$ with H_d. Repetition of this rearrangement with other sets of π and cyclopropane bonds ultimately renders all 10 hydrogens (and carbons) equivalent.

■ **EXAMPLE 10.6** Predict the chemical shift of bullvalene at the fast-exchange limit.

□ *Solution:* Remember that this signal is the weighted average of all exchanging signals. The spectrum of bullvalene at the slow-exchange limit has a 3H multiplet at δ 2.07, 1H at δ 2.13, 3H at δ 5.62, and 3H at δ 5.70. Thus, the average chemical shift of all 10 hydrogens is [3(2.07) + 1(2.13) + 3(5.62) + 3(5.70)]/10 = δ 4.23, exactly as observed!   □

■ **EXAMPLE 10.7** The 25-MHz proton-decoupled ^{13}C spectrum of bullvalene consists of singlets at δ 21.0 (C_a), 31.0 (C_d), 128.3 (C_d), and 128.5 (C_c). (a) Which will coalesce at the higher temperature, the 100-MHz ^1H spectrum or the 25-MHz ^{13}C spectrum? (b) Predict the ^{13}C chemical shift of the fast-exchange spectrum.

□ *Solution:* (a) The maximum $\Delta\delta_0$ in the ^1H spectrum is (5.70 − 2.07 ppm)(100 Hz/ppm) = 363 Hz. In the ^{13}C spectrum the largest $\Delta\delta_0$ value is (128.5 − 21.0 ppm)(25 Hz/ppm) = 2690 Hz. Thus, the ^{13}C spectrum will coalesce at a higher temperature (observed at 40°C). (b) δ = [3(21.0) + (31.0) + 3(128.3) + 3(128.5)]/10 = 86.4.   □

Many examples of this type of fluxional behavior are now known. But had it not been for NMR, most would never have been discovered!

10.5 REVERSIBLE INTERMOLECULAR CHEMICAL PROCESSES

One of the most ubiquitous reactions in chemistry is intermolecular proton (H^+) transfer, that is, the reaction between a Bronsted acid and base. Examples include all reactions that are acid or base catalyzed. In general, proton transfers are among the fastest reactions known, occurring in the fast (or very fast) exchange region of the NMR time scale. This is the reason that hydrogens (such as those in O–H groups) undergoing intermolecular exchange often give rise to broadened signals with concentration-dependent chemical shifts (Section 6.6.1). However, in some cases proton transfer can be accompanied by a rearrangement that slows the overall reaction.

■ **EXAMPLE 10.8** Enols, compounds with an OH group directly bonded to a vinyl carbon, are normally less stable than the corresponding carbonyl ("keto") isomer. [A few enols (e.g., phenols) exist predominantly or exclusively in the enol form.] Though the uncatalyzed equilibration of enols with their keto isomers is usually slow, the rate of isomerization increases dramatically in the presence of a small amount of acid (HA) or base (B):

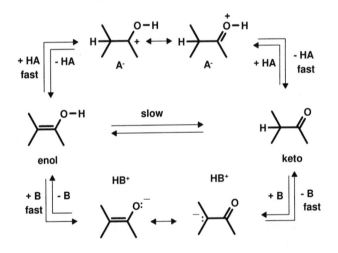

The room temperature ^1H NMR spectrum of symmetrical diketone **10-4**,

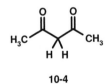

10-4

exhibits the following sharp signals[3]: δ 2.0 (s, 6H), 2.2 (s, 1H), 3.7 (s, 0.16H), 5.7 (s, 1H), 15.3 (broad s, 1H). As the temperature is raised (or if an acid catalysts is added in increasing amount), the signals at δ 2.0 and 2.2 approach coalescence, as do the signals at δ 3.7, 5.7, and 15.3. Explain.

☐ *Solution:* The dynamic behavior, the unexpected number of signals in the spectrum, and the enolic OH signal at δ 15.3 (Section 6.6.1) all suggest that we are dealing with keto–enol equilibration of the β-diketone. The signals at δ 2.2 and 3.7 are in the correct positions (Chapter 6) and ratio (6 : 1) for compound **10-4**. The signals at δ 2.0, 5.7, and 15.3 belong to enol **10-5**, rendered symmetrical by a combination of resonance and intramolecular hydrogen bonding:

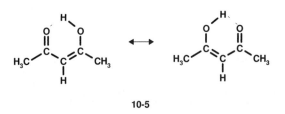

10-5

From the signal integrations it is clear that the enol form constitutes 86% of the equilibrium mixture. Furthermore, at room temperature the rate of equilibration must be much slower than the NMR time scale to observe both sets of sharp signals.   ☐

10.6 REVERSIBLE INTERMOLECULAR COMPLEXATION

One of the most active areas in modern chemical research involves **molecular recognition**, the formation of so-called **host–guest** (or H–G) **complexes** (C) that arise when two molecules "fit" perfectly together, held in place predominantly by hydrogen bonding:

$$H + G \underset{k_{-1}}{\overset{k_1}{\rightleftarrows}} C$$

Such complexes are sometimes called **supramolecules**. The choice of which molecule is the host and which is the guest is somewhat arbitrary, but the larger molecule is usually deemed the host. Enzyme–substrate complexes are a prime biochemical example of such interactions. Cram, Pedersen, and Lehn shared the 1987 Nobel Prize in chemistry for their pioneering work in the area of molecular recognition and supramolecular chemistry. The recent chemical literature is replete with examples in which NMR has been used to study such complexation.[6]

Of particular interest is the strength of the complexation, as measured by equilibrium constant K:

$$K = \frac{k_1}{k_{-1}} = \frac{[C]}{[H][G]} = \frac{X_C}{X_H[G]} = \frac{X_C}{(1 - X_C)[G]} \quad (10.9)$$

where the brackets indicate molar concentration at equilibrium, X_H is the mole fraction of uncomplexed host, and X_C is the mole fraction of complexed host.

10.6.1 Determining K Under Slow-Exchange Conditions

To use NMR methods to study such complexation phenomena, at least one nucleus in the uncomplexed host (or guest) molecule must give rise to a signal whose chemical shift δ_H is significantly different from the same nucleus in the complexed host (or guest) molecule (δ_C). The magnitude of this difference ($\Delta\delta = \delta_H - \delta_C$) often gives information about the structure of the complex and may be as large as several ppm.

With this information, we can determine the value of K under slow-exchange conditions by first rewriting Eq. (10.9) as

$$K = \frac{[C]}{([H]_0 - [C])([G]_0 - [C])} \quad (10.10)$$

where $[H]_0$ and $[G]_0$ are the formal (i.e., precomplexation) concentrations of host and guest; $[C]_0 = 0$). The relative signal integrals at δ_H (I_H) and δ_C (I_C) give X_C by the relationship

$$X_C = \frac{I_C}{I_C + I_H} \quad (10.11)$$

Moreover, since $[C] = X_C[H]_0$, Eq. (10.10) can be rewritten in the form

$$K = \frac{X_C}{[H]_0(1 - X_C)(r - X_C)} \quad (10.12)$$

where r is the ratio $[G]_0/[H]_0$.

10.6.2 Determining K under Fast-Exchange Conditions

Under fast-exchange conditions the observed chemical shift (δ) of the nucleus of interest will be the population average of δ_H and δ_C:

$$\delta = X_H\delta_H + X_C\delta_C = (1 - X_C)\delta_H + X_C\delta_C$$
$$= \delta_H - X_C\Delta\delta \quad (10.13)$$

This leads to the relationship

$$X_C = \frac{\delta_H - \delta}{\Delta\delta} \quad (10.14)$$

Now we can use the observed fast-exchange chemical shift δ, together with δ_H and δ_C from slow-exchange measurements and the known values of $[H]_0$ and $[G]_0$, to evaluate K using Eq. (10.12).

The observed chemical shift under fast-exchange conditions is a complicated function of $[H]_0$, $[G]_0$, δ_H, δ_C, and K:

$$\delta = \delta_H - \frac{\Delta\delta}{2}\left(b - \sqrt{b^2 - 4r}\right) \tag{10.15}$$

where $b = 1 + r + (K[H]_0)^{-1}$ (see review problem 10.4).

Equation (10.15) describes the **complexation-induced shift** (**CIS**) of an NMR signal. The systematic variation of δ with changing values of r (= $[G]_0/[H]_0$) forms the basis of a general technique known as **NMR titration** for the determination of K values for complexation.

Figure 10.3 is a plot of Eq. (10.15) that shows graphically how δ varies with r for several values of K. In all cases δ asymptotically approaches δ_C as r increases. As we might expect, values of K greater than 50 (curves F and G in Figure 10.3) lead to rapid "saturation" of the host sites at relatively low values of r and consequent rapid changes in δ. By contrast, when $K = 0.10$ (curve A), the observed chemical shift changes by only 9% of $\Delta\delta$ when $r = 10$ and by only 50% when $r = 100$ (see Figure 10.4).

For convenience, it is usually easiest to carry out an NMR–CIS experiment by preparing just one solution of known $[H]_0$ and $[G]_0$, then diluting it successively and measuring δ as a function of the dilution. In this case, Eq. (10.15) takes the form

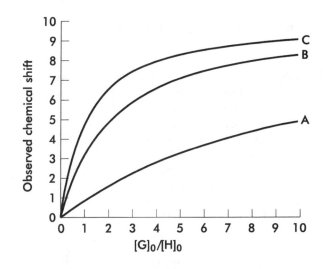

Figure 10.4. Plots of lines A, B, and C from Figure 10.3, extended to larger values of r. [From "An Introduction to NMR Titration for Studying Rapid Reversible Complexation," by R. S. Macomber, *Journal of Chemical Education*, 69, 375–378 (1992). Reprinted by permission.]

$$\delta = \delta_H - \frac{\Delta\delta}{2[H]_0}\left(B - \sqrt{B^2 - 4[H]_0[G]_0}\right) \tag{10.16}$$

where $B = [H]_0 + [G]_0 + K^{-1}$.

When using Eq. (10.16) it is necessary to recalculate values for $[H]_0$ and $[G]_0$ for each dilution. The CIS titration curve can be plotted as δ versus $[G]_0$ (Figure 10.5). It is also possible

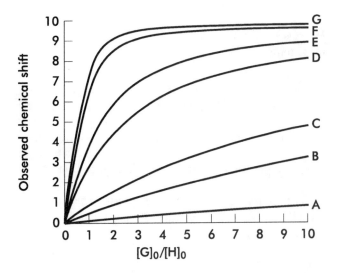

Figure 10.3. Plots of Eq. (10.15) relating δ to r (= $[G]_0/[H]_0$). The values of K are (*a*) 0.1; (*b*) 0.5; (*c*) 1.0; (*d*) 5; (*e*) 10; (*f*) 50; (*g*) 100 M^{-1}. For each line, $[H]_0 = 0.10\,M$, $\delta_H = 0$ Hz, and $\delta_C = 10$ Hz. [From "An Introduction to NMR Titration for Studying Rapid Reversible Complexation," by R. S. Macomber, *Journal of Chemical Education*, 69, 375–378 (1992). Reprinted by permission.]

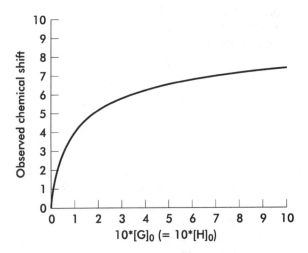

Figure 10.5. Plot of Eq. (10.16), the effect on δ of successively diluting a solution with initial concentration $[H]_0 = [G]_0 = 1.0\,M$. Additional input parameters include $K = 10\,M^{-1}$, $\delta_H = 0$ Hz, and $\delta_C = 10$ Hz. [From "An Introduction to NMR Titration for Studying Rapid Reversible Complexation," by R. S. Macomber, *Journal of Chemical Education*, 69, 375–378 (1992). Reprinted by permission.]

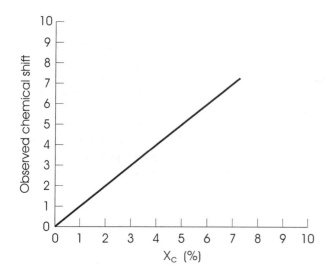

Figure 10.6. Plot of δ versus X_C using the data in Figure 10.5. (review problem 10.7).

to plot δ against X_C [Eq. (10.13)], yielding a straight line with y intercept δ_H (Figure 10.6).

Alas, there is a fly in the ointment. Equations (10.14)–(10.16) all require that both δ_H and δ_C be known. Although the former is easily determined by direct measurement in the absence of a guest, the latter cannot usually be determined directly for two reasons. First, the rate constants for complexation and decomplexation [k_1 and k_{-1}; Eq. (10.9)] are often very large, making it impossible to reach the slow-exchange limit. Second, unless the value of K *is* substantially greater than 10, it may be difficult to approach the δ_C limit of δ at any readily attainable ratio of $[G]_0$ to $[H]_0$ (see Figure 10.3, lines A, B, and C). So, we are left with one equation [Eq. (10.15) or (10.16)] with two unknown parameters, δ_C and K.

But all is not lost. It is possible to use iterative nonlinear curve fitting to determine which values of K and of δ_C best simulate an experimental set of δ-versus-r (or δ-vs-$[G]_0$) data. Programs for doing this are described in ref. 6. The input data include the experimental point-by-point values of δ-versus-$[G]_0$, the values of $[H]_0$ and δ_H, and an estimate of K. One approach is to use the δ-versus-$[G]_0$ data point with the largest $[G]_0$ value, together with the estimate of K, to calculate a value for δ_C using Eq. (10.16). With these estimates of δ_C and K, the program calculates a predicted value of δ (δ_{calc}) for each experimental value of $[G]_0$ using Eq. (10.16). Then it computes the difference ($\delta_{obs} - \delta_{calc}$) for each point. If the sum of these differences is positive, the value of K is incremented; if the sum is negative, K is decremented. The entire process is repeated iteratvely with the updated K value until convergence is achieved.

This curve-fitting approach for the determination of K is most accurate when the experimental data cover a range of r values that is broad enough to allow a significant degree of host saturation. In fact, it can be shown[6] that the optimum

conditions for data collection are $[H]_0 = 0.10/K$, with $[G]_0$ spanning the range $0.10/K$ to $10/K$. In such a case, the resulting titration curve will resemble line C in Figure 10.4. (Another approach to evaluating K is given in review problem 10.6.)

It is important to be aware of certain assumptions that are implicit in the derivation of the above equations. Most importantly, it is assumed that δ_C and δ_H are themselves independent of concentration and temperature effects. While this can often be experimentally verified for δ_H, it is generally impossible to verify for δ_C unless the slow-exchange limit can be attained. Another potential error is the failure to include activity coefficients in the equilibrium expressions, even though concentrations often exceed 1 M. Since iterative nonlinear curve fitting often involves locating a relatively shallow minimum, effects such as these can lead to significant error in derived K.

The bottom line is this: When an NMR spectrum exhibits a shift in δ as a function of the changing concentration of other potential interacting species, this is probably due to complexation-induced shifts under fast-exchange conditions.

■ **EXAMPLE 10.9** Investigators[7] were attempting to use NMR methods to measure the rate for the irreversible reaction of compound **10-6** with trifluoroethanol, forming ether (**10-7**) and trifluoroacetic acid (TFA):

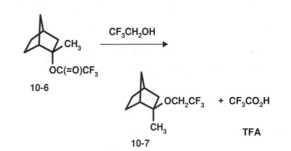

Their initial plan was to follow the rate of disappearance of **10-6** or the rate of appearance of **10-7** by monitoring the methyl signal of each compound as a function of time. To prevented unwanted side reactions catalyzed by TFA, they included in the reaction mixture one equivalent of buffering base 2,6-dimethylpyridine (DMP), to neutralize the acid:

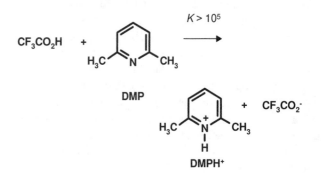

As the reaction proceeded, they noticed something odd. The position of the methyl signal of DMP (initially at δ 2.504) shifted steadily downfield, reaching δ 2.651 at completion. Explain the reason for this shift and how the position of the DMP methyl signal could be used to measure the rate of the conversion of **10-6** to **10-7**.

☐   *Solution:* We are observing the dynamics of a complexation (host–guest) reaction between DMP and H^+ (donated by TFA), forming the complex $DMPH^+$. Because K is large, virtually all the H^+ becomes complexed to DMP as soon as it is formed. But there is also rapid proton transfer from one DMP to another: $DMPH^+ + \textbf{DMP} \rightarrow DMP + \textbf{DMPH}^+$. Thus, once the reaction begins, the DMP methyl signal will be a population average of the signal for DMP (δ 2.504) and that of $DMPH^+$ (δ 2.651, downfield of DMP owing to the deshielding effect of the positively charged nitrogen). The exact relationship is given by Eq. (10.13) with H = DMP and C = $DMPH^+$. Moreover, Eq. (10.14) and the observed chemical shift of the signal can be used to give the mole fraction of $DMPH^+$ at any point during the reaction. Since the mole fraction of $DMPH^+$ must stoichiometrically equal the mole fraction of product **10-7** at any point in the reaction, the rate constant for disappearance of **10-6** (or formation of **10-7**) can be determined by simply monitoring the DMP signal position as a function of time.[7]   ☐

10.7   OTHER EXAMPLES OF REVERSIBLE COMPLEXATION: CHEMICAL SHIFT REAGENTS

10.7.1   Chemical Shift Reagents

In Section 9.9 we discussed how the appearance of a spin-coupled NMR spectrum is determined by the ratio of $\Delta\nu$ (the difference in chemical shifts between the coupling nuclei) to J (the coupling constant they share). For the spectrum to exhibit first-order multiplet intensities (Pascal's triangle; Section 8.5), the value of $\Delta\nu/J$ has to be at least 10. Smaller values of $\Delta\nu/J$ lead to progressively greater complications due to second-order effects.

Although it is possible, in principle, to analyze a second-order spectrum through computer simulation (Section 9.10), it would be easier to interpret such a spectrum if we could somehow "force" it into a first-order appearance. Since the magnitude of J is determined solely by the molecule's structure (and is field independent; Section 8.2), the only way we can increase $\Delta\nu/J$ is to increase $\Delta\nu$. One way to accomplish this is to use a spectrometer with a higher field strength and operating frequency (Section 9.9), but such an instrument may not be readily accessible to us when we need it. In many cases there is a relatively simple way to accomplish the same thing!

There is a family of compounds that, when added to a solution under NMR examination, can lead to large changes in the chemical shifts of the sample's nuclei. These **paramagnetic** (or **chemical**) **shift reagents**[8] (**PSR**s or **CSR**s) are coordination complexes of paramagnetic metal ions of the lanthanide group, such as europium (Eu^{3+}) and praseodymium (Pr^{3+}). One commercially available shift reagent, abbreviated $Eu(fod)_3$ (**10-8**), has an Eu^{3+} ion coordinated to three equivalent β-diketone enolates. This compound is also available in a deuterated version, in which all 30 hydrogens have been replaced with deuterium:

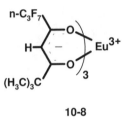

10-8

When a PSR is added to a solution containing the sample (i.e., the host) of interest, there is rapid reversible complexation between the PSR (the guest) and the sample molecules. The sample molecules in the complex are held in relatively close proximity to the metal ion and experience the strong paramagnetic field of the metal ion. As a result, nuclei in the sample molecule usually experience substantial *deshielding*, causing a downfield shift ($\delta_C - \delta_H > 0$) in the position of their signals (Section 6.1). The sample nuclei nearest to the metal ion exhibit the greatest shift. This **pseudocontact shift** interaction operates directly through space (as opposed to through bond), and its magnitude is inversely proportional to the cube of the average distance from the metal ion to the nucleus of interest.

Because of the rapid reversible nature of this complexation, the magnitude of PSR-induced shifts is also a function of the equilibrium constant for complexation as well as the mole ratio of PSR to sample (Section 10.6). But since K is normally greater than 1, δ rapidly approaches δ_C (as in curve G of Figure 10.3) as the ratio of PSR to sample approaches 1 : 1. Beyond that ratio, each sample molecule already has its "own" PSR, so additional PSR normally has little effect.

An example of PSR-induced shifts can be seen in the 80-MHz ^1H spectrum of dibutyl ether:

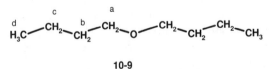

10-9

In the uncomplexed spectrum (Figure 10.7a) the signal for the four H_a nuclei appears as the expected triplet (δ 3.40 ppm), the H_d "triplet" (δ 0.91 ppm) is very distorted by second-order

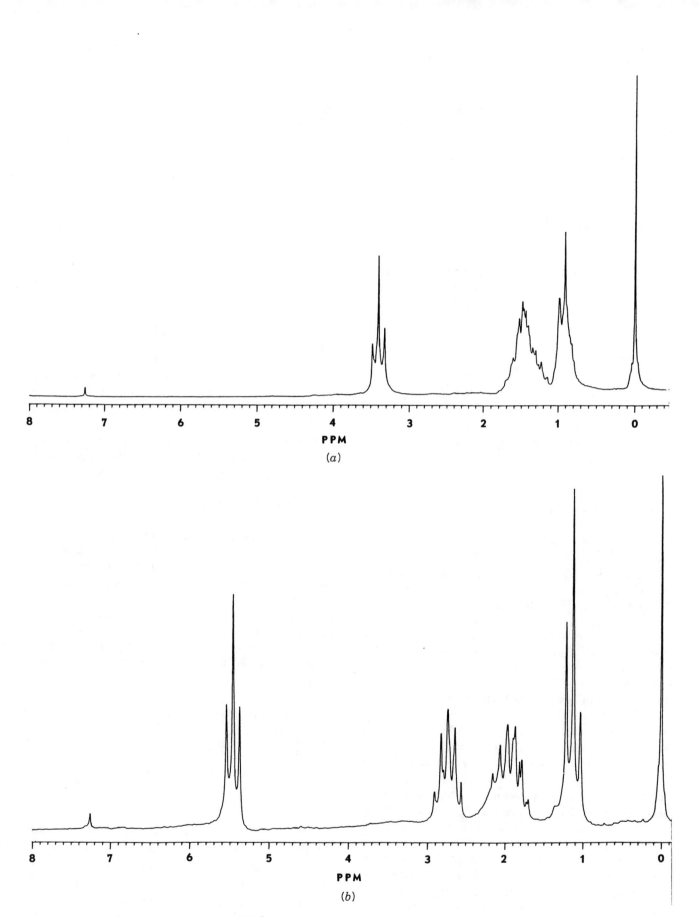

Figure 10.7. (*a*) The 80-MHz ^{1}H spectrum of dibutyl ether (**10-9**) in CDCl$_3$. (*b*) Same sample after addition of 0.72 equivalents of Eu(fod)$_3$.

effects, while the complex H_b and H_c multiplets (δ 1.8 ppm) are not resolved. Addition of deuterated Eu(fod)$_3$ causes the signals of all four sets of hydrogens to shift downfield to varying degrees, with the most downfield signal (H_a) exhibiting the greatest shift (Figure 10.7*b*). When the ratio of Eu(fod)$_3$ to **10-9** reaches 0.72, the four signals are fully resolved, with chemical shifts δ_a 5.46 (triplet), δ_b 2.73 (quintet), δ_c 1.91 (sextet with some additional structure), and δ_d 1.11 (triplet). Note how the multiplicity of each signal is now in accord with first-order expectations.

■ **EXAMPLE 10.10**(a) Calculate the magnitude (in ppm) of the PSR-induced shift for each hydrogen in **10-9**. (b) Where in the structure of **10-9** is the metal ion most likely complexed?

□ *Solution:* (a)

H	PSR-Induced Shift (ppm)
a	$5.46 - 3.40 = 2.06$
b	$2.73 - 1.45^a = 1.28$
c	$1.91 - 1.45^a = 0.46$
d	$1.11 - 0.91 = 0.20$

^{a} Estimated as the center of the multiplet.

(b) Because the PSR-induced shifts follow the order $H_a > H_b > H_c > H_d$, complexation of the metal ion must be nearest to H_a, and furthest from H_d. Complexation therefore must involve sharing of a nonbonding electron pair on the oxygen atom with an empty *d* orbital on europium. Paramagnetic shift reagent complexation takes place at the most electron-rich site of the sample molecule. This site is usually a heteroatom such as oxygen, nitrogen, or halogen.

10.7.2 Solvents as Chemical Shift Reagents

Recall that the chemical shift of a given nucleus is determined by the strength of the external magnetic field *as modified by the neighboring electrons and nuclei through shielding and deshielding effects* (Section 6.1). These neighboring electrons and nuclei include not only those in the same molecule as the nucleus of interest but also those in nearby molecules, including host molecules and even solvent molecules. It is not uncommon for the chemical shift of one or more hydrogens in a molecule to change by a few tenths of a ppm by simply changing the composition of the solvent. In this case the solvent itself is acting as a chemical shift reagent, or a host for the guest solute, because solvation is one more type of rapid reversible complexation (though perhaps weaker than the types we have discussed heretofore).

For example, the 60-MHz ^{1}H spectrum of cyclopropane derivative **10-10** dissolved in carbon tetrachloride exhibits only one sharp signal at δ 1.45 ppm,[9] indicating that the two sets of hydrogens (six methyl and two methylene) are accidentally equivalent (Section 4.4):

10-10

By contrast, when the benzene is used as solvent, two singlets appear [δ 1.11 (2H); 1.17 (6H)], as expected on the basis of symmetry. This effect, known as the **solvent-induced shift (SIS)**, is due to the anisotropic-induced field around the benzene molecules (Section 6.6). Because of the particular way solute molecules are solvated by benzene molecules, all eight hydrogens in **10-10** experience extra shielding by solvent, but the CH$_2$ hydrogens are shielded more than the CH$_3$ hydrogens.

Thus, when confronted with a case of accidental equivalence, it is worthwhile to see if changing the solvent resolves the signals. If not, try a lanthanide shift reagent.

10.7.3 Chiral Shift Reagents

In Section 4.3 we mentioned two important stereochemical terms, enantiomers and enantiotopic nuclei. **Enantiomers** are structures related as the left hand is related to the right: nonsuperimposable mirror images. Any chiral (dissymmetric) molecule can exist in two (and only two) enantiomeric forms. For example, chiral alcohol **10-11** has two enantiomeric configurations, labeled *R* and *S*.[10] **Enantiotopic nuclei** are those related by a plane of symmetry. The methylene hydrogens of benzyl alcohol (**10-12**) are enantiotopic and are labeled **pro-*R*** and **pro-*S***[10]:

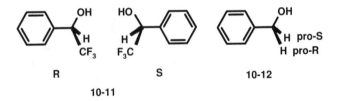

R   **S**   **10-12**

10-11

Here is the most important thing to remember about enantiotopic nuclei: They have identical physical, chemical, and spectroscopic properties and are therefore indistinguishable by NMR *under normal conditions*. Likewise, enantiomers have identical physical, chemical, and spectroscopic properties and are indistinguishable by NMR under normal conditions. (Enantiomers can be distinguished by a technique known as **polarimetry**, which depends on their different absorptivity toward circularly polarized light.[10]) Thus, the two enantiomers of **10-11** exhibit identical NMR spectra, and the methylene hydrogens in **10-12** are magnetically equivalent.

The above proviso, "under normal conditions," is the hedge. In general, enantiomers (as well as enantiotopic nuclei) can be differentiated, provided we place the molecules in an *asymmetric* environment. One way to accomplish this is to use a PSR with ligands that are themselves single enantiomers. The PSR containing ligands with only the *R* configuration will complex differently with the *R* enantiomer of the solute than with the *S* enantiomer. This usually results in greater signal shifts for one of the solute enantiomers. The analogy is this: To differentiate between left hands and right hands, throw in left-handed gloves!

The PSR **10-13**, known as Pr(hfc)$_3$, makes use of three identical ligands related to enantiomerically pure camphor[11]:

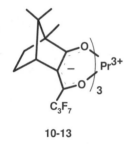

10-13

In the presence of this compound the enantiotopic methylene hydrogens of alcohol **10-12** are no longer equivalent, exhibiting a chemical shift difference up to 0.8 ppm at 100 MHz. Since these hydrogens are no longer equivalent, the coupling constant between them can be measured directly: $^2J = 13$ Hz. Compare this approach with the method discussed in Section 9.4.

Just as solvents can serve as CSRs, it is possible to use enantiomerically pure solvents to provide the asymmetric environment necessary to differentiate enantiomers. This then allows us to determine the relative amount of two enantiomers in a mixture of them, something that is otherwise difficult to accomplish without resorting to polarimetry. Enantiomerically pure alcohol *(R)*-**10-11** has been used as a solvent to differentiate the enantiomers of aminoester **10-14**[12]:

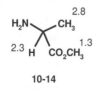

10-14

The numbers above the three sets of hydrogens indicate the differences (in hertz) between the enantiomeric signals.

CHAPTER SUMMARY

1. There are two basic types of chemical processes: reversible (leading to, and maintaining, an equilibrium mixture) and irreversible (proceeding in one direction, i.e., to completion). NMR methods can be used to study both types of processes. A common example of an irreversible process is a chemical reaction with a free energy more negative than ca. 20 kJ/mol. Reversible processes include interconversions of conformations by rotations around single bonds, interconversions of fluxional molecules and valence isomers, and proton transfers, to name a few.

2. To be studied by NMR, a reversible reaction must bring about the exchange in site (molecular environment) of one or more nuclei. The appearance of a spectrum for a compound undergoing exchange depends on the relative magnitude of the exchange rate constant k compared to the difference in chemical shift between the sites (Δv). When $k \ll \Delta v$ (very slow exchange), the spectrum consists of sharp signals for each site in each of the interconverting structures. When $k \gg \Delta v$ (very fast exchange) each set of exchanging nuclei gives rise to one sharp signal whose chemical shift is the average of all the interconverting sites. When $k \approx \Delta v$, each set of exchanging nuclei gives rise to one very broad signal, a condition known as coalescence. The exact lineshapes (equations describing the appearance of an exchanging spectrum) are known. Complete lineshape analysis involves a computer analysis of the spectrum over a wide range in k values.

3. Many rapid reversible chemical reactions involve formation of guest–host complexes between two or more molecules. If a nucleus in the guest molecule exhibits a difference in chemical shift between the complexed and uncomplexed state ($\Delta\delta$), the observed chemical shift (δ) under fast-exchange conditions is related to the magnitude of the equilibrium constant (K) for formation of the complex.

4. Chemical shift reagents are compounds that form complexes with samples of interest, causing changes in the chemical shifts of the sample. Shift reagents that involve paramagnetic lanthanide ions bring about particularly dramatic changes in chemical shifts, but even changing solvent molecules can bring significant changes in chemical shifts.

5. Enantiomers and enantiotopic nuclei normally exhibit identical NMR spectra. However, if such molecules are immersed in an asymmetric medium (e.g., a chiral shift reagent or chiral solvent), the signals for each enantiomer (or each enantiotopic nucleus) will be resolved.

REFERENCES

[1]Rabinowitz, M., and Sievers, R. E., *J. Am. Chem. Soc., 93,* 1522 (1971).

[2]Becker, E. D., *High Resolution NMR,* 2nd ed., Academic, New York, 1980.

[3]For a more detailed treatment of two site exchange, see Gunther, H., *NMR Spectroscopy, An Introduction,* Wiley, New York, 1973.

[4]Kleier, D. A., and Binsch, G. and available from the Quantum Chemistry Program Exchange at Indiana University, Bloomington, IN.

[5]Oth, J. F. M., Mullen, K., Gilles, J., and Schroder, G., *Helv. Chim. Acta, 57,* 1415 (1974).

[6]Macomber, R. *J. Chem. Educ., 69,* 375 (1992).

[7]Creary, X., and Jiang, Z., *J. Org. Chem., 59,* 5106 (1994).

[8]Martin, M. L., Delpuech, J. -J., and Martin, G. J., *Practical NMR Spectroscopy*, Heyden & Sons, Philadelphia, 1980.

[9]Lilje, K. C., and Macomber, R. S., *J. Org. Chem., 39,* 3600 (1974).

[10]The *R/S* and pro-*R/S* nomenclature systems are described in Eliel, E. L., Wilen, S. H., *Stereochemistry of Organic Compounds*, Wiley, New York, 1994, as well as in most organic chemistry textbooks.

[11]Fraser, R. R., Petit, M. A., and Miskow, M., *J. Am. Chem. Soc., 94,* 3253 (1972).

[12]Pirkle, W. H., and Beare, S. D., *J. Am. Chem. Soc., 91,* 5150 (1969).

REVIEW PROBLEMS (Answers in Appendix 1)

10.1 Stan McHardy (with Gary Keck's group at the University of Utah) found that the 500-MHz ^{1}H NMR spectrum of compound **10-15** exhibits temperature dependence, as shown in Figure 10.8:

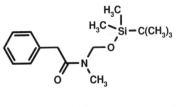

10-15

By assigning each of the signals in the room temperature and 120°C spectra, explain the cause of this temperature dependence.

10.2 The monodeutero derivative of alcohol **10-12** shown below is chiral:

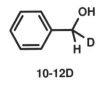

10-12D

The 500-MHz ^{1}H spectrum of a **racemic** (equal) mixture of the R and S enantiomers exhibits a singlet for the benzylic hydrogen, slightly broadened by two-bond H–D coupling. Esterification of this (racemic) alcohol with enantiomerically pure (**R**)-**10-16** gave ester **10-17**, which exhibited two equally intense (and slightly broadened) singlets for the the benzylic hydrogen.

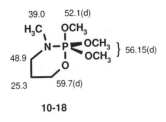

Explain. (This example was contributed by Dhileepkumar Krishnamurthy, also with Keck's group.)

10.3 Alan Sopchik (with Wesley Bentrude's group at the University of Utah) has found that the proton-decoupled 125-MHz ^{13}C spectrum of phosphorane **10-18** is temperature dependent, as shown in Figures 10.9*a* and *b*:

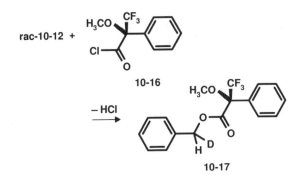

10-18

(The numbers surrounding the structure are the carbon chemical shifts at −90°C; a "d" indicates a doublet due to C–P coupling.) Account for the changes in the spectrum as the temperature is raised from −90 to 20.4°C.

10.4 Derive Eq. (10.15) from Eq. (10.12) and (10.13).

10.5 The 60-MHz ^{1}H NMR spectrum of 1-phenylethylamine **10-19** consists of the following signals: δ 1.29 (d, CH$_3$), 1.47 (s, NH$_2$), 4.01 (q, C–H), and 7.21 (m, ring hydrogens):

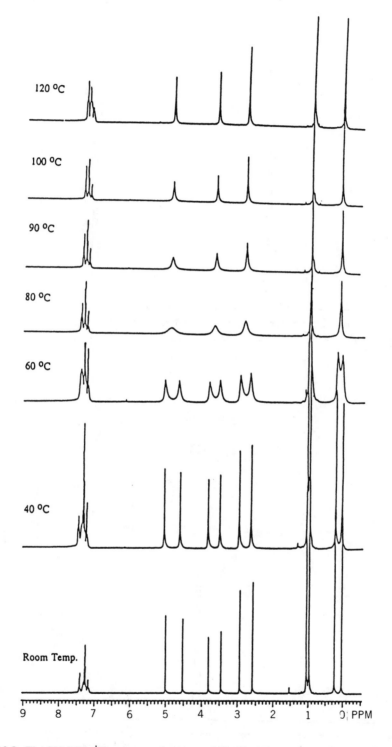

Figure 10.8. The 500-MHz ^{1}H spectrum of compound **10-15** (review problem 10.1) as a function of temperature.

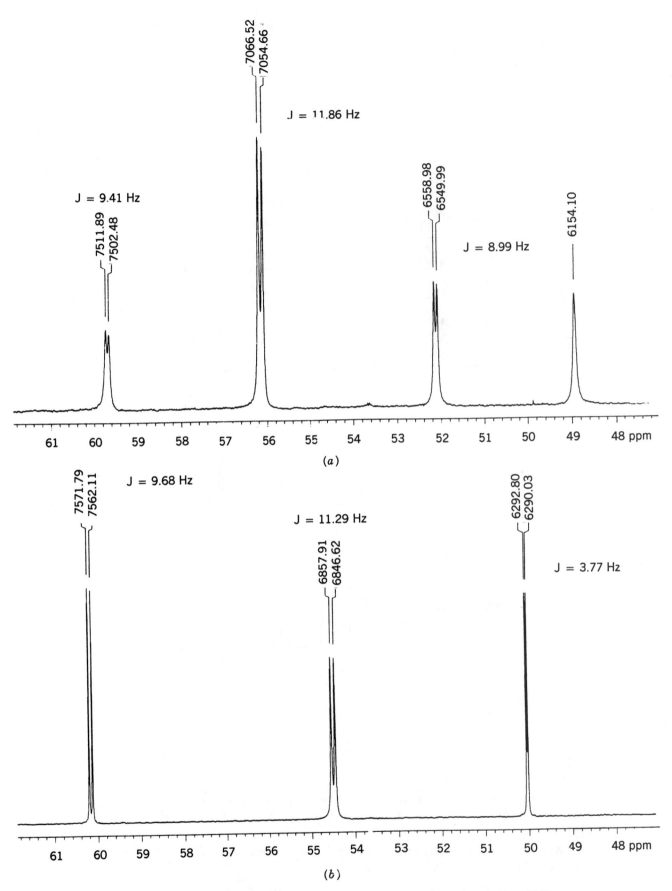

Figure 10.9. (*a*) Partial 125-MHz ^{13}C spectrum of phosphorane **10-18** (review problem 10.3) at −90°C. (*b*) Same spectrum, obtained at 20.4°C.

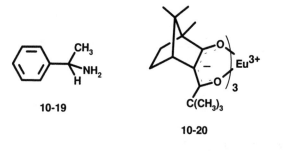

10-19

10-20

Addition of 0.5 equivalents of **10-20** not only causes all five multiplets to shift downfield [CH$_3$ → δ 10.9, C–H → δ 16.9, *ortho*-H → δ 13.1(d), *meta*-H → δ 9.1 (t), and *para*-H → δ 8.8(t)], but also causes each multiplet to separate into two essentially identical multiplets. Explain these observations [See Whitesides, G. M., and Lewis, D. W., *J. Am. Chem. Soc.*, 92, 6979 (1970).]

10.6 Compound **10-21** exhibits some interesting properties. For example, when **10-21** is added to a 0.010 *M* solution of resorcinol (**10-22**), the circled hydrogen

of resorcinol shifts upfield from δ 6.50, approaching δ 3.80 as the ratio [**10-21**]$_0$/[**10-22**]$_0$ approaches 3.0:

(a) Explain specifically what this behavior indicates.

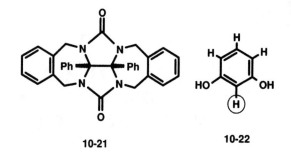

10-21   **10-22**

(b) The chemical shift for the circled hydrogen is δ 5.15 when the ratio [**10-21**]$_0$/[**10-22**]$_0$ is 0.538. Calculate the value of *K*. [See Sijbesma, R. P., Kentgens, A. P. M., and Nolte, R. J. M., *J. Org. Chem.*, 56, 3199 (1991).]

10.7. Verify Eq. (10.4).

11

ELECTRON PARAMAGNETIC RESONANCE SPECTROSCOPY AND CHEMICALLY INDUCED DYNAMIC NUCLEAR POLARIZATION

11.1 ELECTRON PARAMAGNETIC RESONANCE

From Section 2.1.2 you will recall that individual electrons, like protons, have magnetic spin ($I = \frac{1}{2}$) and can therefore adopt either of two precessing spin orientations, $m_s = +\frac{1}{2}$ or $-\frac{1}{2}$ (differing slightly in energy), when immersed in an external magnetic field (Figure 2.1). Thus, in a molecule with an *unpaired* electron, it should be possible to induce transitions between these *electronic* spin energy levels by subjecting the electrons to electromagnetic radiation equal in frequency to their precessional frequency. This fact is the basis of **electron paramagnetic resonance (EPR) spectroscopy**,[1,2] also known as **electron spin resonance (ESR) spectroscopy**, a technique that traces its origin to experiments by Zavoisky in 1945.

The magnetogyric ratio of a free electron is approximately 657 times that of a proton. Modern EPR spectrometers use a microwave generator (klystron) as the source of electromagnetic radiation (i.e., the oscillating magnetic field), with operating frequencies in the range 1–100 GHz (1 GHz = 10^3 MHz = 10^9 Hz); 9.5 GHz (the so-called *X*-band) is perhaps the most common.

■ **EXAMPLE 11.1** What magnetic field strength would be appropriate for a 9.500-GHz EPR spectrometer? For a 35.00-GHz (*Q*-band) spectrometer?

□ *Solution:* From Table 2.1 we know that for a proton the ratio of resonance frequency (in megahertz) to magnetic field strength (in tesla) is 42.576. For an electron, therefore, the ratio is 657(42.576) = 28.0 GHz/T. Therefore, a 9.500-GHz instrument requires a magnetic field of 9.500/28.0 = 0.339 T. The 35-GHz spectrometer needs a

field strength of 1.25 T. Thus, typical EPR spectrometers operate at 20–100 times the frequency of NMR spectrometers using conventional magnets that are about 1/10th as strong. □

The design of EPR spectrometers resembles that of a field-sweep NMR instrument (Section 3.3.2), though pulsed-mode (Fourier transform; Section 3.4) EPR spectrometers are now available. Many of the considerations (such as field stability, lineshape, saturation, relaxation, etc.) that were discussed in Chapters 2 and 3 for NMR are also important in EPR,[1] but there are some significant differences.

11.2 FREE RADICALS

There is a fly in the ointment. EPR spectroscopy can detect only molecules that possess one or more *unpaired* (Sections 2.1.2 and 9.2.1) electrons. Such species are described as **paramagnetic** because their electronic magnetic moments interact strongly with an external magnetic field and they are attracted into the field. Some inorganic ions and organometallic complexes are paramagnetic by virtue of unpaired electrons in the *d* orbitals of certain transition metals; such compounds can be studied by EPR techniques. However, almost without exception, all stable organic molecules (those composed of main-group nonmetallic elements) have an even number of electrons arranged in pairs and are therefore **diamagnetic** (Section 6.1); they are weakly repelled by an external magnetic field. Electron paramagnetic resonance spectroscopy is of no use for studying diamagnetic compounds.

The few organic molecules that do possess an unpaired electron and are paramagnetic are called **free radicals**. They are normally short-lived because of the tendency for pairs of radicals to **recombine**, that is, "pair up" their unpaired electrons, thereby creating a new chemical bond (which, after all, is nothing more than a shared *pair* of electrons). Free radicals are usually encountered as **intermediates** that are formed and then consumed during chemical reactions. These radicals are generated by **homolytic cleavage** of a chemical bond, requiring an amount of energy termed the **bond dissociation energy**. This process leaves one unpaired electron (indicated either by a dot or an arrow) on each fragment radical:

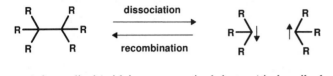

A free radical (with just one unpaired electron) is described as an electronic **doublet** because, in an external magnetic field, the electron can only exist in one of two possible spin states ("up" or "down"). By contrast, a pair or radicals, or a **biradical** (a species with two unpaired electrons in the same molecule) can exist in either of two electronic states: singlet or triplet. In the singlet state the electrons are paired (opposite spin); a singlet radical pair (or biradical) is thus diamagnetic ($M_s = \frac{1}{2} - \frac{1}{2} = 0$) and not observable by EPR. The radical pair above is shown in the singlet state.

In the triplet state the electrons are not paired, so there are three spin combinations: both up ($M_s = \frac{1}{2} + \frac{1}{2} = 1$), both down ($M_s = -\frac{1}{2} - \frac{1}{2} = -1$), or one of each ($M_s = \frac{1}{2} - \frac{1}{2} = 0$). But wait! How is the $M_s = 0$ combination of the triplet different from a singlet? The answer lies in the precessional *phase* of the two

electrons. As shown in Figure 11.1a, the singlet state has zero net magnetization because the two electron magnetic moments are precessing 180° out of phase, completely canceling each other. By contrast, in all three arrangements of the triplet state (Figure 11.1b) the electron magnetic moments are precessing exactly in phase, leading to a nonzero vector sum, even in the $M_s = 0$ state. Thus, a triplet radical pair (or biradical) is paramagnetic and can be studied by EPR.

11.3   THE *g* FACTOR

The precessional frequency (ν) of an unpaired electron is directly proportional to the applied magnetic field strength (**B**) and can be expressed by the equation

$$\nu = g\mathbf{B}\,\frac{\beta_e}{h} \tag{11.1}$$

where β_e is the **Bohr magneton** (9.2740×10^{-24} Joule T^{-1}) and *h* is Planck's constant (Section 1.2), so $\beta_e/h = 1.3996 \times 10^{10}$ Hz T^{-1}. The parameter *g* is a dimensionless proportionality constant called the **electron Zeeman factor**, or ***g* factor**, for short. Compare Eq. (11.1) with Eq. (2.6), which deals with the precession of magnetic nuclei. There is a close parallel between $g\beta_e/h$ and the magnetogyric ratio γ. Just as the value of γ determines the precessional frequency of a magnetic nucleus in a given magnetic field, the value of $g\beta_e/h$ determines the precessional frequency of an unpaired electron in a given magnetic field. The value of *g* for a free electron (g_e) is one of the most accurately known of all physical constants, $2.002319304386 \pm 0.000000000020$.

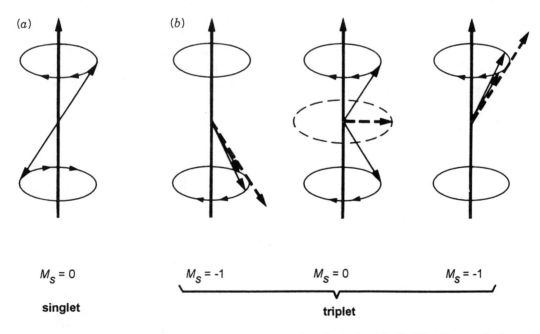

Figure 11.1. (*a*) Singlet and (*b*) triplet electronic states of a radical pair or biradical. Boldface (vertical) arrow indicates the direction of **B₀**; dashed arrows represent the net magnetic vector of the electrons.

Recalling that subtle changes in molecular environment and shielding can cause nuclei to precess at *slightly* different frequencies (Section 6.1), it should come as no surprise that similar environmental effects can bring about small but measurable changes in precessional (and hence, resonance) frequencies of unpaired electrons. So, we will adopt the g factor $[g = \nu/(1.3996 \times 10^{10})\ \mathbf{B}]$ as the signal position parameter ("chemical shift," if you will) in EPR spectroscopy. If a field-sweep instrument is being used, ν will be constant and $\mathbf{B}$ in Eq. (11.1) will be the value of the magnetic field at the center of the signal. In an FT instrument, $\mathbf{B}$ will be fixed, and the value of ν for each signal will be used to calculate the g factor. Moreover, since most of the species we will be studying by EPR have just one unpaired electron, there will only be one signal!

Table 11.1 lists the g factors for a variety of different free radicals, listed in the order of increasing g factor (decreasing shielding). At first glance, it might seem that all the values appear to be quite similar. But, as Example 11.2 shows, these "small" differences in g actually correspond to large and easily measured differences in field strength or frequency.

■ **EXAMPLE 11.2** (a) Assume we are using a 9.41756-GHz instrument. The signal of interest occurs at a field strength of 157.27 mT (1 mT = 10^{-3} T). Evaluate the g factor of this signal. (b) Assume we are using a 0.3400-T FT instrument. Calculate the difference in precessional frequencies between a trichloromethyl radical and an alkylperoxy radical.

TABLE 11.1 Values for g Factors of Several Common Organic Radicals[a]

Radical	g Factor
Hydrogen (atom)	2.0022
Vinyl	2.00220
Free electron	2.00232
Anthracene cation	2.00249
Allyl	2.00254
Methyl	2.00255
Perylene cation	2.002569
Ethyl	2.00260
Perylene anion	2.002657
Naphthalene anion	2.002743
Benzene anion	2.00276
DPPH[b]	2.00354
p-Benzoquinone anion	2.004665
Triphenylphosphine cation	2.00554
Alkylcarboxy	2.0058
Trichloromethyl	2.0091
Alkylperoxy	2.0155

[a]Data from refs. 1 and 2.
[b]2,2-Diphenyl-1-picrylhydrazyl.

□ *Solution:* (a) $g = \nu/(1.3996 \times 10^{10})\ \mathbf{B} = 9.41756 \times 10^9/[(1.3996 \times 10^{10})\ (0.15727)] = 4.2785$. (b) We can recast Eq. (11.1) into the form $\Delta\nu = \Delta g\ \mathbf{B}\ (\beta_e/h)$, where Δ represents the "difference." From Table 11.1, $\Delta g = 2.0155 - 2.0091 = 0.0064$. Therefore, $\Delta\nu = (0.0064)(0.3400)(1.3996 \times 10^{10}) = 3.0 \times 10^7$ Hz = 30 MHz (or about 3000 ppm). □

Because of the huge (compared to NMR) frequency (or field) differences between EPR signals, concerns about resolution, line widths, field homogeneity, and field stability are less worrisome than they were in NMR. Lock signals are not needed; the instrument can be readily calibrated with any of a number of paramagnetic radicals. Therefore, it is not necessary that the sample be dissolved in solution or that it be spun in the probe. Indeed, a large fraction of EPR spectra are run on solid-state samples, either crystalline or in a glassy matrix at very low temperature (to help stabilize the radicals). The g factors listed in Table 11.1 are for solution phase samples, and since these radicals are tumbling freely through the medium, their g factors are averaged over all environments and orientations. Such values are referred to as **isotropic** g factors. When radicals are confined to the crystalline state, such averaging does not take place, so the g factor is **anisotropic**, varying in each of the three Cartesian directions. For the purposes of this chapter, we will confine ourselves to the isotropic g factors.

There are some organic free radicals that are sufficiently stable to last indefinitely, even at room temperature. Such persistent radicals owe their greatly increased stability to one or more of the following structural features: The unpaired electron is formally located on a heteroatom (e.g., oxygen or nitrogen) such that the bond resulting from recombination would be weak (e.g., O–O); the unpaired electron is delocalized over a highly conjugated π system; and/or the molecular structure in the immediate vicinity of the unpaired electron is sterically highly congested, inhibiting recombination. A few examples of such radicals are shown below:

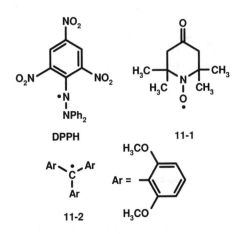

Especially notable are nitroxide radicals such as **11-1**. Molecules of this class are sufficiently stable to survive a variety of chemical reactions, allowing them to be covalently bonded to other molecules (e.g., biopolymers such as proteins and nucleic acids). They can then serve as **spin labels**, transmitting information (via their EPR spectra) about the molecules to which they are attached.

11.4 SENSITIVITY CONSIDERATIONS

In EPR, like ^{1}H NMR, signal intensity is directly proportional to the number of unpaired electrons giving rise to the signal. However, we learned in Chapter 2 that different isotopes have different inherent sensitivity toward generating an NMR signal (Table 2.1). In addition, we saw how the intensity of an NMR signal is very sensitive to saturation effects because of the small differences in spin state populations. Do you expect saturation problems to be as important a consideration in EPR?

■ **EXAMPLE 11.3** (a) What is the energy gap (in joules) between the two spin states of a free electron at 0.34 T? (b) What is the equilibrium population ratio of these states at room temperature?

□ *Solution:* (a) Recall from Section 2.3.1 that the energy gap between two spin states is proportional to the precession frequency, which must be matched by the irradiation frequency:

$$\Delta E = h\nu = (6.63 \times 10^{-34} \text{ J s})(9.50 \times 10^9 \text{ s}^{-1})$$

$$= 6.30 \times 10^{-24} \text{ J}$$

(b) Use the Boltzmann distribution [Eq. (2.8)]:

$$\frac{P_{m=1/2}}{P_{m=-1/2}} = \exp\left(\frac{-\Delta E}{kT}\right)$$

$$= \exp\left[-(6.30 \times 10^{-24}\text{J})/\right.$$

$$(1.38 \times 10^{-23}\text{J K}^{-1})\,(298 \text{ K})] = 0.998486$$

This ratio indicates that at equilibrium 50.0379% of the electrons are in the lower ($m_s = -\frac{1}{2}$) state, while 49.9621% are in the upper state. Or, to put it another way, of one million electrons, there are 758 more in the lower state than in the upper state. Although this may seem like a very small difference, it is 38 times greater than the difference in proton spin state populations, a difference of only 20 nuclei per million at 2.35 T (Example 2.8)! □

The fact that the population difference between spin states is greater for electrons than for nuclei means that EPR spectroscopy is much more sensitive than NMR. Thus, while modern NMR methods (Section 3.3) still require sample concentrations of at least 0.01 M (moles per liter), EPR signals can be detected from radicals in as low a concentration as 10^{-8} M at room temperature (10^{-12} mol in a sample volume of 0.1–0.2 mL).

In Chapter 1 we discussed the concept of spectroscopic time scale. Because EPR involves frequencies on the order of 10^9 Hz (and a resulting time scale of 10^{-9} s), it takes a much faster "snapshot" of dynamic systems than does NMR. As a result, EPR can generate information about chemical processes that are too fast to study by NMR.

You will notice one other difference between an EPR spectrum and an NMR spectrum. All the Lorentzian-shaped NMR signals (Section 3.5.1) encountered so far in this book are displayed in the **absorption mode** (see Figure 3.18). By contrast, EPR signals are usually detected and displayed in the **dispersion mode** (Figure 11.2). You might recognize a dispersion signal as essentially the first derivative (slope) of the corresponding absorption signal (see review problem 11.5). The position ($\Delta\nu_i$ or g) of a dispersion mode signal is the point where the signal line crosses the baseline.

11.5 HYPERFINE COUPLING AND THE *a* VALUE

Just as the NMR signal of one nucleus can be split into a multiplet by other neighboring magnetic nuclei (Chapters 8 and 9), the EPR signal of the unpaired electron in a radical can be split by magnetic nuclei within the radical. This coupling follows *exactly* the same first-order (Pascal's triangle) coupling rules we saw in Section 8.8. Thus, the number of lines in an EPR multiplet is governed by Eq. (8.3). The only differences are that the separation between the lines of a multiplet in an EPR spectrum is called the **hyperfine coupling**, is given the symbol *a* (rather than *J*), and is expressed in the unit of magnetic field strength [millitesla (10^{-3} T = 10 G) rather than hertz].

■ **EXAMPLE 11.4** Predict the position and multiplicity of the EPR signals of (a) a hydrogen atom and (b) a methyl radical ($H_3C\bullet$).

□ *Solution:* (a) The signal will be centered at a *g* value of 2.0022 (Table 11.1) and will be split into a 1 : 1 doublet by the neighboring hydrogen nucleus. (b) The signal will be centered at a *g* value of 2.00255 (Table 11.1) and will be split into a quartet (intensity ratio 1 : 3 : 3 : 1) by the three equivalent neighboring hydrogen nuclei. □

As was true for the sign and magnitude of coupling constant *J* (Chapter 9), the sign and magnitude of a given *a* value depends on the "proximity" of the unpaired electron to the

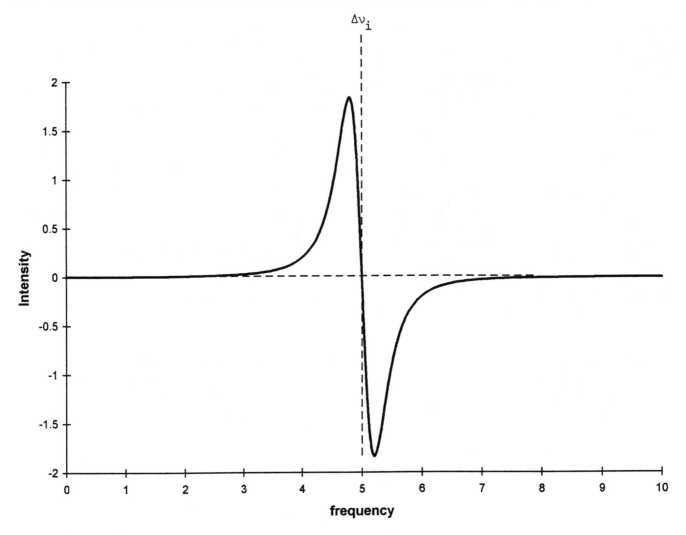

Figure 11.2. Dispersion mode (first-derivative) Lorentzian signal. The position of a dispersion mode signal is where the line crosses its baseline; the halfwidth is the horizontal distance between the maximum and the minimum of the signal curve. Compare this signal with the absorption mode Lorentzian signal in Figure 3.18; both were plotted using the same parameter values (see review problem 11.5).

coupling nuclei. But the word proximity is harder to define for an electron than for a nucleus. Actually, the unpaired electron is to some extent delocalized through the entire molecule by resonance (Sections 7.4 and 9.6), spending a (not necessarily equal) portion of its time near each nucleus. The fraction of time it spends near a given nucleus is called the **electron spin density** (ρ) at that nucleus, and the *magnitude* of the a value for coupling to that nucleus is directly proportional to ρ at that nucleus. In fact, the value of a is the best experimental measure of how the unpaired electron density is distributed in a radical (see review problems 11.1 and 11.2). For example, the magnitude of the a values for the hydrogen atom and the methyl radical (Example 11.4) are 51 and 2.30 mT, respectively. This is because the electron in a hydrogen atom is localized in an orbital centered on the hydrogen

nucleus, while the unpaired electron in the methyl radical is formally located in an orbital centered on the carbon nucleus, one bond away from the hydrogen nuclei. Similarly, the magnitudes (in millitesla) of the a values for the propyl radical (structure **11-3**) are $a_\alpha = 2.21$, $a_\beta = 3.32$, and $a_\gamma = 0.04$,[2] indicating that the unpaired electron has a higher spin density at the H_β's than at the H_α's:

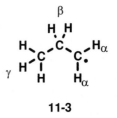

11-3

It may surprise you that a_β is larger than a_α, but this fact is consistent with the way the unpaired electron is delocalized through the molecule. In particular, another contributing "no-bond" (**hyperconjugative**) resonance structure for the propyl radical is written with the unpaired electron directly on H_β:

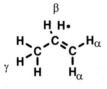

There is no valid resonance structure that places the electron directly on H_α or H_γ.

■ **EXAMPLE 11.5** How many lines would you expect to find in the EPR spectrum of the propyl radical and where would the signal be centered?

□ *Solution:* Because of its similarity to the ethyl radical (Table 11.1), we predict that the signal should occur near $g = 2.00260$. It will be split into a triplet (by the two H_β's) of triplets (by the two H_α's) of quartets (by the three H_γ's), a total of 36 lines! Alternatively, we could have used Eq. (8.3):

$$L = (2 \times 2 \times \tfrac{1}{2} + 1)(2 \times 2 \times \tfrac{1}{2} + 1)(2 \times 3 \times \tfrac{1}{2} + 1) = 36$$

Just as we did for ${}^1H$ NMR, we can neglect hyperfine coupling to the carbons because of the low natural abundance of ${}^{13}C$ (Table 2.1). □

■ **EXAMPLE 11.6** An a value of 3.32 mT at 0.34 T and 9.5 GHz is equivalent to what value in hertz?

□ *Solution:* We can set up a simple proportion:

$$\frac{3.32\text{ mT}}{340\text{ mT}} = \frac{x}{9.5 \times 10^9\text{ Hz}}$$

$$x = 9.3 \times 10^7\text{ Hz} = 93\text{ MHz}$$

Compare this 93 MHz to typical ${}^1H-{}^1H$ coupling constants of ca. 10 Hz! □

As with homo- and heteronuclear coupling constants (Section 9.2), a values have signs. A useful generalization (though not without exceptions) is that the sign of a is a function of the number of bonds separating the unpaired electron from the coupling nucleus. If the separation is zero or an even number of bonds, a will be positive; if the separation is an odd number of bonds, a will be negative. Note that this pattern is exactly the reverse of what we saw for nuclear coupling constants and is due to the negative charge of the electron.

■ **EXAMPLE 11.7** Predict the sign of the a values of the hydrogen atom, the methyl radical, and the propyl radical.

□ *Solution:* For $H\bullet$ (zero-bond separation) the a value is positive; for $H_3C\bullet$ (one-bond separation) the a value is negative, and for the propyl radical, a_α (one-bond separation) is negative while a_β (two-bond separation) is positive. □

Although the sign of a has no effect on the appearance of an EPR spectrum [just as the sign of J has no effect on an NMR spectrum (Section 9.1)], it *does* have an effect on CIDNP behavior (Section 11.7).

11.6 A TYPICAL EPR SPECTRUM

Because resonance interactions play such an important role in determining electron spin density and, therefore, the appearance of EPR spectra, let us consider another example of such effects. The compound **11-4**, BHT (review problem 6.4e), is a common antioxidant that owes its chemical properties to the ease with which it donates a hydrogen atom to an attacking radical ($R\bullet$), thereby forming phenoxy radical **11-5**. This radical exists as a hybrid of five principal resonance structures, **11-5a–e**. From these structures it is apparent that **11-5** has a vertical plane of symmetry (as well as a C_2 axis), relating its left and right halves. The electron spin density is concentrated mainly on the oxygen (in **a**), on the ortho carbons (**b** and **c**), on the para carbon (**d**), and the methyl hydrogens (**e**):

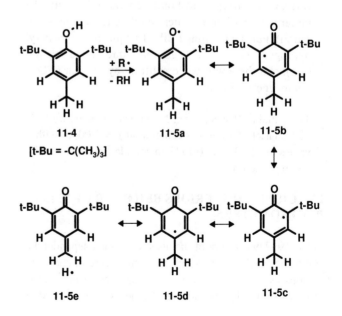

$g = 2.004$

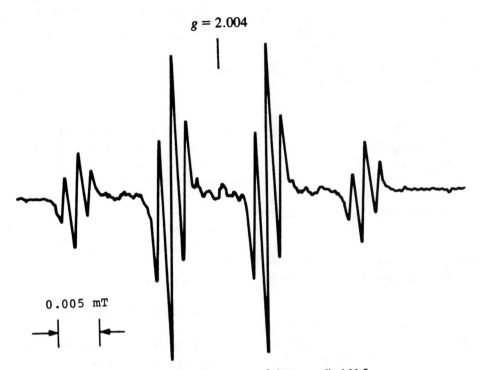

0.005 mT

Figure 11.3. The EPR spectrum of phenoxy radical **11.5**.

■ **EXAMPLE 11.8**   Predict the appearance of the EPR spectrum of phenoxy radical **11-5**.

□ *Solution:* The signal should be centered at the position of resonance-stabilized oxygen radicals (with a g factor in the range 2.0036–2.0058; Table 11.1). It will be split into a quartet by the three methyl hydrogens, and each line of the quartet will be further split into a triplet by the two equivalent aromatic hydrogens. The a value for the former coupling (zero-bond separation) should be larger than that for the latter coupling (two-bond separation). Further splitting by the 18 equivalent *tert*-butyl hydrogens should be negligible (three-bond separation), and there will be no splitting by oxygen (^{16}O is nonmagnetic; Table 2.1).

The actual EPR spectrum of **11-5** is shown in Figure 11.3. The quartet of triplets is centered at $g = 2.0040$, with a values of +1.13 and +0.15 G, in complete agreement with our expectations.   □

11.7   CIDNP: MYSTERIOUS BEHAVIOR OF NMR SPECTROMETERS

Back in 1967 two groups of chemists, working independently, were using NMR spectroscopy to monitor certain chemical reactions as they were occurring. They became quite concerned when the signals portrayed on their spectra seemed to make no sense. Some of the signals were much more intense than they should have been, while other signals were negative (upside down). And in some cases multiplets were composed of both positive and negative signals! Only after the NMR repairman assured them that their instruments were in perfect working order did it become apparent that a new effect had been discovered. This effect has come to be known as **chemically induced dynamic nuclear polarization**, or **CIDNP** (sometimes pronounced "kidnap").[3–6]

11.8   THE NET EFFECT

In order to understand these unexpected observations, we need to review the importance of nuclear spin state populations on NMR signal intensity. As first discussed in Section 2.3 (and again in Section 11.4), the intensity of an NMR signal depends on the ability of the spin system to absorb photons of electromagnetic radiation. The two requirements for this to happen are (a) the frequency of the electromagnetic irradiation must match the nuclear precessional frequency and (b) there must be an *excess* of nuclei in the lower energy spin state. Because at equilibrium the difference in nuclear spin state populations is very small (about 20 ppm; Example 2.8), any factor that affects the relative populations of the two states even slightly can have a profound effect on signal intensity. For example, increasing relaxation rates (Section 2.3), increasing magnetic field strength, or decreasing the temperature of the sample all serve to increase the equilibrium

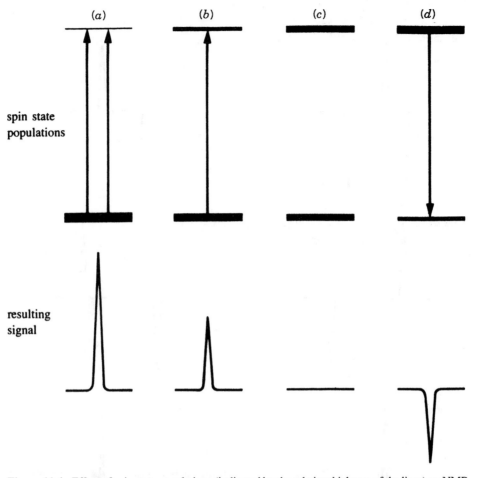

Figure 11.4. Effect of spin state populations (indicated by the relative thickness of the lines) on NMR signal intensity. (*a*) Larger than normal excess in lower state, giving enhanced absorption; (*b*) normal slight excess in lower state, giving normal signal intensity; (*c*) saturation, giving zero signal; and (*d*) inverted population (excess in upper state), giving net emission.

population of the lower state and increase signal intensity. Conversely, if relaxation is slow, the system can become saturated (equal populations in both states; Section 2.3) so that no net absorption is possible. It turns out that certain chemical reactions in which free radicals are involved can form products whose nuclei have non-Boltzmann spin state populations, and these reactions can give rise to the dramatic CIDNP effects. We describe a spin system with nonequilibrium spin state populations as **polarized**.

Suppose, for example, a reaction product is formed in which the nuclei have a polarized spin state distribution, with larger than Boltzmann population in the lower state. This would make the absorption process more favorable, and the resulting signal intensity would be greater than normal. This situation is referred to as **enhanced absorption**, or an *A* **net effect**. On the other hand, suppose for some reason the spin state distribution were polarized in the opposite way, with more nuclei in the *upper* state. Now, not only is absorption

disfavored, but net **emission** of radiation (at the precessional frequency) takes place as the molecules attempt to reestablish equilibrium. This emission of radiation, called an *E* **net effect**, results in a negative NMR signal. These effects are depicted in Figure 11.4.

11.9   THE MULTIPLET EFFECT

As mentioned in Section 11.7, the most baffling aspect of NMR spectra showing CIDNP effects was the occurrence of both positive and negative signals within the same multiplet. Sometimes the left half of the multiplet was positive and the right half negative (an *A/E* **multiplet effect**), while sometimes the reverse was observed (an *E/A* **multiplet effect**). These are shown in Figure 11.5.

As with the net *E* and *A* effects, the multiplet effects can also be rationalized on the basis of nonequilibrium spin state populations. Recall the AB spin system we discussed in

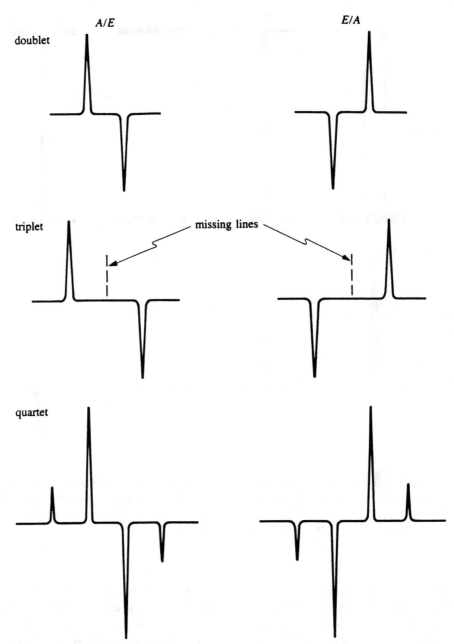

Figure 11.5. Multiplet effects in a doublet, triplet, and quartet. Note how the middle line of the triplet vanishes.

Section 9.1. Consider Figure 11.6, which is derived from Figure 9.2. Suppose a chemical reaction generated a product containing an AB spin system (with $J > 0$), where the polarization increased the populations of the *parallel* spin states (Figure 11.6a). Lines 1 and 2 would exhibit an A effect, while lines 3 and 4 would exhibit an E effect, resulting in E/A multiplet effects in both doublets. Alternatively, if the *antiparallel* states were favored (Figure 11.6c), an A/E multiplet effect would be observed in both doublets.

■ **EXAMPLE 11.9**   How would a negative value of J affect the type of multiplet effects in the CIDNP spectra of an AB system?

☐ *Solution:* You may remember that changing the sign of J changes the ordering of the lines of each multiplet (Figure 9.2c). If the system were now polarized in favor of the parallel states, we would observe A/E multiplet effects, while antiparallel polarization would result in E/A

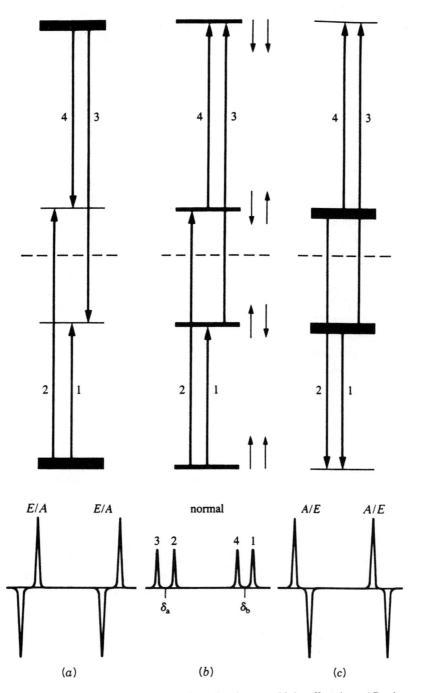

Figure 11.6. Way in which spin state populations give rise to multiplet effects in an AB spin system.

plet effects. Thus, the sign of J is one factor that governs whether the multiplet effect is A/E or E/A.   □

11.10 THE RADICAL-PAIR THEORY OF THE NET EFFECT

Early explanations of CIDNP were based solely on Overhauser effects between an unpaired electron and a nucleus (Section 12.3, Example 12.4), but these early theories failed

to account for several important aspects of CIDNP, such as multiplet effects. Nearly two years passed before a satisfactory theory explaining CIDNP emerged. It is now believed that chemical reactions involving *pairs* of interacting free radicals are responsible for all CIDNP effects.

When the unpaired electrons of two radicals interact within a magnetic field, there are four possible combinations of their precessing spins: the singlet and the three components of the triplet (Section 11.2). If both electrons are equivalent (having the same g factor), their phase relationship will remain constant. That is, a singlet will remain a singlet and a triplet will remain a triplet. But, if the two electrons precess at slightly *different* frequencies (because their g factors are slightly different), their relative phasing will change with time. Although this constantly changing phase relationship will have no significant effect on the $M_s = \pm 1$ components of the triplet radical pair, it *will* cause the $M_s = 0$ component of the triplet state (Figure 11.1*b*) to change periodically into the singlet state (Figure 11.1*a*), then change back again. The rate of this **singlet–triplet mixing** will increase as the difference in precessional frequencies between the two unpaired electrons increases. Therefore, any factor (such as hyperfine coupling to a nearby nucleus) that affects the precessional frequency (by splitting it into several values) will also influence the rate of dephasing and rephasing.

When a radical pair is formed in solution as the result of homolytic bond cleavage, the two radicals remain close together for a short time, held in place by a "cage" of solvent molecules. In order for the radicals to recombine (Section 11.2), the two electrons must be paired, that is, in the singlet state. If the radical pair is in the triplet state, the radicals must stay together long enough for the $M_s = 0$ triplet to mix into the singlet. If this mixing takes too long, the radicals may escape the solvent cage *before* rephasing (and recombination) occurs. Radicals that escape the cage are unlikely to encounter other radicals (because of their exceedingly low concentration), forming instead other types of so-called **escape products**. Thus, recombination products will be formed from radical pairs with nuclear spins that favor the singlet electronic state, while escape products will be formed from radical pairs whose nuclear spins favor the triplet electronic state. This effect has been called **nuclear spin sorting**.

The direction of polarization, and hence the type of net CIDNP effect (A or E), can be predicted if you know the signs of four variables:

1. μ, the original spin state of the radical pair at its "birth" (+ for triplet, – for singlet);

2. ε, the type of reaction leading to the observed product (+ for cage recombination, – for escape);

3. $\Delta g = g_1 - g_2$, the difference in g factors between the two radicals, where g_1 refers to the radical with the nucleus being observed; and

4. a, the hyperfine coupling between the electron and the nucleus being observed.

The type of net effect, given the symbol Γ_n, is determined by the sign of the product of the four variables listed above:

$$\Gamma_n = \mu\varepsilon\,\Delta g\,a \qquad (11.2)$$

In the above equation (attributed to Kaptein) a positive Γ_n corresponds to an A net effect; a negative Γ_n corresponds to an E net effect. The magnitude of the net effect is highly variable and depends on several factors. However, the maximum A net effect can result in an intensity enhancement by a factor of several hundred (see Section 12.3).

■ **EXAMPLE 11.10** Can a net effect be observed with a pair of identical radicals?

□ *Solution:* No! Because $\Delta g = 0$, there will be no dephasing or rephasing of the spins. Thus, in order to generate a net effect, two radicals must have different g factors. □

■ **EXAMPLE 11.11** When the peroxide with structure **11-6** is heated in the presence of iodine, four major products are formed: carbon dioxide and compounds **11-7**, **11-8**, and **11-9**. When the reaction is monitored by ^{1}H NMR, the signal for **11-7** (δ 2.1) exhibits a strong A net effect, while the signal for **11-9** (δ 2.7) shows a strong E net effect:

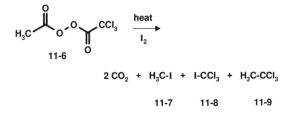

(a) Rationalize these CIDNP results. (b) Why is it that **11-8** does not show a net effect?

□ *Solution:* (a) Heating **11-6** causes the relatively weak O–O bond to break, leading (after expulsion of two CO_2 molecules) to a pair of radicals ($H_3C\bullet$ and $\bullet CCl_3$). The horizontal bars in the reaction scheme below indicate a radical pair formed within a solvent cage:

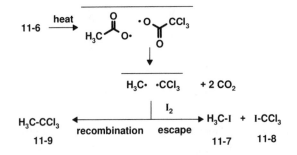

Either the radical pair can undergo cage recombination to form **11-9** or the individual radicals can escape and react with I_2 to form **11-7** and **11-8**.

The signal for recombination product **11-9** shows net emission, so Γ_n must be negative. What does this tell us about μ, the original spin state of the radical pair? We know the g factors for both radicals (Table 11.1), and therefore $\Delta g = 2.0025 - 2.0091$, which is negative. The value of a for $H_3C\bullet$ is negative (Example 11.7), and ε is positive for a recombination product. Thus, expressing each variable in Eq. (11.2) by its sign, we have

$$\Gamma_n = (-) = \mu\varepsilon\,\Delta g\,a = \mu\,(+)\,(-)\,(-)$$

This equation can be true only if μ is also negative, indicating that the radical pair was initially formed in the singlet state. This is eminently reasonable when you consider that precursor **11-6** had all its electrons paired. This result is consistent with the notion that thermal fragmentation of a chemical bond preserves the spin state of the electrons, normally the singlet. Many photochemical fragmentations, on the other hand, proceed by way of electronically excited triplet states and produce triplet radical pairs.

Using the negative value of μ deduced above, let us predict the type of net effect for escape product **11-7**. Everything is the same as before, except that ε is now negative (for an escape product). Thus,

$$\Gamma_n = \mu\varepsilon\,\Delta g\,a = (-)\,(-)\,(-)\,(-) = (+)$$

So, net enhanced absorption is predicted, in agreement with experimental observation.

(b) This is a trick question! Since $ICCl_3$ has no hydrogens, it produces no ^1H NMR signal! However, it might show CIDNP effects in its ^{13}C spectrum (see Example 11.13).   □

From Example 11.11 you can infer that the greatest values of CIDNP are its ability to confirm the involvement of radical pairs in reactions, to determine the spin state of the radical pair precursor at its "birth," and to distinguish between products that arise via recombination and those that arise from escape.

11.11 THE RADICAL-PAIR THEORY OF THE MULTIPLET EFFECT

We saw in Section 11.9 how multiplet effects result from coupled spin states that become polarized. The type of multiplet effect (E/A or A/E) can be predicted in much the same way as the net effect [Eq. (11.2)], except that more parameters must be considered. The multiplet effect (Γ_m) is given by the sign of the product:

$$\Gamma_m = \mu\varepsilon a_i a_j J_{ij}\sigma \qquad (11.3)$$

where μ and ε have the same meaning as before, a_i and a_j are the hyperfine couplings of the electron to nuclei i and j, and J_{ij} is the coupling constant they share. The sign of σ is positive if nuclei i and j are in the same radical, negative if they are in separate radicals. A *positive* Γ_m indicates an E/A multiplet effect; a *negative* Γ_m corresponds to an A/E effect (Figure 11.5). Note that Γ_m does not involve g factors, so even if the two radicals in the pair are identical, they can still generate a multiplet effect.

■ **EXAMPLE 11.12** When symmetrical peroxide **11-10** is thermally decomposed in a chlorinated solvent (abbreviated S–Cl), the major products, beside CO_2, are compounds **11-11** and **11-12**[5]:

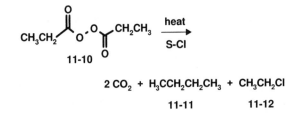

$$2\ CO_2 + \underset{\textbf{11-11}}{H_3CCH_2CH_2CH_3} + \underset{\textbf{11-12}}{CH_3CH_2Cl}$$

Predict the type of ^1H multiplet effect expected for products **11-11** and **11-12**.

□ *Solution:* From our discussion in Example 11.11, we can assume that the product-determining radical pair (**11-13**) is "born" in the singlet state. Furthermore, it is reasonable to expect that **11-11** arises from recombination within the cage, while **11-12** is an escape product:

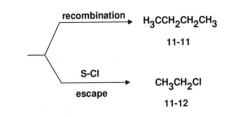

The ^{1}H spectrum of escape product **11-12** consists of the familiar (Chapter 8) quartet and triplet. With regard to the precursor ethyl radical, we know (from Section 11.5) that a_α is negative and a_β is positive, while (from Section 9.2) $J_{\alpha\beta}$ is positive:

Since both H_α and H_β occur in the same radical, σ is positive. Thus, for the CH_2 quartet we predict

$$\Gamma_m = \mu\varepsilon a_i a_j J_{ij}\sigma = (-)\,(-)\,(-)\,(+)\,(+)\,(+) = (-)$$

for an A/E multiplet effect. For the CH_3 triplet,

$$\Gamma_m = (-)\,(-)\,(+)\,(-)\,(+)\,(+) = (-)$$

and again an A/E multiplet effect is predicted. In perfect agreement with our expectations, both multiplets of **11-12** exhibit A/E multiplet effects.[5] In the case of recombination product **11-11**, the only difference is that ε is now positive. Therefore, both multiplets are predicted to show E/A multiplet effects, and they do!   □

11.12   A FEW FINAL WORDS ABOUT CIDNP

In many cases an NMR spectrum exhibits both net and multiplet CIDNP effects. In many of these cases the net E or A effect is superimposed on the multiplet effect of a given signal. For example, when singlet radical pair **11-14** undergoes recombination to **11-15**, the ^{1}H spectrum of the CH_2 group in the product shows an E/A multiplet effect superimposed on an E net effect[3]:

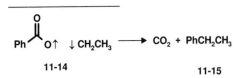

11-14   **11-15**

The E net effect can be rationalized by Eq. (11.2). Recalling that the g factor for ethyl radical is less than that of an alkyl carboxy radical (Table 11.1), we calculate for the CH_2 quartet

$$\Gamma_n = \mu\varepsilon\,\Delta g\,a = (-)(+)(-)(-) = (-)$$

for a net E effect. The E/A multiplet effect in the quartet can be predicted from Eq. (11.3):

$$\Gamma_m = \mu\varepsilon a_i a_j J_{ij}\sigma = (-)(+)(-)(+)(+)(+) = (+)$$

for an E/A multiplet effect.

And finally, as hinted in Example 11.11b, CIDNP effects are not limited to ^{1}H spectra. Below is an example involving ^{13}C NMR.

■ **EXAMPLE 11.13**   The photolytic (light-induced, designated $h\nu$) decomposition of compound **11-16** in solvent CCl_4 is believed to involve singlet carbene (a divalent carbon with two unshared electrons) intermediate **11-17**. This carbene abstracts a chlorine atom from solvent to give singlet radical pair **11-18**, which ultimately undergoes cage recombination to product **11-19**:

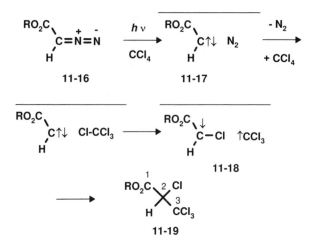

A net effect (E or A) was observed in each ^{13}C signal of product **11-19**. Predict the type of net effect for each carbon.

□ **Solution:**   To calculate Γ_n for each carbon signal, we need to know (a) the sign of Δg (g factors: $\bullet CCl_3$, 2.0091; RO_2C–$C(\bullet)HCl$, 2.003) and (b) the signs of the hyperfine couplings to the carbons (a_1 negative, a_2 positive, a_3 positive). With these data and the fact that we are dealing with cage recombination ($\varepsilon = +$) of a singlet radical pair ($\mu = -$), we predict the following net effects:

Carbon	$\Gamma_n = \mu\varepsilon\,\Delta g\,a$	Net Effect
1	$(-)(+)(-)(-) = (-)$	E
2	$(-)(+)(-)(+) = (+)$	A
3	$(-)(+)(+)(+) = (-)$	E

Believe it or not, this is exactly what was observed![7]   □

The main thing to remember from this discussion is that CIDNP effects are seen in the NMR spectra of the products *as they are being formed* from reactions involving radical pairs. You do not observe the radicals themselves, as you do with EPR spectroscopy. So, the next time your NMR spectrometer shows peaks that are either too big, negative, or strangely constructed, do not automatically assume your instrument is on the blink. Perhaps the spectrum is trying to tell you something!

There is a related effect seen in the EPR spectra of products arising from radical pair reactions. This effect is known as **chemically induced dynamic electron polarization,** or CIDEP. A discussion of CIDEP is beyond the scope of this book, but interested readers can consult ref. 6.

CHAPTER SUMMARY

1. Molecules or molecular fragments that possess one (or more) unpaired electrons are called free radicals. Free radicals are usually formed through homolytic cleavage of a chemical bond; recombination is the process of two free radicals combining to re-form the bond.

2. An unpaired electron, like a proton, can adopt either of two spin orientations when immersed in a magnetic field. An electron in either orientation will precess at a frequency given by Eq. (11.1). The g factor is similar in some respects to the magnetogyric ratio (γ) used in NMR spectroscopy. The value of g, used as the position parameter in EPR spectroscopy, depends on the exact structure of the free radical possessing the unpaired electron.

3. An unpaired electron precesses about 657 times faster than a proton at the same magnetic field strength. The energy gap between the two electronic spin states is therefore 657 times larger than that for protons, which translates into a difference in populations of the two spin states much greater than in NMR.

4. An EPR (or ESR) signal is generated when an unpaired electron in the lower energy spin state absorbs a photon at ν_{prec} and is elevated to the higher energy spin state. The EPR signals are usually displayed in the dispersion (first-derivative) mode.

5. The EPR signals, like NMR signals, can be split into multiplets by coupling between the unpaired electron and neighboring magnetic nuclei. Such splitting is governed by the same type of first-order coupling rules (e.g., the $n + 1$ rule) that govern internuclear spin coupling, except that the hyperfine coupling constant (a value) is expressed in millitesla rather than in hertz.

6. A radical pair (two interacting radicals) can exist in either the singlet state (electronic spins paired) or triplet state (with electronic spins parallel).

7. The NMR spectra of compounds being formed from reactions involving radical pairs often exhibit anomalous signals as the result of chemically induced dynamic nuclear polarization (CIDNP). While EPR signals are due to radicals themselves, CIDNP signals are due to the products arising from radical-pair reactions. These products can be formed with nuclei in polarized (nonequilibrium) spin state distributions.

8. The CIDNP signals are of two basic types: (a) those exhibiting a net effect, that is, highly intense positive (net absorption, or A) signals or intense negative (upside-down, net emission, or E) signals, and (b) those exhibiting a multiplet effect, that is, both positive and negative signals in the same multiplet. Multiplet effects are designated A/E or E/A.

9. The direction (positive or negative) of the net effect (Γ_n) is determined by the sign of the product of the four spectroscopic parameters shown in Eq. (11.2).

10. The type of multiplet effect (A/E or E/A) is determined by the sign of the product of the six spectroscopic parameters shown in Eq. (11.3).

11. The principal value of CIDNP spectra is that they supply information about the spin state (singlet or triplet) of the radical pair at its birth, μ in Eqs. (11.2) and (11.3).

REFERENCES

[1]Weil, J. A., Bolton, J. R., and Wertz, J. E., *Electron Paramagnetic Resonance*, Wiley Interscience, New York, 1994.

[2]Pasto, D. J., and Johnson, C. R., *Organic Structure Determination*, Prentice-Hall, New York, 1969.

[3]Ward, H. R., *Acc. Chem. Res.*, 5, 18 (1972).

[4]Lawler, R. G., *Acc. Chem. Res.*, 5, 25 (1972).

[5]Pine, S. H., *J. Chem. Educ.*, 49, 664 (1972).

[6]Lepley, A. R., and Closs, G. L., *Ed., Chemically Induced Magnetic Polarization*, Wiley, New York, 1973.

[7]Iwamura, H., Imahashi, Y., and Kushida, K., *J. Am. Chem. Soc.*, 96, 921 (1974).

REVIEW PROBLEMS (Answers in Appendix 1)

11.1. One resonance structure of the benzyl radical (PhCH₂•) is shown below:

(a) Draw three additional resonance structures that show how the unpaired electron is delocalized into the aromatic ring. (b) Predict the multiplicity of the EPR spectrum and the sign of each a value. (You need consider only those couplings to hydrogen.) (c) The magnitude of the observed *a* values (in millitesla) are α, 1.5; ortho, 0.49; meta, 0.15; and para, 0.61. What fraction of the unpaired electron density is at the meta carbon? Is your answer consistent with the resonance structures in part (a)?

11.2. The π molecular orbital (π MO) system of 1,3-butadiene consists of four four-center π MO's ($\psi_1 - \psi_4$):

1,3-butadiene

The four π electrons in the system reside as pairs in ψ_1 and ψ_2, as shown below. In each π MO the shading represents orbital phase, while the size of each component *p* atomic orbital represents its relative contribution to that MO:

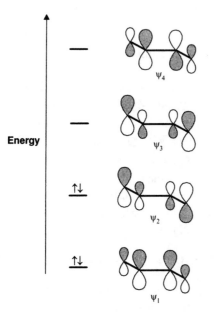

(a) When an additional electron is added to butadiene, the resulting species is a butadiene **radical anion**. Where does the extra electron reside? (b) Predict the multiplicity of the EPR spectrum of this radical anion. (c) The observed EPR spectrum exhibits two *a* values, 0.76 and 0.28 mT. Which *a* value corresponds to which coupling interaction?

11.3. Thiophenol (PhSH) is a reagent capable of intercepting radicals (R•) by the reaction

$$R\bullet + PhSH \rightarrow \rightarrow R–SPh + H\bullet$$

When peroxide **11-10** (Example 11.12) is thermolyzed in the presence of PhSH (rather than S–Cl), one of the products is PhS–CH$_2$CH$_3$. Predict the type of CIDNP effect(s) in the ^{1}H spectrum of both the quartet and triplet of its ethyl group.

11.4. When ethyllithium (CH$_3$CH$_2$Li) and ethyl iodide (CH$_3$CH$_2$I) are combined in an NMR spectrometer cavity, the ^{1}H spectrum exhibits a strong *A/E* multiplet effect (but no net effect) for both the triplet and quartet of the ethyl iodide (see ref. 3). Explain why.

11.5. Using the equation for a Lorentzian absorption signal [Eq. (3.10)], derive the equation of the dispersion curve shown in Figure 11.2.

12

DOUBLE-RESONANCE TECHNIQUES AND COMPLEX PULSE SEQUENCES

12.1 WHAT IS DOUBLE RESONANCE?

In Chapters 2 and 3 we described the basic components of an NMR spectrometer and how an NMR signal is generated. Before proceeding further you might want to review the relevant parts of Sections 2.3–2.5 and 3.4–3.6. To briefly summarize these sections, we can describe the key steps in signal generation as follows:

1. The *external* magnetic field B_0, aligned along the z axis, causes all magnetic nuclei in the sample to assume an equilibrium distribution among spin states precessing around B_0, giving rise to a net nuclear magnetization (M) aligned along the $+z$ axis (parallel to B_0), as in Figure 2.7a.

2. Passage of a brief but powerful pulse of rf current through the transmitter coil (whose axis lies in the xy plane) gives rise to a secondary oscillating magnetic field (B_1) perpendicular to and precessing around B_0. The frequency of B_1 is set to match the precessional frequency of the specific isotope (e.g., ^{13}C) to be observed.

3. When viewed in the rotating frame of reference [Figure 2.9b], B_1 (statically aligned with, for example, the $+x'$ axis) causes M for the observed nuclei to precess by an angle α in the $y'z'$ plane (Figure 2.12). The magnitude of α depends on the strength of B_1, which in turn depends on the power and duration of the pulse.

4. After B_1 is turned off, each nonzero component of M in the $x'y'$ plane (that is, precessing in the xy plane) gives rise to an induced rf signal (which is a summation of all contributing precessional frequencies) in the receiver circuit. This exponentially decaying FID sig-

nal is amplified and collected digitally as a function of time.

5. Fourier transformation of the FID signal produces the conventional frequency-domain spectrum, a plot of signal intensity versus frequency, with a line for each contributing precessional frequency of the observed nuclei.

We will now make a significant addition to our spectrometer: a second transmitter coil that, like the first one, has its axis in the x,y plane. The purpose of the second transmitter coil is to enable us to irradiate a second set of nuclei (e.g., ^1H) while still observing the original nuclei (e.g., ^{13}C). Because these two sets of nuclei undergo resonance simultaneously (though at different frequencies, by virtue of their different γ or δ values), we refer to such experiments as **double-resonance techniques**. To distinguish between the two precessing magnetic fields, we will henceforth refer to the original one as the **observing field** (B_1, with frequency v_1) because it is the one that generates the NMR signal of interest. The other one will be called the **irradiating field** (B_2, with frequency v_2), even though *both* rf fields actually irradiate the sample. Both v_1 and v_2 are normally generated independently by two separately tunable rf oscillators.

When the two sets of nuclei are of different isotopes (e.g., ^{13}C and ^1H), the experiment is described as *heteronuclear* double resonance. When we *observe* the signal from nuclei of isotope A while *irradiating* nuclei of isotope B, we label the resulting spectrum with the shorthand designation A {B}. When the two sets of nuclei belong to the same isotope (e.g., both ^1H), the technique is described as *homonuclear* double resonance.

■ **EXAMPLE 12.1**   Suppose we were to observe the ^{1}H spectrum of CH_2F_2 (Section 8.6.1) at a field strength of 5.87 T, while simultaneously irradiating the fluorines. (a) Does this experiment involve homo- or heteronuclear double resonance? (b) What are the values of ν_1, ν_2, and the difference ($\Delta\nu$) between them?

□ *Solution:* (a) Heteronuclear, because ^{1}H and ^{19}F are different elements. We would designate the resulting spectrum with the label ^{1}H $\{^{19}$F$\}$. (b) The ^{1}H frequency (ν_1) would be 250 MHz (Example 2.6), while the ^{19}F frequency (ν_2) would be 235 MHz, as calculated from Eq. (2.6) and the γ value for ^{19}F in Table 2.1. Therefore, $\Delta\nu$ = 250 MHz – 235 MHz = 15 MHz = 15,000,000 Hz.   □

■ **EXAMPLE 12.2**   Suppose we were to reexamine the 250-MHz ^{1}H spectrum of diethyl ether (Section 8.2 and Figure 8.1). But this time we will observe the methyl triplet while we irradiate the methylene quartet. (a) Does this experiment involve homo- or heteronuclear double resonance? (b) What is the value of $\Delta\nu$ in this case?

□ *Solution:* (a) Homonuclear, because both sets of nuclei are ^{1}H. (b) From Figure 8.1 (or Example 8.2), we can see that the triplet is separated from the quartet by 568 Hz. □

These two examples show that $\Delta\nu$ is very much *smaller* for homonuclear double resonance, suggesting a simple way to generate ν_2. Instead of using a separate rf oscillator, we can use part of the output from the ν_1 channel electronically modulated with an audio-frequency signal to generate output at ν_2. This technique makes it easier to control the exact value of $\Delta\nu$ and to focus ν_1 and ν_2 exactly where we want them.

The family of double-resonance techniques has many members, and in this chapter we will examine some of the most common. We will discover that the different effects of double resonance depend primarily on the exact frequency, timing, duration, phasing, and power of the irradiating field. Furthermore, by careful tuning of the observing transmitter coil, one can send both ν_1 *and* ν_2 through it, obviating the need for the second transmitter coil. It is even possible to add additional irradiating fields, but these higher order multiple-resonance techniques are beyond the scope of this book.

You might wonder what would happen if ν_1 and ν_2 accidentally took on exactly the same value. For example, what would happen to the ^{1}H quartet of diethyl ether (Figure 8.1) if ν_2 were set at its center (870 Hz downfield from TMS)? Under pulse-mode conditions, the Fourier transformation of the multiplet being irradiated normally generates either a beat pattern (Figure 12.1) from the interference between two closely similar frequencies, a broad nonsymmetrical negative peak, or a negative multiplet with sharp lines, depending on the type of experiment employed.

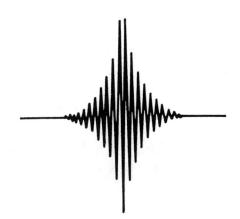

Figure 12.1. Beat pattern resulting from interference between two similar frequencies.

12.2   HETERONUCLEAR SPIN DECOUPLING

When we discussed heteronuclear ^{13}C–H spin coupling (Section 8.6.1), we noted that all the ^{13}C spectra presented in Chapters 5 and 7 had been *proton decoupled*, because in the absence of such decoupling the spectra would have been considerably more complex. Now we are ready to see how the effects of ^{13}C–H spin coupling can be erased when acquiring a ^{13}C spectrum.

Suppose we first generate the undecoupled ^{13}C spectrum of chloroform, $HCCl_3$. We know that, in the absence of decoupling, the carbon magnetization (M_C) will be split into two components (a doublet) by coupling to the "up" and "down" spin states of the lone hydrogen with coupling constant $^1J_{CH}$ (Figure 12.2*a*). We first deliver a $90_{x'}$ B_1 pulse at the precessional frequency of ^{13}C (ν_1) to rotate both components of M_C onto the $+y'$ axis (Figure 12.2*b*). Next, with B_1 turned off, one of the M_C vectors will precess faster than ν_1 (by $J/2$), while the other half precesses slower (by $J/2$). In the *rotating* frame (which itself is precessing at ν_1), these two new vectors will precess, respectively, at frequencies of $+J/2$ (clockwise) and $-J/2$ (counterclockwise), as shown in Figure 12.2*c*. During this time we collect the FID signal (observing at ν_1), perform the FT, and generate the frequency-domain spectrum with two equally intense lines, one at $\nu_1 + J/2$, the other at $\nu_1 - J/2$. Note that the spacing between these lines is J.

Now let us generate the proton-decoupled spectrum of chloroform. Everything is the same as before, until we get ready to collect the FID signal. Just before we start the data acquisition, we begin irradiating the sample with a powerful *continuous* (as opposed to pulsed) B_2 field centered at the average precessional frequency of ^{1}H nuclei (ν_2) and composed of all the frequencies in the ^{1}H region. This **broadband** irradiation of the ^{1}H nuclei causes the net magnetization of the protons (M_H) to precess continuously around B_2 (Figure 12.2*d*). The individual ^{1}H nuclei are no longer distributed

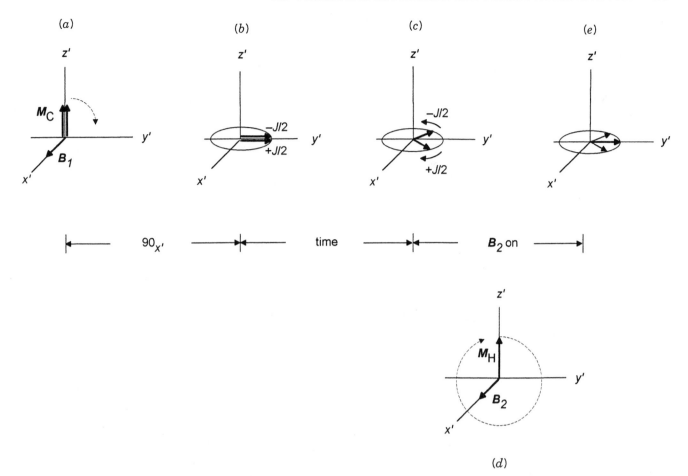

Figure 12.2. Rotating frame depictions of the doublet components of M_C (*a*) before and (*b*) after the $90x'$ pulse, (*c*) divergence of the two components when (*d*) B_2 is off, and (*e*) freezing the components (now shown without boldface) when B_2 is on. The resulting vector sum (boldface) is aligned on the $+y'$ axis and leads to a positive signal.

essentially equally between up ($+z$) and down ($-z$) spin orientations, as required for spin coupling to occur (Section 8.3). Instead, as M_H continuously precesses around B_2, the individual ^1H nuclei are rapidly changing spin from up to down and vice versa. Thus, we have *saturated* the ^1H spin states! Since the ^1H nuclei are no longer aligned solely with and against B_0, they can no longer augment and diminish the magnetic field experienced by the carbon. So the two components of M_C no longer precess at different frequencies; they both precess at ν_1, frozen in the rotating frame, with their vector sum along the $+y'$ axis (Figure 12.2*e*). While B_2 is on, the FID data are collected, and the FT yields a frequency-domain spectrum with just one signal at ν_1. The net result of this technique is that all $^{13}C–H$ coupling interactions disappear, and each ^{13}C multiplet collapses to a singlet! The important thing is that, for decoupling to occur, B_2 *must be on during acquisition of the FID data.*

Example 8.13 described the appearance of the ^{13}C spectrum of 2-chlorobutane in the *absence* of {^1H} decoupling.

Figure 7.1 showed the proton-decoupled ^{13}C spectrum of this compound, with one sharp line for each carbon. Clearly, the main virtue of heteronuclear spin decoupling is its simplifying effect on the appearance of the spectrum, making the task of measuring chemical shifts and assigning signals much easier. But there is an additional benefit as well, an improvement in the signal-to-noise ratio (Section 3.3.3). One reason for this is that the various lines of each multiplet are now collected into a single line with the same total intensity. But there is another, more subtle, intensity-increasing factor, and this is the subject of the next section.

12.3 POLARIZATION TRANSFER AND THE NUCLEAR OVERHAUSER EFFECT

In Section 11.8 we saw how polarization (departure from the equilibrium distribution) of nuclear spin state populations by a nearby unpaired electron can have a profound effect on NMR signal intensity. It turns out that in a ^{13}C {^1H} experiment, irradiation of the hydrogen nuclei can affect the carbon nuclear spin state populations in such a way as to increase the intensity of the ^{13}C signals. This nuclear **Overhauser effect** (**NOE**) was first mentioned in Section 5.4.1. Below is a simplified description of how this special example of a dipole–dipole relaxation process arises.

The NOE interaction between two proton magnetic dipoles takes place predominantly directly through *space* rather than through *bonds* (as is the case in normal spin coupling). This through-space interaction is very sensitive to the distance

(r) between the interacting nuclei, being inversely proportional to r^6. Thus, for two protons to exhibit an NOE, they must be in quite close proximity in the molecular structure. They need not be spin coupled, but the NOE is not precluded if they are.

Consider a molecule that possesses a ^{13}C atom and a nearby (though not necessarily spin-coupled) ^1H atom. The combinations of spin states for this two-spin system, shown in Figure 12.3, are quite similar to the spin states of two ^1H nuclei (Figure 9.1), except that ΔE for the ^1H nuclei is four

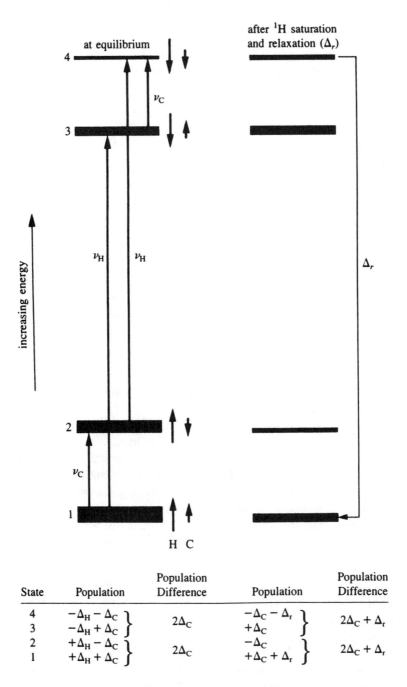

State	Population	Population Difference	Population	Population Difference
4	$-\Delta_H - \Delta_C$		$-\Delta_C - \Delta_r$	
3	$-\Delta_H + \Delta_C$	$2\Delta_C$	$+\Delta_C$	$2\Delta_C + \Delta_r$
2	$+\Delta_H - \Delta_C$		$-\Delta_C$	
1	$+\Delta_H + \Delta_C$	$2\Delta_C$	$+\Delta_C + \Delta_r$	$2\Delta_C + \Delta_r$

Figure 12.3. Spin state populations that result in the nuclear Overhauser effect.

times that for the ^{13}C nuclei (Example 2.5). As we did in Chapter 11, we will depict relative spin state populations by the relative thickness of the spin state line. Recall from Section 2.3.1 that at equilibrium there will be a slight preference for $I = \frac{1}{2}$ nuclei to be in the lower energy spin state (up, parallel to $\mathbf{B}_0$), with slightly fewer in the down state. We will call this population difference $2\Delta_C$ for the carbon and $2\Delta_H$ for the hydrogen, and Δ_H is four times as large as Δ_C (Section 2.3). Therefore, the population of state 1 is augmented by $\Delta_H + \Delta_C$, state 2 by $\Delta_H - \Delta_C$, and so on.

The ^{13}C signal results from transitions between states 1 and 2 and between states 3 and 4. Its intensity is a function of the difference in populations between states 1 and 2 and between states 3 and 4. At equilibrium, this population difference is $2\Delta_C$ in both cases. Now, if the ^{1}H states were saturated in a double-resonance experiment, Δ_H would become zero. This would mean that states 1 and 3 would become equally populated, as would states 2 and 4. At first glance it would seem that the population difference between states 1 and 2 and between states 3 and 4 would still be $2\Delta_C$, but another factor intervenes. Notice that the populations of states 1 and 4 are now polarized. At equilibrium before saturation, the difference between them was $2\Delta_H + 2\Delta_C$, but at saturation it is only $2\Delta_C$. In an attempt to restore equilibrium, a number (Δ_r) of nuclei in state 4 relax to state 1, provided that we allow enough time for this relaxation to evolve. Thus, the population of state 1 will increase (by Δ_r), while that of state 4 will decrease (by Δ_r), so the intensity of both ^{13}C transitions (1 $\rightarrow$ 2 and 3 $\rightarrow$ 4) will increase in proportion.[1]

The NOE is an example of **polarization transfer** (or **cross polarization**), because polarization of one set of nuclear spin states (here, saturation of the ^{1}H nuclei) results in the polarization of another set (here, the ^{13}C nuclear spin states). The maximum Overhauser signal enhancement (η) is given by

$$\eta = \frac{1}{2}\left(\frac{\gamma_{irr}}{\gamma_{obs}}\right) \tag{12.1}$$

where γ_{irr} and γ_{obs} are the magnetogyric ratios of the irradiated (here, ^{1}H) and observed nuclei (here, ^{13}C), respectively, and these can be found in Table 2.1. Notice that an η value of 1.0 indicates a 100% enhancement, to give a signal twice what it was in the absence of the NOE.

■ **EXAMPLE 12.3** Calculate the maximum NOE signal enhancement in a ^{13}C $\{^1\text{H}\}$ experiment.

□ *Solution:* Using Eq. (12.1) and the γ values for ^{13}C and ^{1}H in Table 2.1, we find

$$\eta = \frac{1}{2}\left(\frac{\gamma_{irr}}{\gamma_{obs}}\right) = \frac{1}{2}\left(\frac{267.5 \times 10^6}{67.26 \times 10^6}\right) = 1.988$$

Thus, the signal of a ^{13}C nucleus with a neighboring ^{1}H nucleus can be enhanced by nearly 200%, to give a signal three times as intense as one lacking the NOE, that is, a ^{13}C nucleus with no nearby hydrogens. □

The total NOE enhancement for a given carbon signal tends to increase with the number of nearby hydrogens (though not in direct proportion). This is why we saw in Sections 5.4.1 and 7.2 that proton-decoupled ^{13}C signal intensities often follow the order methyl (CH_3) > methylene (CH_2) > methine (CH) > quaternary carbon (C). However, CH_3 signals are frequently less intense than expected (owing to an efficient spin-rotation contribution to relaxation) while quaternary carbon signals can be more intense than expected if there are enough nearby hydrogen nuclei. Also remember that the enhancement calculated in Eq. (12.1) is the *maximum* NOE; the observed effect falls off with the sixth power of the internuclear distance.

To summarize, the NOE can serve two purposes. First, it provides an added bonus when we generate a proton-decoupled ^{13}C spectrum. By polarization transfer, NOE helps compensate for the low natural abundance and low sensitivity of ^{13}C (Table 2.1). Second, and even more important, NOE can provide a measure of the through-space proximity of one nucleus to another.

Such NOE experiments work best when the *more abundant* (or sensitive) nucleus (e.g., ^{1}H) is irradiated while the effect is observed on the *less abundant* (or sensitive) nucleus (e.g., ^{13}C). It would be fruitless (as well as unnecessary) to carry out the reverse experiment, ^{1}H $\{^{13}\text{C}\}$, and expect to see any enhancement of the hydrogen signal. Furthermore, remember that the NOE requires time to evolve. This is why, during a typical ^{13}C $\{^1\text{H}\}$ decoupling experiment, the ν_2 channel is left on not only during FID acquisition but beforehand as well (during the **presaturation** or **preequilibration delay**).

We will discuss homonuclear NOE in Section 12.7, but before we leave the topic of heteronuclear NOE, try the example below.

■ **EXAMPLE 12.4** Calculate the maximum signal enhancement due to the Overhauser effect between an unpaired electron and a ^{1}H nucleus, as takes place in a CIDNP experiment (Section 11.8).

□ *Solution:* From Section 11.1 we know that the ratio γ_e/γ_H is 657. Therefore,

$$\eta = \frac{1}{2}\left(\frac{\gamma_{irr}}{\gamma_{obs}}\right) = \frac{1}{2}(657) = 329$$

This is in part why CIDNP-enhanced absorptions and emissions can be so intense. □

12.4 GATED AND INVERSE GATED DECOUPLING

It is useful at this point to introduce a simple pictorial representation of what is occurring with both the observing channel (v_1) and the irradiating channel (v_2) during the experiments described above. The normal pulse–FID acquisition–delay sequence will be shown as in Figure 12.4a; compare this with Figure 3.17. A representation of the pulse sequence for a typical proton-decoupled ^{13}C spectrum is shown in Figure 12.4b. Note how the decoupler (v_2) is left on during both FID acquisition (to effect heteronuclear decoupling) and the prior presaturation delay (to generate the NOE).

One relatively simple pulse sequence alteration, called **gated decoupling**, is depicted in Figure 12.5a. Because the NOE requires time for the critical relaxation to evolve (building up or decaying as an exponential function of T_1, Section 12.3), its magnitude depends on the length of time v_2 is on.

Conversely, the decoupling effect of v_2 requires only that v_2 be on during FID acquisition. Therefore, if v_2 is on during the presaturation delay but off during FID acquisition, most of the NOE enhancement would be preserved but all decoupling would be lost. The net result is a *coupled* spectrum with improved signal-to-noise ratio.

Inverse gated decoupling (Figure 12.5b) involves turning v_2 on during data acquisition but off during the presaturation delay. In this case the decoupling effect is preserved, but essentially all of the NOE enhancement is lost. As we will see later, both of these techniques can be used to our advantage in a variety of related applications.

12.5 OFF-RESONANCE DECOUPLING

Every silver lining has a cloud! Although an ^{1}H-decoupled ^{13}C spectrum has the advantages of simplicity and improved

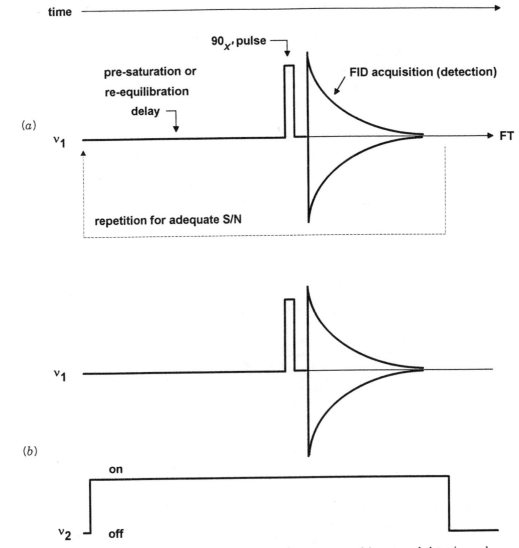

Figure 12.4. Pictorial depiction of two common pulse sequences: (*a*) a normal detection pulse sequence and (*b*) detection with decoupling.

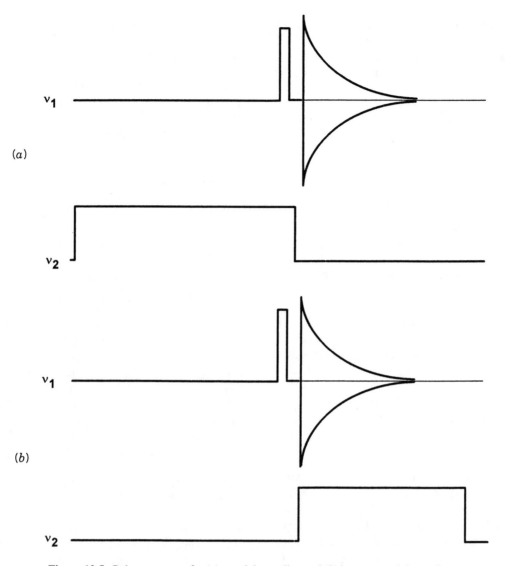

Figure 12.5. Pulse sequences for (*a*) gated decoupling and (*b*) inverse gated decoupling.

signal-to-noise ratio, something has been lost. That "something" is the information coupling provides about the structure of the sample molecules. For example, in a coupled ^{13}C spectrum (considering only one-bond ^{13}C–H couplings), methyl carbons appear as quartets, methylenes as triplets, and so on, making them easier to assign (see, e.g., Figure 8.11). In fully decoupled ^{13}C spectra, all the signals appear as singlets. As we saw in Chapter 7, chemical shift calculations and NOE enhancements do not always allow us to differentiate unambiguously between carbon signals, especially when we study complex molecules. It would be useful if we could turn off all proton–carbon coupling *except* between hydrogens directly attached (one-bond couplings) to carbons.

One way to do this involves a technique called **off-resonance decoupling**. As the term implies, we do not set ν_2 at

the center of the ^1H resonance frequencies, but instead shift it ca. 1000–3000 Hz away. While this is still close enough to preserve most of the NOE enhancement of the ^{13}C signals, all coupling constants (J) are reduced in magnitude according to the equation

$$J_r = \frac{2\pi J \, \Delta\nu}{\gamma_H \, \mathbf{B}_2} \tag{12.2}$$

where J_r is the **reduced** (observed, or *residual*) **coupling constant**, $\Delta\nu$ is the difference between ν_2 and $\nu_{\text{precession}}$ of the coupling ^1H nucleus, and $\mathbf{B}_2$ is the strength of the irradiating rf field (typically 2×10^{-4} T). Notice that if $\nu_2 = \nu_{\text{precession}}$ (and $\mathbf{B}_2$ is sufficiently strong), all ^{13}C–H coupling constants are reduced to zero (complete decoupling).

■ **EXAMPLE 12.5** Typical ^{13}C–H coupling constants for tetrahedral carbons (Sections 9.3–9.5) are ^1J, 125 Hz; ^2J, 5 Hz; and ^3J, 5 Hz. Calculate the reduced coupling constants in an off-resonance decoupled spectrum when $\Delta\nu$ averages 2000 Hz and $\mathbf{B}_2 = 2 \times 10^{-4}$ T.

□ *Solution:* Apply Eq. (12.2):

$$^1J_r = \frac{2\pi J \,\Delta\nu}{\gamma_H \,\mathbf{B}_2}$$

$$= \frac{2\pi\,(125\,\text{Hz})\,(2000\,\text{Hz})}{(267.5 \times 10^6 \,\text{rad T}^{-1}\,\text{Hz})\,(2 \times 10^{-4}\,T)} = 29 \text{ Hz}$$

In like manner, $^2J_r = {}^3J_r = 1$ Hz, which is negligibly small. So, only one-bond C–H couplings are preserved in an off-resonance decoupled ^{13}C spectrum.   □

■ **EXAMPLE 12.6** Predict the appearance of the off-resonance-decoupled ^{13}C spectrum of 2-chlorobutane (Examples 7.3 and 8.13) that would result using the parameters given in Example 12.5:

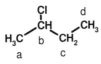

□ *Solution:* Deleting all but one-bond ^{13}C–H couplings leaves C_a–H_a, C_b–H_b, C_c–H_c, and C_d–H_d as the only couplings. Using the chemical shift data in Example 7.3, we therefore predict a quartet at δ 24.85 for methyl carbon C_a, a doublet at δ 60.38 for methine carbon C_b, a triplet at δ 33.33 for methylene carbon C_c, and a quartet at δ 11.01 for methyl carbon C_d. Each multiplet would exhibit a ^1J_r value of ca. 29 Hz. The appearance of this spectrum would be quite similar to that in Figure 8.11, except that the separations between the lines in each of the multiplets would be only one-fourth as large.   □

As useful as off-resonance decoupling once was, there is an even better way to get the same type of information, as we will see later in this chapter and in Chapter 13.

12.6 HOMONUCLEAR SPIN DECOUPLING

It would be useful if we could use a double-resonance approach to generate ^1H spectra free of all homonuclear H–H coupling, so that each set of equivalent hydrogens gave rise to a singlet, as we see in proton-decoupled ^{13}C spectra. However, if we were to subject all the hydrogens to broadband irradiation while simultaneously observing the ^1H signals, ν_1 would equal ν_2 and a horrendous mush of overlapping broad

negative peaks and beat patterns would result. (However, in Chapter 13 we will see a way to accomplish this feat using a 2D NMR technique.)

Nonetheless, we can carry out **partial homonuclear decoupling** by focusing ν_2 on one particular ^1H signal while observing the resulting effects on the rest of the ^1H spectrum. This technique results in partial simplification of the spectrum and confirms which nuclei are coupled to which others. The pulse sequence for this experiment is basically the same as inverse gated sequence shown in Figure 12.5*b*, though in some cases the ν_2 channel output is produced as a rapid series of pulses rather than continuous irradiation.

Take a moment to review the ^1H spectrum of structure **8-2** (Example 8.11). What would be the effect of irradiating the H_b multiplet at δ 1.76? Saturating spin states of H_b by irradiation at that frequency would cause all H–H spin couplings involving that nucleus to disappear. As a result, the doublet centered at δ 0.915 would collapse to a singlet, and the triplet at δ 3.40 would collapse to a doublet.

There are, however, certain operational problems with this technique. First, it can be quite tedious and time consuming to locate ν_2 at exactly the desired frequency, then adjust its power to saturate only *that* signal without disturbing any nearby signals. Second, to confirm *all* couplings in a molecule with *n* sets of mutually coupled nuclei, it would be necessary to generate at least $n - 1$ separate decoupled spectra. Fortunately, there are easier ways to confirm which sets of nuclei are coupled, and we will discuss these in Chapter 13.

12.7 HOMONUCLEAR DIFFERENCE NOE: THE TEST FOR PROXIMITY

Just as saturation of ^1H spin states leads to both proton decoupling and enhancement of signals for nearby ^{13}C nuclei (through NOE cross polarization), homonuclear double resonance can also affect signal *intensities* as well as multiplicities. As before, the effects are significant only if the nuclei are relatively close in space, as the following examples demonstrate. Furthermore, the magnitude of the effect depends in part on the exact relaxation mechanisms operating, for example, whether or not the nuclei are spin coupled (through bonds). If they are *not* spin coupled, a signal *enhancement* can occur; if they *are* spin coupled, the signal can be modestly *diminished* by the NOE. (Interested readers can consult Derome's excellent book listed in the Additional Resources section at the end of this chapter for a more detailed discussion.)

■ **EXAMPLE 12.7** What is the maximum NOE signal enhancement expected in a homonuclear NOE experiment?

□ *Solution:* Because both γ values are equal, Eq. (12.1) reduces to a maximum enhancement of $\eta = 0.5$ (50%). Of course, the further apart the nuclei are, the less will be the enhancement. □

The pulse sequence for **difference-NOE** experiments is similar to the gated decoupling sequence in Figure 12.5a and generates the NOE without decoupling. (Note: The power level used in difference-NOE experiments is much lower than in homonuclear decoupling experiments.) There are, however, two additional wrinkles. First, the sample tube is not spun during the experiment, in order that artifact signals due to second order spinning sidebands (Section 3.3.3) can be avoided. Second, we will initially perform the experiment with ν_2 set exactly on the signal we wish to irradiate, repeating the sequence until the FID has the desired signal-to-noise ratio. (As stated before, if we are testing for proximity between two specific sets of nuclei, we normally irradiate the set comprising the greater number of nuclei.) Then we will repeat the experiment an equal number of times with ν_2 set far away from the frequency of all signals in the spectrum. The FID data for the latter experiment are subtracted from the former, and the remainder FID is Fourier transformed. The resulting difference-NOE spectrum will show a negative signal for the irradiated multiplet and positive signals *only* for

nuclei that experienced NOE enhancements ($\eta > 0$). Customarily the integral of the irradiated signal is set to -100%, so that the integral of all other signals in the difference spectrum are direct measures of the NOE enhancement η.

Figure 12.6 shows a difference-NOE spectrum of the pictured coumarin derivative. Irradiation of the methyl signal at δ 2.23 results in NOE enhancements for only 2 of the 12 hydrogens in the molecule, the circled aromatic hydrogen at δ 6.85 ($\eta = 4.9\%$) and the circled vinyl hydrogen at δ 5.85 ($\eta = 4.8\%$). None of the 10 more remote methylene hydrogens (which are not explicitly shown in the structure) exhibit any enhancement.

■ **EXAMPLE 12.8** At room temperature the ^{1}H spectrum of DMF (Table 8.1 and Section 10.2) consists of two methyl singlets (δ 2.74 and 2.91) and the remaining formyl hydrogen singlet at δ 8.02:

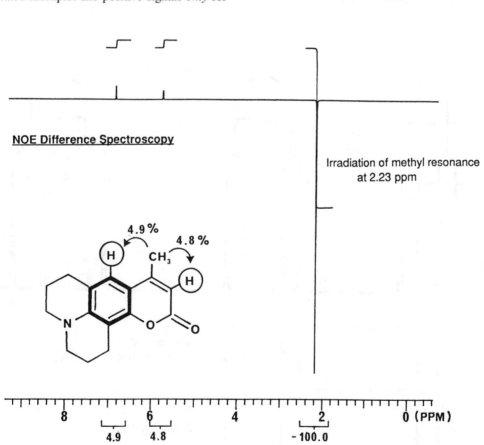

Figure 12.6. Difference-NOE ^{1}H spectrum of the coumarin derivative pictured, resulting from irradiation of the methyl signal. (Contributed by David Lankin.)

There is no observable spin coupling between any of the hydrogens. (The reason the two methyl groups are not rendered equivalent by rotation around the C–N bond was discussed in Section 10.2.) The question is this: Which signal belongs to which methyl group?

□ *Solution:* To carry out this experiment, the NOE enhancement of the formyl hydrogen signal can be observed while first one, then the other, methyl signal is irradiated. When the signal at δ 2.74 is irradiated, the formyl signal is essentially unaffected. By contrast, irradiation of the signal at δ 2.91 leads to an 18% enhancement of the formyl signal.[2] This result indicates that the signal at δ 2.91 belongs to the methyl group closer to the formyl hydrogen, the one that is cis to the formyl hydrogen (and trans to the oxygen). □

12.8   OTHER HOMONUCLEAR DOUBLE-RESONANCE TECHNIQUES

Most other homonuclear double-resonance techniques, with names such as **spin tickling** and **selective decoupling**, provide information about the arrangement of spin states and the

relationships between spin transitions. A comprehensive discussion of all these techniques is beyond the scope of this book, but we will examine one of them in some detail.

In Section 9.1 we noted that each coupling constant has a sign related to the ordering of spin states and spectroscopic lines. The signs of coupling constants usually have no effect on the appearance of a typical NMR spectrum, and the sign cannot be determined from a typical spectrum. (For an exception, see Section 9.12.) However, the parity of the signs (whether they are the same or opposite) can often be determined by the selective decoupling technique. In this method only *part* of a given multiplet is saturated, and decoupling is observed only in the "connected" part of each coupled multiplet.

To understand how this works, consider a three-spin AMX system characterized by three chemical shifts (δ_A, δ_M, and δ_X) and three coupling constants (J_{AM}, J_{AX}, and J_{MX}).[3] The first-order (Section 8.5) spectrum of this system consists of three doublets of doublets, as shown in Figure 12.7. Each line of the A multiplet corresponds to a specific spin orientation of nuclei M and X; each line of the M multiplet corresponds to a specific spin orientation of nuclei A and X; and so on. The

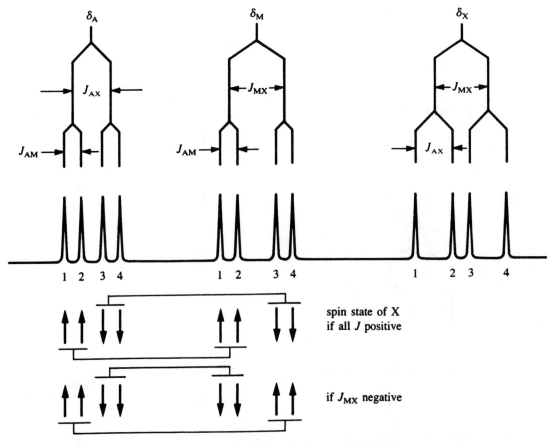

Figure 12.7.   Connected multiplets in a three-spin AMX system.

specific spin orientations that correspond to a certain line are determined by the signs of the coupling constants.

Assume for the moment that all three coupling constants in our AMX system are positive, and focus your attention on the A and M multiplets. Lines A_1 and A_2 correspond to the "up" orientation of nucleus X, while lines A_3 and A_4 correspond to the "down" orientation of X. Similarly, lines M_1 and M_2 correspond to the up orientation of nucleus X, while lines M_3 and M_4 correspond to the down orientation. We describe this situation by saying that the A_1–A_2 doublet is *connected* (via the similar spin state of nucleus X) with the M_1–M_2 doublet, and the A_3–A_4 doublet is connected with the M_3–M_4 doublet. These connections are shown below the spectrum. If we simultaneously saturate only lines A_1 and A_2 (by focusing ν_2 between them and carefully adjusting its power), only the connected lines (doublet M_1–M_2) will collapse to a singlet. Lines M_3 and M_4 would remain intact.

Suppose instead that J_{MX} is negative but the other two coupling constants remain positive. In this case, lines M_1 and M_2 correspond to the down orientation of X, while lines M_3 and M_4 correspond to the up orientation. Now doublet A_1–A_2 is connected with the M_3–M_4 doublet, and the A_3–A_4 doublet is connected with the M_1–M_2 doublet. This time, irradiation of doublet A_1–A_2 results in the collapse of lines M_3 and M_4 to a singlet. Thus, depending on which part of the M multiplet collapses when the A_1–A_2 doublet is irradiated, we can determine which pairs of lines are connected, and from this we can tell whether J_{MX} and J_{AX} have like or unlike signs.

Through a similar series of experiments, we can determine the relative sign of the third coupling constant. Notice that selective decoupling does not establish the *absolute* sign (plus or minus) of the coupling constants, only the relationship between their signs (same or opposite). Usually, however, we can make an educated guess about the absolute sign of at least one of the coupling constants (Chapter 9) and use that to determine the absolute signs of the remaining ones.

■ **EXAMPLE 12.9**   Suppose J_{MX} and J_{AX} have unlike signs, while J_{AM} and J_{AX} have like signs. (a) What is the sign relationship between J_{AM} and J_{MX}? (b) If J_{MX} is a one-bond C–H coupling constant, predict the absolute sign of all three J values.

□ *Solution:* (a) Opposite signs. (b) Since one-bond C–H coupling constants are positive (Section 9.3), J_{MX} is positive, while J_{AM} and J_{AX} are both negative.   □

12.9   COMPLEX PULSE SEQUENCES

Without doubt the biggest advances in NMR technology over the past decade have involved the advent of higher field pulse-mode spectrometers together and the realization that, by careful manipulation of the ν_1 and ν_2 pulse sequences, the net magnetization vector **M** can be redirected at will to provide an almost infinite variety of interesting spectroscopic effects. For example, by adjusting the phase of the ν_1 pulse relative to the detector phase, $\mathbf{B}_1$ can be made to appear not only on the x' axis of the rotating frame but anywhere in the $x'y'$ plane. Also recall (Figure 3.22 in Section 3.6.3) that a component of **M** precessing on the $+y'$ axis is ultimately detected as a positive signal, a component along the $+x'$ or $-x'$ axis gives zero signal, while a component precessing on the $-y'$ axis leads to a negative signal. Before proceeding further, try the example below.

■ **EXAMPLE 12.10**   Recall that **M** will precess around $\mathbf{B}_1$ clockwise (as viewed along the $\mathbf{B}_1$ axis toward the origin) through an angle determined by the duration and power of the pulse. Complete the table below by showing the final orientation of **M** ($\mathbf{M}_f$) resulting from irradiation of **M** (initial alignment $\mathbf{M}_i$) with the indicated pulse.

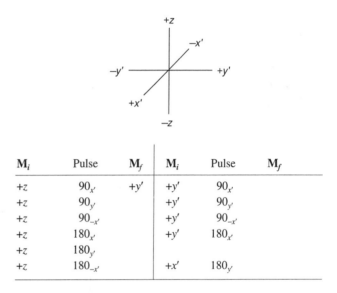

$\mathbf{M}_i$	Pulse	$\mathbf{M}_f$	$\mathbf{M}_i$	Pulse	$\mathbf{M}_f$
$+z$	$90_{x'}$	$+y'$	$+y'$	$90_{x'}$	
$+z$	$90_{y'}$		$+y'$	$90_{y'}$	
$+z$	$90_{-x'}$		$+y'$	$90_{-x'}$	
$+z$	$180_{x'}$		$+y'$	$180_{x'}$	
$+z$	$180_{y'}$				
$+z$	$180_{-x'}$		$+x'$	$180_{y'}$	

□ *Solution:*

$\mathbf{M}_i$	Pulse	$\mathbf{M}_f$	$\mathbf{M}_i$	Pulse	$\mathbf{M}_f$
$+z$	$90_{x'}$	$+y'$	$+y'$	$90_{x'}$	$-z$
$+z$	$90_{y'}$	$-x'$	$+y'$	$90_{y'}$	$+y'$
$+z$	$90_{-x'}$	$-y'$	$+y'$	$90_{-x'}$	$+z$
$+z$	$180_{x'}$	$-z$	$+y'$	$180_{x'}$	$-y'$
$+z$	$180_{y'}$	$-z$			
$+z$	$180_{-x'}$	$-z$	$+x'$	$180_{y'}$	$-x'$

There are two things to note at this point. First, the $180_{x'}$ pulse inverts **M** (rotates it 180° around the x' axis) to the $-z$ axis, giving an excess of spins in the higher energy spin state (a population inversion). The $180_{y'}$ pulse inverts **M** around the y' axis. Such a pulse will become important in the next section.   □

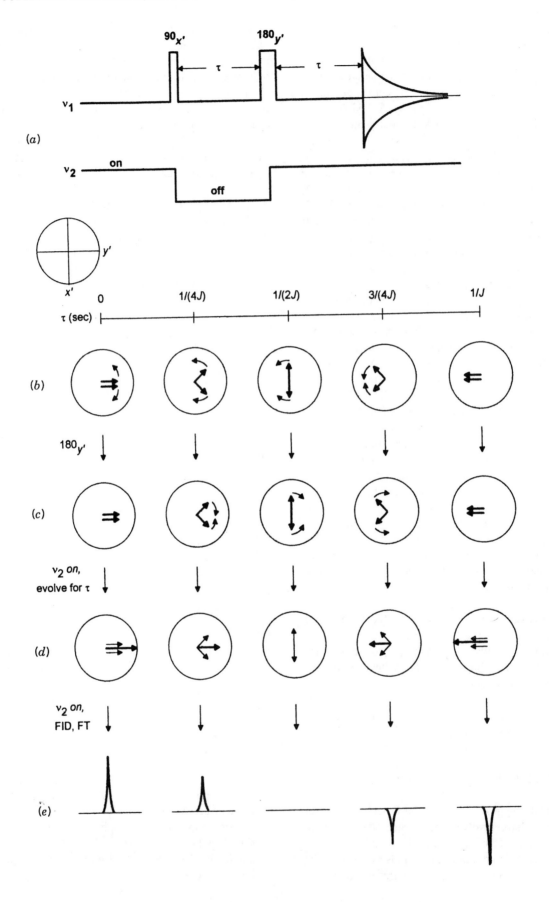

12.10   THE *J*-MODULATED SPIN ECHO AND THE APT EXPERIMENT

The majority of pulse sequence variations focus on the events that precede detection (FID acquisition). The interval preceding detection is divided into a **preparation time** (which may include delays for preequilibration or presaturation), the initiating pulse, an **evolution time** (τ), followed by one or more additional ν_1 pulses and evolution times. During this entire period, plus the subsequent FID collection, the ν_2 (irradiating) channel may also be activated at various times.

Let us investigate what happens when we alter the normal ^{13}C $\{^1$H$\}$ experiment in Figure 12.2 by using the **J-modulated spin echo** pulse sequence shown in Figure 12.8*a*. This pulse sequence is closely related to the spin echo technique described in Section 3.6.3.

We begin with a presaturation delay (ν_2 *on* to produce the NOE), followed by the usual $90_{x'}$ pulse of ν_1, which rotates $\mathbf{M_C}$ onto the +*y′* axis. Next, with the decoupling (ν_2) channel *off*, the two $\mathbf{M_C}$ components of the doublet begin to diverge ("modulated" by *J*), one moving clockwise with frequency +*J*/2 (in the rotating frame), the other counterclockwise with frequency −*J*/2 (Figure 12.2). Note that a frequency of *J*/2 hertz means that one complete revolution of each vector requires 2/*J* seconds, and one-half of a revolution requires 1/*J* seconds. Figure 12.8*b* shows the location of both $\mathbf{M_C}$ components in the *x′y′* plane as a function of time (seconds) from 0 to 1/*J*.

After the desired evolution time (τ) has elapsed, a $180_{y'}$ pulse (Example 12.10) is delivered, reflecting each component of **M** through the *y′z* plane; Figure 12.8*c* shows the result for each of the times shown in (*b*). Next, the decoupler is turned back on, causing all of the components to "freeze" in the rotating frame as well as generating additional NOE enhancement. After a second evolution time (τ) exactly equal to the first (needed to refocus the spin echo; Section 3.6.3), the resulting $\mathbf{M_C}$ vector sum for each of the times is shown in Figure 12.8*d*. With the decoupler still on (so that there is complete proton decoupling and NOE), we collect the FID and carry out the FT, leading to the frequency-domain ^{13}C signals shown in Figure 12.8*e*.

Figure 12.8. (*a*) Pulse sequence for the *J*-modulated spin echo experiment. Both τ values (typically a few milliseconds) are equal and are a thousandfold greater than the pulse widths (typically a few microseconds). (*b*) Position of each doublet component (viewed in the *x′y′* plane) as a function of evolution time τ (from zero to 1/*J* seconds). (*c*) Effect of a $180y'$ pulse on each of the times in part (*a*). Note how the clockwise vector are rotated from bottom to top, while counterclockwise vectors are rotated from top to bottom. (*d*) Effect of resuming decoupling: Each component of the doublet is frozen in the rotating frame (no longer shown boldface), giving a vector sum (boldface) along the ±*y′* axis. (*e*) Frequency-domain signal resulting from each boldface vector in part (*d*).

Note how the sign and intensity of this ^{13}C–H signal is a function of which evolution time τ is selected. An evolution time $\tau = 0$ leads to a normal positive signal for the CH carbon, $\tau = 1/J$ leads to a perfectly inverted signal, and so on.

■   **EXAMPLE 12.11**   How will the appearance of a quaternary carbon signal obtained using the *J*-modulated spin echo technique vary with τ?

□   *Solution:* Because a quaternary carbon is not coupled to any hydrogens, there will be only one component of $\mathbf{M_C}$. The $90_{x'}$ pulse will rotate it onto the +*y′* axis. Since there is no *J* modulation of the signal, it will remain there regardless of the length of τ, and it will be unaffected by the $180_{y'}$ pulse. Subsequent FID acquisition and FT lead to a normal positive signal regardless of the value of τ. □

The reason this pulse sequence is so useful is that ^{13}C signals of different multiplicities (e.g., singlets for quaternary carbons, doublets for CH carbons, triplets for CH$_2$ carbons, and quartets for CH$_3$ carbons) have predictably different responses to the evolution time, as shown in the equations below:

$$\text{CH:} \qquad I = \cos(\pi J\tau) \qquad (12.3a)$$

$$\text{CH}_2\text{:} \qquad I = \tfrac{1}{2}[1 + \cos(2\pi J\tau) \qquad (12.3b)$$

$$\text{CH}_3\text{:} \qquad I = \tfrac{1}{4}[3\cos(\pi J\tau) + \cos(3\pi J\tau)] \quad (12.3c)$$

These equations are plotted graphically in Figure 12.9. Note how the quaternary carbon signal is always positive through the range $\tau = 0$ to $\tau = 1/J$, the CH signal varies from +1 to −1 [compare with Figure 12.8(*e*)], the CH$_2$ signal varies from +1 through a minimum at zero and back to +1, and the CH$_3$ signal goes from +1 through an inflection at zero and on to −1.

Suppose we have a decoupled ^{13}C spectrum consisting of one signal for each carbon but, because of the close proximity of the signals, we are unable to make unambiguous assignments. If all the one-bond C–H coupling constants are approximately equal in magnitude (e.g., $^1J = 125$ Hz if all the carbons have sp^3 hybridization), the *J*-modulated spin echo spectrum with $\tau = 1/J$ [e.g., $\tau = 1/(125 \text{ Hz}) = 8$ ms] will have all signals from quaternary and CH$_2$ carbons positive, while all signals from CH and CH$_3$ carbons will be negative. This experiment is thus known as the **attached proton test**, or **APT**. Put simply, the APT experiment determines the parity (odd or even) of the number of hydrogens attached to a given carbon.

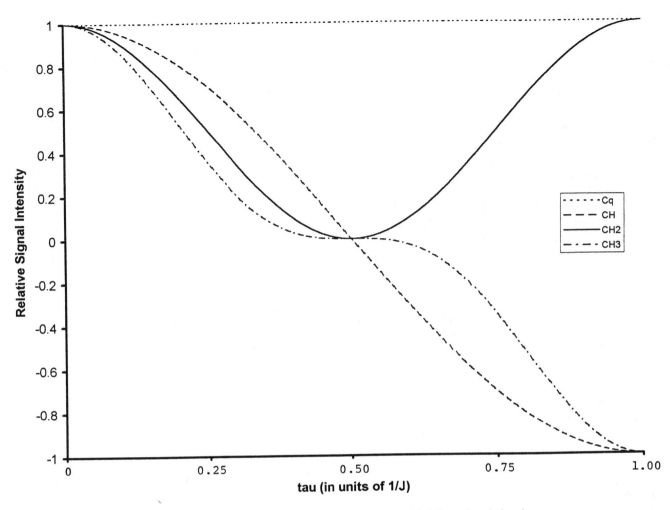

Figure 12.9. APT signal intensity as a function of signal multiplicity and evolution time τ.

■ **EXAMPLE 12.12** Draw the *J*-modulated spin echo ^{13}C spectrum of 2-chlorobutane (Example 12.6) that results from using $\tau = 8$ ms.

□ *Solution:* The 8-ms evolution time correspond to $1/J$ for the 125-Hz one-bond C–H couplings expected for the four tetrahedral carbons in this molecule. According to Figure 12.9, the APT spectrum would exhibit negative peaks at δ 11.01 (CH$_3$), 24.85 (CH$_3$), and 60.38 (CH), and a positive peak at δ 33.33 (CH$_2$). The resulting simulated spectrum is shown in Figure 12.10; compare this to the normal ^{13}C spectrum in Figure 7.1. Note that the signal for TMS (CH$_3$) is negative, while the triplet for solvent CDCl$_3$ is positive, since C–D coupling is unaffected by this experiment. □

■ **EXAMPLE 12.13** Consider the structure of compound **12-1**, in which all 13 carbons are sp^2 hybridized. Ten of the carbons have a hydrogen attached, the other three (shown with dots) are nonprotonated:

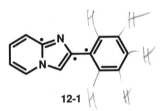

12-1

(a) Predict the number of signals in the normal ^{13}C {^{1}H} spectrum of this compound. In what range of chemical shifts do you expect to find these signals? (b) What is the average value of $^1J_{CH}$ for the protonated carbons in this compound? (c) Predict the appearance of the APT spectrum for this compound using $\tau = 6.4$ ms.

□ *Solution:* (a) Of the 13 vinyl and aromatic carbons in the molecule, there are two sets of 2 (the ortho and meta aromatic hydrogens on the phenyl ring) and 9 additional unique carbons, for a total of 11 signals. From the data in Section 7.4, we expect to find these signals in the δ 100–150 range. (b) From Table 9.1, all the $^1J_{CH}$ values

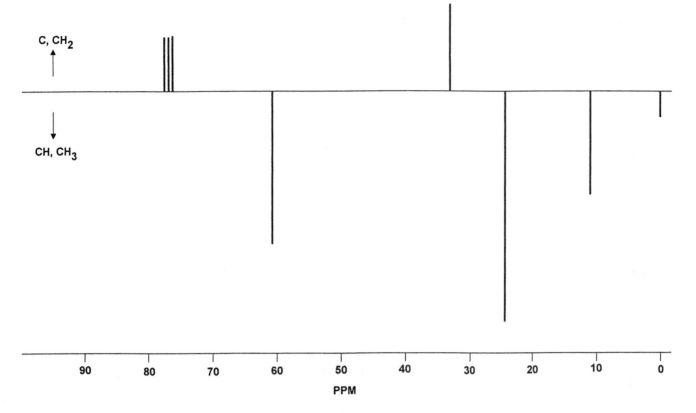

Figure 12.10. Simulated APT ^{13}C spectrum of 2-chlorobutane.

should be approximately 156 Hz. (c) Conveniently, a τ value of 6.4 ms corresponds to 1/156, that is, 1/J. Therefore, we expect the three nonprotonated carbons to appear as positive peaks, while all the CH signals will be negative. Figure 12.11a is the normal proton-decoupled ^{13}C spectrum of **12-1**. Notice that only 10 lines are apparent. This is because the signal at δ 145.5 is actually 2 very closely spaced lines. The APT spectrum, shown in Figure 12.11b, has the unprotonated carbons at δ 145.5 and 133.8 positive and all the remaining CH carbons negative.   □

Although the APT experiment readily differentiates CH and CH$_3$ signals from C and CH$_2$ signals, it is difficult to differentiate CH signals from CH$_3$ signals and quaternary C signals from CH$_2$ signals unless we run a number of experiments with differing τ values. It also requires that all $^1J_{CH}$ values are approximately equal. Fortunately, there is a simpler way to differentiate all four classes of carbons using the DEPT technique, to be described a little later.

12.11   MORE ABOUT POLARIZATION TRANSFER

We began our discussion of internuclear polarization transfer in the context of the NOE in Section 12.3. There, by saturating all the hydrogens by broadband irradiation, we noted en-

hancement of ^{13}C signal intensity for those carbons attached directly to one or more hydrogens. Then (in Section 12.7) we found that with difference-NOE spectra only signals that arise from polarization transfer are observed. For the remaining techniques described in this chapter, as well as many of the experiments to be discussed in Chapter 13, the spectra exhibit *only* those signals that arise by polarization transfer.

12.11.1   Selective Population Inversion

Instead of broadband irradiation of all the hydrogens, let us see what happens to the ^{13}C spectrum of chloroform when we selectively irradiate just one specific hydrogen transition. This experiment resembles selective decoupling (Section 12.8) except that, instead of observing the effect of continuously irradiating (saturating) one signal, we will observe the effects on the ^{13}C spectrum caused by a pulsed *inversion* of just one specific hydrogen signal.

Figure 12.12a depicts the coupled C–H spin system of HCCl$_3$, a composite of Figures 12.3 and 9.2a. There are two ^{13}C transitions ($\nu_{C1} < \nu_{C2}$), each with intensity proportional to population difference $2\Delta_C$, and two ^{1}H transitions ($\nu_{H1} < \nu_{H2}$), each with intensity proportional to population difference $2\Delta_H$ ($\Delta_H = 4\Delta_C$). In the selective population inversion (SPI) experiment, we will irradiate only one specific hydrogen

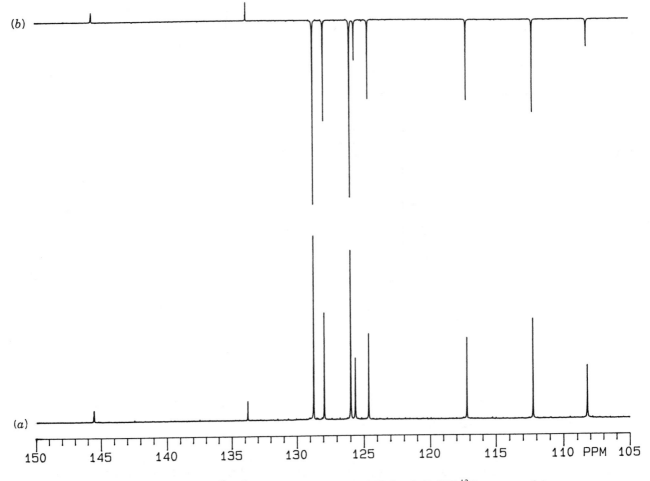

Figure 12.11. (*a*) Normal $^{13}C\ \{^1H\}$ spectrum of compound **12-1** and (*b*) APT ^{13}C spectrum of the same compound. (Contributed by David Lankin.)

transition (for example, ν_{H1}) with a $180_{x'}$ pulse. This has the immediate effect of exchanging the populations of the first and third spin states (Figure 12.12*b*). Although the frequency of each transition remains the same, the intensities vary dramatically. The signal at ν_{C2} now has an intensity proportional to $2\Delta_H + 2\Delta_C$, six times as large as its original intensity, while the signal at ν_{C1} is now inverted with intensity $-2\Delta_H + 2\Delta_C$ (about 60% of the ν_{C2} signal). The signal at ν_{H2} is unaffected, but the signal at ν_{H1} is inverted. The most important result is that both ^{13}C signals have their absolute intensities increased (one positively, one negatively) approximately by the ratio $\gamma_H/\gamma_C = 4$.

Although the SPI experiment is not too useful by itself, we will discover that the technique of polarization by population inversion is central to many of the more advanced pulse sequence experiments.

■ **EXAMPLE 12.14** Describe the result of repeating the above experiment but irradiating at ν_{H2} instead of ν_{H1}.

☐ **Solution:** The second and fourth spin states would exchange so that now ν_{C1} would be positive and ν_{C2} inverted while ν_{H1} is unaffected and ν_{H2} is inverted.   ☐

12.11.2   Insensitive Nuclei Enhanced by Polarization Transfer

Now let us take the SPI experiment one step further. Instead of irradiating just one hydrogen transition we will, in effect, invert all of them simultaneously. This feat is accomplished by the pulse sequence known as Insensitive Nuclei Enhanced by Polarization Transfer (INEPT) depicted in Figure 12.13*a*, which, for a doublet signal results in a spectrum like the one at the bottom of Figure 12.12*b*. The difference in intensity of the two lines ($2\Delta_H + 2\Delta_C$ vs. $-2\Delta_H + 2\Delta_C$; Section 12.11.1) can be eliminated by repeating the INEPT pulse sequence using a $90_{-y'}$ ^1H pulse instead of the $90_{y'}$ pulse, then subtracting the latter data from the normal INEPT data to give the spectrum in Figure 12.13*b*. It is also possible to "refocus"

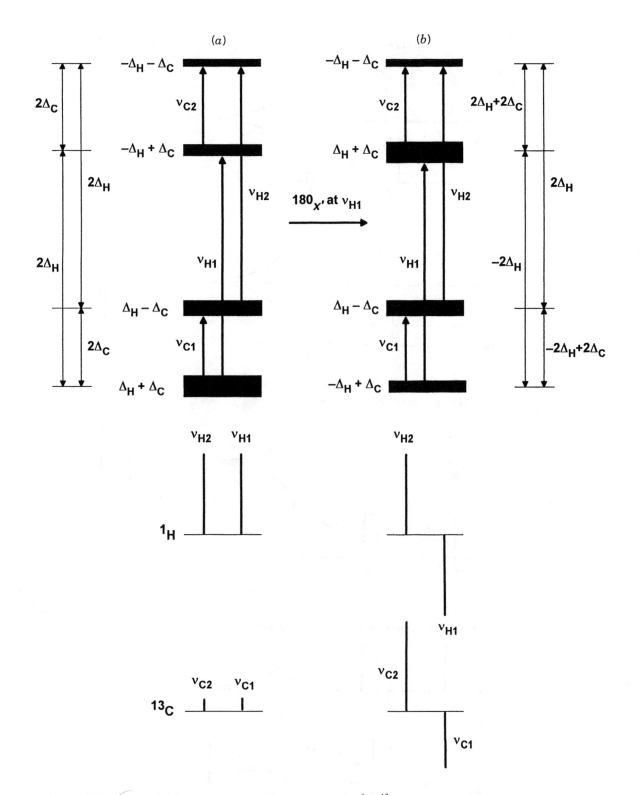

Figure 12.12. (a) Equilibrium spin state population for a coupled ^1H–^{13}C spin system and the two resulting doublets. (b) Effect on spin state populations of selective inversion of the ν_{H1} transition and the resulting two doublets.

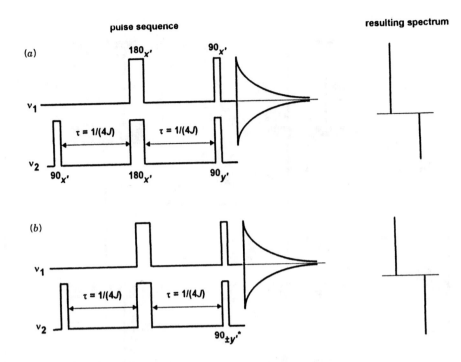

(*a*)

$180_{x'}$ $90_{x'}$

ν_1

$\tau = 1/(4J)$ $\tau = 1/(4J)$

ν_2

$90_{x'}$ $180_{x'}$ $90_{y'}$

(*b*)

ν_1

$\tau = 1/(4J)$ $\tau = 1/(4J)$

ν_2

$90_{\pm y'}^*$

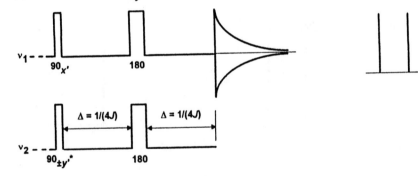

(*c*) **Same as in (*b*) to the $90_{\pm y'}$ pulse; then add**

ν_1 – –

$90_{x'}$ 180

$\Delta = 1/(4J)$ $\Delta = 1/(4J)$

ν_2 – –

$90_{\pm y'}^*$ 180

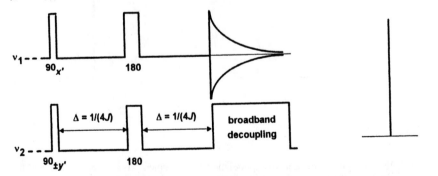

(*d*) **Same as in (*c*), except broadband decoupling is added.**

ν_1 – –

$90_{x'}$ 180

$\Delta = 1/(4J)$ $\Delta = 1/(4J)$ **broadband decoupling**

ν_2 – –

$90_{\pm y'}$ 180

Figure 12.13. Pulse sequences and resulting spectra for (*a*) simple INEPT, (*b*) INEPT with elimination of intensity differences due to natural component ($2\Delta_C$) of M_C, (*c*) refocused INEPT (the evolution time Δ must be adjusted for multiplicities other than doublet), and (*d*) refocused INEPT with decoupling.

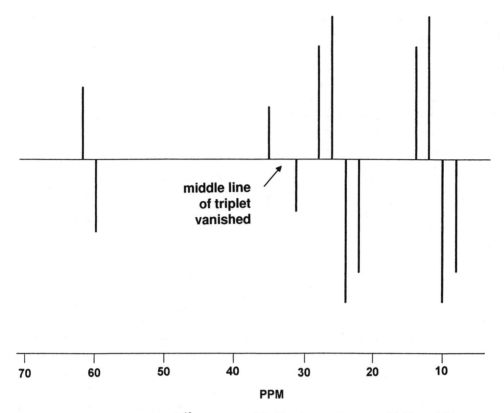

Figure 12.14. Simulated INEPT ^{13}C spectrum of 2-chlorobutane. Compare with Figure 8.11.

(invert) the negative line by the more complicated pulse sequence in Figure 12.13c. The resulting doublet resembles the normal one at the bottom of Figure 12.12a, except that it is four times as intense. Finally, one can add broadband decoupling (Figure 12.13d), resulting in a decoupled singlet that is 33% more intense than one enhanced by a maximum NOE. It should be noted that the evolution time τ must be adjusted somewhat to accommodate multiplets with significantly different multiplicities or J values.[4]

The simulated INEPT ^{13}C spectrum of 2-chlorobutane is shown in Figure 12.14. By comparing this to the corresponding undecoupled ^{13}C spectrum (Figure 8.11), note how each multiplet is both intensified and divided at its center into positive and negative lines that no longer show the same relative intensities as in the undisturbed multiplet. The appearance of these multiplets is quite similar to the multiplet effect CIDNP spectra we saw in Section 11.9, and as was true there, the middle leg of a multiplet with an odd number of lines vanishes.

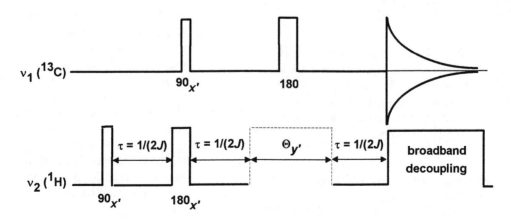

Figure 12.15. Pictorial depiction of the DEPT pulse sequence.

12.12 DISTORTIONLESS ENHANCEMENT BY POLARIZATION TRANSFER

Now we are prepared to combine the APT (Section 12.10) and INEPT (Section 12.11.2) experiments into one of the most useful experiments in modern NMR. Like an APT spectrum, a **DEPT** (distortionless enhancement by population transfer) ^{13}C spectrum is designed to display separate **subspectra** for CH, CH_2, and CH_3 carbon signals. And like an INEPT spectrum, signal intensity (i.e., sensitivity) arises by polarization transfer.

The DEPT pulse sequence is depicted in Figure 12.15. The key addition is the inclusion of a ^1H pulse $\Theta_{y'}$ with a variable tip angle that controls the relative intensity of carbon signals of different multiplicity according to the equations

$$\text{CH:} \quad I = \left(\frac{\gamma_H}{\gamma_C}\right) \sin \Theta \qquad (12.4a)$$

$$\text{CH}_2: \quad I = \left(\frac{\gamma_H}{\gamma_C}\right) \sin 2\Theta \qquad (12.4b)$$

$$\text{CH}_3 \quad I = \left(\frac{3\gamma_H}{4\gamma_C}\right)(\sin \Theta + \sin 3\Theta) \quad (12.4c)$$

These equations are plotted graphically in Figure 12.16. Note that with Θ set to 90° [a DEPT (90) subspectrum] the intensity of the CH signal is at a maximum, while CH_2 and CH_3 signals are zero. Compare these equations with those for the APT experiment [Eq. (12.3) and Figure 12.8].

■ **EXAMPLE 12.15** What subspectrum would result from (a) subtracting the DEPT (135) data from the DEPT (45) data? (b) DEPT (45) + DEPT (135) − 0.707 DEPT (90)?

□ *Solution:* (a) The DEPT (45) gives maximum CH_2 signal intensity and near-maximum CH_3 signal intensity, while a DEPT (135) gives minimum CH_2 signal intensity and near-maximum CH_3 signal intensity. Therefore, a DEPT (45) − DEPT (135) subspectrum exhibits only the CH_2 signals. (b) This subspectrum exhibits only the CH_3 signals. □

This process for separating the various DEPT subspectra by arithmetic manipulation of data resulting from different Θ values is referred to as **editing**. The edited DEPT spectra of the molecule below (artemisinin) are shown in Figure 12.17:

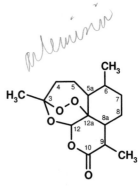

The bottom spectrum is the normal decoupled ^{13}C spectrum consisting of 15 signals. The top three subspectra reveal, respectively, which are the CH_3, CH_2, and CH signals. The fourth subspectrum is the sum of the first three; any signals *not* present in this subspectrum but present in the normal ^{13}C spectrum must belong to nonprotonated carbons.

CHAPTER SUMMARY

1. Double-resonance NMR techniques involve simultaneous irradiation of the sample with two oscillating magnetic fields (**B**$_1$ and **B**$_2$) of frequencies ν_1 and ν_2. This allows us to determine the effect of irradiating one set of nuclei (at ν_2) while observing the NMR signal of another set of nuclei (at ν_1).

2. Homonuclear double resonance indicates that both ν_1 and ν_2 are focused on nuclei of the same isotope, while heteronuclear double resonance indicates that the ν_1 and ν_2 are focused on different isotopes or elements.

3. Depending on the power and exact location of ν_2, a variety of useful effects can be generated during double-resonance experiments. Some of the more common are the following:

   a. Decoupling: Irradiation of nuclei in set B during observation of the signal from those in set A can "erase" the effects of any A–B coupling. For example, irradiation of all ^1H nuclei during acquisition of a ^{13}C spectrum (denoted ^{13}C {^1H}) results in a greatly simplified ^{13}C spectrum free of all C–H couplings. Both homonuclear and heteronuclear decoupling are possible.

   b. Nuclear Overhauser effect: Irradiation of nuclei in set B can have an effect on the spin state populations of other nearby nuclei, causing a change in intensity of the signals of the nearby nuclei (usually an increase in intensity. Thus, in a ^{13}C {^1H} experiment, not only is the spectrum simplified by decoupling but also the intensity of most signals is increased by the NOE [Eq. (12.1)].). The magnitude of the NOE enhancement is highly sensitive to, and therefore a probe of, internuclear distance.

   c. Off-resonance decoupling: When ν_2 is focused somewhat apart from the resonance frequency of

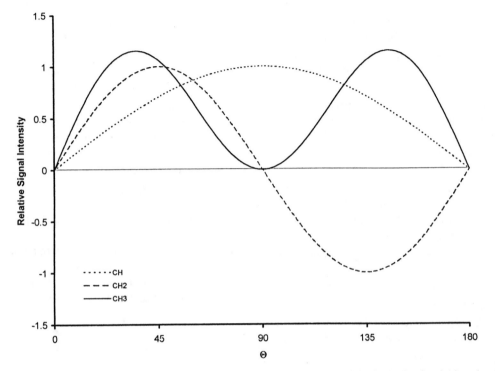

Figure 12.16. DEPT signal intensity as a function of signal multiplicity and the tip angle of variable pulse Θ.

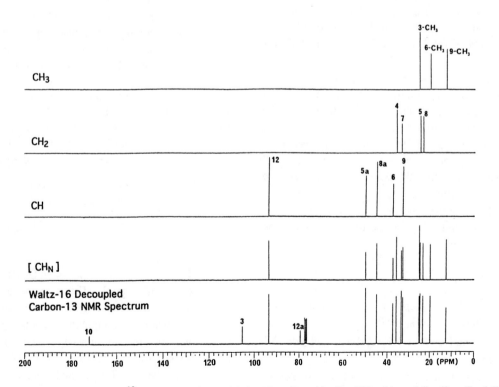

Figure 12.17. Edited DEPT ^{13}C spectrum of artemisinin. (Contributed by David Lankin and Geoffrey Cordell).

set B nuclei while the spectrum of set A nuclei is observed, A–B coupling constants are reduced in magnitude but not eliminated completely. The reduced coupling constant (J_r) is given by Eq. (12.2).

   d. Selective decoupling: By carefully adjusting the power and frequency of ν_2, it is possible to decouple only parts of multiplets and thereby generate information that helps determine the relative signs of coupling constants.

4. Modern pulse-mode NMR spectrometers allow the details of the ν_1 and ν_2 pulse sequence to be altered in an infinite variety of ways to yield useful information about molecular structure. Such pulse sequence variations include the following:

   a. Gated decoupling: If ν_2 is turned on during the presaturation delay but off during FID acquisition, much of the NOE signal enhancement is preserved, though there is no decoupling.

   b. Inverse gated decoupling: If ν_2 is off during the presaturation delay but on during acquisition, the result is a spectrum where there is little NOE but decoupling is preserved.

   c. APT: A ^{13}C pulse sequence by which CH and CH_3 signals can be differentiated from C and CH_2 signals.

5. There is a large and growing class of complex pulse sequences that involve polarization transfer from a set of more sensitive nuclei (e.g., ^1H) to a set of less sensitive nuclei (e.g., ^{13}C), greatly increasing the signal intensity of the latter set. Examples include the following:

   a. SPI, where one particular transition is inverted by a pulse sequence on the ν_2 channel, causing half of the coupled multiplet to be inverted in the observed spectrum;

   b. INEPT, where all ^1H signals are inverted by a pulse sequence on the ν_2 channel and half of all connected multiplets are inverted in the observed spectrum; and

   c. DEPT, where the normal ^{13}C spectrum is separated into individual subspectra for CH, CH_2, and CH_3 signals.

REFERENCES

[1]Cooper, J. W., *Spectroscopic Techniques for Organic Chemists*, Wiley, New York, 1980.

[2]Anet, F. A. L., and Bourn, A. J. R., *J. Am. Chem. Soc.*, *87*, 5250 (1965).

[3]Becker, E., *High Resolution NMR*, 2nd ed., Academic, New York, 1980.

[4]Shoolery, J. N., *J. Nat. Prod.*, *47*, 226 (1984).

ADDITIONAL RESOURCES

1. Derome, A. E., *Modern NMR Techniques for Chemistry Research*, Pergamon, New York, 1987.

2. Benn, R., and Gunther, H., *Angew. Chem. Int. Ed. Engl.*, *22*, 350 (1983).

3. Kessler, H., Gehrke, M., and Griesinger, C., *Angew. Chem. Int. Ed. Engl.*, *27*, 490 (1988).

4. Nakanishi, K., Ed., *One-Dimensional and Two-Dimensional NMR Spectra by Modern Pulse Techniques*, University Science Books, Sausalito, 1990.

5. Duddeck, H., and Dietrich, W., *Structure Elucidation by Modern NMR, A Workbook*, Springer-Verlag, New York, 1989.

REVIEW PROBLEMS (Answers in Appendix 1)

12.1. The ^1H spectrum of 2,2,6,6-tetramethyl-4-heptyn-3-ol (below) consists of two nine-hydrogen singlets (δ 0.96 and 1.20) and two one-hydrogen singlets (O–H at δ 2.25 and C–H at δ 3.94):

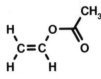

While the intensity of the C–H signal was observed, each *tert*-butyl signal was (separately) irradiated. Irradiation of the signal at δ 1.20 produced no significant change, but irradiation at δ 0.96 gave an 11% increase in the intensity of the signal at δ 3.94. What can you infer from these results?

12.2. Verify Eq. (12.3 b).

12.3. Vinyl acetate, the structure of which is shown below, undergoes addition polymerization to form polyvinyl acetate (PVA), used in paints and adhesives:

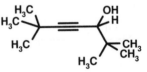

Figure 12.18 shows the edited DEPT ^{13}C spectrum (δ 0–80) of a sample of PVA dissolved in fully deuterated dimethyl sulfoxide. (a) Account for the seven-line pattern (highlighted with dots) near δ 40. Explain why it does not show up in any of the upper four subspectra. (b) Propose a structure for PVA consistent with the DEPT data, and assign all signals in the spectra.

12.4. Figure 12.19 shows the 300-MHz ¹H spectrum of the ester of 3-phenyl-1-propanol shown below:

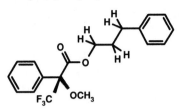

(a) Assign all signals in the spectrum, and account for their multiplicity. *Hint:* see review problem 10.2. (b) Predict the effect on the spectrum of irradiating the multiplet centered at δ 2.04.

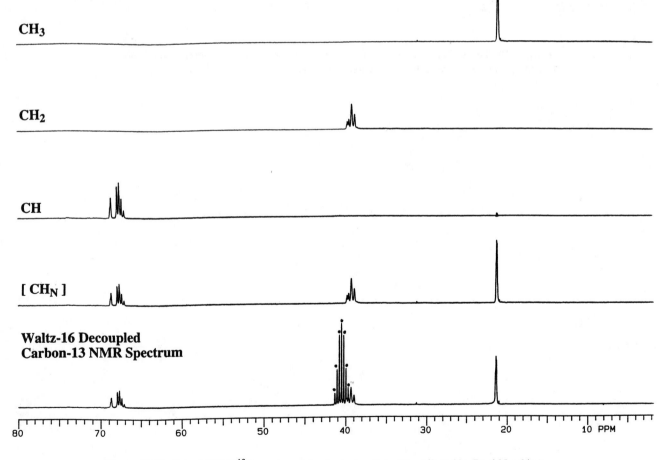

Figure 12.18. Edited DEPT ¹³C spectrum of polyvinyl acetate. (Contributed by David Lankin.)

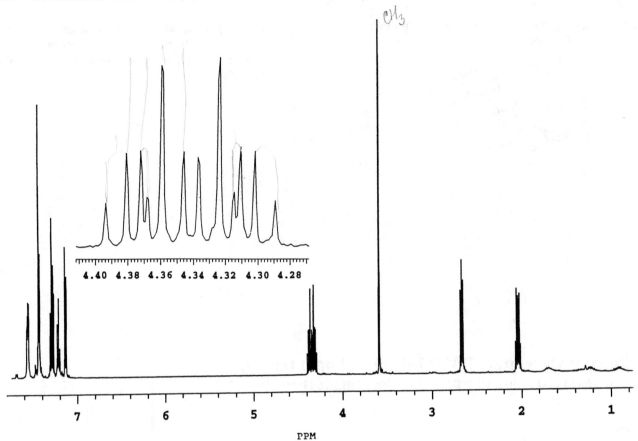

PPM

Figure 12.19. The 300-MHz ^{1}H spectrum of the MTPA ester of 3-phenyl-1-propanol. (Contributed by Dhileepkumar Krishnamurthy, of Gary Keck's group at the University of Utah.)

13

TWO-DIMENSIONAL NUCLEAR MAGNETIC RESONANCE

13.1 WHAT IS 2D NMR SPECTROSCOPY?

By now you have gained an appreciation for the great power of NMR to convey highly detailed information about the structure of molecules and the dynamic processes they undergo. The chemical shift of a nucleus is indicative of its electronic and magnetic environment in a molecule; coupling constants show through-bond proximity between nuclei, difference-NOE spectra provide stereochemical (through-space) information about the arrangement of atoms, and dynamic NMR spectra provide rates of conformational interconversions and other types of fluxional behavior. The problem is that as we graduate to ever more complex molecular structures, their NMR spectra become progressively more difficult to disentangle. The recent advent of 2D NMR techniques now allows us to acquire direct structural information such as connectivity and proximity more efficiently and unambiguously than ever before.

The term **2D NMR**, which stands for two-dimensional NMR, is something of a misnomer. All the NMR spectra we have discussed so far in this book are two dimensional in the sense that they are plots of signal intensity versus frequency (or its Fourier equivalent, signal intensity versus time). By contrast, 2D NMR refers to spectroscopic data that are collected as a function of two time scales, t_1 (evolution and mixing) and t_2 (detection). The resulting data set is then subjected to separate Fourier transformations of each time domain to give a frequency-domain 2D NMR spectrum of signal intensity versus two frequencies, F_1 (the Fourier transform of the t_1 time domain) and F_2 (the Fourier transform of the t_2 time domain). Thus, a 2D NMR spectrum is actually a *three-dimensional* data set!

The F_2 frequency normally corresponds to a chemical shift axis, for example, ^1H or ^{13}C. When the F_1 axis also corresponds to a chemical shift scale, we describe the result as a

2D **shift-correlated** spectrum. When the F_1 axis corresponds to a coupling constant scale (e.g., ^1H–^1H or ^1H–^{13}C), we generate a **2D J-resolved** spectrum.

A typical 2D NMR spectrum is actually composed of a large number of spectra (often several hundred) collected automatically, each with a uniformly incremented value of t_1. For example, consider an unedited DEPT (Section 12.12) experiment on 2-chlorobutane:

2-chlorobutane

The ^{13}C spectrum of 2-chlorobutane, first encountered in Figure 7.1, consists of signals at δ 60.4 (CH), 33.3 (CH$_2$), 24.8 (CH$_3$), and 11.0 (CH$_3$). In our experiment, we will first collect the ^{13}C DEPT spectrum (using the pulse sequence in Figure 12.15), with the variable $\Theta_{y'}$ pulse width set to zero degrees. The process is then repeated 18 more times, with $\Theta_{y'}$ incremented by 10° each time. Finally, the data are subjected to Fourier transformation to give a frequency-domain data set with the F_2 axis corresponding to ^{13}C chemical shift and the F_1 axis to $\Theta_{y'}$.

The resulting 2D NMR data can be plotted in either a **stacked plot** or a **contour plot**. The simulated stacked plot (Figure 13.1) portrays all three dimensions in a perspective view, with positive signal intensity in the upward dimension. (Another example of a stacked plot can be seen in Figure 10.2.) The contour plot (Figure 13.2) represents a *horizontal* cross section of the stacked plot, taken at some selected intensity value above, on, or below the F_1–F_2 plane. In Figure 13.2 the cross section is taken in the F_1–F_2 plane. The size of the circles represents the relative intensity of each signal; closed circles represent positive signals, while open circles

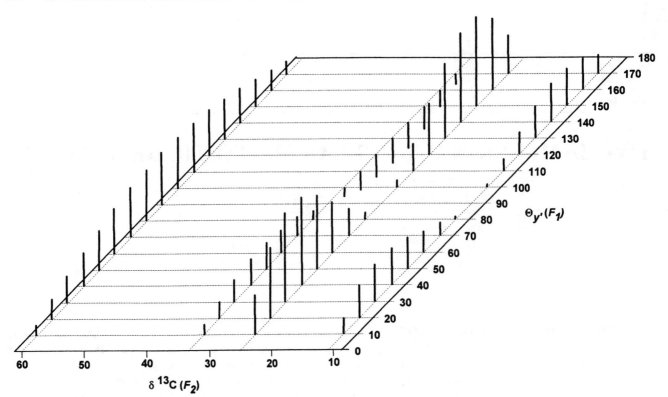

Figure 13.1. Stacked plot of the simulated DEPT spectrum of 2-chlorobutane pulse versus pulse width $\Theta_{y'}$.

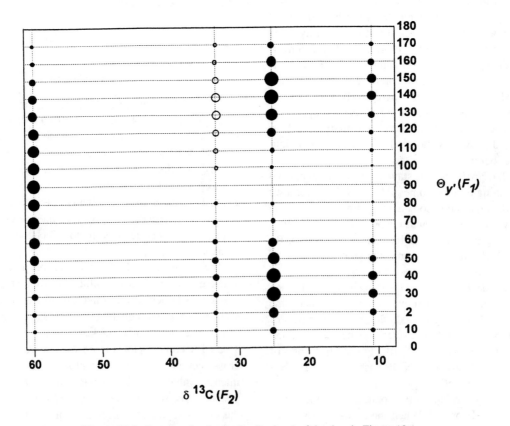

Figure 13.2. Contour plot (in the F_1, F_2 plane) of the data in Figure 13.1.

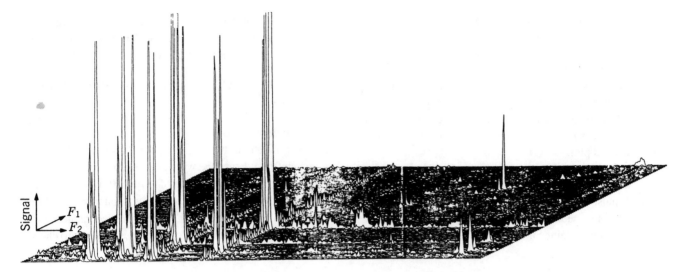

Figure 13.3. Stacked plot of a typical 2D NMR spectrum. (Contributed by R. Marshall Wilson.)

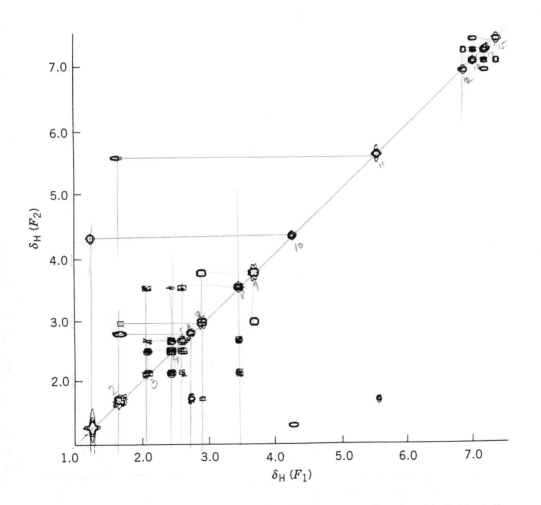

Figure 13.4. Contour plot of a typical 2D hydrogen COSY spectrum. (Contributed by R. Marshall Wilson.)

correspond to negative signals. Compare the two plots; note how the sign and magnitude of each of the four ^{13}C signals vary with $\Theta_{y'}$, in accord with the dependence shown graphically in Figure 12.16.

An actual stacked 2D NMR spectrum is shown in Figure 13.3. Compare this with an actual contour 2D spectrum (of a different compound) in Figure 13.4. In most cases it is easier to derive the desired information from a contour plot than from a stacked plot.

As with the advanced one-dimensional techniques discussed in Chapter 12, there is a large and growing family of 2D NMR pulse sequences.[1-6] In this chapter we will examine some of the most widely used ones, focusing our attention on the information contained in each type of 2D spectrum. How-

ever, we will not describe in any detail the pulse sequences that give rise to each spectrum. Readers interested in these details should consult the references.

13.2   2D HETEROSCALAR SHIFT-CORRELATED SPECTRA

When attempting to unravel the complex structure of an unknown molecule from its ^{13}C and ^{1}H spectra, it is often quite helpful to determine which hydrogens are connected to which carbons. In the parlance of 2D NMR, we wish to find out which ^{1}H signals are *correlated* (associated via one-bond C–H coupling) with which ^{13}C signals. The technique we will use involves heteronuclear (e.g., carbon vs. hydrogen) chemi-

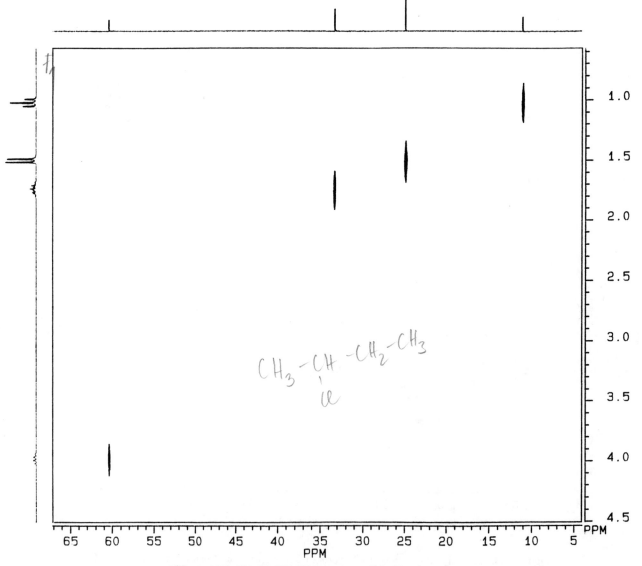

Figure 13.5.   The 2D C,H-HSC spectrum of 2-chlorobutane.

cal shift correlation (**HSC**). The technique is also known as **C,H-HETCOR**, or **C,H-COSY** for carbon–hydrogen correlation spectroscopy.

The contour plot of the C,H–HSC spectrum of 2-chlorobutane is shown in Figure 13.5, where the F_1 (vertical) axis corresponds to ^1H chemical shift and the F_2 axis to ^{13}C chemical shift. Note that the normal ^1H spectrum [δ 1.01 (t, 3H), 1.52 (d, 3H), 1.72, m, 2H), 3.96 (m, 1H)] is plotted vertically along the left side of the 2D spectrum, while the proton-decoupled ^{13}C spectrum is plotted horizontally along the top.

The vertically elongated 2D signals in Figure 13.5 reveal the hydrogen signal(s) to which each carbon signal is correlated by a one-bond coupling constant of ca. 125 Hz. As shown in Figure 13.6, the carbon signals at δ 11.0 (signal **1**,

CH$_3$), 24.8 (signal **2**, CH$_3$), 33.3 (signal **3**, CH$_2$), and 60.4 (signal **4**, CH) correlate, respectively, with the hydrogen multiplets at δ 1.01 (signal **1**), 1.52 (signal **2**), 1.72 (signal **3**), and 3.96 (signal **4**). Note that this roughly diagonal relationship between signals is expected based on our discussion in Section 7.8 (Figure 7.3).

When generating the HSC spectrum of an unknown compound, one with a wide range of one-bond C–H J values and signal multiplicities, it is usually necessary to vary some of the pulse sequence parameters in order to acquire all the desired correlation signals. Another problem involves differentiating weak artifact signals from true correlation signals. These are two aspects of the "art" of 2D NMR, and one becomes an artist only through much practice and experience.

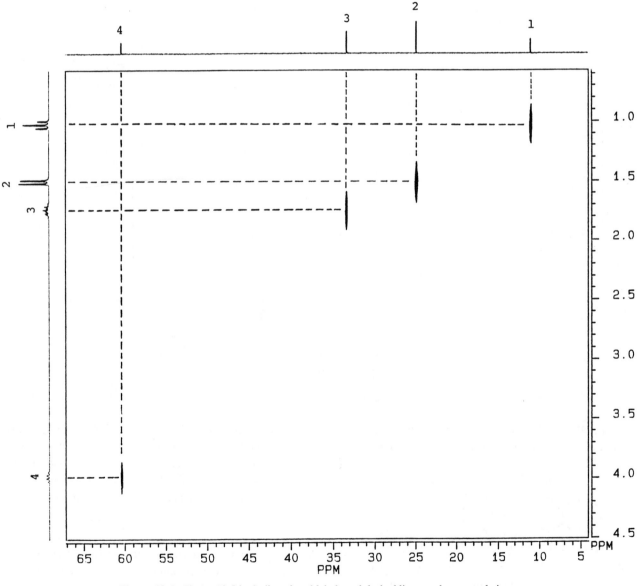

Figure 13.6. Figure 13.5 including signal labels and dashed lines to show correlations.

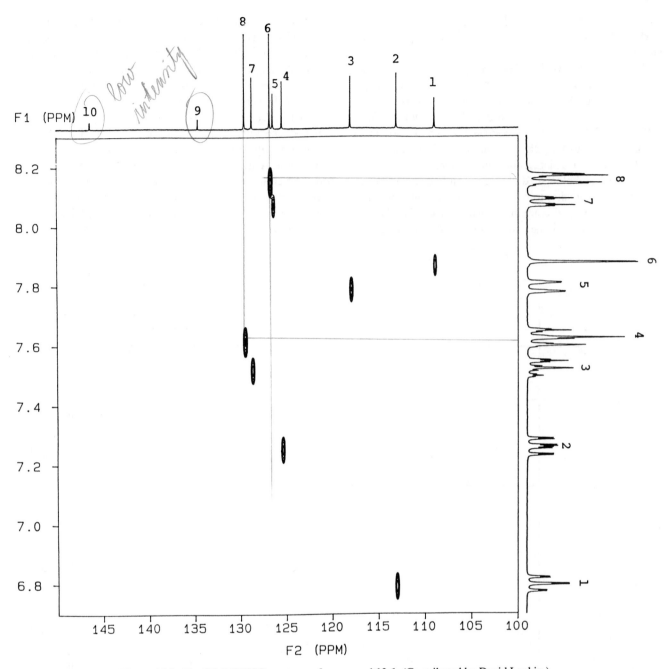

Figure 13.7. The 2D C,H-HSC spectrum of compound **12-1**. (Contributed by David Lankin.)

■ **EXAMPLE 13.1** Figure 13.7 shows the C,H–HSC spectrum of the compound **12-1**, previously encountered in Example 12.13:

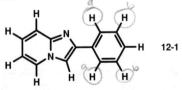

12-1

Recall that there are 11 sets of carbons but only 10 signals (the signal at δ 145.5 is composed of 2 unresolved lines). The hydrogens comprise 8 sets because the 2 ortho phenyl hydrogens are equivalent, as are the 2 meta hydrogens. The 8 hydrogen multiplets are numbered consecutively (in the order of increasing chemical shift), as are the 10 carbon signals. Using these numbers, indicate all H–C correlations.

□ *Solution:* Hydrogen signal number **1** (δ 6.80) correlates with carbon signals number **2** (δ 113), and so on. A complete list of the correlations is tabulated below:

Hydrogen	Carbon	Hydrogen	Carbon
1	2	5	3
2	4	6	1
3	7	7	5
4 (2H)	8	8 (2H)	6

Note that carbon signals **9** and **10** give no correlation signals, so they must correspond to the three unprotonated carbons, a fact that might have been guessed from the low intensity of these signals. Notice also that the roughly diagonal relationship seen in Figure 13.6 is absent here, mainly because the carbon and hydrogen chemical shift ranges (ca. 20 and 1.4 ppm, respectively) are relatively small.   □

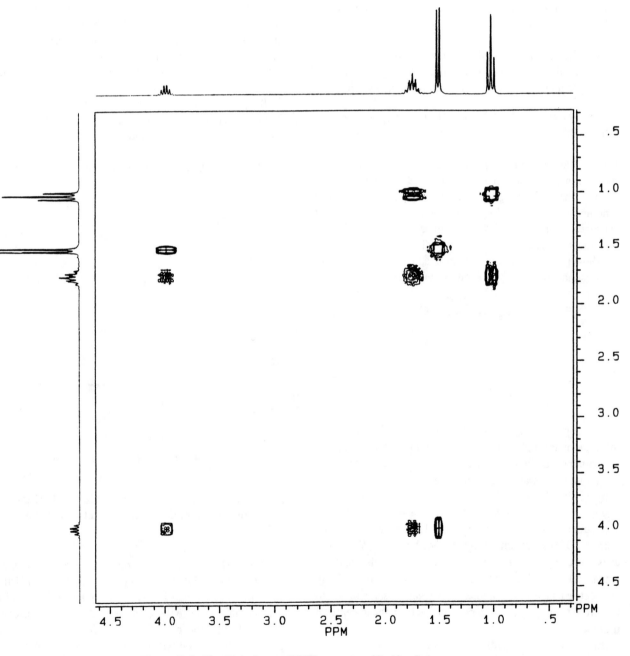

Figure 13.8. The 2D hydrogen COSY spectrum of 2-chlorobutane.

Although the C,H–HSC spectrum indicates all one-bond C–H connectivities, it does not give us any direct information about the more important C–C connectivities. And when complex molecular structures are involved, the HSC spectrum may not help too much in assigning all signals to specific carbons and hydrogens. One approach to resolve this problem is to adjust the HSC pulse sequence to test for longer range C–H couplings. A technique for accomplishing this is known as **COLOC** (correlation spectroscopy via long-range couplings). By comparing a COLOC spectrum with a conventional HSC for the same compound, it is possible (though often with some difficulty) to map out two- and even three-bond C–H coupling interactions. This information can often help in elucidating the carbon–carbon connectivities in the molecule.

Another approach to deduce carbon connectivity patterns from HSC data is to determine which hydrogen signals are coupled together, and that is our goal in the next section.

13.3   2D HOMONUCLEAR SHIFT-CORRELATED SPECTRA

The 2D HSC spectrum in Figure 13.7 indicates, among other correlations, that carbon signals **6** and **8** correlate, respectively, with hydrogen signals **8** and **4**. Suppose you know from a homonuclear decoupling experiment (Section 12.6) that hydrogen signals **8** and **4** are themselves correlated by a three-bond H–H coupling constant. What does this tell you? As shown below, carbon signals **6** and **8** must belong to carbons that are directly bonded to each other!

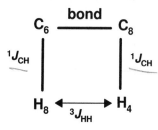

Although we might be able to get all the H–H coupling information through a long series of homonuclear decoupling experiments, there is a much simpler way: the 2D NMR technique known as homonuclear shift correlation spectroscopy, or **COSY**.

Let us take a look at a contour plot of the ^{1}H COSY spectrum of 2-chlorobutane in Figure 13.8. Note first that the normal 1D ^{1}H spectrum is plotted along both axes, since both axes correspond to ^{1}H chemical shift. Notice also that the actual 2D contour signals form a pattern that is perfectly symmetrical about the diagonal, so we get the same information from the upper left triangle as we do from the lower right triangle.

In a homonuclear hydrogen COSY spectrum the signals *along* the diagonal reflect the normal ^{1}H spectrum. It is the

off-diagonal signals (so-called **cross peaks**) that provide the information we need. Since the magnitude of both two- and three-bond H–H coupling constants generally fall in the range 0–20 Hz (Chapter 9), we set the pulse sequence parameters to detect such J values. Thus, in general, each cross peak represents a correlation due to either two- or three-bond H–H coupling. In the case of 2-chlorobutane, all geminally (two-bond) related hydrogens are equivalent, so no two-bond couplings are observed (Section 8.4). Therefore, each cross peak in Figure 13.8 indicates a vicinal (three-bond) relationship.

We again begin by labeling the hydrogen multiplets **1–4** (in order of increasing chemical shift), as shown in Figure 13.9. Then we proceed down the diagonal from the upper right. Signal **1** has a cross peak with signal **3**, signal **2** with signal **4**, and so on. Taken in order, the observed cross peaks indicate the following correlations: **1–3**, **2–4**, and **3–1** (which is the same as **1–3**); **3–4**, and **4–3** (the same as **3–4**); and **4–2** (the same as **2–4**).

Suppose we did not know the structure of 2-chlorobutane, just its molecular formula (C_4H_9Cl). Could we have deduced its structure from the HSC and COSY correlations in Figures 13.6 and 13.9?

Let us begin by tabulating all the known signal correlations as follows:

Carbon	Is Connected To Hydrogen	Which Is Coupled To Hydrogen(s)
1	1	3
2	2	4
3	3	1, 4
4	4	2, 3

From these data we can put together a map of C–C connectivity:

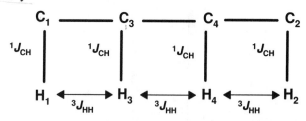

All that remains is to use chemical shift and ^{1}H integration data to put the chlorine on carbon **4** and fill in the hydrogens; *voila*!

■ **EXAMPLE 13.2**   The hydrogen COSY for compound **12-1** (Example 13.1) is shown in Figure 13.10. (a) Using the same hydrogen signal labels as in Figure 13.7, list all H–H correlations. Arrange these, along with all C–H correlations, in a table like the one above. (b) Draw a map of carbon connectivity like the one above. (c) Assign as many ^{1}H and ^{13}C signals to specific nuclei in the molecule as you can.

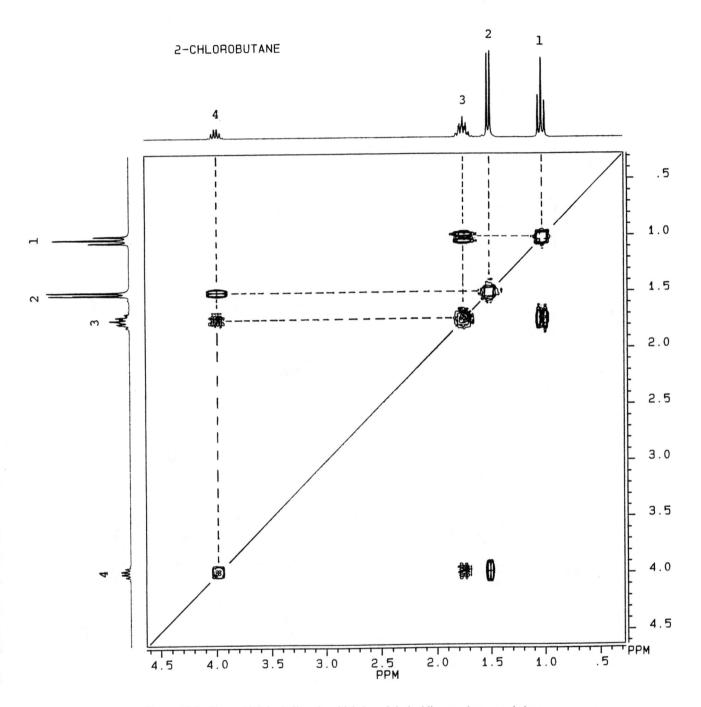

Figure 13.9. Figure 13.8, including signal labels and dashed lines to show correlations.

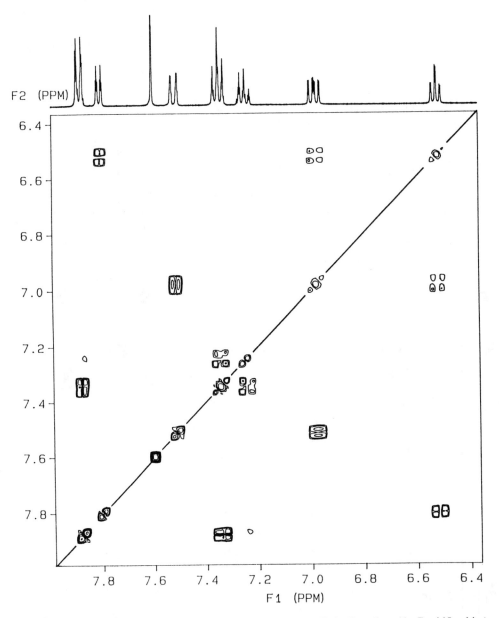

Figure 13.10. The 2D hydrogen COSY spectrum of compound **12-1**. (Contributed by David Lankin.)

☐ *Solution: (a)*

Carbon	Is Connected To	Hydrogen[a]	Which Is Coupled To Hydrogen(s)[b]
1		6	—
2		1	2, 7
3		5	2
4		2	1, 5
5		7	1
6		8	3(w), 4
7		3	4, 8(w)

Carbon	Is Connected To	Hydrogen[a]	Which Is Coupled To Hydrogen(s)[b]
8		4	3, 8
9		—	—
10		—	—

[a]From Example 13.1.
[b](w) indicates very weak (long-range) coupling.

(b) It is generally easiest to start with a carbon whose hydrogen is coupled to only one other hydrogen. Thus,

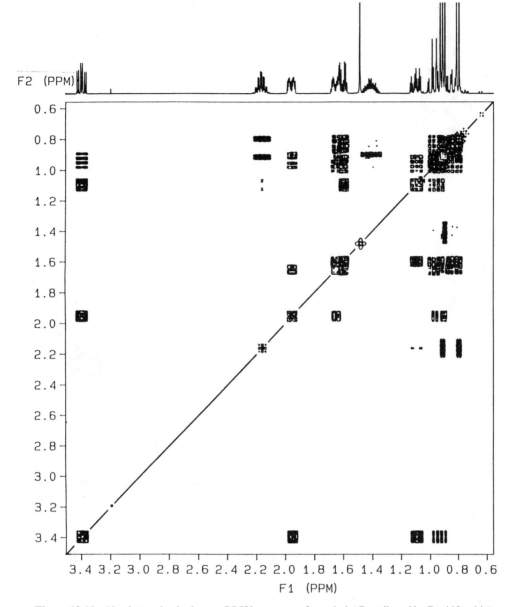

Figure 13.11. Absolute-value hydrogen COSY spectrum of menthol. (Contributed by David Lankin).

hydrogen **7** (attached to carbon **5**) is coupled to hydrogen **1** (attached to carbon **2**); hydrogen **1** is also coupled to hydrogen **2** (attached to carbon **4**); and so on. From this type of reasoning we can generate three pieces of the overall map:

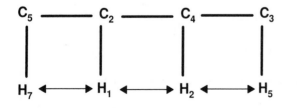

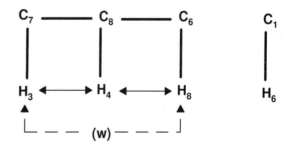

The remaining three nonprotonated carbons (recall that signal **10** represents two accidentally equivalent carbons) must be the connecting links between these three pieces.

(c) Knowing that there are two equivalent hydrogens for carbon signals **4** and **8**, it is logical to assign the three carbon piece to the phenyl group. We can also assign carbon **6** and hydrogen **1**:

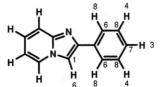

But without making some guesses about chemical shift, we cannot yet be sure which end of the four-carbon fragment belongs where, and we cannot assign carbon signals **9** and **10** unambiguously. But be patient; we will soon have the answer!   □

As is true for many of the advanced 2D NMR techniques, there are several variations of the basic COSY pulse sequence, each of which serves to improve the resolution and signal-to-noise ratio of the 2D signals, remove long-range couplings,

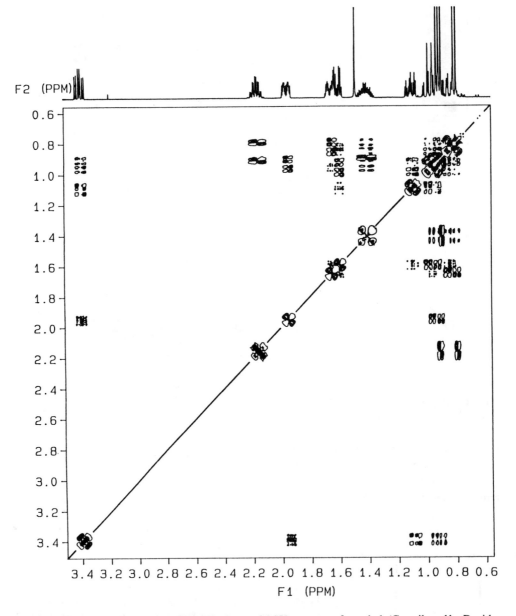

Figure 13.12. Double quantum filtered hydrogen COSY spectrum of menthol. (Contributed by David Lankin.)

or otherwise simplify and improve the quality of the spectrum. For example, Figures 13.11, 13.12, and 13.13 compare three types of hydrogen COSY experiments on menthol, whose ¹H spectrum shows all 14 nonequivalent sets of protons: (**1**) δ 0.80 (d, 3H); (**2**) 0.83 (td, 1H); (**3**) 0.90 (d, 3H); (**4**) 0.92 (d, 3H); (**5**) 0.94 (m, 1H); (**6**) 0.96 (m, H); (**7**) 1.10 (m, 1H); (**8**) 1.41 (m, 1H); (**9**) 1.48 (s, 1H); (**10**) 1.60 (dq, 1H); (**11**) 1.65 (d quintuplet, 1H); (**12**) 1.95 (m, 1H); (**13**) 2.16 (septet d, 1H); (**14**) 3.39 (ddd, 1H).

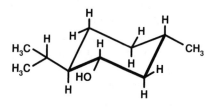

menthol

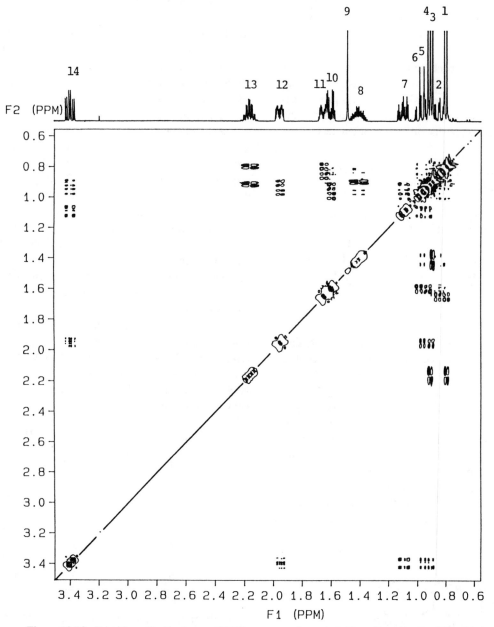

Figure 13.13. Primitive edited hydrogen COSY spectrum of menthol. (Contributed by David Lankin.)

Figure 13.11 shows the "normal" absolute value COSY. Figure 13.12 is the double-quantum-filtered (DQF) phase-sensitive COSY, which removes all uncoupled signals (e.g., the singlet at δ 1.48) from the diagonal. And Figure 13.13 shows the primitive-edited (PE) phase-sensitive COSY, which attenuates long-range couplings (those beyond three bonds). In the phase-sensitive spectra, positive signals are shown as filled contours while negative signals are shown as open contours (as we saw in Figure 13.2). The relative phase of signals can provide information about the relative signs of coupling constants involved (Section 9.12).

■ **EXAMPLE 13.3**(a) Using the numerical labels given in Figure 13.13 for the menthol signals, prepare a table showing all H–H correlations. (b) Assign as many signals as you can to hydrogens in menthol.

□ *Solution:* (a) With a spectrum this complex, the process requires patience and a good set of plastic triangles! Below are the observed correlations; entries in parentheses indicate weak cross peaks. Note that cross peaks involving signals **5** and **6** are somewhat difficult to differentiate.

Hydrogen	Is Coupled To Hydrogen(s)	Hydrogen	Is Coupled To Hydrogen(s)
1	13	**8**	(2), 3, (5)
2	(6), (8), (10), 11	**9**	
3	8	**10**	(2), 6
4	13	**11**	2
5	(8), 12, 14	**12**	5, 14
6	(2), (7), 10	**13**	1, 4
7	(6), 14	**14**	5, 7, 12

(b) Recall that the strongest H–H coupling are the geminal (two-bond) ones and the vicinal (three-bond) ones with anti geometry (i.e., trans diaxial, Example 9.6). We can readily assign doublets **1** and **4** to the methyls of the isopropyl group, since they are both coupled only to multiplet **13**. Similarly, doublet **3** is the lone methyl, coupled only to multiplet **8**. Signal **8** is further (weakly) coupled to **2** and **5**, so these must belong to the two axial hydrogens trans to it. Singlet **9** must be the hydroxyl hydrogen, and multiplet **14** (the most downfield signal) must belong to the hydrogen on the same carbon. Since signal **14** is also coupled to signals **5, 7,** and **12**, we can so far assign all these signals:

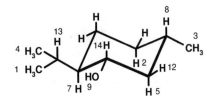

Because signal **7** is further coupled to **6**, which in turn is coupled strongly to **10** and weakly to **2**, we can complete the assignments:

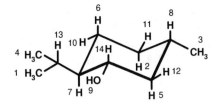

□

13.4 NOE SPECTROSCOPY (NOESY)

By altering the COSY pulse sequence slightly, we can generate a 2D spectrum where the cross peaks reflect homonuclear Overhauser effects (direct through-space polarization transfer, Section 12.7), rather than through-bond coupling. The advantage of a **NOESY** spectrum over the previously encountered difference-NOE spectrum is that, instead of probing just one set of interactions at a time in search of NOE enhancements, we can now probe all possible NOE interactions at once. (Note: The actual experimental details of the steady state difference-NOE are quite different from those of the dynamic NOESY experiment, so that the results of the two experiments are sometimes not comparable.)

The NOESY spectrum usually exhibits most or all of the peaks observed in the corresponding COSY spectrum, plus additional cross peaks from nuclei that are close within the molecule but not directly coupled. It is therefore preferable to compare the COSY and NOESY side by side when making NOESY signal assignments.

To illustrate the complementary value of a NOESY spectrum, compare the NOESY (Figure 13.14) of compound **12-1** (Examples 13.1 and 13.2) with the corresponding COSY (Figure 13.10). The table below lists all cross peaks in the NOESY spectrum, with cross peaks *not* seen in the COSY shown in boldface:

Hydrogen	exhibits NOE with	Hydrogen
1		2, 7
2		1, 5
3		4
4		3, 8
5		2, **6 (w)**
6		**5 (w), 7, 8**
7		1, **6, 8 (w)**
8		4, **6, 7 (w)**

In Example 13.2 we were unable to complete the carbon skeleton of the molecule because we had no information about

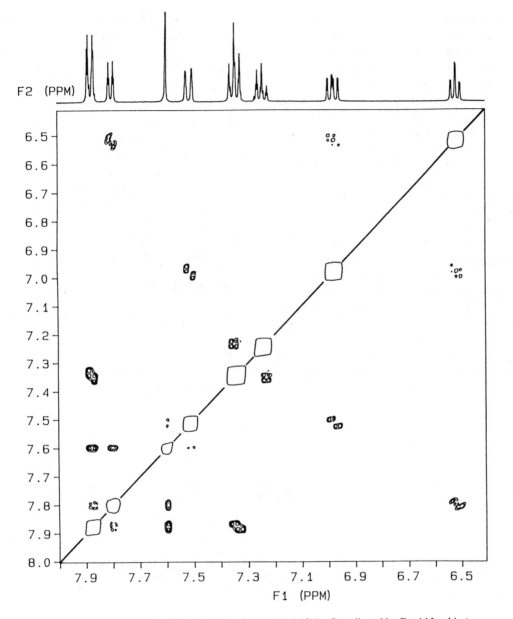

Figure 13.14. The 2D NOESY spectrum of compound **12-1**. (Contributed by David Lankin.)

how the four-, three-, and one-carbon subunits were connected by the remaining three carbons. Now we have the necessary information. Although hydrogen **6** is not directly coupled to any other hydrogens, it is close to hydrogens **7** and **8** (the cross peak with **5** is weak, indicating a longer range interaction). This allows us to assign the sequence of all protonated carbons:

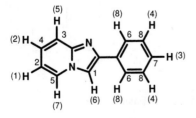

The only assignments that remain are those of carbon signals **9** and **10**.

13.5 HETERO- AND HOMONUCLEAR 2D *J*-RESOLVED SPECTRA

In Section 12.6 we hinted that there was a way to effect complete homonuclear decoupling in a ^{1}H spectrum so that each signal was reduced to a singlet. This would greatly facilitate signal resolution and chemical shift determinations in complex spectra. However, there would be a price for removing all the coupling: All information about through-bond connectivity would be lost.

Fortunately, there is now a 2D technique that accomplishes complete homonuclear decoupling while retaining all the coupling information! It is called the **homonuclear J-resolved 2D spectrum**, or HOM2DJ for short. Here, the F_2 axis is the proton chemical shift, while the F_1 axis is the separation (in hertz) of the component lines of each multiplet. In effect, the HOM2DJ pulse sequence takes the normal ^1H spectrum and rotates each multiplet perpendicular to the chemical shift axis. Thus, all the components of a given multiplet are projected onto a single line along the F_2 axis, but these same components are resolved and displayed along the F_1 axis.

Take a moment to review the rather complex ^1H spectrum of menthol described in Section 13.3 and shown at the top of Figure 13.13. Compare that to the HOM2DJ spectrum of menthol shown in Figure 13.15. Each of the 14 signals long the F_2 axis now appears as a broadened singlet. The multiplicity of each signal is fully resolved in contours along the F_1 axis. For example, symmetrical multiplet signals **7** (12 lines centered at δ 1.10) and **14** (8 lines centered at δ 3.39) of the normal ^1H spectrum (expanded on the right side of Figure 13.15) now appear as the appropriate number of contours (12 and 8, respectively) along the F_1 axis. The magnitude of each coupling constant can be determined from the center-to-center spacing between these contours.

There is also a heteronuclear version of the *J*-resolved experiment, known as **HET2DJ**. The most common examples involve the ^{13}C spectrum displayed as decoupled singlets along the F_2 axis, with the proton coupling (principally the one-bond C–H couplings) displayed along the F_1 axis. Figure 13.16 shows the simulated HET2DJ spectrum of 2-chlorobutane. Note the quartets for methyl signals **1** and **2**, the triplet for methylene signal **3**, and the doublet for methine signal **4**. The magnitude of the one-bond C–H coupling constants can also be ascertained; here all the spacings are approximately 125 Hz. For a complex undecoupled ^{13}C spectrum where the multiplets overlap, the HET2DJ spectrum can confirm the multiplicity (number of attached hydrogens) of each carbon signal, the same information available from an edited DEPT experiment (Section 12.12).

13.6   1D AND 2D INADEQUATE

So far we have discussed homonuclear 2D H,H-shift correlation spectroscopy (H,H-COSY) as well as heteronuclear 2D C,H-shift correlation spectroscopy (C,H-COSY, or C,H-HSC). Let us now consider homonuclear 2D C,C-shift correlation spectroscopy.

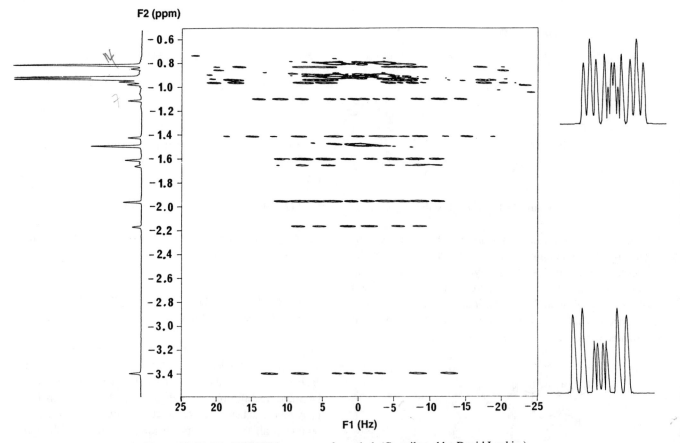

Figure 13.15. The HOM2DJ spectrum of menthol. (Contributed by David Lankin.)

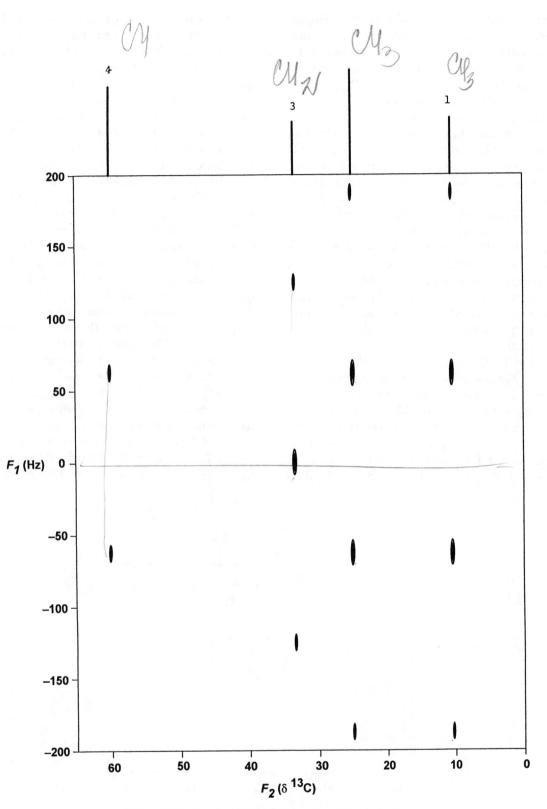

Figure 13.16. Simulated HET2DJ spectrum of 2-chlorobutane.

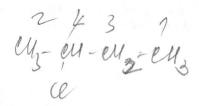

Suppose we could devise a 2D NMR technique to map out all one-bond C–C couplings. What would be the value of such data? A moment's reflection should convince you that this information would provide the carbon–carbon connectivity (the carbon backbone) of the entire molecule! This is the single most useful piece of information in the elucidation of organic structure. But how can we accomplish it?

Recall how we could use the small ^{13}C satellites present in a ^1H spectrum to determine the magnitude of one-bond C–H coupling constants (Section 8.6.1). Now we will need to use ^{13}C satellites *in the proton-decoupled ^{13}C spectrum* to map out all $^{13}C–^{13}C$ one-bond couplings! This will not be easy, however, because the probability of having two ^{13}C nuclei bonded together in a molecule is 1% of 1%, or 1 in 10,000.

There is a 1D NMR technique, not mentioned in Chapter 12, that begins to get at this problem. Known as **1D-INADE-QUATE** (incredible natural abundance double-quantum transfer experiment), this spectrum shows *only* the one-bond ^{13}C satellites of each carbon signal. The main peaks themselves, except for signals due to *isolated* carbons (those not connected to other carbons), are suppressed by the pulse sequence. Thus, for example, the methine carbon of 2-chlorobutane (^{13}C signal **4**) would appear as two overlapping doublets, one due to coupling with the methylene carbon (signal **3**) and one due to coupling with the methyl carbon of signal **2**. Notice that the signal is *not* a doublet of doublets because we are only viewing one pair of ^{13}C nuclei at a time. That is, we see the **4-3** coupling separately from the **4-2** coupling. Figure 13.17*b* shows the simulated 1D INADE-

QUATE multiplet corresponding to signal **4** in the ^{13}C spectrum of 2-chlorobutane. Because the two one-bond coupling constants are nearly equal (ca. 35 Hz between tetrahedral carbons; Section 9.3), the two doublets are barely resolved, appearing instead as one doublet at lower resolution.

If all the individual one-bond $J_{C–C}$ values in a compound could be determined, and the multiplets thereby correlated, we would have the precious C–C connectivity information we are seeking. Unfortunately, for all but the most simple structures, this has proven impossible because of the close similarity in the J values. But do not despair! There is a 2D version of INADEQUATE that provides exactly the correlations we seek, even when all the J values are too close to resolve!

The simulated double-quantum coherence **2D INADE-QUATE** spectrum of 2-chlorobutane is shown in Figure 13.18. The normal ^{13}C spectrum is plotted along the top. Only the cross peaks appear in the contour plot, and each cross peak appears as a doublet (a pair of dots) at this level of resolution. The separation between these dots, about 35 Hz in this case, is $^1J_{CC}$. Each pair of correlated (by one-bond coupling) cross peaks is indicated by a separate dotted horizontal line, with the midpoint of each line on the diagonal. The F_1 axis is the double-quantum frequency, essentially the sum of the δv values of the two coupled nuclei.

Thus, the 2D INADEQUATE spectrum of 2-chlorobutane confirms that the carbon of signal **1** is connected to the carbon of signal **3**, **3** is connected to **4**, and **4** is connected to **2**. Since the DEPT spectrum (or HET2DJ) of 2-chlorobutane already identified signals **1** and **2** as methyls, **3** as a methylene, and **4**

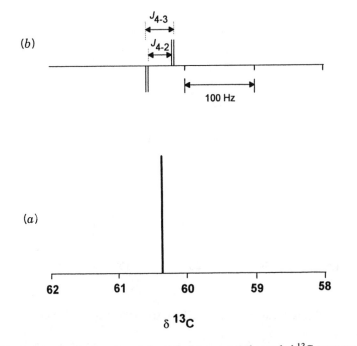

Figure 13.17. (*a*) Simulated expansion of signal **4** of the proton-decoupled ^{13}C spectrum of 2-chlorobutane. (*b*) Simulated 1D INADEQUATE spectrum of the same signal.

$C_2 - C_4 - C_3 - C_1$

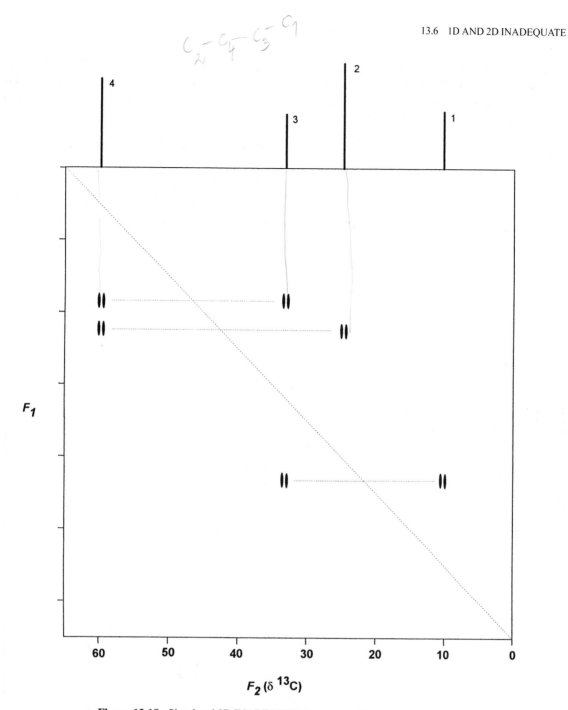

Figure 13.18. Simulated 2D INADEQUATE spectrum of 2-chlorobutane.

as a methine, the structure is unambiguously determined by just these two spectra!

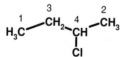

Perhaps the main lesson in this chapter is that a 2D IN-ADEQUATE spectrum combines the information of an H,H-COSY and a C,H-HSC. With only an edited DEPT spectrum and a 2D INADEQUATE spectrum the atomic sequence (structure without stereochemistry) of most molecules can be determined. However, there is a down side: the time and effort required to generate the 2D INADEQUATE spectrum. Because we are looking at very weak sidebands of weak signals, it can take *days* of pulse sequence repetitions to generate the desired information. Moreover, the recycle delay between pulses must be carefully set to exceed (by 1.5- to 3-fold) the longest carbon T_1 value in the molecule. Also, the experiment requires high sample concentrations, careful temperature control, and a nonspinning sample. And most important, one can only probe a limited range of $^1J_{CC}$ values in a given experiment. If a molecule has $C=C$ bonds ($^1J_{CC} = 70$ Hz) or $C\equiv C$ bonds ($^1J_{CC} = 170$ Hz) in addition to C–C bonds, several INADEQUATE spectra, covering the appropriate range of $^1J_{CC}$ values, may have to be collected to acquire a complete carbon–carbon connectivity map.

■ **EXAMPLE 13.4**   Describe the appearance of a 1D and 2D INADEQUATE signal for a methoxy carbon, OCH_3.

□ **Solution:** The 1D INADEQUATE will exhibit a singlet for such a carbon. Since there are no ^{13}C satellites, its signal will not appear in the 2D INADEQUATE spectrum.   □

■ **EXAMPLE 13.5**   A certain compound with molecular formula $C_{10}H_{20}O$ exhibits a proton-decoupled ^{13}C spectrum with signals at δ 16.1, 21.0, 22.2, 23.2, 25.8, 31.7, 34.6, 45.1, 50.1, and 71.3. Figure 13.19 shows the 2D INADEQUATE spectrum of this compound. Numbering the carbon signals from **1** (highest field) to **10**, list the carbon connectivity in the molecule. Then propose a structure for the compound.

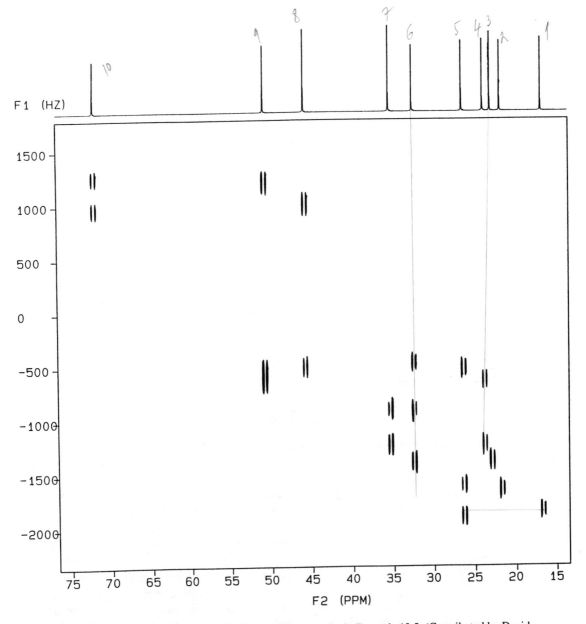

Figure 13.19.   The 2D INADEQUATE spectrum of $C_{10}H_{20}O$, Example 13.5. (Contributed by David Lankin.)

☐ ***Solution:*** The *IOU* for this compound is 1. All 10 carbons are distinguishable. The table below lists the C–C correlations in the INADEQUATE spectrum.

C	Is Connected To	C	C	Is Connected To	C
1		5	6		3, 7, 8
2		5	7		4, 6
3		6	8		6, 10
4		7, 9	9		4, 5, 10
5		1, 2, 9	10		8, 9

From the data in this table we can map out the C–C connectivities:

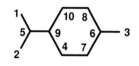

The ring accounts for the IOU value of 1, so the oxygen must be present as OH, and it must be connected to the lowest field signal (**10**). Thus, the structure of the molecule is

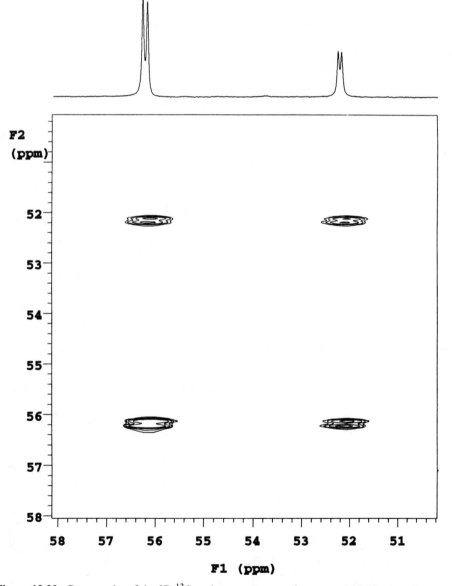

Figure 13.20. Contour plot of the 2D ^{13}C exchange spectrum of compound **10-18**. (Contributed by Alan Sopchik of Wes Bentrude's group at the University of Utah.)

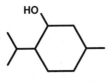

All that remains is to establish the relative stereochemistry of the three substituents, which could be accomplished by NOESY. The compound is our old friend menthol. □

13.7   2D NMR SPECTRA OF SYSTEMS UNDERGOING EXCHANGE

There are many other 2D and higher multidimensional NMR techniques being developed. The common thread of all of them is to show correlations between certain nuclear properties: chemical shifts, couplings, NOE, and so on. We will close our discussion of 2D techniques by mentioning just one more.

Recall how dynamic exchange processes can affect the lineshape of an NMR spectrum (Section 10.3). There is a homonuclear 2D method where the signals of the exchanging system are displayed along the diagonal and the off-diagonal cross peaks indicate which signals are correlated by exchange.[6]

Take a moment to look back at review problem 10.3. The ^{13}C spectrum of compound **10-18** is temperature dependent because of a series of pseudorotations that exchange the equatorial and apical methoxy groups:

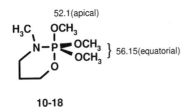

10-18

The relevant part of the 2D ^{13}C exchange spectrum of this molecule (at −90°C in toluene-d_8) is shown in the contour plot of Figure 13.20. Note how the symmetrically placed cross peaks indicate that the signals at δ 52.1 and 56.15 are correlated by exchange at −90°C, even though this exchange is not apparent in the 1D spectrum until the temperature is raised much higher.

CHAPTER SUMMARY

1. The term two-dimensional (2D) NMR spectrum refers to a data set where signal intensity is a function of two frequency domains (F_1 and F_2). The corresponding FID data are collected as a function of two time domains (detection and evolution/mixing) and then Fourier transformed in each dimension. The resulting data are most commonly displayed in a contour format.

2. Two-dimensional NMR techniques in general provide direct information about various types of correlations between nuclei in a molecule. The 2D techniques described in this chapter are characterized in the table below:

Acronym	F_1	F_2	Type of Information Conveyed
C,H-HSC	δ ^1H	δ ^{13}C	Direct C–H connectivity
(H,H-)COSY	δ ^1H	δ ^1H	Two- or three-bond homonuclear H–H coupling
NOESY	δ ^1H	δ ^1H	Proximity of hydrogens through space or through bond
HOM2DJ	J_{H-H}	δ ^1H	H–H multiplets perpendicular to chemical shift axis
HET2DJ	J_{C-H}	δ ^{13}C	C–H multiplets perpendicular to chemical shift axis
INADEQUATE	δ ^{13}C	δ ^{13}C	Direct C–C connectivity

REFERENCES

[1] Derome, A. E., *Modern NMR Techniques for Chemistry Research*, Pergamon, New York, 1987.

[2] Benn, R., and Gunther, H., *Angew. Chem. Int. Ed. Engl.*, *22*, 350 (1983).

[3] Kessler, H., Gehrke, M., and Griesinger, C., *Angew. Chem. Int. Ed. Engl.*, *27*, 490 (1988).

[4] Nakanishi, K., Ed., *One-Dimensional and Two-Dimensional NMR Spectra by Modern Pulse Techniques*, University Science Books, Sausalito, 1990.

[5] Duddeck, H., and Dietrich, W., *Structure Elucidation by Modern NMR, A Workbook*, Springer-Verlag, New York, 1989.

[6] Ernst, R. R., Bodenhausen, G., and Wokaum, A., *Principles of Nuclear Magnetic Resonance in One- and Two-Dimensions*, Oxford University Press, London, 1987, Chapter 9.

REVIEW PROBLEMS (Answers in Appendix 1)

13.1. Figure 13.21 is a stacked plot corresponding to one of the contour plots encountered earlier in this chapter. To which figure does it correspond?

13.2. (a) Predict the number of ^1H signals in the spectrum of alkaloid **13-1**:

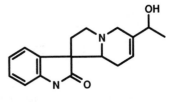

13-1

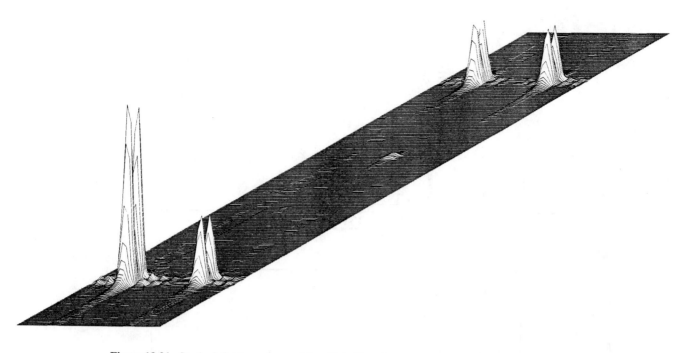

Figure 13.21. Stacked plot for review problem 13.1. (Contributed by Alan Sopchik of Wes Bentrude's group at the University of Utah.)

(b) The 300-MHz ^{1}H spectrum of **13-1** consists of the following 17 signals: δ 1.28 (d, 3H), 1.67 (br. s, 1H), 1.76 (br. m, 2H), and a series of 1H signals at 2.13 (dt), 2.47 (td), 2.62 (q), 2.78 (dd), 2.92 (d), 3.48 (td), 3.73 (d), 4.29 (q), 5.56 (br. s), 6.89 (d), 7.03 (t), 7.20 (t), 7.39 (d), and 8.15 (s). A D$_2$O exchange experiment indicates that the signals at δ 1.67 and 8.15 can be assigned respectively to the O–H and N–H hydrogens. Numbering the remaining signals from **1** to **15**, prepare a table of H–H COSY correlations, given that Figure 13.4 is the ^{1}H COSY of **13-1** (with the chemical shift scales *increasing* from left to right).

(c) Assign as many of the signals in the ^{1}H spectrum of **13-1** as you can.

13.3. The ^{1}H spectrum of 9-benzylanthracene consists of the following six signals: δ 5.35 (s, 2H), 7.50 (m, 5H), 7.80 (m, 4H), 8.38 (m, 2H), 8.57 (m, 2H), and 8.78 (s, 1H). The 2D COSY and NOESY spectra for this molecule

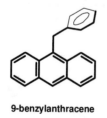

9-benzylanthracene

are reproduced in Figures 13.22 and 13.23. Assign all ^{1}H signals to hydrogens in the molecule.

13.4. A natural product with molecular formula C$_{10}$H$_{14}$O exhibited the following spectroscopic data (abstracted from ref. 5):

^{1}H: δ 0.72 (s, 3H), 1.22 (s, 3H), 1.74 (d, *J* = 1 Hz, 3H), 1.77 (d, 1H), 2.16 (m, 1H), 2.35 (m, 1H), 2.53 (m, 1H), 5.42 (q, *J* = 1 Hz, 1H).

^{13}C (DEPT multiplicity): δ 21.8 (CH$_3$), 23.3 (CH$_3$), 26.3 (CH$_3$), 40.5 (CH$_2$), 49.4 (CH), 53.6 (C), 57.3 (CH), 120.8 (CH), 169.9 (C), 203.3 (C).

The H,H-COSY and C,H-HSC correlations for this compound are summarized below:

C	Is Connected To H	Which Is Coupled To H(s)
1	1	
2	3	8
3	2	
4	4	7
4	7	4, 5, 6
5	5	6, 7
6		
7	6	
8	8	3
9		
10		

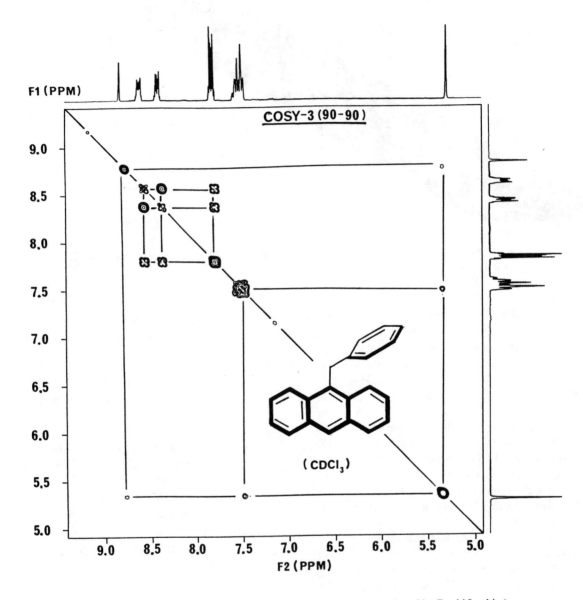

Figure 13.22. The COSY spectrum of 9-benzylanthracene. (Contributed by David Lankin.)

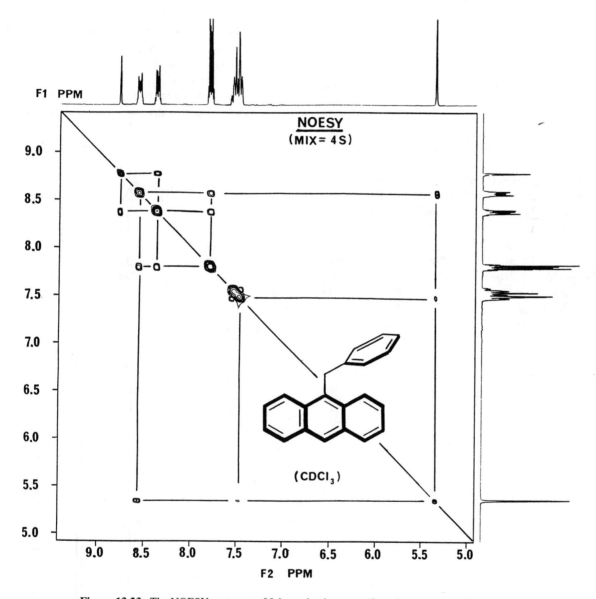

Figure 13.23. The NOESY spectrum of 9-benzylanthracene. (Contributed by David Lankin.)

A series of 1D NOE experiments confirmed the following proximities: H1 to H2; H2 to H5, H6, and H7; and H3 to H8. (a) Propose as complete a structure as you can for this compound and assign all NMR data. (b) Confirm or reconstitute your structure so it is consistent with the 2D INADEQUATE spectrum of the compound tabulated below:

C	Is Connected To	C	C	Is Connected To
1	6		6	1, 3, 5, 7
2	9		7	4, 6, 10
3	6		8	9, 10
4	5, 7		9	5, 8
5	4, 6, 7		10	7, 8

13.5. The ^1H spectrum (Figure 13.24) of a certain natural product with molecular formula $C_{15}H_{22}O_5$ exhibits 16 signals, all of which are 1H multiplets except for 3H doublets at δ 0.99 and 1.21, a 3H singlet at δ 1.44, and a 1H singlet at δ 5.82. The ^{13}C spectrum of this compound consists of 15 signals, the 11 shown at the top of the HSC spectrum in Figure 13.25, plus signals at δ 80, 95, 106, and 172. The COSY and NOESY spectra for this compound are shown in Figures 13.26 and 13.27, respectively. (a) From the data in Figure 13.25, prepare a table showing which hydrogen(s), as designated by their signal number, are attached to each carbon. From this, determine the multiplicity (CH_3, CH_2, CH, or C) of each carbon. (b) From the data in Figure 13.26, prepare a table showing which hydrogen(s) are coupled to each other. (c) From a comparison of Figures 13.26 and 13.27, determine which hydrogens, though not directly coupled through bonds, are nonetheless close in space. (d) See if you can integrate the above information into a unique structure for the compound. (e) The table below lists the C–C correlations from the 2D INADEQUATE spectrum of the unknown compound. Can you now complete the structure of the compound?

C	Is Correlated With	C	C	Is Correlated With	C
1	6		9	2, 7, 11	
2	9		10	3, 6, 12	
3	7, 10		11	4, 9, 12	
4	8, 11		12	10, 11, 13	
5	14		13	12	
6	1, 10, 15		14	5, 8	
7	3, 9		15	6	
8	4, 14				

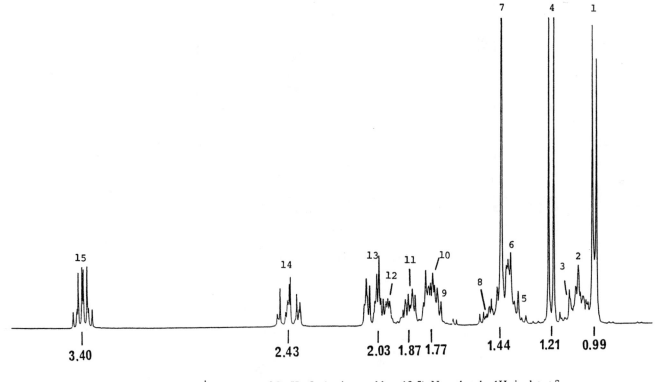

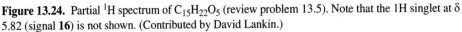

Figure 13.24. Partial ^1H spectrum of $C_{15}H_{22}O_5$ (review problem 13.5). Note that the 1H singlet at δ 5.82 (signal **16**) is not shown. (Contributed by David Lankin.)

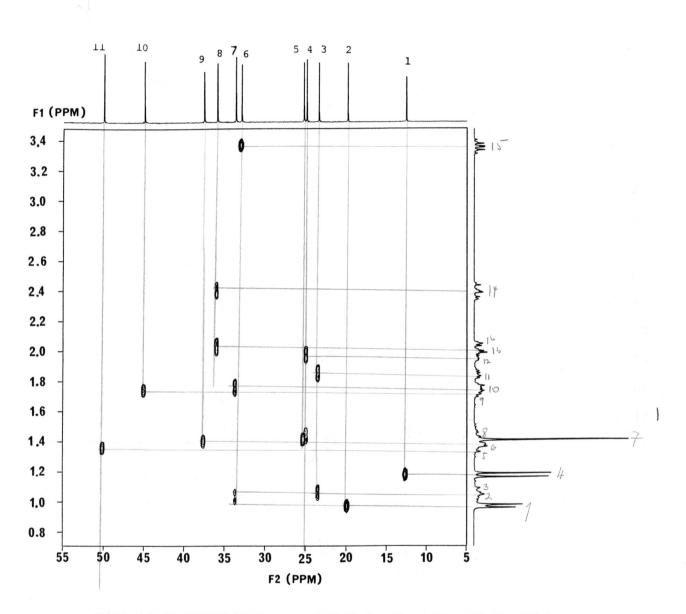

Figure 13.25. Partial 2D H,C-HSC spectrum of $C_{15}H_{22}O_5$ (review problem 13.5). Note that the correlation between carbon signal **13** and hydrogen signal **16** is not shown. (Contributed by David Lankin and Geoffrey Cordel.)

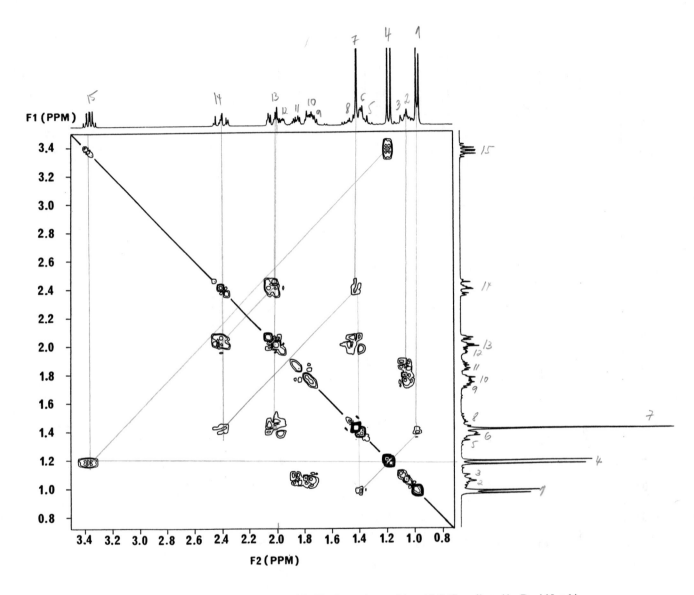

Figure 13.26. The COSY spectrum of $C_{15}H_{22}O_5$, review problem 13.5 (Contributed by David Lankin and Geoffrey Cordel.)

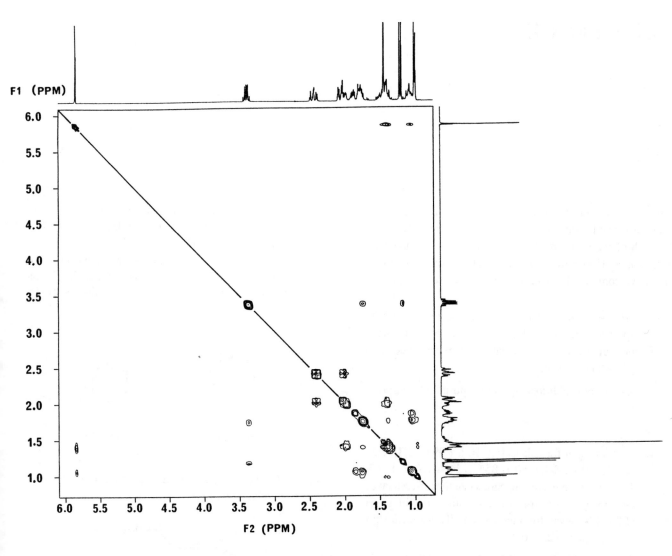

Figure 13.27. The NOESY spectrum of $C_{15}H_{22}O_5$, review problem 13.5. (Contributed by David Lankin and Geoffrey Cordell.)

SELF-TEST II

The following questions range in difficulty from relatively straightforward to quite difficult; the answers are in Appendix 1. Each of the first three problems consists of a molecular formula, a ^1H spectrum, and a ^{13}C {^1H} spectrum of an unknown compound. For each of these three sets of spectra:

(a) Propose a molecular structure consistent with both spectra; then assign all signals in both spectra to nuclei in your structure by calculating the expected position of each resonance.

(b) Answer any additional questions posed in each problem.

1. (a) Molecular formula C_9H_8; ^1H: δ 1.99 (s, 3H), 7.30 (m, 5H); ^{13}C: δ 4.1, 80.1, 85.9, 124.4, 127.6, 128.3, 131.7.[1] (b) Predict the appearance of the off-resonance decoupled ^{13}C spectrum of this compound, with ν_2 set at 2000 Hz away from the average ^1H chemical shift and B_2 equal to 2×10^{-4} T.

2. (a) Molecular formula $C_7H_{14}O_2$; ^1H: δ 0.95 (t, $J = 7$, 3H), 1.22 (d, $J = 7$, 6H), 2.18 (t, $J = 7$, 2H), 4.92 (m, $J = 7$, 1H); ^{13}C: δ 13.7, 18.7, 22.0, 36.6, 67.1, 172.7.[1] (b) Predict the appearance of the APT ^{13}C spectrum of this compound with τ set to 8 ms.

3. (a) Molecular formula $C_7H_6ClNO_2$; ^1H: δ 4.63 (s, 2H), 7.56 (d, $J = 11$, 2H), 8.19 (d, $J = 11$, 2H); ^{13}C: δ 44.6, 123.8, 129.8, 144.9, 147.5.[1] (b) By drawing a contour plot similar to Figure 13.6, predict the appearance of the C,H-HSC spectrum of this compound. You may assume that $^1J_{CH}$ values of ca. 125–160 Hz can be detected.

4. Predict the position and intensity of each line in the *coupled* ^{13}C spectra of (a) acetone-D_6 [$D_3C-C(=O)-CD_3$] and (b) F–$^{13}CH_3$.

5. Cyclooctatetraene exhibits only one signal in its ^1H spectrum, at δ 5.75. Compare this value to the chemical shift of benzene (Section 6.5) and discuss its significance:

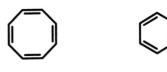

cyclooctatetraene   **benzene**

6. (a) Predict the appearance of the first-order ^1H spectrum of compound A[2]:

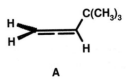

A

(b) The actual 60-MHz ^1H spectrum of A is shown in Figure ST2.1. Account for any differences between it and your predictions. What are the small signals flanking the singlet at δ 1.03?

(c) Next, consider the related molecule B.[3] By labeling each hydrogen, indicate which, if any, are symmetry equivalent. You may assume free rotation around all single bonds.

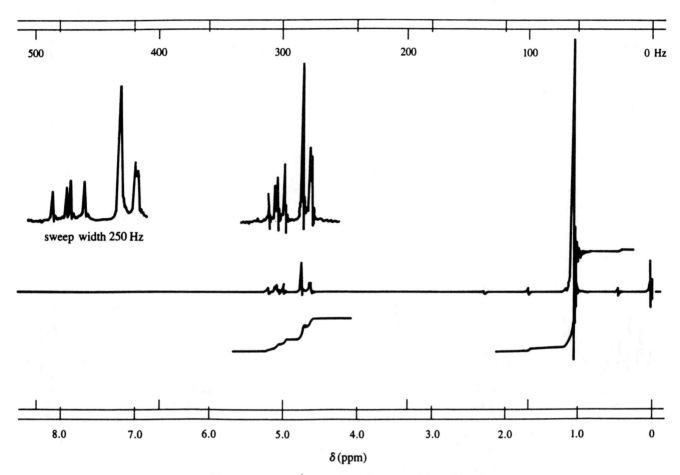

sweep width 250 Hz

δ (ppm)

Figure ST2.1. The 60-MHz ^{1}H spectrum of compound **A**, problem 6(a).

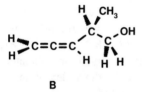

B

(d) How many δ and J values would be needed to generate an accurate computer simulation of the ^{1}H spectrum of **B**? Estimate the magnitude of each of these quantities.

(e) Figure ST2.2 is the 90-MHz ^{1}H spectrum of **B**. Assign all signals. According to first-order rules, what is the maximum number of lines in the multiplet at δ 2.3?

(f) Predict the result on the rest of the spectrum in Figure ST2.2 of irradiating the signal at δ 2.3.

(g) What is unique about the ^{13}C signal from the middle carbon in a C=C=C linkage?

(h) Section 6.6 states that O–H and N–H resonances are broadened and decoupled by exchange processes. Explain how this exchange broadens the signals.

7. Consider structures **C** and **D**. The ^{1}H spectrum of **C** is *independent* of temperature, showing (among other signals) two singlets for the OCH$_3$ groups [δ 4.04 (6H), 4.10 (3H)]. By contrast, the ^{1}H spectrum of **D** is reversibly temperature dependent. At +38°C there are two OCH$_3$ singlets [δ 3.44 (6H), 3.79 (3H)], at –23°C only one is observed (δ 3.79), and at –67°C there are three [δ 2.76 (3H), 3.79 (3H), and 4.13 (3H)].[4] Account for these observations:

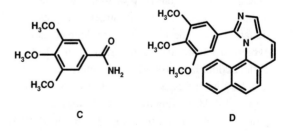

C   **D**

8. In Example 2.2 (and Table 2.1) we encountered the two isotopes of chlorine. Each of these isotopes has an *I* value of $\frac{3}{2}$ and a substantial γ value as well as an

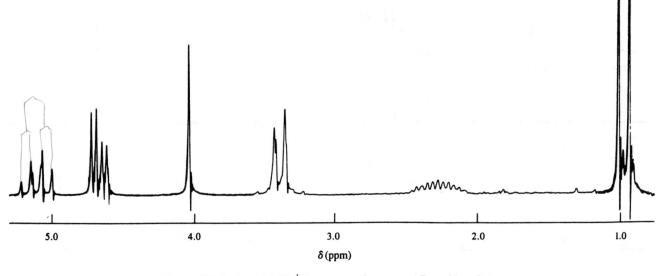

Figure ST2.2. The 90-MHz ^{1}H spectrum of compound **B**, problem 6(c).

electric quadrupole moment. Yet in all the NMR spectra of chlorine-containing compounds we have discussed there has never been any evidence of coupling between the chlorine and the other hydrogens and carbons in the molecules. In contrast, hyperfine coupling between the chlorine and the unpaired electron is commonly observed in EPR spectra of chlorine-containing radicals. Why is coupling to chlorine observed in EPR spectra but not in NMR spectra?

9. (a) Figure ST2.3 is the 500-MHz ^{1}H spectrum (in $CDCl_3$) of a highly purified compound with molecular formula $C_{11}H_{14}O$. Deduce its structure and assign all signals. *Hints*: The J values in the doublets at δ 2.13 (which disappears in D_2O) and 2.44 are matched in the triplet of doublets at δ 4.82. Furthermore, the "singlet" at δ 1.80 is very weakly coupled to the singlets at δ 4.86 and 4.93. (b) Figure ST2.4 shows the effect of adding $Eu(hfc)_3$ (the Eu analog of compound **10-13**; Section 10.7.3) to the solution. The signals in the ranges $\delta < 2.0$ and $\delta > 10$ are due to the hfc ligand (see structure **10.13**). Explain the changes in the spectrum. (These spectra were contributed by Dhileepkumar

Krishnamurthy with Gary Keck's group at the University of Utah.)

10. A novel marine natural product with molecular formula $C_{28}H_{39}O_4Cl$ was recently isolated and characterized by 2D NMR.[5] The ^{13}C, DEPT, and C,H-HSC data for this compound are tabulated below (signals are numbered in the order of increasing chemical shift).

C	δ	DEPT	C,H-HSC with H(s)
1	11.5	CH_3	1
2	17.7	CH_3	9
3	20.9	CH_2	11, 12
4	21.7	CH_3	3
5	21.8	CH_3	4
6	23.4	CH_2	8, 15
7	27.7	CH_2	10, 19
8	27.8	CH	25
9	28.9	CH_2	21, 27
10	31.0	CH_2	6, 18
11	34.0	CH_2	22
12	34.4	CH_2	16, 20

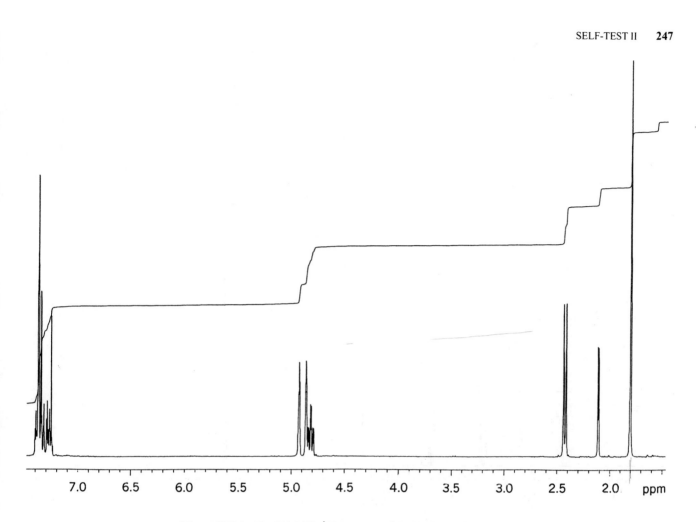

Figure ST2.3. The 500-MHz ^{1}H spectrum of $C_{11}H_{14}O$, problem 9(a).

C	δ	DEPT	C,H-HSC with H(s)
13	35.1	CH	13
14	36.8	CH$_2$	7, 17
15	40.9	CH$_2$	24, 26
16	41.4	C	
17	42.1	CH	23
18	42.2	C	
19	52.1	CH	14
20	53.9	CH	2
21	55.3	CH	5
22	121.6	CH$_2$	28, 29
23	127.2	C	
24	154.9	C	
25	164.8	C	
26	179.8	C	
27	190.8	C	
28	200.2	C	

The ^{1}H and COSY data for the compound are tabulated below.

H	δ (m, #H)	COSY with H(s)
1	0.86 (s, 3)	
2	0.95 (td, 1)	**11, 12, 13**
3	0.99 (d, 3)^a	**25**
4	0.99 (d, 3)^a	**25**
5	1.02 (m, 1)	**8, 13, 15**
6	1.05 (m, 1)	**13, 18, 21, 27**
7	1.08 (m, 1)	**11, 12, 17**
8	1.17 (m, 1)	**5, 10, 15, 19**
9	1.22 (s, 3)	
10	1.31 (m, 1)	**8, 14, 15, 19**
11	1.40 (qd, 1)	**2, 7, 12, 17**
12	1.55 (m, 1)	**2, 7, 11, 17**
13	1.57 (m, 1)	**2, 5, 6, 18**
14	1.60 (q, 1)	**10, 19, 23**

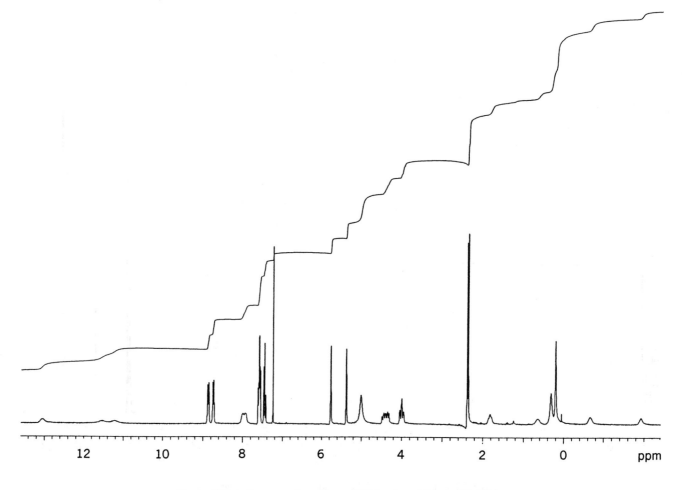

Figure ST2.4. Spectrum from Figure ST2-3, after addition of Eu(hfc)$_3$.

H	δ (m, #H)	COSY with H(s)
15	1.66 (m, 1)	**5, 8, 10, 19**
16	1.68 (td, 1)	**20, 22**
17	1.89 (m, 1)	**7, 11, 12**
18	1.90 (m, 1)	**6, 13, 21, 27**
19	1.91 (m, 1)	**8, 10, 14, 15**
20	2.01 (dt, 1)	**16, 22**
21	2.17 (td, 1)	**6, 18, 27**
22	2.55 (m, 2)	**16, 20**
23	2.76 (td, 1)	**14, 24, 26**
24	2.85 (dd, 1)	**23, 26**
25	2.85 (sept, 1)	**3, 4**
26	3.08 (dd, 1)	**23, 24**
27	3.23 (ddd, 1)	**6, 18, 21**
28	5.71 (d, 1)	**29**
29	5.96 (d, 1)	**28**
30	11.0 (exch, 1)	

^{a} Signals with the same chemical shift could be differentiated in 2D spectra.

Finally, a 2D COLOQ experiment showed the following long range C–H correlations:

H	C
1	**14, 18, 19, 21**
14	**7, 17, 18, 26**
24, 26	**24, 28**
3, 4	**8, 24**
28, 29	**8, 24, 28**
9	**12, 16, 20, 25**

Using these data establish the structure of the compound and assign all NMR signals.

11. Another steroidal natural product has molecular formula $C_{30}H_{54}$. The C,H-HSC and COSY spectra (contributed by David S. Watt at the University of Kentucky) are shown in Figures ST2.5 and ST2.6, respectively; asterisks indicate signals with the same chemical shift that were differentiated in the 2D spectra. To help identify the individual signals, the ^{1}H and

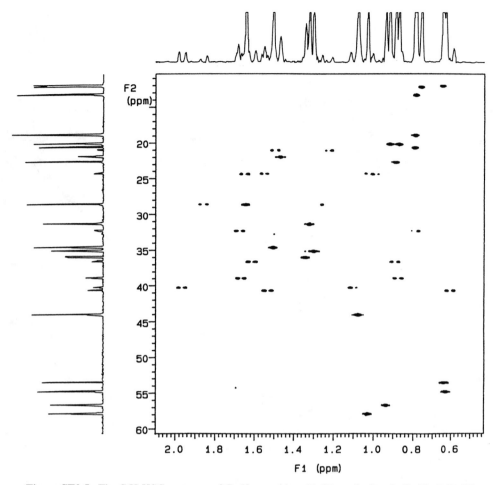

Figure ST2.5. The C,H-HSC spectrum of $C_{30}H_{54}$, problem 11. [From Stoilov, I., Smith, S. L., Watt, D. S., Carlson, R. M. K, Fago, F. J., and Moldowan, J. M., *Mag. Res. Chem., 32*, 101 (1994). Copyright 1994 by John Wiley & Sons.

^{13}C spectra for this compound are tabulated below;[6] Identify the structure of this compound, and assign all resonances.

C	δ	DEPT
1	12.0	CH_3
2	12.1	CH_3
3	13.3	CH_3
4	18.8	CH_3
5	20.1*	CH_3
6	20.1*	CH_3
7	20.6	CH_3
8	21.0	CH_2
9	21.9	CH_2
10	22.6	CH_3
11	24.2	CH_2
12	24.3	CH_2
13	28.5	CH_2

C	δ	DEPT
14	28.6	CH
15	31.3	CH
16	32.3	CH_2
17	34.6	CH
18	35.1	CH
19	35.9	CH
20	36.4	C
21	36.6	CH_2
22	38.9	CH_2
23	40.2	CH_2
24	40.6	CH_2
25	42.6	C
26	44.0	CH
27	53.5	CH
28	54.8	CH
29	56.6	CH
30	57.9	CH

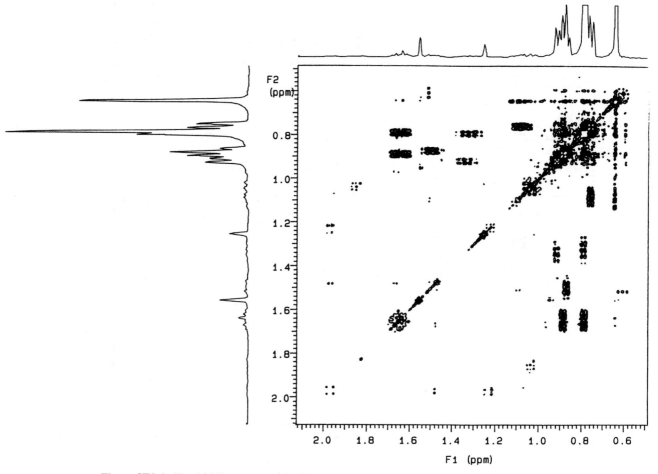

Figure ST2.6. The COSY spectrum of $C_{30}H_{54}$, problem 11. [From Stoilov, I., Smith, S. L., Watt, D. S., Carlson, R. M. K., Fago, F. J., and Moldowan, J. M., *Mag. Res. Chem.*, *32*, 101 (1994). Copyright 1994 by John Wiley & Sons. Reprinted by permission.

H	δ	#H if 1
1	0.60	
2	0.62	
3	0.64	
4	0.65	3
5	0.76	3
6	0.79*	3
7	0.79*	3
8	0.79*	3
9	0.79*	
10	0.87*	3
11	0.87*	
12	0.89*	3
13	0.89*	
14	0.92	3
15	0.94	
16	0.99	
17	1.03*	
18	1.03*	

H	δ	#H if 1
19	1.08	
20	1.10	
21	1.22	
22	1.25	
23	1.30*	
24	1.30*	
25	1.32	
26	1.34	
27	1.46	2
28	1.49	
29	1.53	
30	1.55	
31	1.62	
32	1.64	
33	1.65	
34	1.67*	
35	1.67*	
36	1.86	
37	1.97	

(b) The idealized 2D INADEQUATE data for this compound are tabulated below. Can you now complete the structure?

C	is bonded to C
1	26
2	25
3	20
4	14
5	17
6	19
7	15
8	23, 28
9	21, 22
10	14
11	13, 29
12	16, 27
13	11, 30
14	4, 10, 26
15	7, 21, 27
16	12, 18
17	5, 24, 26
18	16, 28, 29
19	6, 24, 30
20	3, 22, 27, 28
21	9, 15
22	9, 20
23	8, 25
24	17, 19

C	is bonded to C
25	2, 23, 29, 30
26	1, 14, 17
27	12, 15, 20
28	8, 18, 20
29	11, 18, 25
30	13, 19, 25

REFERENCES

[1]The ^{1}H data are abstracted from *The Sadtler Standard NMR Spectra*, Sadtler Research Laboratories, Division of Bio-Rad Laboratories, Philadelphia, 1972. The 13C data are abstracted from *The Sadtler Guide to Carbon-13 NMR Spectra*, Simons, W. W., Ed., Sadtler Research Laboratories, Division of Bio-Rad Laboratories, Philadelphia, 1983.

[2]Macomber, R. S., and Lilje, K. C., *J. Org. Chem., 39*, 3600 (1974).

[3]Macomber, R. S., *J. Org. Chem., 36*, 999 (1971).

[4]Schmidt, J. C., Benson, H. D., Macomber, R. S., Weiner, B., and Zimmer, H., *J. Org. Chem., 42*, 2003 (1977).

[5]Carney, J. R., Scheuer, P. J., and Kelly-Borges, M., *J. Org. Chem.,58*, 3460 (1993).

[6]Stoilov, I., Smith, S. L., Watt, D. S., Carlson, R. M. K., Fago, F. J., and Moldowan, J. M., *Mag. Res. in Chem., 32*, 101 (1994).

14

NMR STUDIES OF BIOLOGICALLY IMPORTANT MOLECULES *

14.1 INTRODUCTION

The utilization of NMR spectroscopic techniques to study biologically important molecules began in the 1960s. Biochemists realized that the same methods being used to determine the structure of small organic molecules could also be applied to much larger biomolecules. NMR techniques could be used to determine (1) the structures of unknown molecules, (2) the absolute stereochemical configuration of known molecules, (3) the modes of interaction of small molecules with larger ones, (4) the kinetics and thermodynamics of such interactions, and (5) the dynamics of molecules in supramolecular assemblies.

Due to their larger size, biomolecules do present problems that are not encountered in the structural analysis of smaller organic molecules. Biomolecules are polymers comprising a large number of monomers with similar structures, resulting in considerable overlap of individual resonances. Therefore, unambiguous assignment of these resonances to individual atoms becomes more difficult. Additionally, as these molecules increase in size, their relaxation times decrease, leading to increasingly broad peaks. These difficulties are being overcome with the development of higher field instruments and more sophisticated hardware and software.

14.2 NMR LINE WIDTHS OF BIOPOLYMERS

As we saw in Section 3.5, the line width (peak width at half height) of a Lorentzian line is determined by the reciprocal of the effective spin–spin relaxation time (T_2^*). For small organic

molecules, T_2^* is long and the intrinsic line width is small, on the order of a few hertz or less. For spherical molecules undergoing isotropic motion, increasing molecular weight causes a decrease in molecular motions and averaging of field inhomogeneities and a consequent broadening of lines. This effect is reflected in the dependence of T_2 on the motional correlation time, τ_c, which is the time required for a molecule to rotate through 1 rad (Section 2.5). This relationship is shown in Figure 14.1. The line width can range from a few hertz for small oligopeptides ($n = {\sim}10$) to tens of hertz for proteins with up to ~100 amino acids. For large proteins with molecular weight (MW) > 50,000 daltons, the line widths are very large and the signals are usually broadened beyond detection.

Not all large molecules are spherical; they can also be long random coils with a certain degree of flexibility along the chain. If the chain is flexible, there are rapid local fluctuations between monomers and the dipole–dipole interactions are effectively averaged. An example of this is polyuridylic acid (poly U), which has a molecular weight of 100,000 daltons but is nonetheless very flexible. The line widths for poly U as a random coil are only a few hertz. If the chain is incorporated into a rigid double or triple helix, these local motions are lost and the resonances are broadened beyond detection (see Section 14.7.2).

Many small molecules have a tendency to form large aggregates in solution. It is no longer the molecular weight of an individual monomer that determines the line width but the overall mobility of the molecule in the aggregate. An extreme example of this phenomenon is lipid molecules in membranes (Section 14.8).

---

*Contributed by George P. Kreishman and Elwood Brooks.

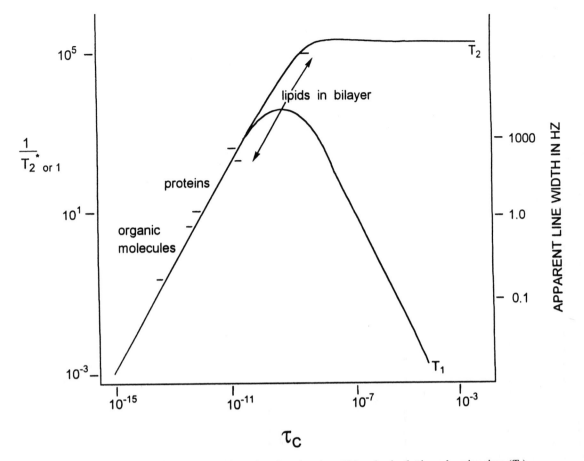

Figure 14.1. Dependence of the spin–spin relaxation time (T_2) and spin–lattice relaxation time (T_1) on the motional correlation time τ_c. Approximate values of the expected line widths for small organic molecules, proteins, and lipids in a bilayer are indicated.

14.3 EXCHANGEABLE AND NONEXCHANGEABLE PROTONS

The spectra of small biomolecules can be obtained in organic solvents such as DMSO-d_6. The ^{1}H spectra of adenosine in DMSO-d_6 and D_2O are shown in Figures 14.2 (page 254) and 14.3 (page 255), respectively. In DMSO-d_6, all of the protons are observed. The base protons appear at ~δ 8, the NH$_2$ protons at δ 7.4, H$_{1'}$ of the sugar at δ 5.9, the hydroxyl protons between δ 5.5 and 4.5, and the remaining ribose protons between δ 4.3 and 3.0. In D_2O, the amine and hydroxyl protons have been exchanged with deuterons and no longer appear in the ^{1}H spectrum. Comparison of the two spectra facilitates the assignment, but this comparison is limited to small monomers and oligomers of biomolecules.

With larger biopolymers, the solvent of choice is an aqueous medium since, in most cases, water is required to maintain the native conformation in solution. In addition, the exchangeable protons are of importance because they are involved in such things as hydrogen bonding and salt bridges, which are critical in stabilizing the native structure. The total spectrum

of the native conformation of a biomolecule can be obtained in 90%/10% H$_2$O/D$_2$O solutions. The small amount of D$_2$O is used for the lock signal. Since the majority of the solvent is H$_2$O, the exchangeable protons will not be significantly deuterated and can be observed in the spectrum. To eliminate the extremely large solvent peak, a pulse sequence called **presaturation** (Figure 14.4, page 255) can be used. Prior to excitation, a "soft" pulse is applied at the solvent's chemical shift. This pulse saturates the solvent peak before the excitation (observing) pulse, and therefore it does not contribute to the spectrum. Because of the low amplitude and long pulse length (several seconds), only a small spectral region (usually a few hertz) is saturated. When the "hard" excitation pulse of short duration (several microseconds) and high amplitude is applied, all of the remaining protons except for the solvent protons are excited. With this pulse sequence, the spectrum of both the exchangeable and nonexchangeable can be obtained. A portion of the presaturated spectrum of horse heart cytochrome c in 90%/10% H$_2$O/D$_2$O and the spectrum in 99.9% D$_2$O are compared in Figure 14.5 (page 256). As can be seen, the resonances for the exchangeable amide protons can be

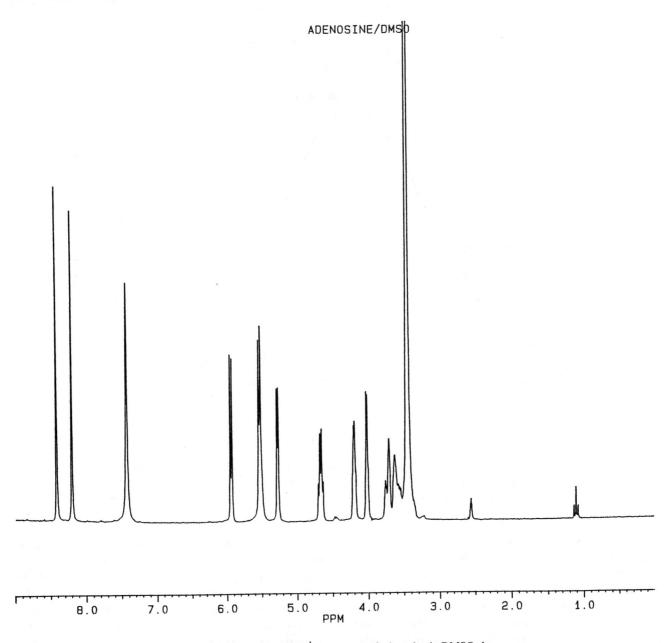

ADENOSINE/DMSO

Figure 14.2. The 250-MHz ^{1}H spectrum of adenosine in DMSO-d$_6$.

observed in the H$_2$O sample but do not appear in the D$_2$O sample.

14.4 CHEMICAL EXCHANGE

In the earlier discussion of the effects of dynamic processes on NMR spectra of small molecules (Chapter 10), the NMR time scale for the exchange process was determined by the relationship between the preexchange lifetimes (or the inverse rate constants) and the chemical shift difference between the two states (or sites). This is also true for many systems of biological importance when the line widths of the resonances

are small and observable. However, it is not the case for biological systems where one component is very large and the T_2^* values of the protons are extremely short. In the last case, the NMR time scale is determined by the relationship between the preexchange lifetime and T_2^* of the bound state. An example of this is the binding of a substrate to a large enzyme. If the preexchange lifetime is long compared to T_2^* of the large molecule, the magnetization of the small molecule will have relaxed before departing the large molecule and no signal will be observed. If, however, the preexchange lifetime in the bound state is short compared to T_2^*, the observed spectrum will be a weighted average of the individual free and bound

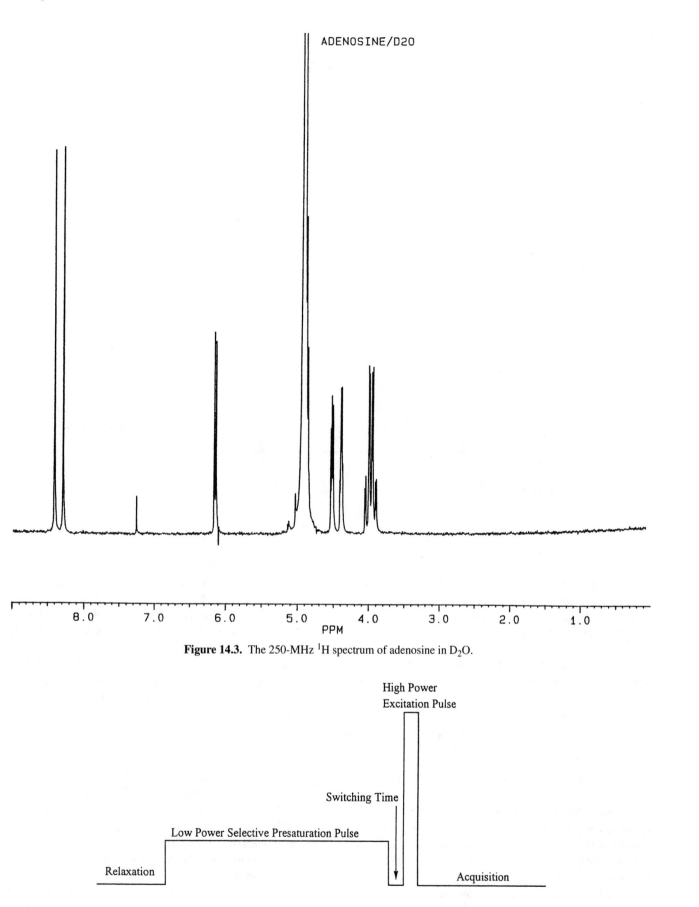

Figure 14.3. The 250-MHz ^1H spectrum of adenosine in D_2O.

Figure 14.4. Pulse sequence of a solvent suppression experiment.

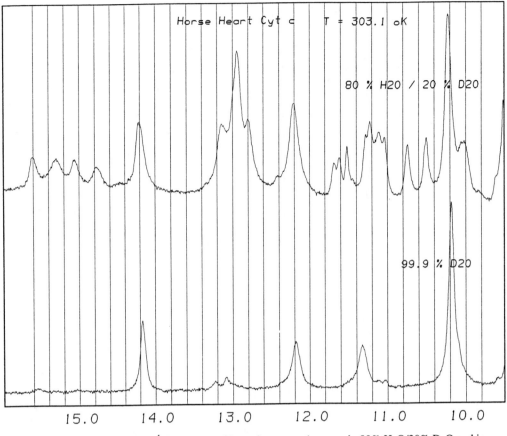

Figure 14.5. The 400-MHz ^{1}H spectrum of horse heart cytochrome c in 80% H$_2$O/20% D$_2$O and in 99.9% D$_2$O.

states, characteristic of a fast-exchange process (Section 10). Thus, even though the resonance of a small molecule bound to a large molecule cannot be detected directly, the chemical shift can be deduced from the weighted average spectrum as compared to that in the free state.

14.5   THE EFFECTS OF pH ON THE NMR SPECTRA OF BIOMOLECULES

The acid and base groups found in biomolecules can be protonated or deprotonated depending on the pH of the medium and the pK value for that group. Since the electron density around a nucleus will change as the charge on the molecule changes, the chemical shift will also be influenced. The chemical shift of the H$_2$ resonances of two histidine residues of ribonuclease are shown in Figure 14.6.[1] With increasing pH, the resonances are shifted upfield as the ring is deprotonated. The point of maximum inflection is the pK_a value for that specific group in a given environment. Titration curves are useful in assigning resonances to particular amino acids because of the characteristic pK_a values. In the native protein both histidine residues are exposed to the solvent and

their pH dependencies are approximately the same. Upon addition of an inhibitor that binds to the active site of the protein, the titration curve for His-12 is markedly shifted because it is being shielded from the solvent by the inhibitor molecule. This simple experiment confirmed that His-12 is in the active site of the protein.

Examples of NMR studies of the four classes of biomolecules (proteins, nucleic acids, lipids, and carbohydrates) are described in the next sections. Each class has unique characteristics, so they will be discussed separately.

14.6   NMR STUDIES OF PROTEINS

Proteins constitute a diverse class of polymers made up of 20 amino acids linked by peptide bonds. They range in size from small oligomers to large molecules with molecular weights in the millions of daltons. The primary structure (the sequence of amino acids) of a protein also determines its secondary and tertiary structure. The general structure of the peptide backbone is shown in Figure 14.7, and the structures of the R side chains of the individual amino acids are shown in Figure 14.8 (page 258). The approximate chemical shifts (in D$_2$O) of the

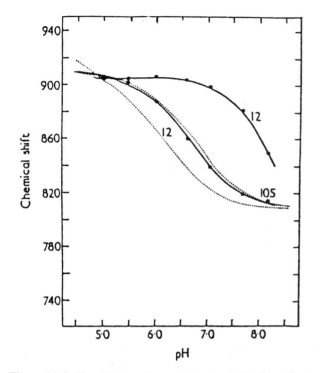

Figure 14.6. The pH dependence of the chemical shift of the H_2 resonances of several histidine residues in ribonuclease: ($\cdots\cdots$) in the absence of inhibitor; (———) in the presence of inhibitor. (From ref. 1. Copyright 1969 by Academic Press. Reprinted by permission.)

various protons in these isolated amino acid are also indicated in Figure 14.8. These chemical shifts will be affected by the local environments of the amino acids in the native protein.

As an example, consider the ^{1}H spectrum of ribonuclease, a small protein of 124 amino acids (MW = 12,000 daltons). This protein can be reversibly denatured either by the addition of denaturants such as urea or guanadinium hydrochloride or by high temperatures. Upon denaturation, the protein unfolds from its tertiary structure to a random coil with four disulfide bridges. These disulfide bridges can be cleaved with the addition of β-mercaptoethanol and the protein becomes a completely random coil,[2] as depicted in Figure 14.9. There are four histidine residues in the protein. In the native state, each of the four occupies a unique environment, and the NMR spectrum consists of four resolved resonances for the H_2 protons (Figure 14.10, page 260).[3] With the addition of increasing amounts of guanadinium hydrochloride, the protein begins to unfold and the four resonances for the H_2 protons

coalesce into a single resonance. The NMR spectrum of the native protein does reflect the unique structure of the protein, and this demonstrates how powerful NMR is as a tool for studying tertiary structures of proteins.

14.6.1   Strategies For Assigning Individual Resonances

The complete assignment of individual resonances for a protein can, in principle, be achieved by using multidimensional NMR spectroscopic techniques. For simplicity, the following three 2D NMR ^{1}H techniques will be discussed. They are COSY (Section 13.3), NOESY (Section 13.4), and TOCSY. These techniques allow for the identification of resonances for nuclei that are connected through bonds, those that are in close proximity in space, and those that are within a given spin system, respectively.

The initial assignment begins with the resonances of the protons in the peptide backbone. In Figure 14.11 (page 260), a generalized peptide backbone is illustrated. Beginning with the amino protons on residue 1 (H_1), this proton is spin coupled to the alpha proton on the same residue ($H_{\alpha1}$) and would exhibit a cross peak in the COSY spectrum. The $H_{\alpha1}$ is in close proximity to the amide proton of the second residue (H_2), and a cross peak for these protons would be exhibited in the NOESY spectrum. The H_2 is spin–spin coupled to $H_{\alpha2}$ and would show a cross peak in the COSY spectrum. This pattern of alternating COSY–NOESY correlations continues down the chain. In the assignment process for the backbone protons, it is customary to combine a COSY spectrum and a NOESY spectrum into a single spectrum where the upper left half is the NOESY spectrum and the lower right half is the COSY spectrum. A simulated spectrum for the backbone described above is shown in Figure 14.12 (page 260). Some characteristic information about the starting amino or alpha proton must be ascertained independently. This could be the unique shift of the terminal amino group as compared to the amide proton of the peptide bond or some correlation with a unique side-chain group that will be discussed later. Starting with the unique amino proton, one looks for a correlation with the alpha proton in the COSY spectrum. This defines the chemical shift of the first alpha proton. Next, correlation with an amino proton is found in the NOESY spectrum, and the chemical shift of the second amide proton is assigned. The second alpha proton connectivity sequence is found in the COSY spectrum. Its proximity to the third amide proton is found in the NOESY spectrum, and so on. This process can be continued sequentially down the chain as long as the individual cross peaks can be resolved. In practice, this is usually limited to about 30 residues; thereafter, ambiguities may arise.

With the backbone assigned, it is necessary to continue with the assignment of the side-chain protons. Commonly, another 2D spectroscopic technique, **TOCSY** (total correlation spectroscopy), is employed in addition to COSY. This

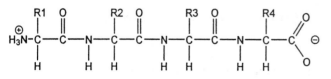

Figure 14.7. Generalized structure of the backbone of a protein.

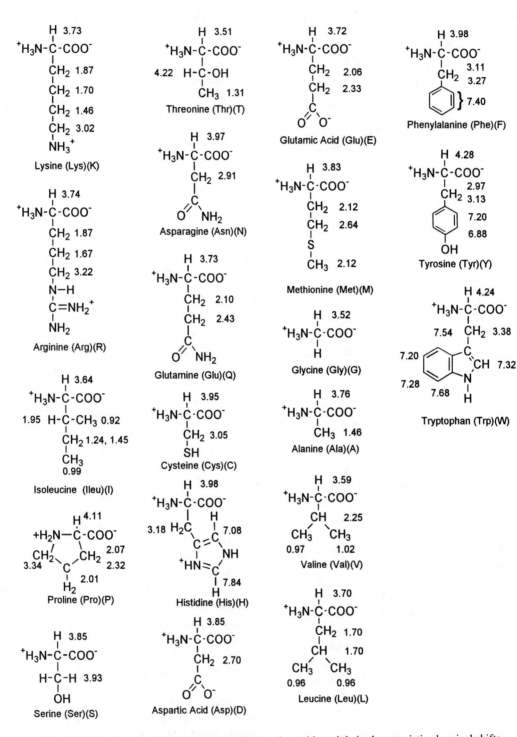

Figure 14.8. Structures of the side chains of the amino acids and their characteristic chemical shifts in D_2O.

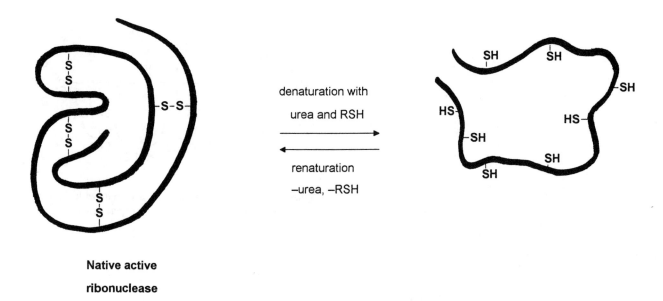

Native active

ribonuclease

Figure 14.9. Pictorial depiction of the native and denatured structures of ribonuclease.

technique is similar to a COSY spectrum, but the cross peaks arise from all of the connectivities within the spin system, not only those through three bonds. The pattern of these connectivities is unique for protons within a given spin system. Taking isoleucine as an example, its structure, COSY spectrum, and TOCSY spectrum are shown in Figure 14.13 (page 261). From the previously assigned connectivities of the alpha protons, the individual side-chain protons can be assigned. The TOCSY pattern can be used initially to assign an alpha proton of a unique amino acid residue, and this can then be used to assign sequentially the remaining backbone protons.

One of the first total assignments of a protein spectrum using these techniques was for the DNA binding domain (or head piece) of the *Escherichia coli lac* repressor. This protein fragment contains 51 amino acid residues (MW 5,500 daltons).[4] As can be seen in Figure 14.14 (page 262), the 2D spectrum for even such a small protein is quite complex. Using only 2D techniques, complete assignment can be achieved for proteins with molecular weights below ~10,000 daltons. With the application of 3D and higher dimensional techniques, as well as superhigh magnetic fields (800 MHz), the upper limit for molecular weights is now ~40,000 daltons.

In addition to the use of NOESY spectra for the assignment of individual resonances, they can also be used for the determination of the three-dimensional structure of the protein in solution. Up to now we have only dealt with the connectivities along the backbone between protons that are close to each other because of chemical bonding. There are also correlations between amino acids that are far apart in the primary structure but are in close proximity because of the folding found in the tertiary structure. For example, a portion of the chain near the amino terminus can be near the carboxy termi-

nus. These correlation patterns can be used to determine the overall structure of the protein.

14.6.2 Binding of Small Molecules to Proteins

In a classic study utilizing NMR titration techniques (Section 10.6.2), the mode of interaction of an inhibitor, 5'-cytidine-monophosphate (5'-CMP) with ribonuclease (RNAase) was ascertained.[1] In RNAase, the individual resonances for the H_2 resonances of the four histidines can be resolved and have been assigned. Of the four histidines, two are in the active site. Upon addition of 5'-CMP, two of the histidine resonances are shifted downfield with increasing inhibitor concentration (Figure 14.15, page 262). In addition to the histidine resonances, a resonance from a phenylalanine group of the protein is shifted upfield with increasing 5'-CMP concentration (Figure 14.16, page 262). The aromatic resonances of the cytosine base of 5'-CMP are shifted downfield compared to their chemical shifts at low concentrations in the absence of RNAase (Figure 14.16).

These data can be interpreted as follows. Since the chemical shifts of both resonances from the protein and the inhibitor are concentration dependent, the exchange rate between the free and bound states is fast on the NMR time scale. The observed chemical shifts are a weighted average of the two states [Eq. (10.13)]. For the protein resonances, the fraction bound increases with increasing inhibitor concentration. At zero concentration of the inhibitor, the observed chemical shift is that of the unbound state, while at concentrations above ~0.06 M inhibitor, the observed chemical shift is that of the bound state. For the *inhibitor*, the highest fraction bound occurs at the lower concentrations of the inhibitor and the lowest fraction bound at the higher concentrations. There-

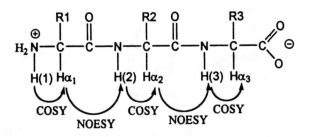

Figure 14.11. Peptide backbone of a protein with the expected correlations from a COSY and NOESY experiments.

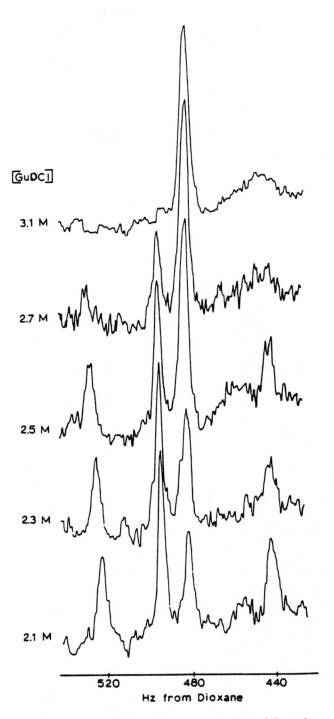

Figure 14.10. The H_2 resonances of the histidines of ribonuclease as a function of added guanadinium hydrochloride. (From ref. 3. Reprinted with permission of Elsevier Science-NL, Sara Burgerhartstraat 25, 1055 KV Amsterdam, The Netherlands.)

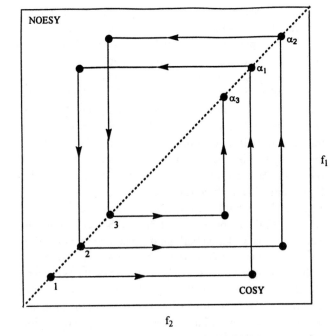

Figure 14.12. Simulated combined COSY/NOESY spectrum of a peptide backbone.

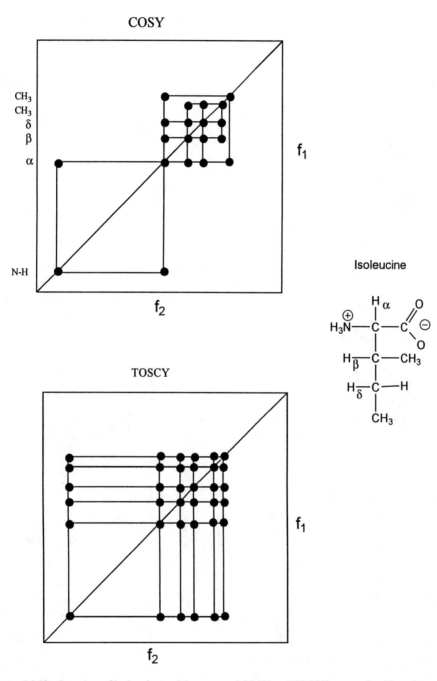

Figure 14.13. Structure of isoleucine and the expected COSY and TOCSY spectra for this amino acid residue in 90% H_2O/10% D_2O.

ω_2 (ppm)

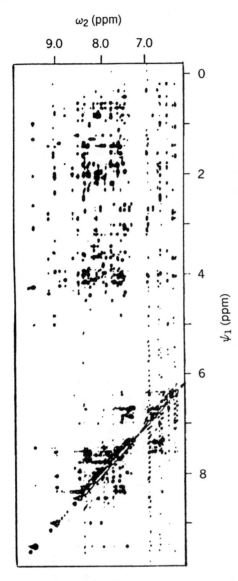

Figure 14.14. The 500-MHz ^{1}H 2D NOESY spectrum of the *lac* repressor headpiece in H$_2$O. (From ref. 4. Copyright 1985 by John Wiley & Sons, Inc. Reprinted by permission.)

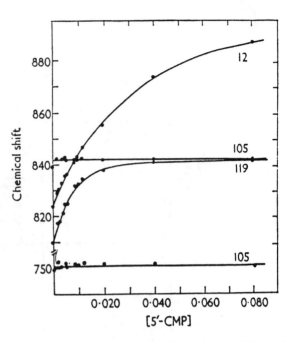

Figure 14.15. Chemical shifts of the H$_2$ resonances of ribonuclease as a function of 5′-CMP concentration at pH 7.0. [RNAase] = 0.0065 *M*. (From ref. 1. Copyright 1969 by Academic Press. Reprinted by permission.)

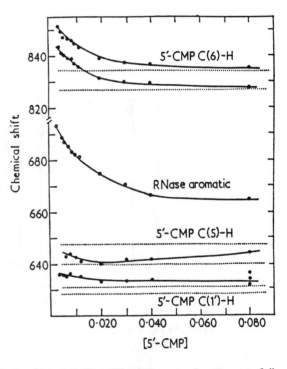

Figure 14.16. Chemical shifts of the aromatic resonance of ribonuclease and the H$_{1'}$, H$_5$, and H$_6$ resonances of ribonuclease as a function of 5′-CMP concentration at pH 5.5 [RNAse] = 0.0065 *M*. The dotted lines indicate the chemical shifts of the 5′-CMP resonances in the absence of RNAase. (Copyright 1969 by Academic Press. Reprinted by permission.)

fore, the histidine resonances of the protein are shifted downfield and the phenylalanine resonances are shifted upfield in the bound state. For the inhibitor, the aromatic resonances are shifted downfield in the bound state. These observed shifts are in agreement with the proposed configuration of 5′-CMP in the active site (Figure 14.17). The downfield shift of the histidines results from the deshielding effects of the nearby negatively charged phosphate group. The upfield shift of the phenylalanine resonances arises from their close proximity to the shielding region of the cytosine ring. The cytidine ring resonances are shifted downfield because they are in the deshielding region of the phenylalanine ring.

NMR techniques can thus be used to ascertain the nature of the binding of small molecules to large proteins. Although initial studies involved the simplest of systems, similar analy-

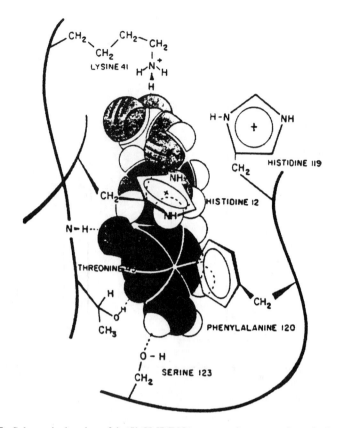

Figure 14.17. Schematic drawing of the 5′-CMP/RNAase complex as seen from the back of the active site cleft. (Copyright 1969 by Academic Press. Reprinted by permission.)

ses have since been applied to a wide variety of more complex systems.

14.7 NMR STUDIES OF NUCLEIC ACIDS

Nucleic acids, the molecules that transmit all hereditary information, are polymers comprised of nucleosides and phosphoric acid. Each nucleoside consists of a nitrogenous base and a sugar. The structures and ^{1}H spectra of several nucleosides are shown in Figures 14.18a–d (pages 264–265). The sugar resonances are found at ~δ 3–4 for the $H_{2'}$, $H_{3'}$, $H_{4'}$, $H_{5'}$ and $H_{5''}$ protons and at ~δ 5 for the $H_{1'}$ proton. Since all of the nucleosides contain the same sugar group, even small oligomers ($n < 5$) exhibit considerable signal overlap, so this region of the spectrum can be very complex. The base protons of the pyrimidine and purine rings are downfield at ~δ 5–9 and do not exhibit as much overlap as the sugar resonances.

As with proteins, 2D techniques can be applied to elucidate assignments and structure for nucleic acids. Since there is no spin coupling between sugar protons and ring protons, COSY spectra are not normally used, but NOESY spectra are still useful. For a single nucleoside such as uridine, there are distinct NOE interactions between the sugar protons and the ring protons (Figure 14.19, page 266):

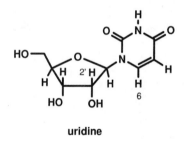

uridine

The most prominent cross peak is that between the $H_{2'}$ of the sugar and H_6 of the base. In solution, uridine must be in a conformation where $H_{2'}$ is in close proximity to H_6. Such a conformation is shown in Figure 14.20 (page 266).

14.7.1 Conformation of a Dinucleotide in Solution

The relative conformation of the bases in a dinucleotide such as adenylyl 3′→5′ adenosine (ApA) can be ascertained from the differential shifts of the aromatic protons of the purine ring.[5] For ApA, the downfield region of the ^{1}H spectrum consists of two resonances for the nonequivalent H_2 protons and two resonances for the nonequivalent H_8 protons (Figure 14.21, page 267). The H_8 proton resonances can be differen-

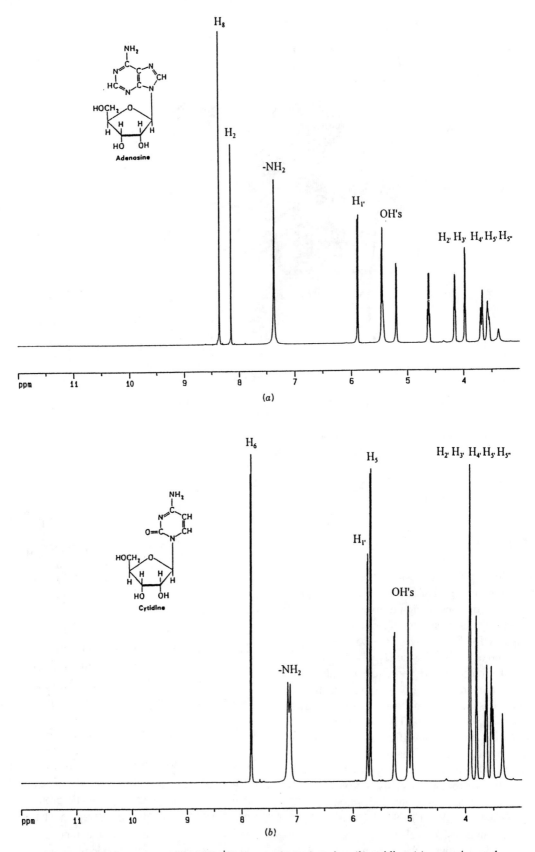

Figure 14.18. Structures and 250-MHz ^{1}H spectra of (a) adenosine, (b) cytidine, (c) guanosine, and (d) uridine in DMSO-d_6.

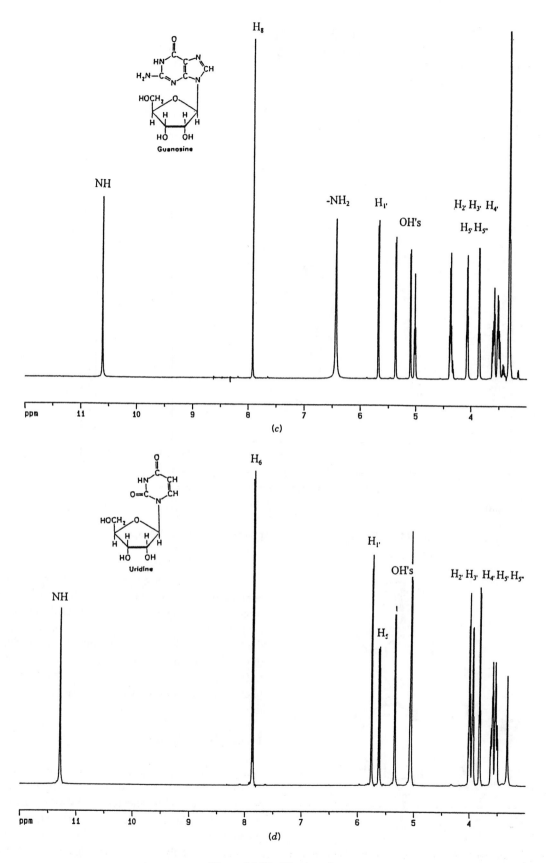

Figure 14.18. (Continued)

Uridine in DMSO-d6
NOESY spectrum
500 msec mixing time

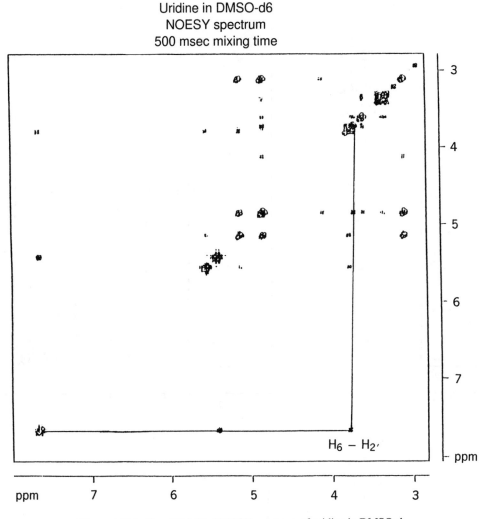

$H_6 - H_{2'}$

Figure 14.19. The 400-MHz NOESY spectrum of uridine in DMSO-d_6.

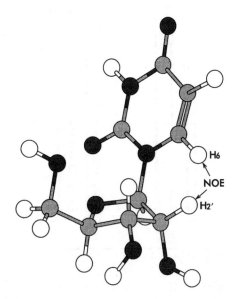

H6

NOE

H2'

Figure 14.20. Conformation of uridine in DMSO-d_6.

tiated from the H_2 proton resonances by elevating the temperature of the sample in D_2O to ~90°C for an hour; the H_8 protons will exchange with the deuterated solvent and disappear. In the conformation of ApA in solution (Figure 14.22), the $H_{2(3')}$ proton is in close proximity to the shielding region of the neighboring adenine base and is shifted upfield relative to $H_{2(5')}$. Due to its close proximity to the negatively charged phosphate group of the backbone, $H_{8(5')}$ is deshielded relative to $H_{8(3')}$.

14.7.2   Helix Formation

Nucleic acid polymers are known to undergo transitions between single strands and double and triple helices. These transitions are readily detected by NMR because of the large differences in the line widths of a flexible random coil compared to a rigid helix. A mixture of adenosine, a mononucleoside, and poly U, a large single-stranded polymer, is known to form a triple helix at low temperatures.[6] The down-

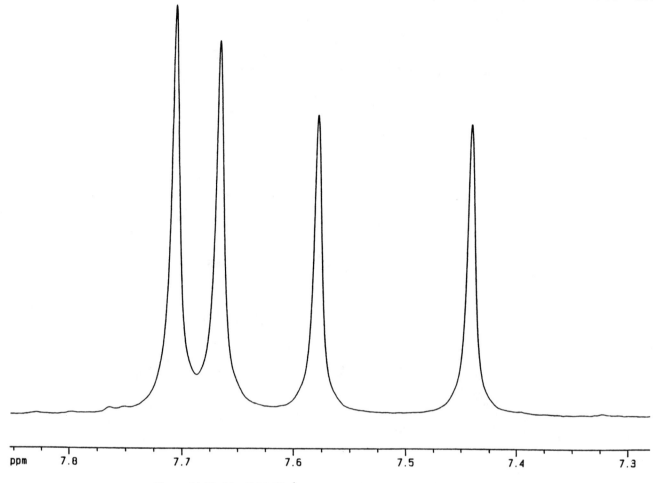

Figure 14.21. The 400-MHz ¹H spectrum of ApA in the aromatic region.

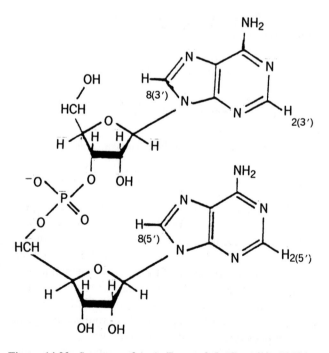

Figure 14.22. Structure of ApA. From ref. 5. (Copyright 1969 by the American Chemical Society. Reprinted with permission.)

field region of the ¹H spectrum for a 1 : 2 base ratio of a mixture of adenosine and poly U is shown in Figure 14.23, where the H_2 and H_8 proton resonances of the adenosine ring and the H_6 proton resonance of the uridine ring are observed. At room temperature, there is no significant interaction between the adenosine and poly U. The line widths of all of the resonances are fairly sharp because adenosine is a small molecule and poly U is a flexible single strand (Section 14.2). As the temperature is lowered, broadening in all of the resonances is observed. This is indicative of interaction between the adenosine and the poly U, which results in partial immobilization of the aromatic bases. Below ~20°C, the resonances are broadened beyond detection. At these temperatures, the adenosine and poly U form a rigid helix with little flexibility. This results in very short T_2 values and extremely broad lines. Since all of the resonances are broadened beyond detection, the ratio of the bases must be one adenosine to two uridine in a triple helix.

14.7.3 Structural Analysis of Small Helices

The structure of small helices of 10 base pairs or less can be obtained from their NOESY spectra and subsequent computer

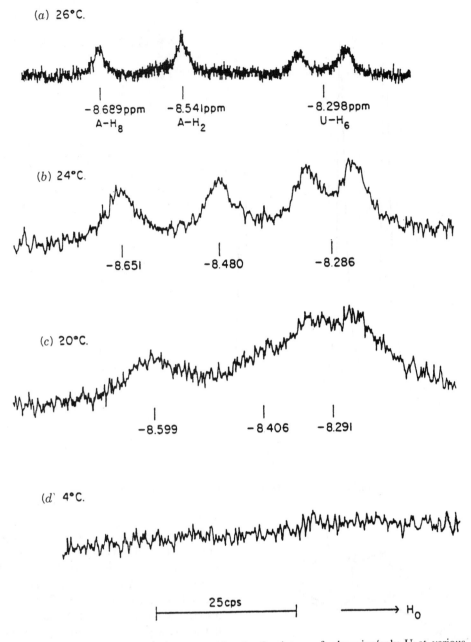

Figure 14.23. The 100-MHz ^{1}H spectrum of a 1 : 2 mixture of adenosine/poly U at various temperatures. (From ref. 6. Reprinted with permission of S. I. Chan.)

analysis. The self-complimentary decanucleotide, CATAGCTATG, forms a double helix at room temperature.[7] In addition, it forms a complex with cisplaten, a popular antitumor drug. The expected NOESY correlations of H_6 of the pyrimidines and H_8 of the purines with $H_{1'}$ of the deoxyribose sugars are depicted in Figure 14.24, where the backbone of the helix is depicted. Each $H_{1'}$ proton is in close proximity to the H_6 or H_8 proton of a base on its own nucleotide, as well as the one on the base of its neighbor. The actual NOESY correlations can be seen in Figure 14.25. For the complimentary decanucleotide, the assignment of resonances begins

with H_6 of cytidine-1 at δ 7.4, which gives rise to a cross peak with the $H_{1'}$ signal at δ 5.3. This $H_{1'}$ exhibits a cross peak with the H_8 signal of the neighboring adenosine at δ 8.2, which in turn shows a cross peak with the $H_{1'}$ signal δ 6.0, and this correlates with the H_6 signal of the neighboring thymidine at δ 6.8. Continuing in this manner, the entire chain can be assigned sequentially. With the assignment of the base protons and $H_{1'}$ protons complete, the $H_{2'}$ and $H_{2''}$ protons can be made from correlations made with the $H_{1'}$ protons.

The chemical shift and cross-peak intensities are very sensitive to the secondary structure of the helix. Each of the

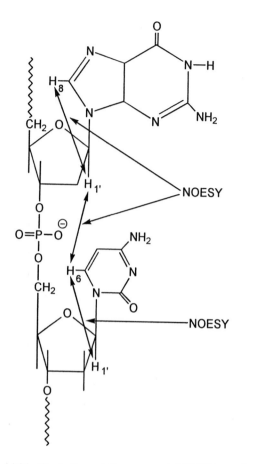

Figure 14.24. Illustration of the expected NOESY correlations along the polynucleotide backbone.

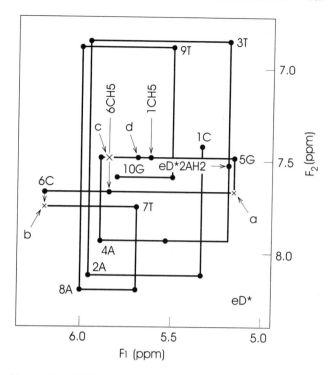

Figure 14.25. The NOESY spectrum of CATAGCTATG in the region where correlations between the $H_{1'}$ protons of the deoxyribose sugar and the H_6 and H_8 protons of the nitrogenous bases are observed. (From ref. 7. Copyright 1995 by the American Association for the Advancement of Science. Excerpted with permission.)

known types of helices, the A form, the B form, and the Z form, yield a distinctive pattern. From the analysis of the data it can be ascertained that the majority of the bases, four at each end, are in a B-type helix. The interior G and C bases are not in the helix but are complexed with the cisplatin or turned out. Using the distance constraints from the NOE measurements and those from covalent linkages, computer analysis generated the three-dimensional structure of the complex shown in Figure 14.26.

14.7.4 Hydrogen Bonding between Nucleosides

In nonaqueous solvents such as DMSO-d_6, complementary nucleosides will form Watson–Crick base pairs that can be monitored by NMR. The spectra of guanosine and cytidine, both separately and together, are shown in Figure 14.27. For guanosine, the observed resonances include H_8 at δ 8.6, NH_2 at δ 7.2, and the $H_{1'}$ at δ 6.4. For cytidine, the observed resonances are H_6 at δ 8.5, NH_2 at δ 7.8, $H_{1'}$ at δ 6.5, and H_5 at δ 6.45. In a 1 : 1 mixture, the resonances for the amine protons of both the guanosine and cytidine bases are shifted downfield due to hydrogen bond formation. In addition, the

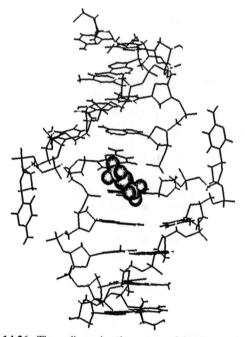

Figure 14.26. Three-dimensional structure of the decanucleotide–cisplatin complex. The cisplatin is shown in CPK format and the turned-out cytidines are shown in black. The coordinates were supplied by B. P. Hoskins. (From ref. 7. Copyright 1995 by the American Association for the Advancement of Science. Excerpted with permission.)

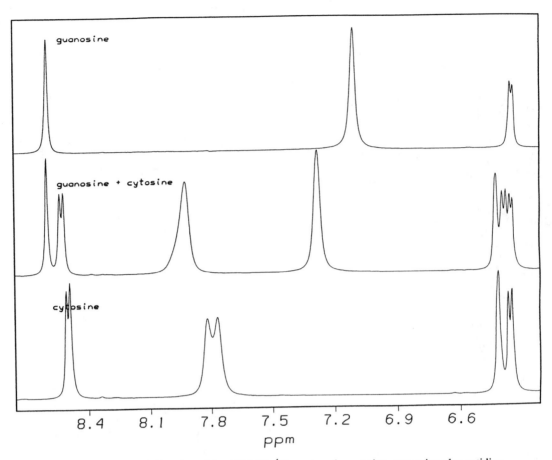

Figure 14.27. Downfield region of the 400-MHz ^{1}H spectra of guanosine, guanosine plus cytidine, and cytidine in DMSO-d$_6$.

H$_5$ and H$_6$ resonances are shifted downfield because each proton lies in the deshielding region of the neighboring guanosine base, and the two bases are in the same plane when they are hydrogen bonded.

14.8   LIPIDS AND BIOLOGICAL MEMBRANES

The two major components of biological membranes are lipids and cholesterol. One major class of lipids are the phospholipids. The ^{1}H spectrum of one such phospholipid, dipalmitoylphosphatidylcholine (DPPC), in chloroform is shown in Figure 14.28. The ^{1}H spectrum of cholesterol in chloroform is shown in Figure 14.29 (page 272).

In nonaqueous solvents such as chloroform, such molecules do not aggregate and their intrinsic line width of several hertz is controlled primarily by their molecular weight. The spectrum of DPPC consists of the resonances for the choline head group at 3.3 ppm, those for the ethyl group and glycerol group in the region δ 4–5.3, the methylene groups from δ 2.3 for C$_2$ to δ 1.2 for the others, and the terminal methyl group at δ 0.8. The spectrum of cholesterol consists of the vinyl proton at δ 5.3, the methine hydrogen on the hydroxyl-bearing

carbon at δ 3.5, the various ring protons from δ 0.7 to 2.4, and the three singlets for the methyl groups near δ 0.7.

In aqueous solutions, lipids form huge supramolecular aggregates, the major one being the bilayer structure shown in Figure 14.30 (page 272). This is a sheet that is only two molecules thick but extends large distances in a two-dimensional plane. The bilayer structure can undergo a phase transition from gel to liquid crystalline, the former being almost solid and the latter exhibiting liquidlike diffusion in a constrained manner. In the gel phase, the motions of the lipids are so constrained that the intrinsic line widths of the protons are thousands of hertz, broadened beyond detection. Above the phase transition temperature, the lipids are more mobile and an NMR spectrum can be detected (Figure 14.30). The motion of the lipid molecules is anisotropic. There is rapid rotation around the long axis of the molecule and slower lateral diffusion within the plane of the bilayer. In addition, the mobility of the fatty acid residues increases toward the terminal methyl group. There is very little motion in the glycerol backbone. Due to this asymmetry of motion of the lipid molecules, the line widths of the various components vary markedly. The terminal methyl group at ~δ 1.0 and some of

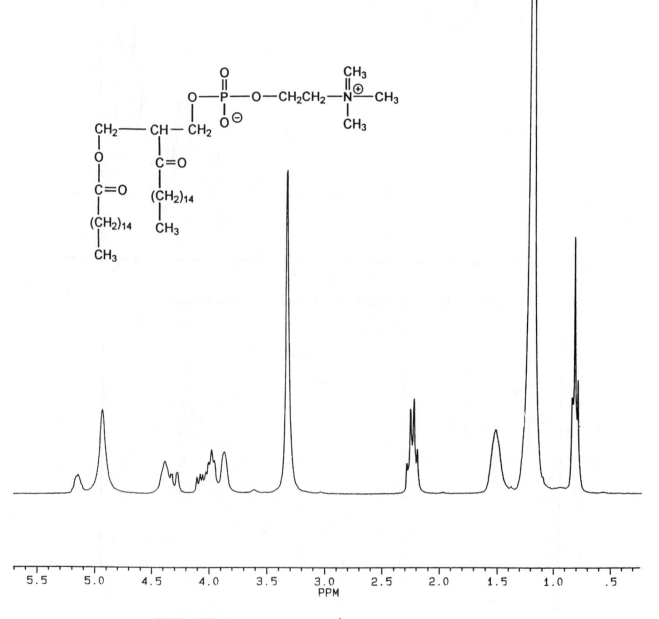

Figure 14.28. Structure and 250-MHz ^{1}H spectrum of DPPC in CDCl$_3$.

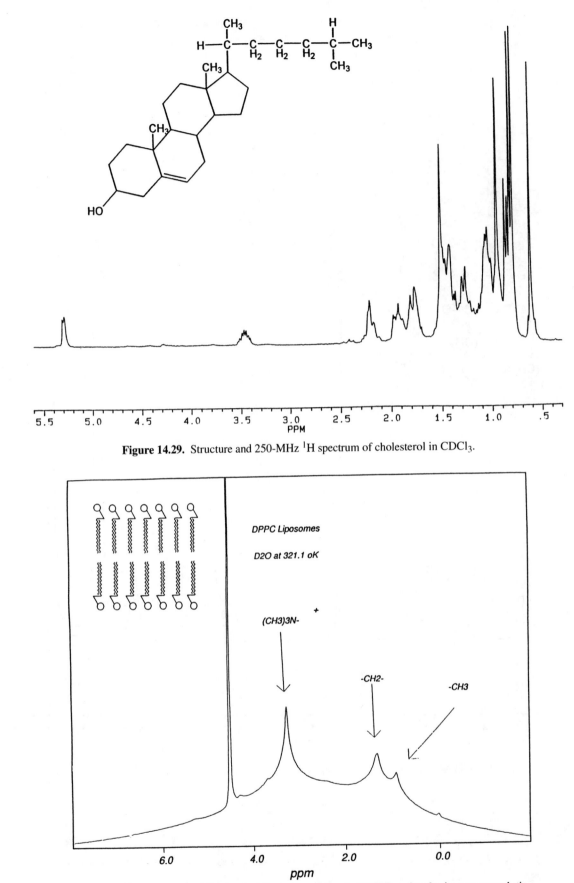

Figure 14.29. Structure and 250-MHz ^{1}H spectrum of cholesterol in CDCl$_3$.

Figure 14.30. Pictorial depiction of the bilayer arrangement of DPPC molecules in aqueous solution and the 400-MHz ^{1}H spectrum of DPPC at 45°C.

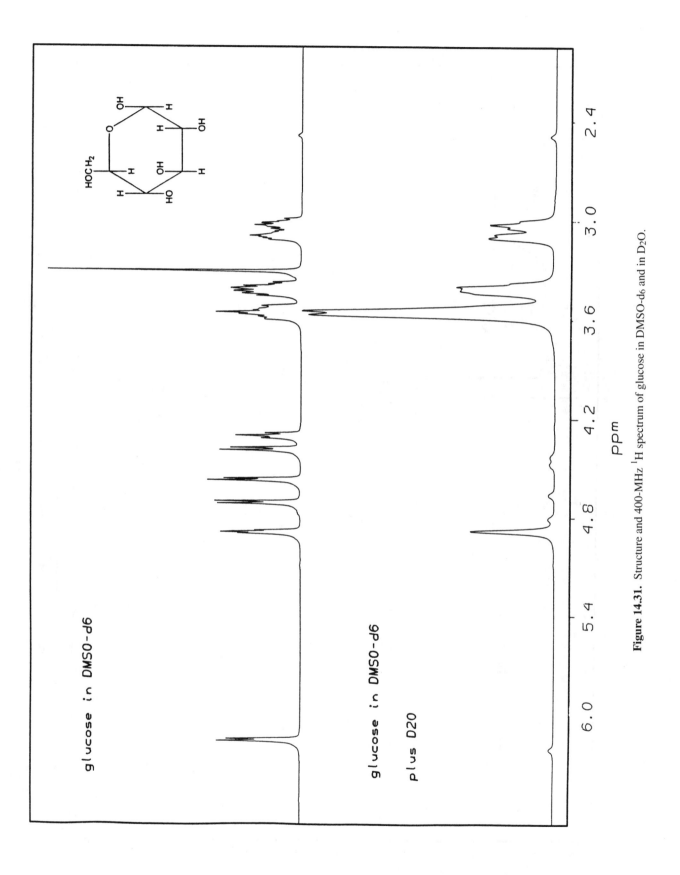

Figure 14.31. Structure and 400-MHz ^{1}H spectrum of glucose in DMSO-d_6 and in D$_2$O.

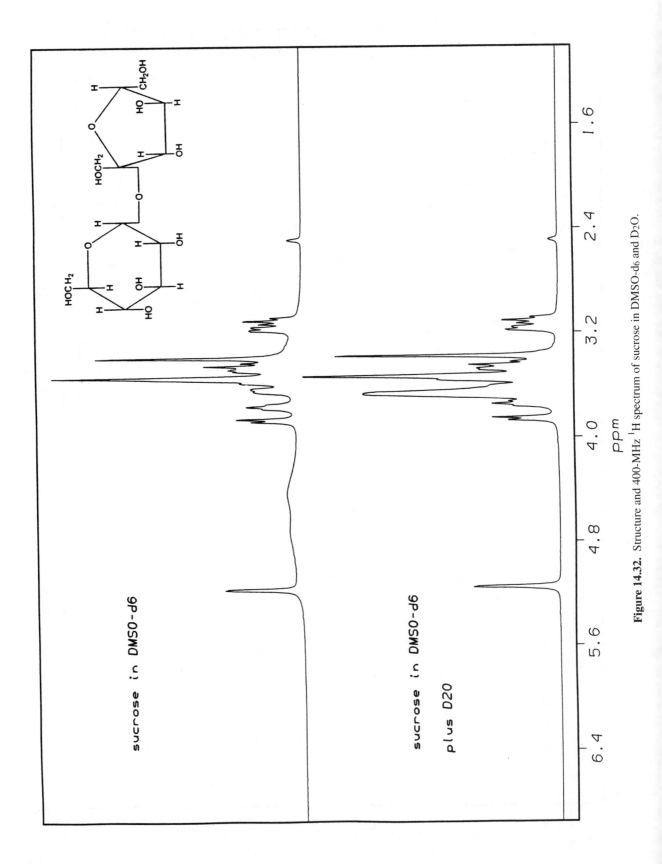

Figure 14.32. Structure and 400-MHz ^{1}H spectrum of sucrose in DMSO-d$_6$ and D$_2$O.

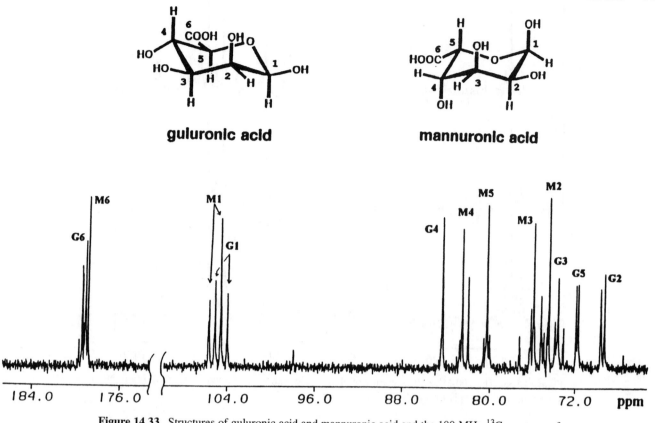

guluronic acid   **mannuronic acid**

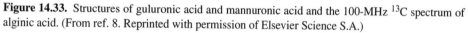

Figure 14.33. Structures of guluronic acid and mannuronic acid and the 100-MHz ^{13}C spectrum of alginic acid. (From ref. 8. Reprinted with permission of Elsevier Science S.A.)

the nearby methylene groups at ~δ 1.5 are reasonably sharp (~50–200 Hz). The same is true for the methyl groups of the choline head group at ~δ 3.5. The remaining protons form a very broad (> 1000 Hz) non-Lorentzian line centered at approximately δ 2.5. The analysis of the line widths and lineshapes is complex but can be used to ascertain the dynamic properties of lipid molecules in the bilayer. Similarly, the effects of cholesterol on the mobilities of various portions of the lipid molecule can be analyzed.

14.9 CARBOHYDRATES

Carbohydrates have the general formula $(CH_2O)_n$, where $n = 3, \ldots, 8$. Because of the presence of stereoisomers, there are literally hundreds of monosaccharides. The monosaccharides can combine to form disaccharides and higher polymers. There is a high heterogeneity of carbohydrates in nature. The ^{1}H spectra of two common carbohydrates, D-glucose and sucrose (in DMSO-d$_6$), are shown in Figures 14.31 (page 273)

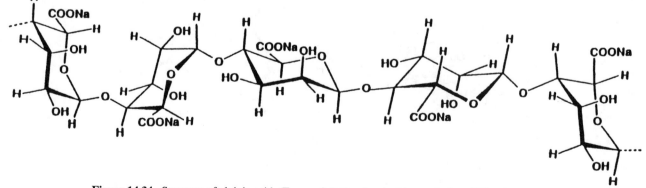

Figure 14.34. Structure of alginic acid. (From ref. 8. Reprinted with permission of Elsevier Science S.A.)

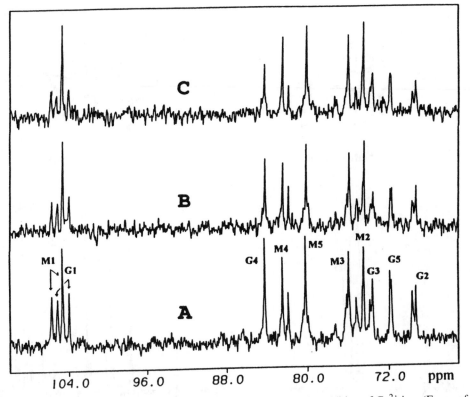

Figure 14.35. The 100-MHz ^{13}C spectrum of alginic acid with the addition of Ca^{2+} ion. (From ref. 8. Reprinted with permission of Elsevier Science S.A.)

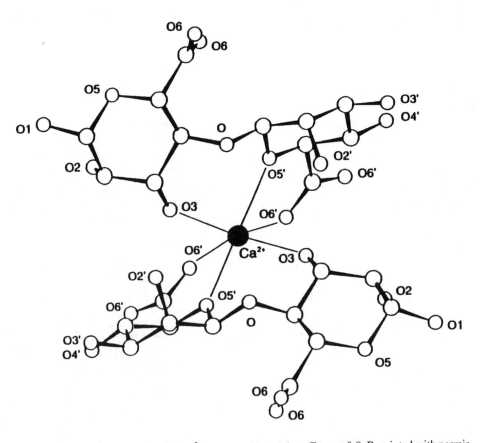

Figure 14.36. Proposed structure of Ca^{2+}/alginic acid complex. (From ref. 8. Reprinted with permission of Elsevier Science S.A.)

and 14.32 (page 274), respectively. The spectra consist of the anomeric proton at δ 5–6, the hydroxyl protons at δ 4–5, and the other protons at δ 3–4. The hydroxyl protons can be identified by exchange in D_2O.

In contrast to ^{1}H spectra, where there is considerable overlap of resonances, ^{13}C spectra of carbohydrates show greater chemical shift dispersion. The ^{13}C spectrum for alginic acid is shown in Figure 14.33 (page 275).[8] Alginic acid is a naturally occurring linear, binary copolymer of 1,4-linked α-L-guluronic acid and β-D-mannuronic acid found in brown algae (Figure 14.34, page 275). The sequence of residues is random, which leads to three types of blocks: homopolymeric sequences of mannurate (MM blocks) and guluronate residues (GG blocks) and a region where the two residues alternate (MG blocks). In solution, alginates behave like flexible coils, as evidenced by the relatively sharp lines. Some divalent cations such as Ca^{2+} ions are known to bind to the carboxylic acid groups of mannuronic acid. Titration of alginic acid with Ca^{2+} is shown in Figure 14.35 (page 276). With increasing Ca^{2+}, there is a decrease in the intensities of the mannuronic acid resonances. For these groups, there are two states, free and bound. In the bound state, it is postulated that the mannuronic acid residues form a chelatelike structure around the Ca^{2+} (Figure 14.36, page 276). The binding of the ions immobilizes these residues, and the effective averaging of the dipole–dipole interactions is lost. These resonances are broadened beyond detection. Since exchange between the free and bound states is slow on the NMR time scale, only the resonances from the free state are observed. Integration of the intensities of these resonances can be used in determining the fraction bound.

CHAPTER SUMMARY

1. Nuclear magnetic resonance spectroscopy can be applied to the study of biomolecules for structural and conformational elucidation, the modes of interaction between and within molecules, and the dynamics in supramolecular complexes.

2. The NMR line widths are affected by the molecular weight of biomolecules, their flexibility, and the degree of association of the molecules.

3. Exchangeable (hence, hydrogen-bonding) protons are important in biomolecules, and special techniques may be used for their observation.

4. Dynamic processes between molecules of biological importance can be characterized by NMR studies.

5. The NMR parameters of certain groups in biomolecules are sensitive to the pH of the medium and the presence of other complexing agents.

6. The NMR studies of proteins lead to information about their primary, secondary, and tertiary structure as well as information about their interactions with substrates.

7. The NMR studies of nucleic acids can yield information about base sequences, conformations in solution, helix formation, and hydrogen bonding between base pairs.

8. The NMR studies of lipids can be used to probe the structure and function of biological membranes.

9. The NMR studies of carbohydrates provides information about structure, conformation, and complexation with small molecules and ions.

REFERENCES

[1]Meadows, D. H., Roberts, G. C. K., and Jardetzky, O., *J. Mol. Biol.*, *45*, 491 (1969).

[2]Zubay, G. L., Parson, W. W., and Vance, D. E., *Principles of Biochemistry*, Wm. C. Brown: Dubuque, IA, 1995.

[3]Bebz, F. W., and Roberts, G. C. K., *FEBS Lett.*, *29*, 263 (1973).

[4]Zuiderweg, E. R. P., Scheek, R. M., and Kaptein, R., *Biopolymers*, *24*, 2257 (1985).

[5]Chan, S. I., and Nelson, J. H., *J. Am. Chem. Soc.*, *91*, 168 (1969).

[6]Bamgerter, B. W., and Chan, S. I., *Proc. Natl. Acad. Sci., U.S.A.*, *60*, 1144 (1968).

[7]Huang, H., Zhu, L., Reid, B. R., Drobny, G. P., and Hopkins, P. B., *Science*, *270*, 1842 (1995).

[8]DeRamos, C. M., Irwin, A. E., Nauss, J. L., and Stout, B. E., *Inorg. Chim. Acta*, *256*, 69 (1997).

ADDITIONAL RESOURCES

1. James, T. L., *Nuclear Magnetic Resonance in Biochemistry: Principles and Applications*, Academic, New York, 1975.

2. Schraml, J., and Bellama, J. M., *Two-Dimensional NMR Spectroscopy*, Wiley, New York, 1988.

3. Yarger, J. L., Nieman, R. A., and Bieber, A. L., *J. Chem. Educ.*, *74*, 243 (1997).

REVIEW PROBLEMS (Answers in Appendix 1)

14.1. Draw the structure of arginine. Predict the appearance of its ^{1}H spectrum. Number these signals in order of increasing chemical shift, then use the numbers to list all the correlations expected in the ^{1}H COSY spectrum of arginine. Draw the COSY spectrum.

14.2. Gramicidin S is a cyclic decapeptide antibiotic with the following sequence:

D-Phe - L-Leu - L-Orn - L-Val- L-Pro
    | |

L-Pro- L-Val - L-Orn - L-Leu - D-Phe

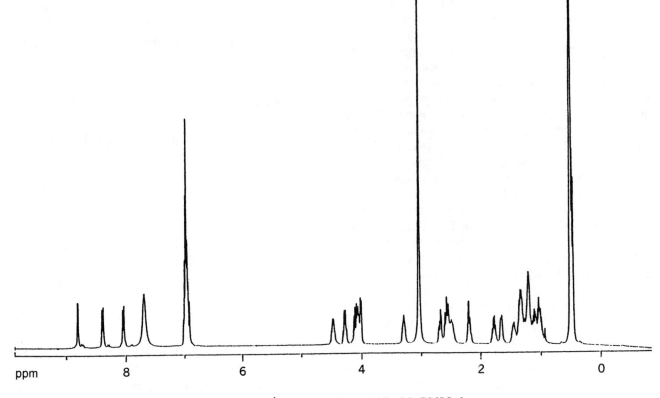

Figure 14.37. The ^{1}H spectrum of gramicidin-S in DMSO-d$_6$.

L-Orn is ornithine, an amino acid found in metabolic pathways and is one methylene shorter than lysine. (a) Draw the structure of gramicidin S. Is there an axis of symmetry? Are the pairs of like amino acids equivalent? (b) Figure 14.37 shows the ^{1}H spectrum of gramicidin-S. The full COSY spectrum is shown in Figure 14.38, and expansions of the upfield and downfield halves are shown in Figures 14.39 and 14.40. The combined NOESY/COSY spectrum of the correlations between the amino and α protons is shown in Figure 14.41. How many amino proton signals are expected and where do they appear in the COSY spectrum? (c) The resonance at δ 8.1 is the amino group of the L-Orn side chain. Beginning there, map the correlations in the COSY spectrum of L-Orn to its backbone amino group. Locate the other COSY correlations for the other side chains. Map the remaining COSY–NOESY correlations along the backbone of the entire peptide.

14.3. Figure 14.42 (page 281) shows the temperature-dependent ^{1}H spectrum of a 1 : 1 mixture of adenosine and Poly U. At 30°C, the spectrum is identical to that of the separate components. With decreasing temperature, broadening of all the signals is observed. Below 21°C, the H$_6$ resonance of Poly U is broadened beyond detection. The H$_2$ and H$_8$ resonances of adenosine are half as intense as they were at higher temperatures, and *sharpen* with decreasing temperature. Account for this behavior.

14.4. Benzaldehyde dehydrogenase (BDH), a hypothetical enzyme, catalyzes the oxidation of benzaldehyde to benzoic acid. The enzyme has a flavin prosthetic group that accepts electrons from the substrate upon oxidation:

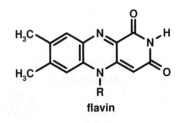

flavin

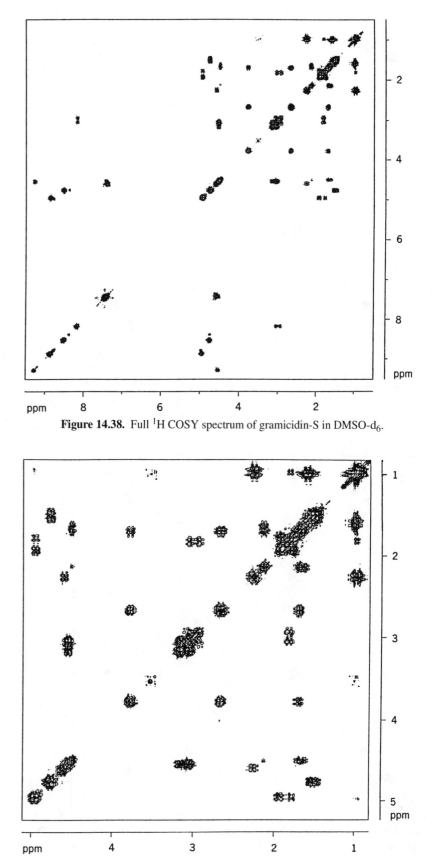

Figure 14.38. Full ¹H COSY spectrum of gramicidin-S in DMSO-d_6.

Figure 14.39. Upfield region (δ 6.0 to 0.5) of the ¹H COSY spectrum of gramicidin-S in DMSO-d_6.

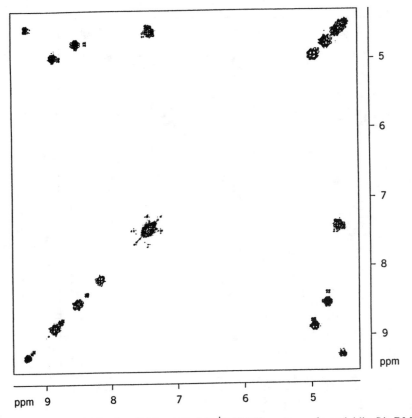

Figure 14.40. Downfield region (δ 9.5 to 4.0) of the ^{1}H COSY spectrum of gramicidin-S in DMSO-d$_6$.

NOESY

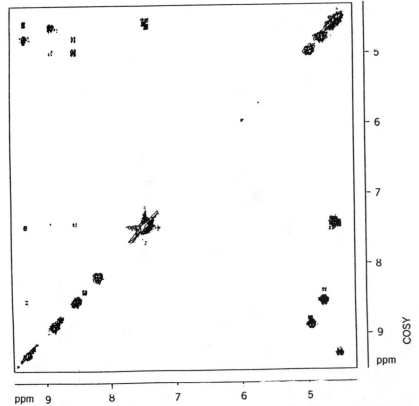

Figure 14.41. Downfield region (δ 9.5 to 4.0) of the combined ^{1}H NOESY/COSY spectrum of gramicidin-S in DMSO-d$_6$.

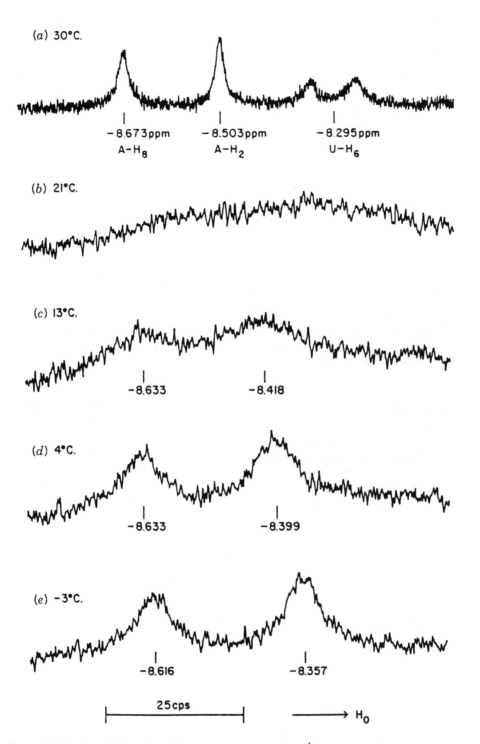

Figure 14.42. Downfield region of the temperature-dependent ^1H spectrum of a 1 : 1 mixture of adenosine and poly U. The H_8 and H_2 resonances of adenosine and the H_6 resonance of uridine are visible in this region. (From ref. 6. Reprinted with permission of S. I. Chan.)

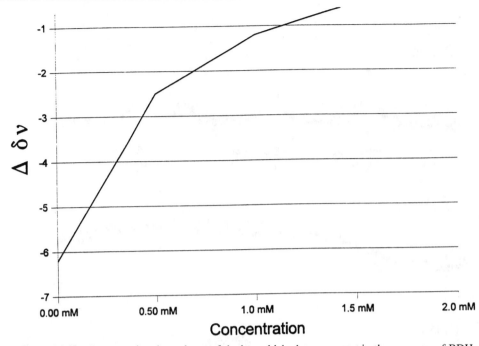

Figure 14.43. Concentration dependence of the benzaldehyde resonances in the presence of BDH. [BDH] is constant. Chemical shifts are given relative to benzaldehyde resonances at the same concentration in the absence of BDH.

The concentration dependence of the resonances of benzaldehyde are shown in Figure 14.43, where the BDH concentration is kept constant. Is exchange fast or slow on the NMR time scale? If positive $\Delta \delta \nu$ is downfield, is the plane of benzaldehyde ring parallel or perpendicular to the flavin ring?

15

SOLID-STATE NMR SPECTROSCOPY*

15.1 WHY STUDY MATERIALS IN THE SOLID STATE?

Most NMR spectroscopy is performed on fluid samples, typically solutions. But there are materials for which requiring a fluid sample to analyze in the spectrometer just will not do. (Have you ever tried ordering a 2.0 M solution of Illinois No. 6 coal?) The word "material" is used here in the very specific sense employed by materials scientists: A **material** is an aggregation of matter that might consist of many compounds and in which things such as phase (solid, liquid, gas), microstructure (crystallinity, grain size, fiber orientation), and molecular motions (diffusion, entangling of polymer chains, exchange of a small molecule between the gas phase and the adsorbed state on a silica gel surface) all figure importantly in the properties of the material. In such cases, it would make no sense to dissolve the material for the purpose of NMR analysis because the properties of interest may then be destroyed. Nuclear magnetic resonance is one of a group of analytical techniques called **nondestructive**, and the use of these analytical methods in the characterization of materials is known in the materials science community as **nondestructive testing** (NDT) or **nondestructive evaluation** (NDE). Examples of **destructive** techniques include atomic absorption spectrophotometry (the sample is dissolved and burned) and wet chemical analysis (the sample is dissolved and altered chemically).

How NMR spectroscopy can work on solid samples without dissolution (sometimes even without grinding to a powder)—and how it differs from the fluid phase NMR with which most chemists are familiar—is the subject of the next

section. Here we can list some materials that are commonly studied with *solid state* NMR spectroscopy.

An early impetus for the development of solid-state NMR was the study of polymers. The NMR of polymer solutions was (and still is!) an important means of characterizing the chemical structure of the monomer segments, their stereochemistry (for a polymer, this includes the monomer sequencing along the polymer chain), the branch points in polymer chains, and so forth. However, many polymers are highly **crosslinked** (the chains are covalently bonded together into a tight network) and may be **filled** (contain dispersions of small particles for added strength). Even pure polymer **melts** may have many of their properties, such as viscosity, determined by the extent of molecular **entanglements** (compare swimming through a bowl of soup versus a bowl of spaghetti). Dissolving the polymer would require destroying the crosslinks, leaving the fillers behind, and making the entanglement effects disappear. You would not be studying the material you had presumed to analyze.

The other early driving force for the development of solid-state NMR techniques was the study of fossil fuels such as coal. Coal represents the fossil remains of thick beds of decaying plant matter, such as that produced in bogs. It consists of an agglomeration of innumerable organic molecules, many of which are highly crosslinked, and often contains various minerals and elemental carbon. Dissolving coal requires breaking chemical bonds and therefore amounts to destroying the material. Solid-state NMR has proved invaluable for quantitative analysis of the aromatic and aliphatic content of coals.

Solid-state NMR is used in the analysis of ceramics (they are neither soluble nor meltable), biological cell membranes, wood, cellulose and other plant materials, bone, and even gemstones (any volunteers for solution NMR spectroscopy of

*Contributed by Jerome L. Ackerman

your diamond ring?). One can even study highly reactive intermediates such as free radicals if they are isolated and frozen solid in an inert matrix (such as argon) at very low temperature. In all these cases, fluid-state NMR would be inappropriate.

Another application of solid-state NMR is in the study of solid-state conformation. In solution molecules tumble rapidly, and rotations about single bonds are fast; recall that in solution spectra the three methyl protons are normally equivalent (Section 4.2). In the solid state many of these motions are restricted or suppressed entirely, and many of the chemical shift degeneracies to which we have become accustomed are now split. For example, *para*-dimethoxybenzene, structure **15-1**, exhibits a solution ^{13}C spectrum in which the two methoxy carbons are equivalent to each other, as are the two substituted aromatic carbons, and each set gives rise to a single resonance:

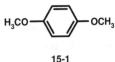

15-1

The four unsubstituted aromatic carbons also are equivalent and yield a single resonance. In the solid, crystal packing forces (the **crystal field**) create slightly different microenvironments for each carbon in the molecule, so crystalline **15-1** gives rise to eight distinct ^{13}C signals. Often, for many compounds, packing effects create different environments for molecules that are otherwise identical but occur in different orientations or positions in the **unit cell** of the crystal structure. (The unit cell of a crystal is the smallest part of a crystal that can be used to construct the entire crystal by successive replications along the unit cell axes.) The result is a spectrum with separate resonances for each environment.

Lastly, molecular motion affects solid-state spectra just as chemical exchange does in solution (Chapter 10 and Section 14.2). Nuclear magnetic resonance provided the proof that benzene molecules, structure **15-2**, rotate in place about their sixfold axes (Example 4.3) in the crystal above 90 K (Kelvins absolute temperature; 0°C = 273 K):

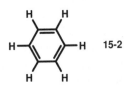

15-2

Below that temperature this thermally activated motion is frozen out. There is a characteristic change in the solid-state lineshape that clearly demonstrates the effect. Nuclear magnetic resonance can in such cases provide important information about molecular structure and dynamics in solids.

15.2 WHY IS NMR OF SOLIDS DIFFERENT FROM NMR OF FLUIDS?

For nuclear spins, the difference between a solid and a fluid depends on the extent and type of molecular motion and how the motion affects the various interactions experienced by the spin. Let us clarify this by looking at motions and interactions in more detail. By now we are familiar with the *J* coupling interaction, which is a magnetic interaction between pairs of spins that is transmitted through the covalent chemical bonding network of a molecule (Chapters 8 and 9). But that is only a small part of the entire story. The maximum coupling that can exist between a pair of protons can be on the order of 50 kHz (50,000 Hz) or greater, far larger than the few hertz found in solution spectra. Why is it that we never see such an enormous coupling in solution?

Figure 15.1 shows a pair of protons in the magnetic field B_0. Because each spin possesses a magnetic moment, each is surrounded by a magnetic field that is experienced by the other. This is the **direct**, or **through-space**, **dipole–dipole** (or **dipolar**) **coupling**. The equation that describes the direct splitting $\Delta\nu$ (in hertz) in the spectrum is

$$\Delta\nu = \frac{3\mu^2}{hr^3}(3\cos^2\theta - 1) \tag{15.1}$$

Notice two things about this equation. First, the splitting depends very strongly upon the distance r between the nuclei, varying as the inverse third power. Second, the splitting also depends upon the angle θ the internuclear vector makes with the static field B_0. This is both a blessing and a curse. By measuring the splitting caused by the direct dipole–dipole coupling, we can in principle measure internuclear distances

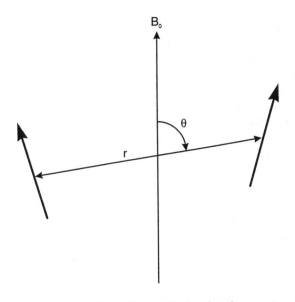

Figure 15.1. Dipole–dipole spin pair.

with great accuracy. However, this measurement is complicated by its **anisotropic** nature: It is dependent upon the orientation of the spin pair.

■ **EXAMPLE 15.1** At what orientation is the dipolar coupling at its maximum? Its minimum? Exactly zero? What is the maximum splitting for two protons separated by 1 Å (10^{-8} cm)?

□ *Solution:* The orientation is given by the angle θ. The function $3 \cos^2 \theta - 1$ is equal to 2 (maximum splitting) when $\theta = 0°$ ($\cos \theta = 1$), which corresponds to the spin pair aligned along the static field. The function equals -1 when θ equals $\pi/2$ (90°; $\cos \theta = 0$), which corresponds to the spin pair lying in the plane perpendicular to the static field. And it equals zero at the so-called **magic angle**, which can be found as follows:

$$3 \cos^2 \theta - 1 = 0$$

$$\cos \theta = (\tfrac{1}{3})^{1/2}$$

$$\theta \cong 54.74° \cong 54°44'$$

To compute the size of the maximum splitting for $r = 1$ Å, we need the magnetic moment of the proton, 1.41×10^{-23} erg/G, and we will use $\theta = 0$ in Equation (15.1):

$$\Delta \nu = \frac{3(1.41 \times 10^{-23} \text{ erg/G})^2}{(6.63 \times 10^{-27} \text{ erg s})(10^{-8} \text{ cm})^3} (3 - 1)$$

$$= 180 \text{ kHz!}$$

(Do not worry about making the units come out, or about doing the calculation in MKS units; electromagnetic units can be very complicated and are beyond the scope of this book.) □

That is a rather large splitting, far larger than the entire chemical shift range for protons at any reasonable field strength. For an isolated spin pair, this size and variability in the splitting would be quite manageable. But now consider the approximately 10^{21} spins in a millimole-sized crystal. *Each pair* of spins (there are 10^{42} possible pairs!) experiences the direct dipole–dipole interaction. For each pair there is a particular distance and angle, giving a huge number of possible splittings spanning the continuum between the minimum splitting and the maximum. What is actually observed in the proton spectrum of a solid (if no special experimental technique is employed) is a featureless broad line (since this many splittings could never be separately resolved), typically approximating a Gaussian (review problem 3.3) shape, with a width at half height on the order of half of the splitting given by Equation (15.1) evaluated for the nearest-neighbor interproton distance. Because the width of solid-state proton lines is so large compared with the chemical shift range, resolving and measuring chemical shifts in this situation is hopeless without special techniques. You are more familiar with the magic angle than you may realize at this point. It is actually one-half the tetrahedral angle, of course familiar to chemists. Additional—and equally familiar—geometric interpretations will show up later.

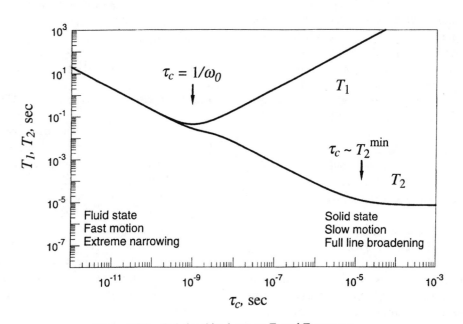

Figure 15.2. Relationships between T_1 and T_2 versus τ_c.

Let us reexamine the relationship between the correlation time τ_c and the relaxation times T_1 and T_2 introduced in Section 2.5. Remember that τ_c is a characteristic time for molecular motion: The shorter the value of τ_c, the faster the motion. The precise definition of τ_c depends upon the details of the specific type of motion being discussed, but for our purposes, we may again regard it as characterizing the average time it takes for a nuclear spin pair (say, the protons in a methylene group) to become completely randomized in orientation with respect to any given starting time. (Correlation times are statistical quantities because molecular motions are random and do not actually occur as, for example, simple, smooth rotations.)

We may plot Equation (2.12), which approximately describes T_1, and also plot a similar relation for T_2 in Figure 15.2. The axes are both logarithmic because the variables cover large dynamic ranges. On the left side of the graph, τ_c is short, and therefore this side represents fast molecular tumbling and low-viscosity fluids (**extreme narrowing** of the dipole–dipole interaction that underlies this relaxation mechanism). On the right side of the graph, τ_c is long, which corresponds to slow motions and high viscosity. We may therefore regard the right-hand side of the plot as the limit of a fluid of very high viscosity: essentially a solid with no free molecular tumbling.

We notice two features of the graph. First, in the limit of low viscosity and fast motion, T_1 and T_2 are equal and vary inversely with τ_c. Second, in the limit of high viscosity and restricted motion T_2 becomes much shorter than T_1, with T_1 varying directly with τ_c and T_2 approaching a very small and constant value, the **rigid-lattice** limit. We therefore expect T_1 in solids to be very long and T_2 to be very short. The T_1 minimum discussed in Chapter 2 is quite apparent from Figure 15.2. The minimum occurs when the correlation time is equal to the inverse of the Larmor frequency in radians per second and is about where T_1 and T_2 diverge. This theory was expounded in a famous 1948 paper by Nikolaas Bloembergen, Edward Purcell, and Robert Pound (the former are two of the seven Nobel Prize winners who have influenced NMR greatly) and is known affectionately as **BPP theory**.

These relaxation time effects usually are bad news for the spectroscopist attempting to analyze solid materials. Since the T_1 is long, the delay time t_w between rf pulses must be long to allow the spins to repolarize. This reduces the signal-to-noise ratio that can be accumulated in a given time. Because T_2 is short, the resonance is spread over a large spectral bandwidth (range of frequencies), which reduces the peak amplitude of the line with respect to the noise. Another way of understanding this effect is to realize that the noise power acquired is proportional to the spectral bandwidth of the measurement. Measuring a signal that occupies a larger bandwidth necessitates acquiring more noise along with the signal. Finally, the broad lineshapes reduce the spectral resolution.

It is worth including a final note on molecular motions in systems other than simple fluids or fully rigid solids. Some compounds may appear solid macroscopically but may experience substantial motion at the molecular level. For example, the hydrocarbon adamantane and the molecule camphor are both roughly spherical in shape:

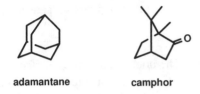

adamantane camphor

At room temperature, these molecules occupy well-defined locations in their respective crystal lattices. However, they tumble freely and **isotropically** (equally in all directions) in place at their lattice positions. As a result, their solid phase NMR spectra show features highly reminiscent of liquids. We will see an illustration of this point shortly. Other molecules may reorient anisotropically (as in solid benzene). Polymer segmental motions in the melt may cause rapid reorientation about the chain axis but only relatively slow reorientation of the chain axes themselves. Large molecular aggregates in solution (such as surfactant micelles or protein complexes or nucleic acids) may appear to have solidlike spectra if their tumbling rates are sufficiently slow. There are numerous other instances in which our macroscopic motions of solid and liquid may be at odds with the molecular dynamics. Nuclear magnetic resonance is one of the foremost ways of investigating these situations.

To summarize the features of solid-state NMR spectroscopy, spin–lattice relaxation is generally inefficient (long T_1's), spin–spin relaxation is extremely efficient (short T_2's), and spectral lines are quite broad due to the anisotropic nature of the interactions.

15.3 CHEMICAL SHIFTS IN SOLIDS

15.3.1 Revisiting the Origin of the Chemical Shift

We learned in Section 6.1 that the chemical shift effect results from shielding of the nucleus by the electron cloud of the molecule. Imagine the simplest case, for example, a helium atom (^{3}He has a spin of $\frac{1}{2}$ and gives a strong NMR signal, although its natural abundance is zero; it is obtained as a byproduct of tritium production). This is a perfectly spherical atom, and when placed in a magnetic field, the electrons in the atomic orbitals circulate about the field in a direction so as to produce diamagnetic shielding; thus, the Larmor frequency is slightly lower than it would be if only a bare ^{3}He nucleus were present. The spherical atomic symmetry assures that no matter how the atom is reoriented in the magnetic field, only a single well-defined chemical shift value is observed.

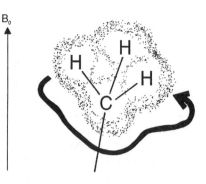

Figure 15.3. Electronic circulation in the nonspherical molecular orbitals of a methyl group give rise to an anisotropic chemical shift.

There is probably not a great demand or future for ^{3}He NMR spectroscopy, so let us examine a slightly more interesting situation: the ^{13}C resonance of a methyl group, diagrammed in Figure 15.3. At room temperature, the methyl protons reorient rapidly about the local C_3 axis, imposing **axial** symmetry C_∞ (continuous rotational symmetry about an axis). Picture the electron cloud of the carbon atom and how the B_0 field might drive the electronic circulation. If the local C_3 axis is parallel to B_0, we might expect the electrons to follow circular paths about the field direction, just as in the isolated spherical atom. But suppose the methyl group axis is tipped away from B_0 by some angle. The paths of electronic circulation can no longer be circular; in any case it is clear that the shielding effect must be altered in some way and that the value of the chemical shift may depend upon the orientation of the molecule with respect to the magnetic field. This effect is known as **chemical shift anisotropy**. Just like the direct dipole–dipole interaction, chemical shielding is anisotropic.

15.3.2 Chemical Shift Anisotropy: The Chemical Shift Tensor

This anisotropy is just what is observed when we obtain spectra from molecules in specific orientations, for example, in a single crystal of a compound or salt. There are normally just a few discrete possible molecular orientations in a crystal, and if the spectrum of the compound is not too complicated, it is possible to see resonances for all the molecular orientations. The salt calcium formate, structure **15-3**, contains 16 different formate ions in the unit cell of its crystal structure:

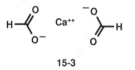

15-3

Remember that the entire crystal may be built up by replicating the unit cell of a crystal structure via translation along all three crystal axes. This is an example of a **space symmetry**,

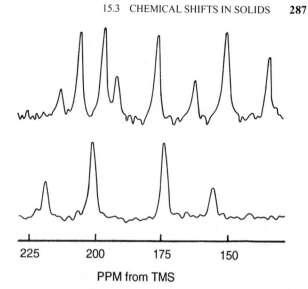

PPM from TMS

Figure 15.4. The ^{13}C spectra of a single crystal of calcium formate in two orientations with respect to B_0. In the top spectrum the crystal is oriented such that all eight distinguishable chemical shift tensors give different chemical shift values. In the lower spectrum the crystal is oriented such that B_0 lies in a symmetry plane of the crystal structure, causing pairwise degeneracies of the chemical shift values.

as opposed to the **point symmetries** discussed in Chapter 4. The calcium formate unit cell contains a center of inversion (a point symmetry) that relates eight of the formate ions to the other eight. Nuclear magnetic resonance spectroscopy cannot

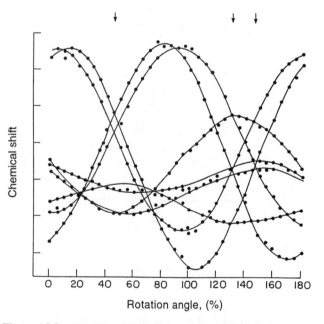

Rotation angle, (%)

Figure 15.5. Rotation pattern of a single crystal of calcium formate, showing the ^{13}C resonance positions as a function of orientation. The three arrows at top indicate orientations at which B_0 lies in a symmetry plane, causing pairwise chemical shift degeneracies as in the lower spectrum of Figure 15.4.

distinguish structures related by an inversion, and so there can in principle be up to eight lines in the ^{13}C spectrum. Figure 15.4 shows two such spectra of a crystal of calcium formate in two arbitrary orientations. In the top spectrum, all eight magnetically distinct formate resonances are resolved. In the bottom spectrum, the crystal is turned such that pairs of ions occupy similar orientations with respect to B_0, and only four (degenerate) lines are resolved. This can occur if B_0 lies within a mirror plane (point) symmetry element in the unit cell. Figure 15.5 plots the resonance positions as spectra are obtained at successive angles of rotation of the crystal about a particular axis. All eight resonances are now apparent. We are used to ascribing a single chemical shift value to a nucleus (or a set of equivalent nuclei) in any given compound. How are we now to describe this complex situation of orientation-dependent chemical shifts? The answer is provided by the mathematics of **tensors**, the details of which are beyond the scope of this book. However, the principles are easily understood if we refer to Figure 15.6. The figure shows a formate

ion in two views, one in the molecular plane and one from above. The ^{13}C **chemical shift tensor** is represented by an ellipsoidal surface. If we were to draw the direction of B_0 passing through the center of the ellipsoid representing the chemical shift tensor, the "radius" from the center to where B_0 intersects the surface is a measure of the chemical shift for that particular orientation of the molecule with respect to B_0. We see that there are three special directions, the **major axes**, or **principal directions**, or **eigenvectors**, of the ellipsoid (and of the chemical shift tensor), with corresponding **principal values**, or **eigenvalues**. The principal values are usually labeled in order of increasing shielding as $\delta_{11} < \delta_{22} < \delta_{33}$. (The double-subscript notation comes from tensor algebra.) We see also that the chemical shift tensor is not aligned exactly along the symmetry axes of the formate ion. With some reflection, this should not be surprising, because we are not observing an isolated gas phase ion but rather one that is packed into a crystal lattice. Crystal field effects "push" the electron density around, distorting the molecular orbitals from the shapes they would have if the ion were free of packing forces.

We can interpret the principal values of the chemical shift tensor in similar ways to the use of liquid-state chemical shifts and make use of the directional information the tensor provides. In the formate ion, the most shielded direction lies approximately perpendicular to the O–C–O plane, and the least shielded direction lies approximately along the σ bond. This turns out to be a general trend for carboxylate ^{13}C tensors.

What is the relationship between the chemical shift tensor and the shifts measured in liquids? The *average*, or *isotropic*, chemical shift of the tensor, that is, the mean chemical shift of a large collection of crystals distributed in all orientations in space, is given simply by the average of the three principal values:

$$\delta_i = \overline{\delta} = \tfrac{1}{3}(\delta_{11} + \delta_{22} + \delta_{33}) \tag{15.2}$$

The isotropic shift is the equivalent to the shift we might observe if the sample were truly liquid. In the case of calcium formate, we could dissolve the crystal in water and measure the ^{13}C spectrum again. Rapid and isotropic molecular tumbling in solution would create an averaging, precisely in the same manner as does chemical exchange (Chapter 10) between sites, yielding the isotropic average of the tensor. In the case of calcium formate, there are two slightly, but measurably, different tensors, corresponding to the two ions in the **asymmetric unit** (the smallest possible part of the structure that can be related to the rest of the crystal by all the symmetry operations of the **space group** of the crystal structure; the asymmetric unit is always part of or equal to the unit cell). There is no requirement that the two ions of the asymmetric unit be superimposable. The principal values for

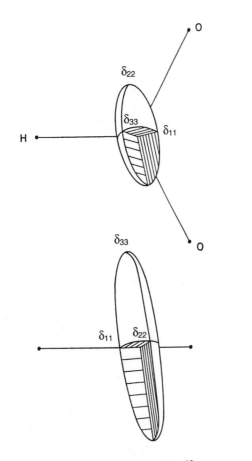

Figure 15.6. Graphical representation of the ^{13}C chemical shift tensor in calcium formate; views from above and from within the OCO plane. (Reprinted with permission from Ackerman JL, Tegenfeldt J, Waugh JS. ^{13}C chemical shielding anisotropies in single crystal calcium formate. J Amer Chem Soc. 1974; 96: 6843–6845. Copyright 1974 American Chemical Society.)

δ_{11}, δ_{22}, and δ_{33} are respectively δ 234, 189, and 104 ppm for tensor 1 and δ 239, 185, and 104 ppm for tensor 2.

■ **EXAMPLE 15.2** Compute the isotropic chemical shifts for solid calcium formate.

□ *Solution:* Use Eq. (15.2).

For tensor 1: $\delta_i = \frac{1}{3}(234 + 189 + 104) = 176$ ppm

For tensor 2: $\delta_i = \frac{1}{3}(239 + 185 + 104) = 176$ ppm

Since the two ions are essentially the same chemically but merely differ slightly in how they pack into the crystal, it is not surprising the average shifts do not differ significantly. For comparison, the chemical shift of a 1 *M* solution of calcium formate is δ 172 ppm. Again, we expect similarity, but not identity, with respect to the isotropic solid-state shift. In the solution case the ion is surrounded by a solvation sphere, whereas in the crystal, the "solvent" is other ions. It is not unreasonable to expect a moderate shift due to that effect. It is also very often the case that principal values are individually more sensitive to intra- and intermolecular effects than are isotropic shifts. □

It is important to take note of the large 135-ppm span in chemical shift that occurs in the spectra of just this single compound. This span is more than half of the typical range of liquid shifts for all carbon compounds! Such large chemical shift anisotropies are the norm in solids!

15.3.3 The Powder Pattern

The most common applications of solid-state NMR do not involve simple single crystals. What does a spectrum of a powder or other disordered solid look like? Let us grind up our crystal of calcium formate and observe the spectrum. (Do not worry, calcium formate crystals are relatively easy to grow from water solution!) If we have done a thorough job of grinding, the powder will contain numerous tiny crystallites, and all possible orientations will be equally represented. Each crystal will exhibit its own spectrum similar to those in Figure 15.4, but with the resonance positions determined by the orientation of the chemical shift tensor with respect to $\mathbf{B}_0$. Since the number of crystals is large, and the resonance lines are not particularly narrow, we will not be able to resolve the spectra of individual crystals. Instead we see a **powder pattern**, a broad line with a characteristic shape (Figure 15.7). Despite our lack of knowledge about the orientation of any one of the crystallites, we can still read from the spectrum the three principal chemical shifts, which correspond to the upper and lower limits and the peak of the powder pattern.

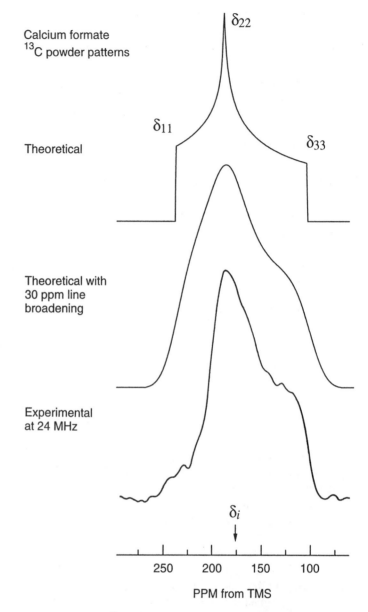

Figure 15.7. The ^{13}C powder pattern spectrum of calcium formate.

Each spectral position in the powder pattern arises from crystals with a particular orientation or family of orientations. The characteristic shape of the pattern arises from the mathematics of tensors and is entirely predictable. Although we need not be concerned with the exact mathematical details, we can use the shapes to learn about the symmetry environment of molecular fragments, simply by inspection of the spectra. Figure 15.8 presents ^{13}C spectra of adamantane and frozen benzene and the ^{31}P spectrum of monetite, CaHPO$_4$, along with the theoretical powder patterns.

Adamantane, as noted previously, has a diamondlike structure but overall is spherical in shape. At room temperature adamantane molecules reorient rapidly and isotropically in

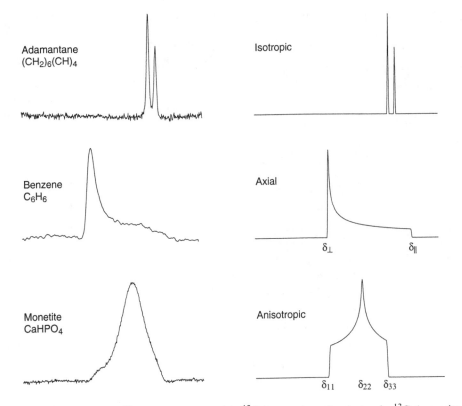

Figure 15.8. Isotropic (^{13}C adamantane), axial (^{13}C benzene), and anisotropic ^{13}C (monetite, CaH^{31}PO$_4$) powder spectra and theoretical powder patterns.

their places in the crystal lattice. The bulk material appears solid by any normal measure, but the NMR spectrum reveals that this molecular motion is able to average the methylene and methine chemical shift tensors completely to their isotropic values. The result is a spectrum with two relatively sharp lines and no powder patterns. Molecular motion has imposed isotropic symmetry on the spectrum. In effect, the motionally averaged chemical shift tensor can be described as having all three principal values (as well as the isotropic value) equal, and the graphical representation of the tensor would be a sphere.

Benzene, in contrast, is a disk-shaped molecule and can reorient rapidly, *but not isotropically*, in the solid between 90 K and its melting point. Because of the sixfold symmetry, all six chemical shift tensors in the isolated molecule should be *congruent* (identical except for the rotations needed to convert one into the other) and probably nearly so in the solid. The rapid molecular rotation partially averages the tensors to axial symmetry. Polymers often show axial symmetry when well above their **glass transition temperature**, where reorientation about the chain axis is rapid. Below the glass transition, the chains are locked into rigid but disordered positions and would generally possess symmetry lower than axial. It turns out that strict threefold C_3 molecular symmetry about a particular direction is sufficient to impose axial symmetry on a chemical shift tensor. This is why most methyl groups (to the

extent that their inherent symmetry is preserved when packed into a crystal) tend to exhibit axial powder patterns. In axial symmetry, two principal values of the chemical shift tensor are **degenerate** (equal), and the graphical representation is a **prolate** (cigar-shaped) or **oblate** (fast rotating-planet-shaped) **spheroid**. These are both examples of **ellipsoids of revolution**. The two distinguishable principal values may be determined by inspection of the spectrum. They are often denoted $\delta_\perp$ and $\delta_\parallel$ because they correspond to $\mathbf{B}_0$ being oriented respectively perpendicular and parallel to the symmetry axis.

■ **EXAMPLE 15.3**   Explain why $\delta_\perp$ and $\delta_\parallel$ are assigned as shown in Figure 15.8.

□ *Solution:*   Imagine $\mathbf{B}_0$ superimposed on the graphical representation of an axial tensor and passing through a small patch of ellipsoid surface. Let us consider how many ways $\mathbf{B}_0$ can be oriented parallel to the axis. In how many ways (by comparison) can $\mathbf{B}_0$ be oriented perpendicular to the axis? We see that there is only a single "small patch" at each of the poles of the axially symmetric ellipsoid, whereas there are many small patches along its equator. We conclude that there are many more relative orientations of the tensor with respect to $\mathbf{B}_0$ in the perpendicular direction, and the spectrum should be correspondingly larger in amplitude at the $\delta_\perp$ position. In all axially symmetric powder patterns, the taller feature is therefore $\delta_\perp$. Whether $\delta_\perp$ is upfield or downfield with respect to $\delta_\parallel$ depends of course upon which orientation is more shielded. □

In analogy to the well-established empirical rules for isotropic chemical shift correlations with molecular structure, there is a series of generalized empirical rules for predicting the relative magnitudes of the principal values of the chemical shift. The least shielded direction generally is found to be perpendicular to the ring for aromatic carbons, parallel to the C_3 axis for methyl carbons, and perpendicular to the sp^2 plane for carbonyls and carboxylates. The intermediate principal direction is typically tangent to the ring for aromatic carbons, perpendicular to the C_3 axis for methyl carbons (and in the mirror plane if the methyl does not rotate), and in the sp^2 plane but perpendicular to the C–C bond for carboxylates. The most shielded direction is typically in the plane of the ring and oriented radially for aromatic carbons, perpendicular to the C_3 axis for methyl carbons (and perpendicular to the mirror plane if the methyl does not rotate), and upright in the sp^2 plane for carbonyls and carboxylates. Methyl groups normally have rather small chemical shift anisotropies ($\delta_{11}-\delta_{33}$) of about 20–50 ppm, methylenes about 30–60 ppm, aromatics about 170–230 ppm, carboxylates about 130–160 ppm, and carbonyls about 150–200 ppm. Carbon monoxide has an anisotropy around 400 ppm.

15.3.4 Eliminating Powder Patterns: Magic Angle Spinning

The information provided by the solid-state spectra we have seen so far can be enormously useful for understanding molecular structure and dynamics in systems that are chemically relatively simple. What if we wish to study a molecule with 10, 20, or 30 different types of carbons? In any but the simplest compounds there will be a sizable number of powder patterns that will mostly overlap, leading to complex, broad, featureless, and largely uninterpretable spectra. What is needed is a means to convert our spectra to something akin to those of fluids, namely spectra containing sharp resonances with one resonance per distinguishable nuclear site. In 1959 two research groups independently came up with the same trick to accomplish this. One group was led by Irving Lowe at the University of Pittsburgh and the other by E. Raymond Andrew at the University of Cambridge in the United Kingdom. The solid sample was placed in a small container, which formed the rotor of a very high speed gas turbine. A jet of compressed gas impinging on flutes in the rotor caused the sample to spin. When the spinning axis was rotated so that it made an angle of 54.74° with $\mathbf{B}_0$, the broad powder patterns "magically" scaled to narrow lines!

Magic angle spinning, or **MAS**, relies on the very fortunate mathematical and physical equivalence between the iso-

tropic averaging of tensor interactions such as the chemical shift performed by Brownian motion in liquids and the averaging performed by mechanical spinning about an axis. Rather than delve into the mathematics, the principles may be understood with reference to Figure 15.9. As chemists, we are quite familiar with atomic orbitals and with the concept that particular orbitals (electron wavefunctions) may be represented as linear combinations of other orbitals. A simple example is the representation of sp^3 hybrid orbitals in terms of linear combinations of the s and p atomic orbitals (the **basis set**). The entire collection of $s, p, d, f, \ldots$ orbitals has particular significance for any function of orientation: They form a complete basis set. This means that *any* function that is solely dependent upon orientation variables (for instance, the angles θ and ϕ of a spherical polar coordinate system) may be expanded in terms of a series of the functions (known as the **spherical harmonics**) that describe the $s, p, d, f, \ldots$ orbitals. This is in rough analogy to the use of the Fourier transformation to represent a function of time as an expansion in terms of the linear harmonic functions (sines and cosines).

Since the anisotropic chemical shift depends on orientation, it is a good candidate for expansion in terms of orbitals, or spherical harmonics. It turns out that this approach yields a particularly simple expansion. We know from our graphical representation of chemical shift tensors that they are invariant under inversion, which eliminates p (and $f, h, \ldots$) orbitals from the expansion. Furthermore, it can be shown that only terms to second order need be retained, which means that we need go no higher than d orbitals. As shown in the top half of Figure 15.9, the orientation dependence of any chemical shift tensor may be expanded in terms of the s and the five d orbitals. Another way of viewing this picture is to say that the angular dependence of the chemical shift tensor is represented as the angular dependence of a function made up of these six orbitals in combination, each multiplied by an appropriate coefficient, or mixing contribution. The s contribution represents the isotropic shift (it does not vary with orientation), while the d contributions express the anisotropic part (with the expected invariance under inversion).

How does magic angle spinning affect this picture? Rapid spinning imposes motional averaging, just as nature might through the random motions of individual molecules. In the MAS case, we as experimenters average *every* tensor in the sample to an axial tensor, *with every tensor axis now being aligned with the spinner axis*. Which orbitals can survive this averaging process? There are now only two: s and d_{z^2}, as indicated in the lower half of the figure. The s orbital still carries the isotropic part, while the entire anisotropy (which is now purely axial) is represented by the d_{z^2} term. At this point we consult our favorite textbook on atomic orbitals to discover that the functional form of the d_{z^2} orbital is given by (lo and behold!)

$$d_{z^2}(\theta) = \tfrac{1}{2}(3\cos^2\theta - 1) \qquad (15.3)$$

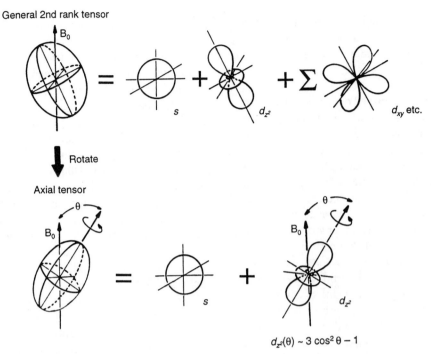

Figure 15.9.   Chemical shift tensor expansion in spherical harmonics (atomic orbitals); top: second-rank tensor with no special symmetry; bottom: axially symmetric tensor.

All tensor interactions in the sample are scaled down by this factor $d_2(\theta)$, where θ is the angle between the spinning axis and the field. For $\theta = 0°$, $d_2(\theta) = 1$; for $\theta = 90°$, $d_2(\theta) = -\frac{1}{2}$; and for $\theta \cong 54.74°$ (our old friend, the magic angle), $d_2(\theta) = 0$. Therefore, if the spinning axis is tilted at $54.74°$ relative to $\mathbf{B}_0$,

all tensor interactions are scaled to zero: The powder patterns narrow to sharp liquidlike lines. A diagram of one type of magic angle spinner appears in Figure 15.10.

Figure 15.11 shows ^{13}C spectra of hexamethylbenzene (structure **15-4**),

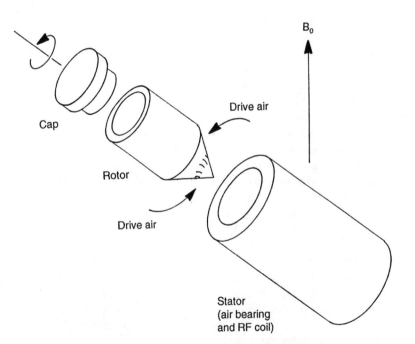

Figure 15.10.   Diagram of a magic angle spinner.

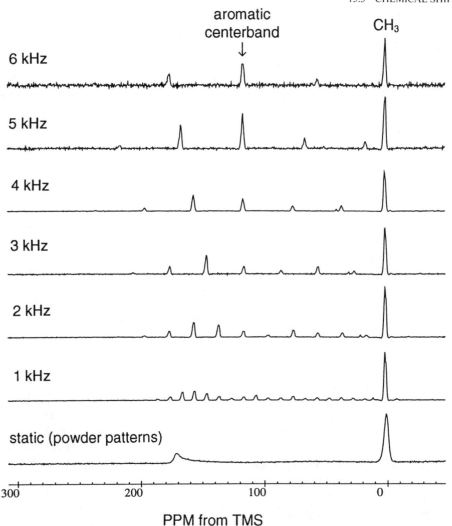

Figure 15.11. Hexamethylbenzene ^{13}C spectra at various spinning speeds. The rotation of the molecule about its sixfold axis imposes axial symmetry on its chemical shift tensors. The double rotation of the methyl group (first about its own C_3 axis and then about the molecular D_6 axis) averages its chemical shift anisotropy to an exceedingly small value.

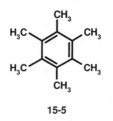

15-5

nonspinning, and spinning at several different speeds at the magic angle. The aromatic and methyl resonances are readily apparent. Magic angle spinning makes a dramatic difference in spectral resolution!

How fast should one spin the sample to achieve line narrowing? Just as with chemical exchange narrowing, the criterion for motion to narrow the spectrum is that the rate of

motion—the inverse of the correlation time in the case of molecular motion, the spinning speed in the case of MAS— must be greater than the width (in the corresponding frequency units) of the spectral feature to be narrowed. That may turn out to be a rather fast spinning speed. On a 200-MHz spectrometer, ^{13}C nuclei resonate at about 50 MHz. An anisotropy of 100 ppm will span 5 kHz of spectral bandwidth. To get a better feel for this number, multiply by 60 s min^{-1} to get rpm: (5000 s^{-1})(60 s min^{-1}) = 300,000 rpm. The acceleration at the inner wall of a 5-mm-inside-diameter rotor at this speed is about *one-quarter of a million* times the force of gravity! Nevertheless, modern MAS probes work well at these speeds and in some cases can go much higher. The sample must be carefully prepared and packed into the rotor so that it is mechanically balanced in order to reach the required speed.

It turns out that the requirement of the spinning rate being greater than the anisotropy can be relaxed to a fair extent, making life on very high field spectrometers and with large anisotropies somewhat easier. If we examine the spectra in Figure 15.11, we find that the effect of MAS is not merely to collapse the powder pattern to a single sharp line but rather to produce a series of **spinning sidebands**. These spinning sidebands are analogous to those produced by spinning a liquid sample in a standard NMR tube when the field is not fully shimmed. There is a centerband at the isotropic shift position that is flanked by spinning sidebands spaced at the spinning frequency. The envelope of the spinning sideband pattern approximately traces out the shape of the nonspinning powder pattern. For a relatively slow spinning rate, there are many sidebands, and as the spinning rate increases, the sideband separation increases, yielding a reduced number of sidebands. Because each sideband is relatively sharp, MAS is still quite useful even if the "fast spinning" criterion cannot be fully met. Sideband patterns from overlapping powder patterns may overlap, but it is often the case that the centerbands (peaks at the isotropic shift positions) can still be resolved and identified to permit analysis of the spectrum. If the centerbands cannot be unambiguously identified by inspection, the spectrum can be acquired again at a slightly different spinning rate. Only the sidebands will move, and the centerbands will remain at the same spectral positions.

We might note in passing that there is a 2D NMR MAS technique known as **variable angle spinning** (**VAS**) in which spectra are obtained as a function of spinning angle. Many other variations on MAS (including spinning at two angles simultaneously: **double rotation**, or **DOR**) are possible.

■ **EXAMPLE 15.4**   A sample of brushite, $CaPO_4 \cdot 2H_2O$, with a ^{31}P anisotropy of 121 ppm yields sidebands of 2 ppm width when spun at 2.5 kHz on a 400-MHz (proton frequency) spectrometer. Estimate the effective increase in the signal-to-noise ratio.

□ *Solution:*   From Table 2.1, we find that the ^{31}P Larmor frequency on this spectrometer is (400 MHz)(17.235)/(42.5759) = 161.92 MHz. The powder anisotropy therefore covers (161.92 MHz)(121 ppm) = 19.6 kHz. At a spinning rate of 2.5 kHz, there will be approximately (19.6)/(2.5) = 7.8 ≈ 8 sidebands within the span of the powder pattern. The total signal intensity (area under the powder pattern) will be distributed among eight sidebands of about 2 ppm width each, or a total of about 16 ppm, when the spinner is running. This reduction in signal bandwidth of 121/16 = 7.6 will result in an increase in the height of the resonance by the same factor of 7.6 (the total area under the resonance is preserved). If no other changes are made, the signal-to-noise ratio will increase by 7.6, which is equivalent to a factor of 58 in signal averaging time.   □

15.4   SPIN–SPIN COUPLING

We discussed direct dipole–dipole coupling in fair detail in Section 15.2. In this section we will explore ways of reducing its effects on solid state spectra, so that chemical shifts may be resolved. Spin–spin couplings fall into two broad classifications: homonuclear (between like spins, for instance proton–proton) and heteronuclear (between unlike spins, for instance proton–carbon). We are already familiar with ways to eliminate the effects of heteronuclear couplings: To obtain a ^{13}C spectrum free of the effects of proton couplings, we may apply strong continuous rf at the proton resonance frequency (proton decoupling; Section 12.2). This is particularly important in solid-state spectroscopy because direct heteronuclear couplings can be very strong, on the order of 50 kHz for methylene groups. This would totally obscure any chemical shift information if proton decoupling were not employed. The couplings would not occur in discrete multiplet patterns as they do in liquids. Rather, the couplings would be manifest as extreme line broadening, since in the solid the couplings are anisotropic (giving powder patterns) and have both inter- and intramolecular contributions. The rf power level required for solid-state proton decoupling is much higher than for liquids because the interactions to be averaged are much stronger (compare 150 Hz J couplings in liquids with 50 kHz through space couplings in solids). Because of this, solid-state proton decoupling is always performed in the gated mode (just during the sampling of the FID; Section 12.4) and is never applied continuously. Continuous application of proton rf power of this magnitude would quickly destroy the sample and the probe. Since motions in solids are highly restricted, nuclear Overhauser enhancements are extremely weak or nonexistent, and so there is no point to keeping the decoupler running continuously anyway. (It is these strong direct couplings, which are made time dependent by molecular tumbling and are averaged out of the liquid spectrum, that provide the relaxation pathway necessary for NOEs to work in liquid spectroscopy.) To differentiate the two types of decoupling, the terms **dipolar decoupling** (for solid-state interactions) and **scalar decoupling** (for liquid-state interactions) are sometimes used. Figure 15.12 shows a gated decoupling pulse sequence.

The familiar liquid-state J couplings also play a role in the solid state. Unlike direct through-space coupling, the electron-mediated J coupling has a nonzero mean that survives motional averaging in the liquid state. It is this isotropic, or scalar, part of the J coupling interaction that gives rise to the familiar multiplet patterns in liquid-state spectroscopy. The J coupling interaction has an anisotropic part that is manifest in solids and cannot in fact be easily distinguished from the through-space coupling, because both follow the same tensorial behavior with respect to rotations. However, the J coupling is normally much smaller than the through-space coupling for the most commonly studied nuclei. For nuclei in

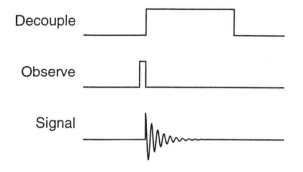

Figure 15.12.   Gated decoupling pulse sequence.

the lower rows of the periodic table, where the very large atomic polarizabilities lead to very large J interactions (1–10 kHz in some cases) and the large atomic radii reduce the through-space interactions, it is possible for the anisotropic J coupling to dominate.

In 1965 a unique experiment was reported by the physicists Lee and Goldberg. They applied continuous rf power to a solid-state proton system and found that the normally extremely short T_2 became longer during the pulse. In fact, the spin–spin relaxation rate followed the relation

$$\frac{1}{T_2} \propto \tfrac{1}{2} (3 \cos^2\theta - 1) \tag{15.4}$$

In this equation, θ is the effective field angle (the angle between the effective magnetic field and the z axis) in the rotating frame (Section 2.4) and is set by adjusting the strength of the rf and its resonance offset. This was the first indication that homonuclear spin–spin couplings could be manipulated with rf pulses in a manner that might be useful for chemical shift spectroscopy in solids. (This report was also the first to use the term "magic angle.") At the magic angle, the solid-state FID, dominated by the direct spin–spin coupling, was lengthened by about a factor of 10. The key to the potential usefulness of the experiment was that resonance offsets, chemical shifts in particular, were scaled by

$$\delta \propto \cos\theta \tag{15.5}$$

At the magic angle, chemical shifts were scaled by $\cos(54.74°) = 0.577$.

In the Lee–Goldberg experiment, spin–spin couplings were scaled nominally to zero, while chemical shifts were reduced only by about half, suggesting that this might be a means to eventually see solid-state chemical shift spectra in the presence of attenuated spin–spin couplings. As it turns out, various higher order effects combined with the difficulty of assuring that the magic angle condition was met throughout the entire sample volume (because of the inhomogeneity of

the field produced by the rf coil) prevented this experiment from allowing practical spectroscopy, but it pointed out the way forward. As opposed to the use of mechanical motion (magic angle spinning) to achieve scaling of an interaction (the anisotropic part of the chemical shift) in **real space**, the Lee–Goldberg experiment is an example of averaging in **spin space**, that is, by manipulating the spins in the rotating frame with rf pulses.

Shortly afterward, in 1968, a now famous experiment by Waugh, Huber, and Haeberlen, the **WAHUHA multiple-pulse sequence**, was published. It built on the Lee–Goldberg concept of scaling the dipolar coupling as close to zero as possible while retaining the chemical shift. Instead of using continuous rf, it employed a rapid but precisely timed sequence of 90° pulses, which provided two critical advantages. First, the FID could be sampled during the pulse sequence, in "windows" between pulses. The use of continuous rf in the Lee–Goldberg experiment meant that only a single point of the FID could be sampled (at the end of the pulse) in any single repetition. Second, small adjustments of the timing, phase, and amplitude of the rf pulses in the WAHUHA sequence could compensate for various experimental errors and inefficiencies, leading to much superior performance. This was the first practical sequence for performing high-resolution NMR spectroscopy in homonuclear dipolar broadened solids.

A full analysis of this pulse sequence requires the use of quantum theory, but the general idea follows easily by extending our now well developed notions of averaging and scaling to the rotating frame. In the Lee–Goldberg experiment the effective rf magnetic field is placed at the magic angle in the rotating frame, which scales both chemical shift and dipolar interactions by the formulas above. Another way of achieving the same effect is to place the effective field successively and rapidly along each of the coordinate axes x, y, and z in the rotating frame. If done properly and with careful attention to the quantum mechanical analysis (beyond our scope here), a much more accurate and robust scaling of the dipolar coupling can be achieved. This brings up yet another interpretation of the magic angle: It is the angle between the edge and the body

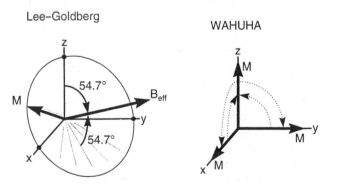

Figure 15.13.   Rotating-frame diagrams for the Lee–Goldberg and WAHUHA pulse sequences.

Digitize one FID point

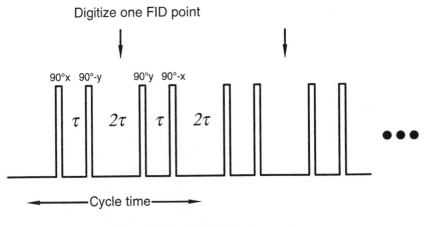

Figure 15.14. WAHUHA pulse sequence.

diagonal of a cube or between the z axis and the 111 direction of the rotating frame. Figure 15.13 shows the rotating-frame diagram for the Lee–Goldberg and WAHUHA pulse sequences. It is clear how WAHUHA, on average, accomplishes the equivalent averaging as Lee–Goldberg. The pulse sequence timing diagram for WAHUHA is shown in Figure 15.14.

A flurry of improved multiple-pulse sequences, compensating for various higher order effects and experimental errors, followed the introduction of WAHUHA. The improved sequences are built up of successively nested cycles and subcycles of rf pulses, in some cases amounting to as many as 128 pulses per sampled FID point, and thousands of rf pulses per FID. The most popular sequence became known as **MREV-8** (Mansfield–Rhim–Ellman–Vaughan eight-pulse cycle). Magic angle spinning was added to narrow the proton powder patterns in a technique that is known as **CRAMPS** (combined rotation and multiple-pulse spectroscopy). This achieves averaging in spin and real space simultaneously. In exceptional cases, dipolar broadening can be narrowed by about four orders of magnitude, down to a few tens of hertz. Most typically, however, no more than about a 1–2 ppm limiting line width can be achieved routinely. Because the isotropic chemical shift range for protons is only about 10 ppm, this achievable spectral resolution may not always be useful. Multiple-pulse spectroscopy remains difficult because of the need to produce long, tediously adjusted, and exceptionally stable trains of very high power rf pulses, and not all spectrometers are properly equipped to perform the measurement. Nevertheless, multiple-pulse spectroscopy can be a powerful tool for studying solid-state chemistry.

A special case occurs if the solid contains nuclear spin pairs that are sufficiently far apart from each other such that intrapair coupling dominates interpair coupling. In this situation, instead of a featureless broad lineshape, we observe a well-defined powder pattern known as a **Pake doublet** (Figure 15.15). Since the dipolar coupling between two nuclei

depends upon the same function of orientation as does the axial chemical shift tensor [compare Eqs. (15.1) and (15.3)], it would not be surprising to observe an axial powder pattern for each component of the dipolar splitting doublet. The Pake doublet is simply two back-to-back axial powder patterns. The separation of the "horns" may be used to determine the internuclear separation directly from Eq. (15.1) evaluated with θ = 90°. (Exercise: Why 90°?)

The *very first* application of NMR to a problem in chemistry was just this measurement. As a physics graduate student working in the Harvard laboratory of Edward Purcell (one of the codiscoverers of NMR who shared the 1952 Nobel Prize in physics with Felix Bloch for this achievement), George Pake measured the proton spectrum of gypsum, $CaSO_4 \cdot 2H_2O$, in 1948. The separation between the peaks in the Pake doublet (he probably was not the one who named it the Pake doublet) gave a value of 1.58 Å for the interproton distance in the waters of crystallization. Chemists immediately began to take notice of this new physical phenomenon called NMR. Hydrogens are notoriously difficult to locate in crystals by means of X-ray diffraction, because their very low electron density yields negligible X-ray scattering. They suspected that NMR might be useful in determining hydrogen positions in com-

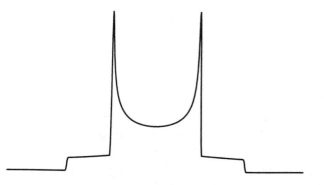

Figure 15.15. Pake doublet quadrupole powder pattern.

pounds, and indeed that has turned out to be the case. Pake later went on to become a vice president of research at Xerox Corporation.

As a final note, it may already have occurred to you that the homonuclear dipolar interaction may be averaged to zero by means of magic angle spinning alone. That is in principle true, except that the typical safe upper limit of the order of 5 kHz spinning speed is usually insufficient to give narrowing in most organic compounds of interest. Unlike the anisotropic chemical shift, "slow" spinning does not produce sharply resolved spinning sidebands in homonuclear dipolar broadened systems. However, there are now advanced "supersonic" MAS probe technologies that enable spinning speeds of around 20 kHz (1.2 million RPM) or higher to be reached routinely. The design of MAS probes revolves around the same high-speed aerodynamic technology used in designing modern aircraft. The extremely high spinning speeds require that the rotor be suspended on a bearing film of compressed gas, usually air; liquid lubrication would introduce unacceptable frictional losses and prevent reaching high speed. The rotational velocity is normally limited to that at which the linear surface speed equals the speed of sound in the bearing gas. This is the same limitation imposed by ideal gas thermodynamics for aircraft reaching the sound barrier: The gas flow no longer conserves entropy, a "shock" front is created, and the drive energy requirements rise drastically.

15.5 QUADRUPOLE COUPLING

An interaction that is never directly seen in liquid spectra but that, if present, always dominates solid-state spectra is the quadrupole interaction. In Figure 15.16 we see the energy level scheme for a spin $I = 1$ nucleus, such as deuterium, ^{2}H. Recall that a system with an angular momentum I has $2I + 1$

levels, and in the absence of other effects these levels will be equally spaced. Nuclei with $I > \frac{1}{2}$ behave as if their electric charge distribution is nonspherical. In effect, we may regard these nuclei as being shaped like oblate or prolate spheroids (again, the respective fast-spinning planet or cigar shapes). Atomic nuclei, as well as the electrons, are subject to enormous electrostatic forces. These forces give atoms and molecules their shapes and keep nuclei in their nominal positions. The nuclear positions are the places where the net electrostatic forces on nuclei are balanced; otherwise any net nonzero forces would displace the nuclei elsewhere. However, in most locations in space occupied by matter there exist **electrostatic field gradients**, which are simply regions of varying electrostatic field. Although a given atomic nucleus may reside at a balance (null) point with respect to the net electric field (or force), that point in most cases will experience an electric field gradient. This simply means that if the nucleus were to be displaced in either direction along the gradient, there would be a restoring force pushing it back.

Nuclei with $I = 0, \frac{1}{2}$ do not care about electric field gradients: Their charge distribution is spherical. But nuclei with higher spin do care, as the figure shows. The prolate nuclear charge distribution is seen immersed in an electrostatic field with a nonzero gradient. For some orientations, the extended reaches of the nuclear charge distribution are slightly closer to the higher potential regions, whereas for other orientations, the nuclear charge distribution is a bit closer to regions of negative potential. Note that the *net* force on this nucleus remains zero, so there is no tendency for the nucleus to be displaced. However, the nucleus is certainly "happier" in certain orientations than in others; in other words, there is an electrostatic energy associated with varying nuclear orientation. As NMR spectroscopists, we are in the business of reorienting nuclei. Therefore, when we deal with quadrupolar

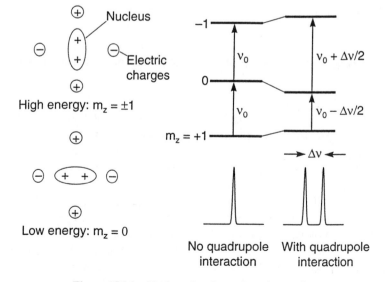

Figure 15.16. Nuclear electric quadrupole coupling.

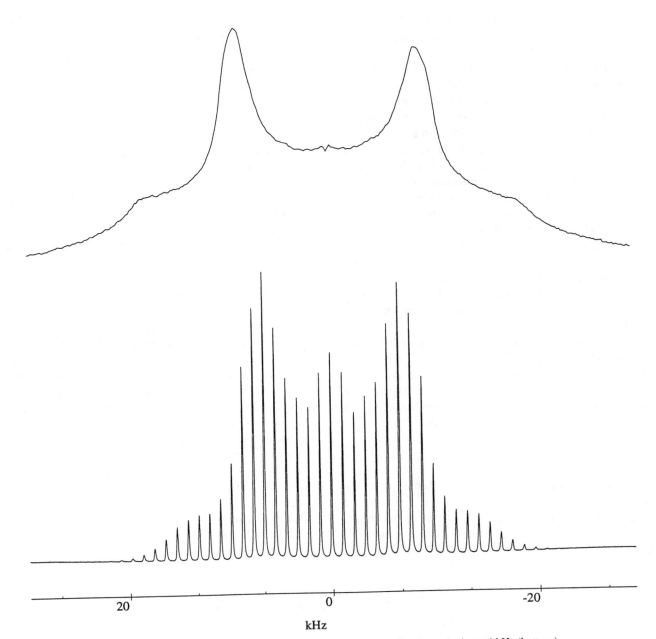

Figure 15.17. Deuterium quadrupole spectra static (top) and magic angle spinning at 1 kHz (bottom).

nuclei (and this includes *most* nuclei in the periodic table, since only a few have $I = \frac{1}{2}$) we can expect there to be electric quadrupole interactions.

Nuclei with $I > \frac{1}{2}$ have an **electric quadrupole moment** q that is a measure of the deviation of the nuclear charge distribution from spherical symmetry. The intensity of the electrostatic field gradient is given the symbol Q, and the **electric quadrupole interaction** is given by

$$\Delta\nu = \frac{3e^2qQ}{4hI(2I-1)}(3\cos^2\theta - 1) \qquad (15.6)$$

In this equation e is the charge on the electron, I is the spin quantum number of the nucleus, and $\Delta\nu$ is the separation between adjacent lines of the $2I$-component multiplet. This spectrum arises because the quadrupolar interaction pushes some energy levels up and some down; the transitions between adjacent energy levels (recall that the only allowed transitions are between adjacent levels) are shifted by amounts proportional to m_z. The center of gravity of the spectrum is not affected to first order. Equation (15.6) applies only to axially symmetric electric field gradient tensors, such as those experienced by ^{2}H in a typical C–D bond. The expression for nonaxial electric field gradient tensors is far more complicated, but the general ideas remain the same.

Figure 15.17 shows ^{2}H spectra of powdered hexamethylbenzene-D_{18}. Since we are dealing with a powder, we expect powder patterns, and the back-to-back superimposition of a pair of asymmetric powder lineshapes is clearly seen. Quadrupolar interactions can get quite large, and in most cases they will dominate the chemical shift spectrum. For deuterium, which has a rather tiny quadrupole moment, the quadrupolar interaction varies from about 25 kHz in hexamethylbenzene-D_{18} to about 200 kHz in carboxylate deuterons. In the heavier nuclei, such as ^{127}I, the quadrupolar interaction can reach 2000 MHz! No one is ever likely to observe a spectrum that wide.

There is a redeeming instance that often makes partial spectra of quadrupolar nuclei observable. For odd-half integral nuclei, such as ^{23}Na ($I = \frac{3}{2}$), the $m_z = \frac{1}{2}$ to $m_z = -\frac{1}{2}$ transition is not affected by the quadrupolar interaction to first order and remains relatively sharp, while the satellite transitions, such as $m_z = \frac{3}{2}$ to $m_z = \frac{1}{2}$ and $m_z = -\frac{1}{2}$ to $m_z = -\frac{3}{2}$, are broadened by the powder distribution. There is a substantial second-order effect of the quadrupole coupling that broadens even the central transition, but this broadening is often only a few kilohertz to a few tens of kilohertz—large but much smaller than Megahertz.

Another special case occurs if the nucleus is at a site of cubic point symmetry (actually tetrahedral symmetry or higher will do). If this condition holds, the electric field gradient must by symmetry disappear, and the quadrupole coupling vanishes. Common examples are tetrahedral ions such as SO_4^{2-}, MoO_4^{2-}, TcO_4^-, and simple salts, such as NaCl, in which the ions reside in sites of cubic symmetry. Unfortunately, this situation almost never occurs for the quadrupolar nuclei, such as ^{17}O, ^{14}N, and ^{35}Cl, that one might wish to study in organic compounds. The only possible exceptions are symmetrically tetrasubstituted ammonium salts.

Of course, in single crystals the spectrum consists of discrete lines whose frequencies move with crystal orientation. In some cases involving large couplings it is possible to track the positions of these discrete lines as the crystals are rotated by retuning the spectrometer and probe at each crystal orientation. In that manner a spectral width of several megahertz may be covered in smaller, more practical pieces of bandwidth. This method has been used to determine the sequence and tertiary structure of crystalline proteins by observing the ^{14}N resonances of the amide backbones.

Magic angle spinning can be used on quadrupole couplings as well as on the other interactions. It works only to first order for the quadrupole coupling, but that is often good enough to achieve satisfactory spectral resolution. The catch with using magic angle spinning has to do with the size of the interaction to be scaled. A 1° error in the angle setting makes the scaling factor about 2.5% [try comparing $\theta = 55.74°$ with $\theta = 54.74°$ in Eq. (15.3) for the scaling factor]. If a 10-kHz powder width

is being studied, this gives a residual line width of 250 Hz. If we are attempting to narrow a 1-MHz powder width, the same angle error gives a line width of 25 kHz, not narrow at all. For good deuteron solid-state spectroscopy (the chemical shifts are the same as for protons) extreme accuracy and stability (within hundredths of a degree) of the angle setting are essential.

15.6   OVERCOMING LONG T_1: CROSS POLARIZATION

Until the early 1970s solid-state NMR was the province of physicists. The phenomenal power of NMR spectroscopy to reveal molecular structure to chemists was primarily dependent upon the high spectral resolution afforded by liquid-state spectroscopy. In 1962 Sven Hartmann and Erwin Hahn (the inventor of the spin echo and the first to perform pulsed NMR routinely) published an experiment in which they transferred the magnetization of one spin system—an **abundant** spin species that was concentrated and gave a strong signal—to another, **rare** spin species, one that would be extremely difficult to observe because of low signal-to-noise ratio. The sample they used was calcium fluoride, CaF_2, sometimes known in NMR circles as the physicists' crystal, since this salt was used in many early solid-state NMR experiments. Fluorine-19 is 100% abundant and gives signals almost as strong as protons. On the other hand, ^{43}Ca is a total loser in terms of its NMR visibility: Larmor frequency only 6.7% that of the proton, 0.143% natural abundance, $I = \frac{7}{2}$ (fortunately its quadrupole moment is unusually small). Flourine-19 is easy to detect with NMR, while ^{43}Ca is extraordinarily difficult.

To understand **cross polarization** and its many variants, we need several new concepts. These concepts include **spin-locking** (one element of this double-resonance pulse sequence), **rotating-frame relaxation times**, **spin temperature** and **Curie's law**, and finally, establishing **contact**, or **mixing**, between spin systems by means of applied rf.

Spinlocking is best understood by reference to the rotating frame (Figure 15.18). Following a 90° pulse, the rf phase is shifted by 90°, bringing the magnetization of the sample into alignment with the applied rf field. Recall that there is an

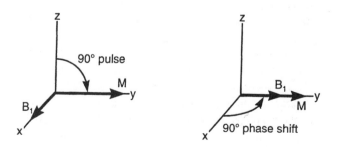

Figure 15.18.   Spinlocking in the rotating frame.

energy E associated with the interaction of a magnetic moment **M** and an applied magnetic field **B** given by

$$E = -\mathbf{M} \cdot \mathbf{B} = -MB \cos \theta \qquad (15.7)$$

In general, the magnetic moment **M** and the magnetic field **B** are **vector** quantities, since they have both magnitude and direction. Equation (15.7) states that the energy is given by the **dot product** of the two vectors, which is equal to the product of the magnitudes times a factor that depends upon how the vectors are aligned. If the two vectors are parallel (and pointed in the same direction), $\cos \theta = 1$, and the energy is at a minimum, the most stable situation. If the two vectors are antiparallel (parallel but pointed in opposite directions), $\cos \theta = -1$, and the energy is at a maximum, the most unstable situation. The energy varies smoothly between these limits as the vectors are rotated relative to each other from one extreme to the other. If the vectors are at right angles, $\cos \theta = 0$, and the energy is zero.

In the laboratory frame of reference, then, any component of magnetization along $\mathbf{B}_0$ is associated with an energy. We also know that this longitudinal component relaxes back to its equilibrium value $\mathbf{M}_0$ with a time constant T_1. Spin–lattice relaxation, which changes the magnitude of the component of magnetization along the field, requires a change in the energy of the spin system; that energy is exchanged with the lattice, and this process can be a bottleneck (it may be slow). On the other hand, any component of magnetization perpendicular to $\mathbf{B}_0$ has no energy of interaction with the field, and we are familiar with the fact that this transverse component relaxes back to its equilibrium value of zero with a time constant T_2. Spin–spin relaxation, which does not change the magnitude of the component of magnetization along the field, requires no exchange of energy with the lattice.

Now we are ready to understand the concept of spinlocking. Equation (15.7) holds in any frame of reference. Let us jump into the rotating frame (hold on tight!) and perform the same analysis. Any component of magnetization aligned with the $\mathbf{B}_1$ field must exchange energy with the lattice, and we might expect this process to be slow: a spin–lattice type of relaxation. Any component of magnetization perpendicular to the $\mathbf{B}_1$ field need not exchange energy with the lattice, and we might expect this spin–spin relaxation process to be faster. In solids, T_1 is usually much longer than T_2. Therefore, a spin-locked magnetization, although it is transverse to the static laboratory field $\mathbf{B}_0$, is "locked;" that is, it is prevented from decaying at the normal T_2 rate because it is forced to exchange energy with the lattice. These rotating-frame relaxation times (more precisely, relaxation in the rotating frame with respect to the $\mathbf{B}_1$ field) are given the symbols $T_{1\rho}$ and $T_{2\rho}$ (ρ for rotating frame). They need not be equal to the familiar laboratory frame time constants, and in fact they *must* be equal to or less than the laboratory frame values.

Next we need an expression for the size of the equilibrium magnetization $\mathbf{M}_0$. This is given by Curie's law:

$$\mathbf{M}_0 = \frac{C\mathbf{B}_0}{T} \qquad (15.8)$$

In Eq. (15.8) C is the Curie constant, which is specific to a given type of spin, and T is the absolute temperature measured in kelvins. This equation holds quite generally for magnetic systems of any kind. It states that the magnetization when fully relaxed (after an infinite number of T_1 intervals; about 3–5 times T_1 works for our purposes) will be proportional to the field and inversely proportional to the absolute temperature. The equation expresses the effect of two counterbalancing forces: that of the external field tending to align atomic (or nuclear) magnets and that of thermal motions tending to randomize their alignment. For a given field, hot spins are weakly magnetized, cold spins strongly magnetized. (Pierre Curie was a physicist who studied magnetism; he later joined his wife Marie in her studies on radioactive elements and the isolation of polonium and radium, for which the two shared the 1903 Noble Prize in physics.) Now we can apply Curie's law to develop the concept of spin temperature. We start with spins equilibrated at a laboratory temperature T_L of about 300 K (27°C) in the laboratory magnet at field $\mathbf{B}_0$ (say 10 T to make the example concrete), with a magnetization $\mathbf{M}_0$. These three quantities are related by Curie's law. Next we quickly jump into the rotating frame and spinlock the magnetization. Before any relaxation can take place (we need not rush anyway, since the magnetization is locked) let us take another look at Curie's law. We have a magnetization $\mathbf{M}_0$ that is aligned along a field $\mathbf{B}_1$. We forgot to bring along a thermometer so we might choose to use Eq. (15.8) to find the temperature knowing the magnetization and the field. If we happen to be using a typical rf field strength of 10 G (0.001 T), which corresponds to a 90° pulse length of 5 μs, we find that the apparent absolute **rotating-frame temperature** T_ρ is based on the proportionality of Eq. (15.8):

$$T_\rho = T_L \frac{\mathbf{B}_1}{\mathbf{B}_0} = (300\ \text{K})\frac{0.001\ \text{T}}{10\ \text{T}} = 0.03\ \text{K} \qquad (15.9)$$

That is quite close to absolute zero, and intensely cold! The interpretation is simple: We have a laboratory-frame-sized magnetization locked along a rotating-frame-sized field: The spins seem to be behaving as if they are in the deep freeze. If left to themselves, the spins will begin to warm up to the lattice temperature. What is the time constant for this warming process? It is none other than $T_{1\rho}$. In essence, the spins can be thought of as a separate **thermal reservoir**, or **spin reservoir**, permeating the material and intermixed with the other components of the lattice. Thermally speaking, the spins are rather well "insulated" (isolated) from their surroundings, the de-

Hartmann-Hahn matching

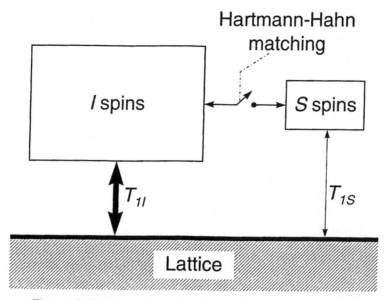

Figure 15.19.   Thermal spin reservoirs involved in cross polarization.

gree of insulation or isolation being expressed by the size of the spin–lattice relaxation time (that is, whichever spin–lattice relaxation time, T_1 or $T_{1\rho}$, is appropriate to the experimental situation).

If these cold spins should come in "contact" with spins at a higher temperature, the cold spins will warm while the hot spins will cool; equivalently, the magnetization of the initially cold spins will shrink and that of the initially hot spins will grow. That is the essence of cross polarization: Magnetization may be essentially transferred from one reservoir of spins to another by means of a thermal contact. The process may be understood in terms of heat transfer between hot and cold thermal reservoirs. These concepts are summarized graphically in Figure 15.19. In Hartmann and Hahn's original experiment, the abundant spins, labeled I in the figure, are ^{19}F (lots of spins, each with large magnetic moments, constituting a thermal reservoir of high **heat capacity**), and the rare spins, labeled S in the figure, are ^{43}Ca (a much smaller number of spins, each of which has a small magnetic moment, constituting a thermal reservoir of low heat capacity). Our only missing element is the "switch" in the figure, the part that breaks or establishes thermal contact between the reservoirs.

How do we connect the reservoirs by means of an NMR spectrometer? This is done through the direct dipole–dipole coupling between the rare and abundant spins. Under normal conditions, these spins have differing Larmor frequencies in any given magnet. Although each type of spin species may cause splittings in the spectrum of the other, there is no way for them to undergo mutual spin flip flops, thereby interchanging (transferring) their magnetizations, because this process would not conserve energy. We require a way to give them the identical Larmor frequencies, which is equivalent to their having identical Zeeman splittings. This is obviously

impossible in the laboratory magnetic field, but it is simple to do in the rotating frame of each spin. Just as the laboratory frame Larmor frequency is given by $\omega_0 = \gamma \mathbf{B}_0$, the rotating-frame Larmor frequency is given by

$$\omega_1 = \gamma \mathbf{B}_1 \qquad (15.10)$$

(One quarter of the period of the rotating-frame Larmor frequency is just the 90° pulse length, right?) If we simultaneously apply rf at both Larmor frequencies, we can control the relative rotating-frame Larmor frequencies merely by adjusting the relative outputs of the two rf amplifiers. To make the two spin systems mutually resonant—that is, have equal Zeeman splittings—we simply adjust the power of each of the two transmitters to match the rotating-frame Larmor frequencies (the **Hartmann–Hahn condition**):

$$\omega_{1I} = \omega_{1S} \text{ or } \gamma_I \mathbf{B}_{1I} = \gamma_S \mathbf{B}_{1S} \qquad (15.11)$$

This famous relation is equivalent to adjusting the rf powers such that the lengths of the 90° pulses of the two spin systems are equal in duration. Now, within their rotating frames, the two types of spins have identical Larmor frequencies and energy splittings and may undergo mutual spin flip flops, thereby exchanging magnetization.

Some students have difficulty envisioning just how this doubly rotating frame (simultaneously rotating at two different Larmor frequencies) works. If this concept bothers you (it apparently does not bother the spins!), just imagine that although the transverse components of magnetization rotate at different frequencies and do not readily have an effect on each other, the longitudinal component of each spin's magnetization is directly felt by the others in its vicinity. If an I spin is driven by its own applied rf to precess about its $\mathbf{B}_1$ field, the longitudinal component will oscillate at a radian frequency of ω_{1I}. An S spin in the vicinity will see this field oscillating longitudinally in *its own* rotating frame, because the z axes of *all* rotating frames always coincide. If the S spin is spin locked, we now have the same situation as plain old nuclear resonance in the laboratory frame: a spin polarized along a static field (except that the static field is $\mathbf{B}_{1S}$ in the S rotating frame) with a transverse oscillating field at the Larmor frequency (which is now ω_{1S}). One spin system "does" NMR on the other; mutual exchange of magnetization (or energy, or heat) can now freely occur. The key element for this to be possible is that there be a dipolar coupling between the spins that can be modulated by the rf. Our switch is complete: It is simply the matching or mismatching of the Hartmann–Hahn condition.

Now we are equipped to analyze Hartmann and Hahn's original experiment. They spin locked the abundant (or I) ^{19}F spins and then irradiated simultaneously at the rare (or S) ^{43}Ca spin Larmor frequency for a short **mixing**, or **contact**, period. They then shut off both rf transmitters and immediately observed the fluorine FID. If the matching condition had not been achieved, there would be no effect on the ^{19}F signal. If, however, matching had been achieved, there would be a cooling of the ^{43}Ca system and a corresponding heating of the ^{19}F system, which would cause loss of the ^{19}F signal when its FID was observed. By searching for the ^{43}Ca Larmor frequency and the correct rf amplitude to achieve Hartmann–Hahn matching, they could map out the otherwise unobservable ^{43}Ca spectrum by **indirect detection** via the ^{19}F FID. They thus achieved the detection of the rare spin spectrum at the sensitivity of the abundant spin system!

This technique did not attract much attention among chemists until about a decade later. The fly in the ointment was that the rare spin spectrum was obtained with the dipolar couplings intact, which means the spectrum was of low resolution: Chemical shifts were not resolvable. However, in 1972 Alex Pines, Michael Gibby, and John Waugh figured out how to modify the Hartmann–Hahn pulse sequence to obtain resolved chemical shifts. Following spinlocking of the abundant spins and the thermal contact, they turned off the rare spin rf while maintaining the abundant spin rf. The rare spin FID was then observed directly. Figure 15.20 diagrams the pulse sequence. In this variation, as the abundant spins warm, the rare spins cool and therefore develop a large magnetization that can be more easily observed than their normal thermal equilibrium magnetization [which would be given as usual by Eq. (15.8)]. During the observation of the rare spin FID the abundant spin rf served to achieve dipolar decoupling. Thus, the extremely large sensitivity enhancement afforded by indirect detection is traded away for a more modest enhancement of the magnetic moment over the thermal equilibrium value, but at least the spectrum is obtained free of

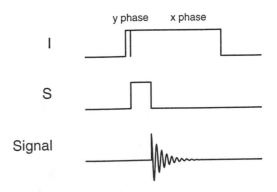

Figure 15.20. Cross-polarization pulse sequence.

abundant spin coupling. The single-contact CP enhancement of the rare spins is given by γ_I/γ_S. The enhancement is actually substantially greater than this analysis would suggest, since it is usually the case that abundant spins such as protons have much shorter T_1's than rare spins. It is not unusual to have in a typical organic solid proton T_1 values of a few seconds and carbon T_1 values of minutes or hours. The ability to recycle at the proton T_1 rather than at the carbon T_1 usually represents a dramatic sensitivity enhancement in itself.

This cross-polarization (**CP**) pulse sequence became the standard for solid-state NMR, because it made ^{13}C solid-state spectroscopy practical for the first time. Shortly thereafter, magic angle spinning was added to make **CP/MAS** a powerful technique for studies of materials in the solid-state. This is by far the most commonly encountered solid-state NMR experiment. Figure 15.21 illustrates some CP/MAS spectra. Many variants of the method have been since developed. For example, the contact time can be varied in successive steps to map out the buildup of rare spin magnetization, yielding the value of the **cross-polarization time constant** T_{CP}. Because cross polarization requires the existence of a dipolar coupling, which we know is strongly distance dependent, the value of T_{CP} can often diagnose which carbons are strongly coupled (that is, are close) to protons and which are not. By inverting the phase of the rare spin rf midway in the contact period, certain rare spin resonances may be "depolarized," or nulled (since the polarization, after proceeding forward for a time, is shifted back the other way), thereby achieving a form of **spectral editing**; this is called **polarization inversion** or **differential cross polarization** (**DCP**). In the **dipolar dephasing** experiment a small gap is inserted into the abundant spin decoupling to let the strongly coupled rare spins dephase, retaining only the less strongly coupled spins in the spectrum. In **reverse cross polarization** (**RCP**) the roles of rare and abundant spins are interchanged to observe the spectrum of only those abundant spins near certain rare spins. Many two-dimensional versions of these variants may be (and have been) constructed.

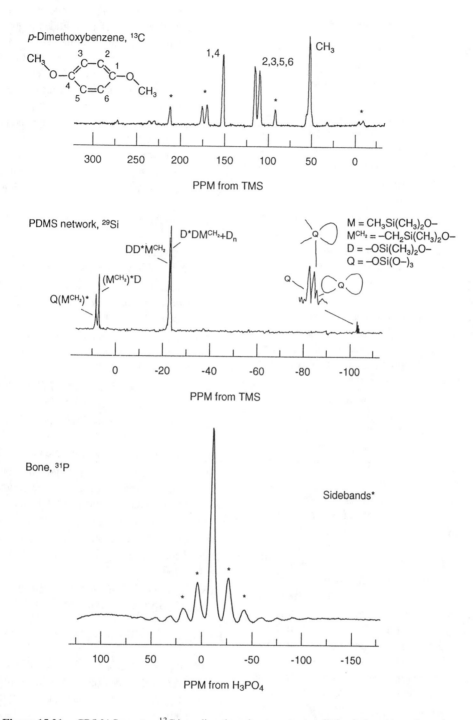

Figure 15.21.   CP/MAS spectra: ^{13}C in p-dimethoxybenzene (note splitting induced in ortho carbons due to the asymmetry of the methoxy group), ^{29}Si in poly-[dimethylsiloxane] (PDMS), and ^{31}P in bone mineral [roughly equivalent to $Ca_5OH(PO_4)_3$]. (^{29}Si spectrum reprinted with permission from Beshah K, Mark JE, Himstedt A, Ackerman JL. Characterization of PDMS model junctions and networks by solution and solid state silicon-29 NMR spectroscopy. J Polymer Sci B: Polymer Phys. 1986; 24: 1207–1225. Copyright 1986 John Wiley & Sons.)

CHAPTER SUMMARY

1. Solid-state NMR spectroscopy is useful for studying materials that cannot be brought into solution or when the properties of interest would be perturbed by dissolving.

2. In the solid state, all interactions of nuclear spins exhibit anisotropic behavior, leading to complex spectra and broad lineshapes, the latter being the source for extremely short T_2 and T_2^*. Because of reduced molecular motion, T_1 typically becomes very long.

3. The most commonly encountered broadening interactions in solids are the chemical shift anisotropy, the direct dipole–dipole interaction, and the quadrupolar interaction. The chemical shift and quadrupolar interactions are inhomogeneous: They involve spins individually, whereas the dipole–dipole interaction involves pairs of spins.

4. Each interaction is mathematically described as a second-rank tensor, which can be graphically represented by an ellipsoid and mathematically expanded as a series of spherical harmonic functions in analogy to the atomic orbitals of the hydrogen atom.

5. Tensors may be isotropic (no orientation dependence, as in a sphere), axial (two principal values equal, as in an oblate or prolate spheroid), or asymmetric (no special symmetry). Axial tensors vary as $\frac{1}{2}(3\cos^2\theta - 1)$, with θ being the angle between the tensor axis and $\mathbf{B}_0$.

6. Magic angle spinning averages all tensors to axial, with their symmetry axes all coincident with the rotor's axis. Then the anisotropic parts of all tensors will scale to zero when the rotor is tilted to the magic angle $\cos^{-1}\sqrt{\frac{1}{3}} \cong 54.74° \cong 54°44'$, and the spectral powder patterns are narrowed to sharp liquidlike lines.

7. Multiple-pulse sequences such as WAHUHA and MREV-8 achieve magic angle averaging in spin space, that is, by rf pulses. Dipole–dipole and quadrupole couplings are scaled to zero, and the isotropic and anisotropic chemical shifts are scaled to some factor, on the order of 0.5, which depends upon the particular pulse sequence. CRAMPS is a combination of a multiple-pulse sequence and magic angle spinning and scales all anisotropies to zero.

8. The central transition, $m_z = \frac{1}{2}$ to $m_z = -\frac{1}{2}$, of odd-half integral spin quadrupolar nuclei is not broadened to first order and may often be used for spectroscopy even when the full quadrupolar powder width is too large to be observable.

9. Cross polarization is an rf-mediated transfer of magnetization from an abundant spin species I, most often protons, to a rare spin species S, for example ^{13}C, ^{31}P, ^{15}N, or ^{29}Si. In CP, both nuclear spin species are spin locked and the rf amplitudes adjusted so that their Larmor precession frequencies in their respective rotating frames are equal: the Hartmann–Hahn condition. The single-contact CP enhancement of the rare spins is given by γ_I/γ_S. Additional enhancement normally occurs because the CP experiment may be recycled at a rate limited by the usually much faster spin–lattice relaxation rate of protons rather than that of the rare spins.

10. Cross polarization is usually combined with magic angle spinning to give the CP/MAS experiment.

ADDITIONAL RESOURCES

1. Abragam, A., *The Principles of Nuclear Magnetism*, Oxford University Press, London, 1970.

2. Haeberlen, U., *High Resolution NMR in Solids, Advances in Magnetic Resonance, Supplement 1*, Academic, New York, 1976.

3. Fukushima, E., and Roeder, S. B., *Experimental Pulse NMR: A Nuts and Bolts Approach*, Addison-Wesley, Reading, PA, 1981.

4. Mehring, M., *Principles of High Resolution NMR in Solids*, 2nd ed., Springer-Verlag, Berlin, 1983.

5. Gerstein, B. C., and Dybowski, C. R., *Transient Techniques in NMR of Solids, An Introduction to Theory and Practice*, Academic, New York, 1985.

6. Axelson, D. E., *Solid State Nuclear Magnetic Resonance of Fossil Fuels: An Experimental Approach*, Multiscience Publications, Canada, 1985.

7. Ernst, R. R., Bodenhausen, G., and Wokaun, A., *Principles of Nuclear Magnetic Resonance in One and Two Dimensions*, Oxford University Press, New York, 1987.

8. Slichter, C. P., *Principles of Magnetic Resonance*, 3rd ed., Springer-Verlag, Berlin, 1990.

REVIEW PROBLEMS (Answers in Appendix 1)

15.1. The α-diketone 1,2-cyclodecadione undegoes a base-catalyzed reaction to provide a high melting solid, $C_{20}H_{32}O_4$, which was incapable of being dissolved in any solvent:

1,2-cyclodecadione

The 25.2-MHz CP/MAS ^{13}C spectrum of this compound exhibited sharp signals at δ 217.6, 83.5, and

52.5 and an envelope of broader lines with peaks discernable at δ 29.1, 26.1, and 21.6. Suggest a structure for this product.

15.2. The cumulatriene **A** exhibits a ^{13}C spectrum with signals at δ 151.4, 116.8, 35.3, 28.2, and 26.6:

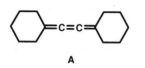

A

When heated in degassed tetrahydrofuran, **A** provides in high yield an insoluble microcrystalline material with *empirical* formula C_7H_{10}. The 68-MHz CP/MAS ^{13}C spectrum of this compound exhibited sharp signals at δ 88.8 and 48.2 and an envelope of broader lines with peaks discernable at δ 33.7 and 25.8. Suggest a structure for this product.

16

NMR IN MEDICINE AND BIOLOGY: NMR IMAGING*

16.1   A WINDOW INTO ANATOMY AND PHYSIOLOGY

One of the most remarkable developments in the history of NMR is its application to the study of living organisms. Certainly, the most visible example of this development is **NMR imaging**, often called **magnetic resonance imaging**, or **MRI**. This part of our field is familiar to the public and is becoming widely used by the medical community, most intensely by **radiologists**, physicians who assist in the diagnosis of disease by the use of radiation that penetrates the body. Three key factors have led to NMR's current status in biomedicine. The first factor is the discovery of methods that permit the creation of **noninvasive** (the biological equivalent of nondestructive) images of the internal **anatomy** (physical structure) of living organisms. The second factor is NMR's exquisite sensitivity to molecular structure and dynamics, which provides a unique window into **physiology**, the study of the functioning of cells, tissues and organs, as opposed to their structure. The third factor is the complete absence in NMR of **ionizing radiation** (ultraviolet, X-ray, and higher energy electromagnetic radiation and subatomic particles), which contributes to its relative safety when used on humans.

It is without doubt that one of the foundations of modern medical diagnosis was the discovery of X-rays by the German physicist Wilhelm Roentgen (pronounced "rentghen"), about a century ago in 1895. So taken with this discovery were the scientific world and public at that time that the first Nobel Prize in physics was awarded to Roentgen. One of the most famous pictures in the history of science is Roentgen's X-ray photograph (a **radiograph**) of his wife's hand, showing the many bones in delicate detail as well as her wedding band. The impact on medicine was immediate and explosive. Within

months of the announcement, the first medical X-rays were taken at Massachusetts General Hospital in Boston. By 1896, physicians around the world were constructing and using their own X-ray devices. A new branch of medicine—radiology—came into being and experienced phenomenal growth. Never before in the thousands of years of the practice of medicine had there been an opportunity to peer into the body without the use of the scalpel.

The tragic consequences of the uncontrolled exposure to ionizing radiation, unfortunately, became clear over the next few decades, with the suffering and death of many scientists, physicians, workers, and patients from radiation sickness. Nevertheless, the overwhelming value of the now numerous and varied techniques of radiologic practice became firmly established. The increasing impact of visual information on the internal workings of the body grows unabated even today, although safety issues are foremost in the modern use of radiation in medicine. Injury resulting from medical imaging procedures is now an exceedingly rare occurrence.

Nuclear magnetic resonance, in the form of MRI, is an important tool for the diagnosis of disease today and has become the method of choice (the "gold standard") in certain diagnoses such as multiple sclerosis and degeneration of intervertebral disks (the pads of cartilage separating the bones of the spine). Although research in NMR spectroscopy of the human body has been conducted for many years and shows great potential for the diagnosis of some conditions, it has not received the same level of acceptance among practicing clinicians as has MRI.

16.2   BIOMEDICAL NMR

16.2.1   The Sample Is Alive!

What is different about biomedical NMR? The title of this section says it all! As chemists, we are used to having rela-

*Contributed by Jerome L. Ackerman

tively "cooperative" samples. We can grind, mix, and dissolve them, put them in 5-mm glass NMR tubes, and spin them; in most cases they are stable. They never breathe, complain, squirm, sneeze, bite, urinate, litigate, or try to crawl out of the spectrometer. In biomedical NMR, we may encounter all of these possibilities. Even the most complex of chemical samples, proteins or fossil fuels for example, pale in comparison to the complexity of even the simplest living creature. The outcome of an individual's life could very well hinge on our analysis of the data. In addition to all of the aspects of magnetic resonance we have learned so far, a whole host of new issues come into play in biomedical NMR.

Let us examine some of these issues in more detail.

16.2.2 Diagnosis of Disease

Disease may be most generally defined as abnormal functioning of the body. The origin of a disease (its **etiology**) may be contagious (caused by microorganisms or viruses), physical (trauma—physical damage—or radiation for example), chemical (toxins), genetic (inherited through our **genome**; most cancers are now thought to involve genetic as well as environmental factors), or psychological (involving brain function). Remarkably, NMR has relevance in *all* of these areas! The physician makes a diagnosis of a disease by considering many separate clues and then drawing a conclusion, often based on incomplete or ambiguous information (much like detectives, managers, and scientists!). One class of information is anatomic: an enlarged heart wall, a broken bone, a blocked artery, a swollen patch of skin. Such features are obviously revealed in images. A second class of information is physiologic: the volume of blood ejected into the aorta in a single heartbeat, the volume flow rate of blood through a given mass of tissue (the tissue **perfusion**), or the presence of excess lactate ion (a product of **anaerobic metabolism**, that which is carried out in the absence of sufficient oxygen) in muscle or brain tissue. These may also be gleaned from the appropriately measured images.

16.2.3 Biochemistry in Living Systems

Analyzing for constituents such as lactate, phosphate, and other metabolites in living tissue sounds a lot like the spectroscopy with which we are so familiar. And this is why NMR has become such a powerful diagnostic tool: Cellular-level physiology is tightly integrated with biochemistry, and NMR is a superb analytical technique for organic chemistry. In particular, NMR, whether in the *imaging mode* or the *spectroscopic mode*, is highly sensitive to in situ (literally, "in the undisturbed position") chemistry. The term in situ normally is used in reference to a nonbiological system, that is, to noninvasive and nondestructive analysis of chemistry within a functioning system. For living systems, the corresponding descriptive term is in vivo, "in the living body." When samples of living systems (*specimens* in medical jargon) are studied in isolation from the original organism, the analysis is said to be ex vivo, "out of the living body," or in vitro, "in glass."

The enormous advantage of noninvasive biochemical analysis cannot be overstated. For example, in many research projects involving animals it is necessary to measure the concentration of some metabolite. The conventional way to do this is the **freeze clamp** method. The animal is sacrificed (killed), and the tissue of interest is rapidly frozen in liquid nitrogen (at 77 K), ground to a powder at that temperature, dissolved, and then analyzed by whatever means is appropriate. If, however, it is possible to detect the compound by its in vivo NMR spectrum, the measurement can be performed noninvasively without harm to the animal. This obviously makes the animal much happier and may be far less complicated and expensive (assuming an NMR spectrometer with large enough magnet is available!). Noninvasive measurement has numerous scientific advantages as well. There is no uncertainty about whether the freeze clamp or subsequent operations have introduced artifacts into the analysis (it is quite a major perturbation on a living system!). Measurements such as intracellular pH are easily accomplished by NMR but are impossible by the freeze clamp method (microelectrodes can be inserted into cells, but that is invasive and certainly a major perturbation). Finally, **longitudinal** studies are possible with NMR, in which an animal or human subject is measured repeatedly over an extended period of time.

One potential drawback to in vivo measurements is the complexity of living organisms. Most chemists' samples are relatively simple, often a single compound dissolved in a solvent. Cells contain untold thousands of constituents: solvents (water, fat), inorganic ions (Na^+, CO_3^{2-}, PO_4^{3-}), organic compounds (too varied and numerous to even begin listing), polymers (proteins, polysaccharides, nucleic acids), and so on. They are highly complex multiphase systems with very specific microstructures. Most of these substances are not detectable by in vivo NMR because their concentrations are too low, their spectra are unresolvable or obscured by other resonances, or their physical state (solidlike, adsorbed, associated, or otherwise immobilized) makes their spectroscopy problematic. However, in a few selected (but exceptionally important) cases in vivo NMR is easy and quite useful.

16.2.4 Maintaining Living Creatures in NMR Magnets

What if you are asked to be the sample (the **subject**) for an NMR experiment? What sorts of comforts would you hope for? The most obvious is that you will need to fit into the magnet. Today, NMR magnets come in room temperature bore sizes (the internal diameter of the dewar) up to a little over 1 m in diameter. If you are going to be subjected to some signal averaging, it might be nice to be able to lie down comfortably. Magnets for in vivo NMR have horizontally aligned bores. You probably would not agree to be spun at 10

Hz for highest resolution, so the spectroscopist will have to be satisfied with at most 10–20 Hz for the water line width. You would probably be willing to lie there for at most 30 minutes (an hour if you are really determined), so overnight signal acquisitions are out of the question. Most NMR equipment for humans is equipped with internal illumination, a fan for ventilation, music, an intercom to speak with the operator, sometimes a periscope to allow seeing outside the magnet, and a comfortable bed that can be slid in and out of the magnet. The rf coil that fits over your head has openings to allow you to breathe and see freely.

Animal subjects in research projects may not be so cooperative. They are normally sedated or anesthetized (put to sleep) for the duration of the measurement. They may need to have tubes or electrical wires attached to monitor or control physiological functions such as breathing, pulse rate, temperature, blood pressure, and so forth. This constitutes the physiological preparation (the *prep*) of the animal. There may be instances in which some of these are required for human subjects as well. For example, infants normally need to be sedated, and an adult's heart rate may need to be monitored so that the data acquisition is synchronized with the beating of the heart (we will discuss **cardiac gating** later). Some people are subject to **claustrophobia**, the fear of being in confined spaces. They may not be able to enter an NMR magnet or to complete an exam. Other individuals may have **ferromagnetic** (strongly magnetic, like iron) implants, such as steel surgical clips. Depending upon how large these objects are (which determines the physical forces on them when passing from a low to a high field at the magnet bore entrance) and how rigidly they are held in the body, it may not be safe to subject such a person to NMR. A small steel pin firmly implanted in a tooth is relatively innocuous; a clip confining an aneurysm (a ballooning) in an artery is life threatening. These are all aspects of **life support** that must be considered for living subjects.

Figure 16.1 is a photograph of a typical 1.5-T clinical NMR magnet suitable for humans, and Figure 16.2 is a photograph of a 4.7-T magnet used for animal research.

Often the question is asked, "Is NMR harmful?" There is a story told by Nobel Prize winner and Professor of Physics Edward Purcell of Harvard that about the time the university's cyclotron was being constructed, he wondered whether a person could sense the tipping and forced precession of proton spins in his or her own brain. The perfect opportunity arose just as the huge cyclotron magnets had been completed but before the vacuum chamber had been inserted into the magnet gap. He and a colleague constructed an rf coil in the gap; one of them placed his head in the coil, while the other connected an oscillator and tuned it to the proton Larmor frequency. Purcell reports that neither physicist could tell when his brain passed in and out of resonance. That undoubtedly marks the first NMR experiment on human subjects, and no untoward

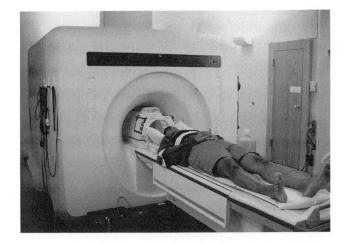

Figure 16.1. A 1.5-T, 100-cm-diameter bore NMR system for human clinical use. The table automatically moves the subject into the magnet.

consequences apparently ensued (the experiment was never published).

As discussed above, NMR uses no ionizing radiation, so that particular element of risk to subjects is eliminated. Just as it is possible to damage a sample (and the probe!) by

Figure 16.2. A 4.7-T, 30-cm-diameter bore NMR system for animal research. (Courtesy of Bruker Instruments.)

applying too much rf power (such as by accidentally specifying, say, a high-power rf pulse of 10 s duration), it is similarly possible to heat and burn tissue. The heating effect is always greater for electrically conductive samples such as ionic solutions (including living tissue) and increases with rf frequency; this is why precautions must always be taken when performing experiments that use proton decoupling on ionic solutions to avoid undue heating. (Microwave ovens work by heating tissues at a frequency of 2.45 GHz.) All clinical NMR equipment has software- and hardware-imposed rf power limits to protect subjects. There seem to be no ill effects on humans caused by magnetic fields up to about 2 T (a bit higher than the most common field used for humans, 1.5 T). Of course, flying steel objects are always a hazard around NMR equipment. Above that strength, it becomes easier to detect physiological effects caused by $\mathbf{B}_0$. Moving an electrical conductor in a magnetic field generates an electric potential (a voltage), which will cause a current to flow in a conductive medium (Section 3.1). If you wag your head in a 4-T NMR magnet (the most intense field available for humans as of this writing), you can sense taste effects on your tongue and see phosphenes (subtle glow effects at the periphery of your visual field) if your eyes are closed. These effects are caused by stimulation of nerves by the electric potentials created by movement in the field. Some individuals experience nausea with frequent and repeated entry into the very highest field magnets; this may be caused by electrical effects on the vestibular system (the semicircular canals of your inner ears, which are involved in sensing balance). There is no evidence of any long-term or harmful effects of high static field. Rapid switching of fields, such as when the magnetic field gradients of NMR imaging systems are pulsed (we will learn about these shortly), can induce nerve stimulation. Clinical NMR systems have controlled switching rates to preclude these effects, although there are no reports of harm to humans caused by field switching (in principle, it might be possible to precipitate heart or brain malfunction, but these have never been reported).

16.3 PICTURES WITH NMR: MAGNETIC RESONANCE IMAGING

16.3.1 The Projection Method

As chemists performing spectroscopy, we are usually looking for the highest spectral resolution possible. Generally, one of the factors that limit the spectral resolution is the uniformity of $\mathbf{B}_0$ (the **homogeneity** of the static field) over the volume of the sample. If the field strength differs from point to point within the sample, there will be a similar point-to-point variation of Larmor frequency, and the width of the resonance line will reflect that distribution in frequency. The shim system (Section 3.2), which is used in a spectrometer to improve $\mathbf{B}_0$ homogeneity, consists of a set of electrical coils within the

magnet bore. Electric currents are passed through the coils, each coil generating a small magnetic field over the sample region. Each coil in the set is wound in a particular geometric configuration to produce a specific variation of its field with position. The currents in individual coils are adjusted so as to compensate as much as possible for the inherent spatial variation of $\mathbf{B}_0$. The labels given to each shim coil (x, y, z, xy, etc.) represent (to an approximation) the spatial variation produced in $\mathbf{B}_0$ by that particular coil. A large enough set of coils constitutes a nearly geometrically complete set that can compensate most any inherent $\mathbf{B}_0$ variation. (Note that the labels of the shim coils correspond to the labels for atomic orbitals, both representing complete sets of expansion terms for functions of orientation.)

It was in 1973 that a physical chemist and NMR spectroscopist then at the State University of New York at Stony Brook, Paul Lauterbur, made an interesting discovery while he shimmed his spectrometer. Most of the shims were reasonably well adjusted, but one of them, perhaps the x shim, was way off. Let us re-create that observation by imagining (not imaging!) the situation depicted in Figure 16.3. The spectrometer contains a 5-mm tube of water, and we are recording the proton spectrum. The axis of the tube is along the y direction of the magnet (it was an iron magnet and cw spectrometer in those days). The x shim, being misset to a very large value, produces a linear **magnetic field gradient** across the sample tube, as shown by the plot of $\mathbf{B}_0$ versus the x position. It is the job of the x coil to produce a field, the z component of which varies linearly with position; the setting of the shim value on the spectrometer console determines the magnitude and sign of the gradient (the slope of the plot). The z component is the only shim field component the spins "see" as long as $\mathbf{B}_0$ is reasonably large.

It is not hard to understand how the spectrum of a tube of water in a field gradient arises. From Figure 16.3 we see that the value of $\mathbf{B}_0$ increases as we move from left to right across the sample, and from the proportionality between Larmor

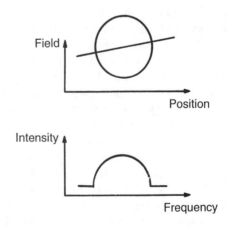

Figure 16.3. Proton spectrum of a cylindrical tube of water in a field gradient.

frequency and field we therefore also expect the spectral frequency to rise linearly (we are ignoring the usual reversal of the chemical shift frequency scale to keep the argument simple). Now let us superimpose on this picture an imaginary movable plane parallel to the tube axis and perpendicular to the paper. We will start with the plane near the left edge of the tube. Water in the plane will have a spectral frequency lower than the central frequency. Because of the circular cross section, there will not be very much water in the plane, and so the height of the resonance will be small. As we shift the plane rightward, toward the axis of the tube, the spectral frequency creeps upward, moving the resonance position of water within the plane to the right. Also, the amount of the water contained in the plane increases to a maximum as the plane passes through the tube axis. As the plane continues to sweep rightward, the spectral frequency continues to increase but the water content falls until the plane passes out of the tube and the spectral intensity drops to zero.

The linear transverse magnetic field gradient produces a spectrum that appears to be a cross-sectional **projection** of the tube along the gradient direction, somewhat like what we would expect from scanning an X-ray beam across the tube (along the gradient direction) and plotting the beam attenuation on the paper. The width of the spectrum or projection is proportional to the intensity of the field gradient, while the overall vertical scaling of the projection is proportional to the concentration of the compound being detected. You may think of the gradient intensity as controlling the magnification.

The relationship between the geometrical projection of an object and its NMR spectrum in the presence of a linear magnetic field gradient holds quite generally for all possible object shapes and for all relative orientations of the object and the gradient direction. Figure 16.4 shows a slightly more complicated phantom/object consisting of two columns of water. Imaging test objects are often called **phantoms**. We can impose on the phantom field gradients in multiple directions within a plane, or in arbitrary directions in three-dimensional space if we wish. In each case we obtain the corresponding projection of the object along that direction. If we can measure a sufficient number of such "views" of the object, we might expect to be able to guess what it actually looks like from only the spectra without ever seeing the object itself. This idea in fact is quite old in medicine and is known as **tomography**. Physicians have long used two or three radiographs (X-ray images) taken in different (usually perpendicular) directions to detect or estimate the shape or size of features not easily discernible in a single view. Using the principles of tomography, automated devices have been constructed to produce images of a desired isolated plane deep within the body while avoiding image "contamination" from the overlying or underlying structures. This is known as **analog tomography**. In the 1960s it was discovered that a sufficiently large number of digitally recorded projections could be used to reconstruct an accurate representation of the

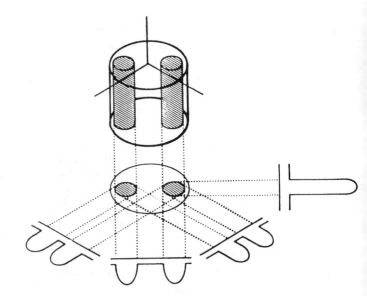

Figure 16.4.   Phantom (test object) containing two columns of water and a set of projections.

object by a computer algorithm, based on mathematical principles discovered early in this century. Over the following decade, **computed tomography** (CT) led to a revolution in diagnostic radiology almost akin to the discovery of X-rays. "CAT scanners" (for computed axial tomography) are now common in larger hospitals.

Lauterbur realized that, as in CT, the multiple views created by field gradients could likewise be reconstructed to form an image of the object in the rf coil. Nuclear magnetic resonance imaging was born! The very first magnetic resonance images were obtained by using standard cw NMR spectrometers and rotating the specimens contained in 5-mm NMR tubes with respect to a magnet-fixed gradient to obtain the multiple views. Soon pulsed methods were employed to obtain the signal-to-noise advantage of Fourier transform NMR, and the gradients were reoriented electronically rather than by moving the specimen. The gradient fields combine like vectors, so that fields from three gradient coils x, y, and z may be combined to produce a gradient field in any desired direction. An example of a simple projection imaging pulse sequence is shown in Figure 16.5.

In order to accept human-sized specimens, NMR instruments underwent substantial development. Magnets and rf coils were engineered to be large enough for human bodies. The earliest whole-body magnets consisted of a set of four or six large coils of water-cooled wire at room temperature, continuously powered. The coils were sized and positioned to fall on the surface of an imaginary sphere, which gives a reasonable degree of $\mathbf{B}_0$ homogeneity (in the early days about 100–200 ppm over a 20-cm-diameter volume was considered good!). The clear bore (the straight-through opening) was about 60 cm, and the typical fields achieved were 0.15 T (6.4

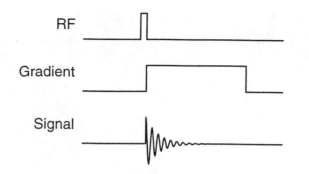

Figure 16.5. Pulse sequence for projection imaging using the free induction decay.

MHz for protons). Later, whole-body superconducting magnets were built. Today, a typical imaging magnet has a field strength of 1.5 T (64 MHz proton frequency; only superconductors are practical at this field), a clear bore of 100 cm, a mass of about 7000 kg, and a complete set of superconducting and room temperature shim coils yielding a homogeneity of 1–2 ppm over a spherical volume of 30 cm. The transmitter power necessary to produce a 100-μs 90° pulse over the diameter of the torso is around 15 kW. The amplifiers that generate the magnetic field gradient pulses are typically capable of about 25 kW for each axis. For comparison, conventional high-field spectrometers have rf transmitters typically capable of 0.1–1 kW and amplifiers that power the shim coils capable of at most 50 W per channel. The magnetic energy stored in a clinical MRI magnet is substantial. The superconducting coil of a clinical magnet may carry a typical current of 500 A and have an inductance on the order of 80 Hy. The magnetostatic energy E_M stored in an inductance L at a current i is given by

$$E_M = \tfrac{1}{2} L i^2$$

$$= \tfrac{1}{2}(80 \text{ Hy})(500 \text{ A})^2$$

$$= 1 \times 10^7 \text{ J} \qquad (16.1)$$

This is the same as the kinetic energy of a 100-ton locomotive moving at 14 m s^{-1} (31 mph). It is also the energy that would be converted to heat if the superconducting magnet were to **quench** (accidentally revert to a normal conductor with finite resistance, a result of the helium bath evaporating to a point too low to keep the coil cold). This does not happen often, but when it does, the remaining helium is instantly vaporized and blows out the top of the magnet in a dramatic display most people would rather not witness close up.

Figure 16.6 shows a cross-sectional view of a human head obtained by a standard MRI method. It is important to always keep in mind that with this image we are not looking at an optical photograph of what a slice of the head might look like but rather a spatially dependent "map" of NMR signal intensities. The true visual image has been "filtered" by the NMR properties of the material at each position represented in the image. Each **voxel** (volume element) in the subject contains a vast array of substances. Each substance yields some resultant signal intensity depending upon the nature of the pulse sequence and the various timing parameters. The composite signal is then represented as brightness in the image **pixel** (picture element, or **pel**) that corresponds to the voxel in the subject. Concentration is only one factor that affects image brightness; relaxation parameters such as T_1 and T_2 also affect brightness, often more so than concentration. In Figure 16.6 the skin surrounding the skull is relatively bright because of its high fat content (the methylene resonances in lipids in vivo tend to have short T_1's), while the **cerebrospinal fluid** surrounding the spinal cord appears relatively dark because of its long T_1. The actual proton concentration throughout most body tissues varies in a range of only a few percent. It is the much larger variation in NMR relaxation parameters that creates most of the **contrast** (light/dark variation) in the image. We will discuss more about image contrast later.

Lauterbur's imaging method, as well as that used CT, is known as **projection reconstruction**. There are several mathematical techniques for accomplishing image reconstruction from a set of projections (for example, back projection, filtered back projection, algebraic reconstruction, and Fourier back projection), but their details are beyond the scope of this book. These different reconstruction techniques vary in their computational efficiencies, their response to noise and errors in the data, and their propensity to create various types of **artifacts** (systematic errors) in the reconstructed image. The relative importance of these factors may change depending on whether the data and image are two or three dimensional.

16.3.2 Resolution and Field of View

Let us look at the effect of a magnetic field gradient in more detail. A gradient G_x along the x direction may be expressed in units of gauss per centimeter (a convenient and popular

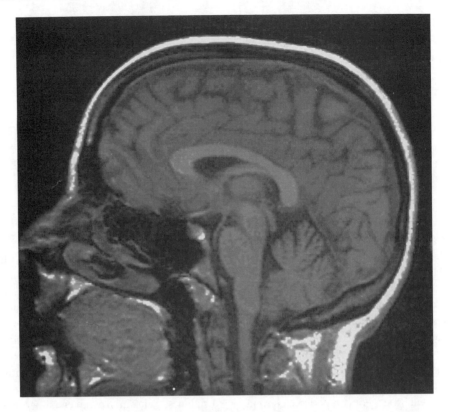

Figure 16.6. Sagittal (side) proton NMR image of a human head. (Courtesy of Terrance A. Campbell.)

unit) or the proper SI unit tesla per meter. A typical value of G_x is 1 G cm^{-1}, or 0.01 T m^{-1}. We may express the Larmor frequency ω (in radians per second) of a spin with magnetogyric ratio γ at position x as the product of the field B at x and the magnetogyric ratio γ, or

$$\omega = \gamma B = \gamma x\, G_x \qquad (16.2)$$

If we have an object that extends for a length Δx (the **field of view**, or **FOV**) along the x axis and the object occupies a spectral bandwidth $\Delta\omega$ (in radians per second) or $\Delta\nu = \Delta\omega/2\pi$ (in hertz), the relation between these quantities is

$$\Delta x = \frac{\Delta\omega}{\gamma G_x} \qquad (16.3)$$

Just as when acquiring a chemical shift spectrum, we would set the acquisition spectral width, or sweep width (SW) (in hertz), to the field of view $\Delta\nu$ (again in hertz). (Actually, we would set the spectral width a bit larger to avoid having the spectrum or image bump into the upfield and downfield Nyquist spectral limits.) The **spatial resolution** obtained is limited by the spectral resolution of the acquisition. If we

acquire N_x *complex* points (N_x point pairs, or $2N_x$ ordinary points) at a rate of SW complex points per second, we will spend a time $t = N_x/\text{SW}$ seconds sampling the FID, giving a spectral resolution $\delta\omega$ (in radians per second) or $\delta\nu = \delta\omega/2\pi$ (in hertz) of

$$\delta\nu = \frac{1}{t} = \frac{\text{SW}}{N_x} = \frac{\Delta\nu}{N_x}$$

$$= \frac{\gamma\, \Delta x\, G_x}{2\pi N_x} = \frac{\gamma\, \delta x\, G_x}{2\pi} \qquad (16.4)$$

The spatial resolution δx is therefore given by

$$\delta x = \frac{\Delta x}{N_x} = \frac{2\pi\, \delta\nu}{\gamma G_x} \qquad (16.5)$$

(To keep our notation consistent, we have used Δ to denote "macroscopic" quantities that relate to the field of view and δ to denote "microscopic" quantities that relate to individual pixels.) We see that the size of the image produced (the magnification) is proportional to the gradient and to the magnetogyric ratio. The resolution obtained (smaller is better) is proportional to the field of view and inversely proportional to the gradient, the magnetogyric ratio, and the number of samples.

It turns out that one of the more important artifacts that affect projection imaging is any effect that produces an offset in the Larmor frequency other than the field gradient. For example, suppose that the rf frequency is set off resonance downfield by 100 Hz. This shifts all the projections upfield by the same amount, leading to an error in the apparent center of rotation and to a reconstruction artifact in the final image. If we image a compound, for example methanol, with a complex chemical shift spectrum (more than one line is complex enough!), each line will have its own resonance position and will contribute a separately distinguishable "ghost" image to the final reconstruction (a **chemical shift artifact**). Finally, the resonance position may drift during the data acquisition, either because we are using an electromagnet with limited field stability or perhaps because the subject is moving (a **motion artifact**). All of these effects contribute to artifacts in the image.

In addition to considering parameters such as the desired spatial resolution, the field of view, the number of samples, and the gradient strength, we must also consider the intrinsic spectral line width of the resonance to be imaged. This line width sets a lower limit to the possible spectral resolution. Therefore, when setting up an imaging measurement, we first estimate the intrinsic spectral width (which is actually the full width of a complex spectrum containing several spectrally resolvable lines, not the width of a single line). We then set the remaining parameters to force that intrinsic spectral width to occupy one or at most a few pixels. We do this by deciding how many spatial resolution elements (pixels) we need to cover the field of view, or alternatively, what the linear extent of a single pixel is to be. Then by means of Eqs. (16.4) and (16.5) we determine the required spectral width and gradient strength.

■ **EXAMPLE 16.1** A subject's head is to be imaged at a field of view of 30 cm. The water resonance has a width of 30 Hz in this magnet. An image matrix of 256×256 pixels is to be obtained. What should be the gradient strength in each axis?

□ *Solution:* The number of samples along the x axis is 256. Inverting Eq. (16.4) gives

$$SW = \Delta v_x = N_x \, \delta v_x$$

$$= (256)(30 \text{ Hz})$$

$$= 7.68 \text{ kHz}$$

Equation (16.5) yields the x spatial resolution and the x gradient strength:

$$\delta x = \frac{\Delta x}{N_x}$$

$$= \frac{30 \text{ cm}}{256}$$

$$= 0.12 \text{ cm}$$

$$G_x = \frac{2\pi \, \delta v_x}{\gamma \delta x}$$

$$= \frac{(6.2832)(30 \text{ Hz})}{(26{,}753 \text{ rad s}^{-1} \text{ G}^{-1})(0.12 \text{ cm})}$$

$$= 0.059 \text{ G cm}^{-1}$$

The same parameters would normally be used for the y axis. □

There is nothing to prevent us from setting the magnification differently along the two coordinate axes, and occasionally there may be good reason for doing so. Imaging the skin at high magnification might be most usefully performed with intense gradients perpendicular to the skin surface to resolve the thin layers while keeping the magnification and gradients in the plane of the skin surface more modest. This ability to choose independent magnification factors for each axis is a unique feature of MRI.

Why not magnify ad infinitum? There are limits to the magnification, the most severe of which is signal-to-noise ratio. As the gradient is increased, the projection occupies a larger receiver bandwidth and is therefore accompanied by additional noise power, which is proportional to bandwidth. Therefore, the signal-to-noise ratio decreases with increasing magnification. The signal averaging time required to achieve a fixed signal-to-noise ratio varies with the square of the magnification for each axis; these factors multiply for each dimension. In three dimensions, the sixth-power relation between linear spatial resolution and the time required to obtain a given signal-to-noise ratio per pixel imposes a severe barrier to arbitrarily increasing the magnification.

16.3.3 Pulsed Gradients

To partially address the problem of resonance offset artifacts, we can first separate the nuclear spin phase evolution induced by the gradient from the evolution induced by all other sources of resonance offset. Nobel Prize winner Richard Ernst introduced a form of imaging that uses pulsed gradients, in analogy

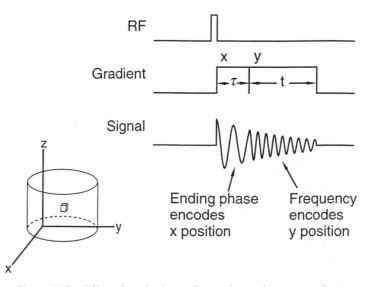

Figure 16.7. Effect of a pulsed y gradient and a continuous x gradient.

with two-dimensional spectroscopy. (Interestingly, Ernst published this early form of 2D spectroscopy before his landmark paper on true 2D spectroscopy.) Examine Figure 16.7. If, following a 90° pulse, we apply a pulse of gradient G_y of duration τ to the spins and then sample the FID during the time interval t (but with no x gradient applied yet), we obtain a spectrum that has accumulated some extra phase according to where the spins were along y. This extra phase ϕ is given by

$$\phi = \gamma y G_y \tau \qquad (16.6)$$

The phase is proportional to the y position, gradient, and duration of the pulse. We could now image by stepping the duration τ (or alternatively the amplitude G_y) in equal increments, acquiring the FID each time, while the x gradient is on continuously during the data acquisition. The pulsed y gradient is known as the **phase-encoding** gradient, while the continuous x gradient is known as the **frequency-encoding** gradient. By analogy to any 2D spectroscopic experiment (τ becomes t_1, t becomes t_2) a 2D Fourier transform of the time-domain data yields a 2D spectrum that represents the 2D image in x and y. The field of view and resolution in the y dimension can be calculated by asserting that two pixels separated by a distance δy (the y resolution) acquire a relative phase difference of $\delta \phi = \pi$ at the maximum amplitude G_y and fixed duration τ:

$$\text{Resolution:} \quad \delta y = \frac{\pi}{\gamma G_y \tau}$$
$$\qquad (16.7)$$
$$\text{Field of view:} \quad \Delta y = N_y \, \delta y = \frac{\pi N_y}{\gamma G_y \tau}$$

Similar to the frequency-encoding case, the spatial resolution and field of view of a phase-encoding gradient pulse are inversely proportional to the magnetogyric ratio and to the gradient strength. They are also inversely proportional to the duration of the pulse. Equations (16.7) are actually identical to Eqs. (16.3) and (16.5) for the respective frequency-encoding field of view and resolution. This can be seen if we identify the duration of the pulse τ with the inverse of the spectral resolution $1/\tau = \delta\nu$ and note that the phase-encoding gradient would be stepped over a range of $\pm G_y$ (i.e., the phase-encoding gradient *range* is twice G_y).

16.3.4 Spin Warp Imaging

This two-dimensional scheme does not yet completely solve the problem of resonance offsets. We can do this partially by using a spin echo during the frequency-encoding period. Recall that at the peak of a spin echo all resonance offsets are refocused. The **spin warp**, or **2D Fourier transform (2DFT)**, imaging pulse sequence is pictured in Figure 16.8 (unfortunately the rather general term 2DFT got co-opted as the name of the most common form of imaging pulse sequence). It looks complicated, but we can analyze it with the tools we now have under our belts.

The rf part of the pulse sequence generates a Hahn spin echo at time 2τ. In imaging jargon the time from the center of the 90° pulse to the center of the echo is called the **echo time** (or time to echo) **TE** $= 2\tau$ (where τ is the spectroscopist's usual symbol for the time from the 90° pulse to the 180° pulse). The time from the center of the 90° pulse to the center of the next 90° pulse is called the **repetition time** (or time to repeat) **TR**. Spectroscopists know TR as the time equal to the recycle delay plus the time taken by the pulsing and data sampling.

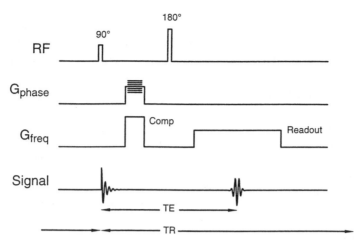

Figure 16.8. Spin warp or 2DFT pulse sequence.

At the center of the echo all resonance offsets from interactions linear in the spin quantum number are canceled *as long as these interactions operate for the full duration of TE.* Linear spin interactions include chemical shifts, heteronuclear dipolar couplings, field inhomogeneity, field gradients, and transmitter frequency offsets but do not include quadrupolar and homonuclear dipolar couplings. There will however be a net phase evolution induced by an interaction to the extent its duration or intensity is not balanced with respect to the two halves of TE (that is, the balance with respect to amount of phase evolution on either side of the 180° pulse).

If we examine the TE interval closely, we see that all interactions operate in a balanced manner on either side of the 180° pulse except for the field gradients. The phase-encoding gradient operates only during one-half of TE, so we can expect the echo of the signal from every voxel to acquire a phase evolution ("phase warp"—hence the name "spin warp") that depends on the voxel's y position. The echo is "read out" in the presence of a frequency-encoding x gradient, so the spectrum obtained by Fourier transform of the echo will be a projection along x (phased, of course, by the y-gradient evolution). For this reason the frequency-encoding gradient is also known as the **readout gradient**. To ensure that the net readout-gradient-induced phase evolution is exactly canceled at the Hahn rf spin echo position, an x-gradient pulse of duration equal to half of the readout is placed in the first half of the TE interval. This pulse is known as the **compensation gradient** ("**comp**" gradient). The comp gradient may overlap in time with the phase-encoding gradient; their effects are additive.

The samples of the echo readout correspond to the t_2 interval, and the phase encode steps to the t_1 interval (not a true time interval of course), of a 2D spectroscopic acquisition. A 2D Fourier transform yields an image in which the two frequency axes correspond to the x and y spatial axes. This method is a distinct improvement over the projection method

with respect to frequency offset artifacts. There is no spin evolution other than that caused by the y position in the phase-encoded domain because it is not a real-time domain, so there are no frequency shift artifacts at all. In the frequency-encoded domain, the spin echo refocuses most frequency shifts, making their effects small.

16.3.5   Slice Selection

The next refinement to the spin warp sequence is the introduction of **frequency-selective** rf pulses. Normally, we use pulses of sufficient intensity to "cover" the entire spectral width. By using a weaker rf pulse, we cover a smaller spectral bandwidth. If a field gradient is present during the weak, limited bandwidth rf pulse, that bandwidth is translated into the excitation of a band of spins along the gradient (a "**slice**"). The shape of the pulse (the amplitude as a function of time) determines the shape (**slice profile** or cross section) of the slice (the pulse shape and slice profile are approximately related by the Fourier transform). We may therefore incorporate into the pulse sequence shaped 90° and 180° rf pulses, during which gradient pulses are applied to excite only those spins within selected slices of the subject. This is known as **slice selection**. Slice selection avoids having to resolve the entire subject in three dimensions, which is generally an extremely time, memory, and computationally intensive operation. Actually, the acquisition of multiple slices may be interleaved, since an acquisition of one slice leaves the others untouched. Therefore, most applications of spin warp imaging operate in the **multislice** mode. After the first phase-encode step of the first slice is performed, the first phase-encode step of the next slice is acquired during the TR of the first slice acquisition, and so forth. The number of slices that may be acquired is largely determined by the number of copies of the sequence in Figure 16.8 that fit into a single TR interval. Typical acquisitions may use 1–30 slices. The slice thickness is determined by the amplitude of the slice selection gradient pulse (usually the z gradient) and the duration, amplitude, and shape of the rf pulse. Typical values are 0.5 cm thickness, 0.1 G cm^{-1} slice selection gradient, and 4 ms 90° "sinc"-shaped pulse [$\sin(x)/x$ shape, which Fourier transforms to a rectangular slice profile]. The precise relationship between these variables depends upon the details of the specific rf pulse shape function. The selective 90° pulse causes a dispersion in the phase of the spins that is refocused by a reversal of the slice selection gradient for a short period equal to half the selective 90° pulse time; the selective 180° pulse is self-refocusing.

16.3.6   Putting it All Together

The complete spin warp pulse sequence is shown in Figure 16.9. A series of slices from a multislice sequence is shown in Figure 16.10. In setting up a sequence, we would first decide on the parameters such as field of view, resolution, slice thickness, slice spacing, and number of slices. We would then set the selective pulse by finding the rf *amplitude* (not the duration as in conventional spectroscopy) that nulls the signal in the absence of the slice selection gradient. This is the 180° pulse amplitude. The 90° pulse is half this amplitude. The duration of the 90° pulse determines its frequency selectivity. The amplitude of the slice selection gradient is next adjusted to achieve the desired slice thickness. If the spin echo is read out in the presence of the slice selection gradient, rather than the readout gradient, we obtain a profile of the slice itself (by Fourier transforming the echo) for setup purposes. Next we adjust the slice selection compensation gradient to give maximum echo signal. Then, with the readout and compensation frequency-encoding gradients in place, we adjust their amplitudes together to obtain the desired frequency-encoding

field of view. At this stage, the projection of the entire object in the frequency-encoding direction is obtained by Fourier transformation of the echo. Next the compensation gradient is adjusted to obtain the maximum signal. Finally, the phase-encoding gradient is added, and a short image acquisition is performed with minimal signal averaging (a **scout** image, often used for a quick look at the general features in the subject). Small adjustments of the frequency- and phase-encoding gradient amplitudes may be required to correct the x and y magnification so that objects in the image have the proper shape.

Practical sequences may have other features built in to compensate for various errors in pulse and time interval settings. Especially on clinical imagers, most of the adjustments, including those described above (setting the pulses, adjusting the receiver gain, etc.) would automatically be performed by the imager for speed, accuracy, and safety. Because of the numerous parameters that need to be calculated, adjusted, and checked (and the many ways to go wrong), the source codes that program these pulse sequences are very complex and have numerous internal consistency checks. They compute for the operator various parameters such as gradient strengths, maximum number of slices, and so on, and check for errors. They often allow parameters such as field of view, image center, and image orientation (allowing for arbitrary rotation of the coordinate system) to be graphically prescribed with respect to a scout image. Clinical imagers, as part of the pulse sequence code, automatically determine the 90° pulse length, water center frequency, and other calibrations needed for proper operation and then set the imaging parameters accordingly. The code for a multislice **multiecho** sequence (a separate image may be produced for each echo of a Carr Purcell echo train) may run to many tens of pages.

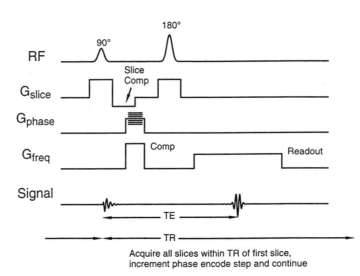

Figure 16.9.   Spin warp or 2DFT pulse sequence with slice selection.

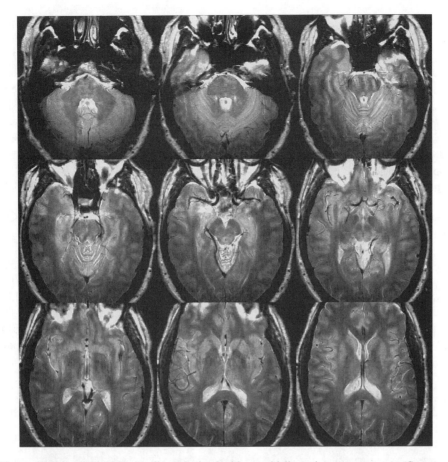

Figure 16.10. Multiple image slices obtained with a multislice spin warp sequence. (Courtesy of Terrance A. Campbell.)

16.4 IMAGE CONTRAST

We have already encountered the notion of image contrast. The image brightness is a direct measure of NMR signal intensity obtained under the particular pulse sequence used in the image acquisition. As spectroscopists, we are accustomed to analyzing resonances on the basis of spectral frequency, amplitude, lineshape, and the presence of familiar spectral patterns (for example, multiplets). These are all forms of contrast, in the sense that these features differentiate one resonance from another. It is normally impractical to perform such analyses on image data because there are so many data elements (pixels), and the data may not have been recorded in a manner appropriate to permit this type of analysis. Often, we know that some NMR parameter, such as the water spin lattice relaxation time, will differentiate one region from another in an informative manner given the problem at hand. For example, we may be searching for degenerating cartilage between spinal vertebrae and know that degenerated intervertebral disks have a higher content of water with longer T_1 than do healthy disks. We may not even need to measure the value

of T_1, but merely need to visualize regions that differ by certain T_1 values. In this case, we may adjust the parameters of the pulse sequence to weight the signal intensities by the predominant T_1 values within each of the voxels in the subject. The brightness of image pixels will then reflect to a moderate extent the T_1 differences. Other factors will influence pixel brightness as before, but the importance of T_1 differences will be enhanced. If, for example, we wish to discriminate between regions of 100-ms T_1 material from regions of 1-s T_1 material, an easy way to obtain a T_1–**weighted** image is to set the TR to some intermediate value such as 300 ms. The TE would be set short to prevent attenuating any signals with short T_2. Now pixels representing short T_1 material will appear with near full brightness, while those representing long T_1 material will be largely saturated and will be relatively dark. Watery body fluids such as blood and cerebrospinal fluid (**CSF**, the fluid that bathes the brain and spinal cord) tend to have T_1's around 1 s, while water in cells tends to exhibit T_1's about an order of magnitude shorter.

Alternatively, we may obtain a T_2–**weighted** image by setting TE to a value in the middle of the range we are trying to discriminate. The TR would be set long to prevent saturating any signals with long T_1. Watery bodily fluids have T_2's of a few hundred milliseconds, while tissue water has T_2 values in the range 10–100 ms. The actual values of these relaxation times vary widely and are dependent on the composition of the tissue in a highly complex and not completely understood manner. For example, the $1/T_1$ and $1/T_2$ of the water in blood are both approximately linearly proportional to the **hematocrit** (red blood cell content). This is due to a number of factors, among which are the presence of large amounts of protein in **plasma**, the watery fraction of blood outside the cells. Water molecules in the bulk experience chemical exchange with protein-bound water, which has different relaxation times and pathways. Bulk plasma water also exchanges across red cell membranes. Inside the red cell, water can approach **hemoglobin** molecules (the protein that transports oxygen and carbon dioxide), which in the deoxy state contain a strongly paramagnetic Fe(II) center. In this state, water relaxation is highly efficient. Analogous situations hold in all tissues. (We promised complexity!) Many physical theories of tissue water relaxation have been proposed and refined over the years.

Although we now have a substantial level of understanding, it is far from complete.

Figure 16.11 shows an examples of T_1 and T_2 contrast, in which a spine image is obtained with (on the left) short TR and short TE (T_1 weighting) and (on the right) long TR and long TE (T_2 weighting). On the left, the short T_2 signals from marrow in the vertebrae, disks, and spinal cord come through brightly, while the CSF (surrounding the cord), with its long T_1, is attenuated. The anatomy of the spine is revealed, especially the herniated disk compressing the cord (ouch!). On the right, the situation is reversed, and the CSF, which has long T_2, is imaged brightly.

If we acquire an image with long TR and short TE, there is little weighting on the basis of T_1 and T_2; the resulting image reflects concentration of protons and is called a **proton density** image. If we acquire a series of images in which TR or TE is systematically varied, we can perform a least-squares fit of the pixel intensities just as we would in simple spectroscopy to obtain the corresponding T_1 or T_2 value in each pixel. This gives a computed image of the corresponding parameter. This method is applicable to any NMR parameter that can be used in weighting magnetic resonance images.

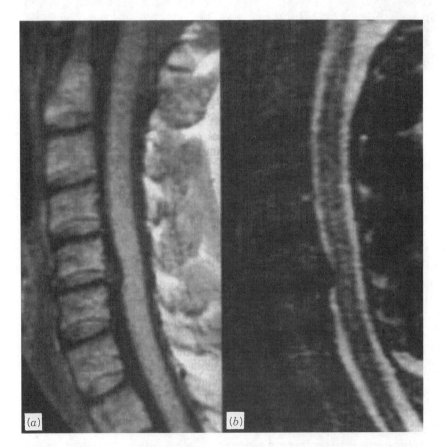

Figure 16.11. Successive proton images of the spine of the same subject showing two different types of contrast. (Reprinted with permission from Wehrli FW, Shaw D, Kneeland JB, eds. *Biomedical Magnetic Resonance Imaging. Principles, Methodology and Applications.* VCH Publishers, Inc., Weinheim, Germany, 1988. Copyright 1988 VCH Publishers, Inc.)

We do not need to depend on the inherent relaxation differences in tissues to develop image contrast. We may also introduce an **exogenous** (from outside the body) **contrast agent** that produces some observable effect on an image. For example, we may inject a small amount of a strongly paramagnetic compound into the blood. Gadolinium diethylene-triaminepentaacetic acid (Gd-DTPA), structure **16-1**, has high T_1 **relaxivity**:

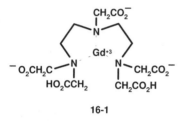

16-1

The relaxivity is a coefficient, expressed in units of $s^{-1}\,mM^{-1}$, which describes the contribution of a species to the water relaxation rate. The additional contribution to $1/T_1$ is given by the product of the relaxation agent's concentration and the spin–lattice relaxivity. A similar coefficient describes the spin–spin relaxivity. Most relaxation agents contain a high (electron) spin transition or lanthanide metal ion strongly chelated by some ligand or ligands to reduce toxicity and tailor the agent's **biodistribution** (distribution among various organs in the body) properties. The Gd^{3+} aquo ion provides the strongest water proton relaxivity of any element, and most of its relaxivity is preserved in the DTPA complex. This **magnetopharmaceutical** is used in humans at doses around $0.1\ mmol\ kg^{-1}$ of body mass.

Figure 16.12 shows T_1-weighted **transverse** (plane perpendicular to the body axis) head images pre- (left) and post- (right) contrast-enhanced. A tumor is detectable in both images, but the contrast-enhanced image outlines the tumor margins clearly for the surgeon and delineates what is probably a **necrotic** center (dead tissue, common in larger tumors) which has no circulation and does not take up the contrast agent. The dark nonenhancing region on the right side of the brain below the tumor is **edema** (swelling with excess water) and appears with reduced signal intensity because its T_1 is long due to the higher water content. The dark oblong region on the left side of the brain is the patient's right **ventricle**, a space in the brain filled with CSF. It is compressed because of the pressure exerted by the growing tumor but is otherwise normal. (In the conventional radiologic presentation the subject's right side appears on the left, as if the subject is facing you.)

Another form of contrast is based on physical displacement of spins. The spin echo works perfectly only as long as the interactions that cause dephasing of the FID (chemical shifts, for example) operate in both halves of the TE interval to the same degree (again, they must be balanced with respect to the 180° pulse). Imagine a spin in a particular position x_1 along the x gradient when the spin warp sequence is applied. If the spin is stationary, its Larmor frequency, determined by its position and the gradient strength, is fixed throughout the entire TE interval. Any dephasing is exactly refocused at the echo peak. Now suppose the spin is displaced to some new position x_2 (with a new Larmor frequency) at some time during TE. The phase evolution during the first half of TE is no longer exactly compensated by the phase "rewinding" during the second half of TE. The spin echo no longer works

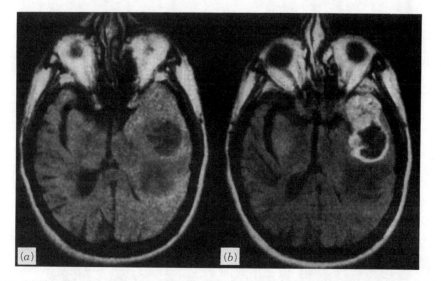

Figure 16.12. The T_1-weighted images of a brain tumor patient before (left) and after (right) administration of a paramagnetic contrast agent. (Reprinted with permission from Wehrli FW, Shaw D, Kneeland JB, eds. *Biomedical Magnetic Resonance Imaging. Principles, Methodology and Applications.* VCH Publishers, Inc., Weinheim, Germany, 1988. Copyright 1988 VCH Publishers, Inc.)

perfectly, and there is some net dephasing and signal loss at the echo peak. The displacement may have been caused by molecular diffusion or by fluid flow. Therefore, motion may be a contrast mechanism.

There are many ways to develop motion contrast. Images may be sensitized (or desensitized) to random molecular diffusion; to steady, turbulent, or pulsatile flow; or to the elastic vibrations induced by the application of acoustic energy (ultrasound). The procedure may make use of the value of the net phase induced by the displacement, or the extent of signal attenuation at the echo peak, or merely the fact that some dephasing has taken place. Precisely engineered sequences of gradient pulses and particular timings of rf and gradient pulses are used to achieve this magic. The simplest way to do it is to introduce a matched pair of intense gradient pulses centered about the 180° pulse. These **motion-sensitizing** gradient pulses may be chosen along any direction, irre-

spective of the image orientation, to sensitize the contrast to displacement along that direction. You may realize at this point that there are always frequency-encoding gradients surrounding the 180° pulse in any spin warp sequence, so that this sequence always has some motion sensitivity.

In Figure 16.13 we see a magnetic resonance **angiogram** (picture of the vascular system) that highlights the moving blood in the arteries of the brain. This image was obtained by subtracting two images that were sensitized to the excess phase induced by flowing blood. One image was obtained during **systole**, when the powerful *left ventricle* of the heart has just contracted, and there is a surge of blood flow. The other image was taken during **diastole**, when the left ventricle is relaxed and the flow rate is lower. The subtraction cancels out all stationary spins, leaving only those that experience a significantly different phase evolution between the two states. The images were not actually obtained in their entirety at

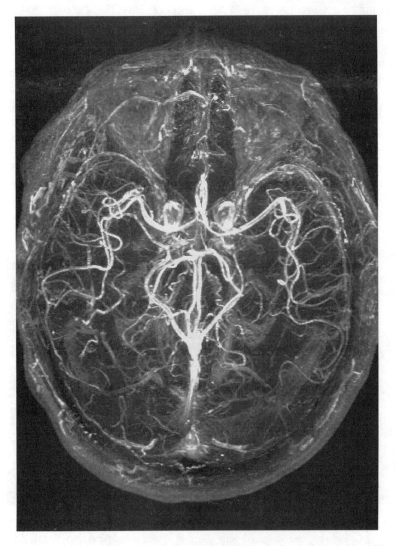

Figure 16.13.   Magnetic resonance angiogram highlighting flowing blood. (Courtesy of Terrance A. Campbell.)

these particular instances in time, since a single heart beat cycle occurs in only a fraction of a second. Rather, the images were acquired stroboscopically in synchrony with the heart beat. An **electrocardiograph** monitors the electrical activity of the heart by detecting small voltages on the chest by means of electrodes. This provides an electrical signal that can be used to pace the acquisition of each phase-encoding step. The pulse program timing is adjusted to acquire the spin echo at a precise point in the cardiac cycle, either systole or diastole, and the image is obtained in 128 or 256 heart beats (whatever the number of phase-encoded steps happens to be). This is known as **cardiac gating**.

Gating is essential when imaging the heart itself, because the heart's motion produces severe motion artifacts in images. The more or less random phase shift introduced into the spin echo at each phase-encoding step as the data acquisition comes into and out of synchronization with the heart motion causes some of the signal intensity to be "splattered" across the final image, along the phase-encoding direction. By synchronizing the acquisition with the heart cycle, this motion artifact is much reduced. Figure 16.14 demonstrates the image improvement with cardiac gating.

A clinically very important relaxation mechanism related to both relaxation agents and motion is due to **magnetic susceptibility** effects. Magnetic susceptibility is the shielding a material exerts in the presence of a magnetic field. It is measured in ppm and is essentially the same physical phenomenon as chemical shielding, except that magnetic susceptibility represents the bulk shielding effect on macroscopic objects, rather than the shielding effect of electronic distributions in individual atoms or molecules on single nuclei. Most materials are slightly **diamagnetic** (net shielding, equivalent to a negative value of magnetic susceptibility) by a few ppm,

while some materials (notably some metals) are slightly **paramagnetic** (net deshielding, positive susceptibility) by a few ppm. Some compounds with unpaired electrons are strongly paramagnetic. The O_2 molecule, possessing a ground-state triplet electronic configuration, is a highly effective relaxer in solution at normal atmospheric partial pressure; bulk liquid oxygen adheres to a magnet because of its strong paramagnetism. Finally, a very few substances are ferromagnetic (extreme positive values of magnetic susceptibility) and may possess an intrinsic magnetization even in the absence of an external field. Familiar examples of ferromagnetics are iron, nickel, and cobalt. Ceramics known as **ferrites**, having the generic formula $MOM_2'O_3$, where M and M′ are metal ions, are sometimes ferromagnetic. The most familiar example is the mineral known originally as ferrite, Fe_3O_4, or lodestone, which is the material most commonly used as a magnetic recording medium.

The magnetic susceptibility of plasma (the watery component of the blood) is very close to that of the surrounding tissue. When the susceptibility of the blood is altered, for example by injecting a contrast agent possessing high susceptibility (Gd-DTPA is a good susceptibility agent as well as a spin–lattice agent), there is now a small shift in the value of B_0 between blood and tissue, typically on the order of 1 ppm. That tiny shift, however, represents a huge magnetic field *gradient* over the dimensions of small structures, such as the thickness of a capillary vessel wall. When water molecules diffuse (or flow) in and out of such gradients, substantial dephasing occurs and spin echo signal intensity is lost.

■ **EXAMPLE 16.2** What is the magnetic field gradient created by a 1-ppm magnetic susceptibility difference across a capillary wall of thickness 1 μm in a 1.5-T imager?

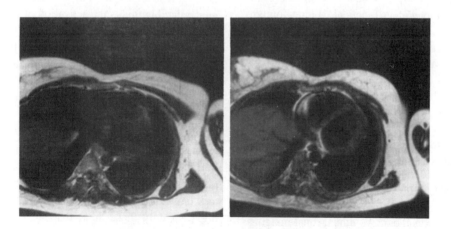

Figure 16.14. Images of the chest obtained without (left) and with (right) cardiac gating to synchronize the data acquisition with motion of the heart. (Reprinted with permission from Potchen EJ, Siebert JE, Haacke EM, Gottschalk A, eds. *Magnetic Resonance Angiography. Concepts & Applications.* Mosby-Year Book, Inc., St. Louis, 1993. Copyright 1993 Mosby-Year Book, Inc.)

□ *Solution:* The gradient is given by the magnetic field difference divided by the distance:

$$G = (1 \text{ ppm})(1 \times 10^{-6}/\text{ppm})(1.5 \text{ T})/(1 \times 10^{-6} \text{ m})$$

$$= 1.5 \text{ T m}^{-1}$$

$$= 150 \text{ G cm}^{-1} \qquad \qquad □$$

This is an enormous field gradient. Its dephasing effect is emphasized in T_2-weighted images. However, the effect depends on the susceptibility agent being *compartmentalized* within the blood and excluded from the surrounding tissue to maintain a high field gradient. If the normally tightly sealed *blood–brain barrier* (probably an evolutionary adaptation to protect the central nervous system from damage by infectious or chemical agents) becomes damaged, say as a result of **stroke** (an interruption of blood flow in the brain), the agent diffuses into the surrounding brain tissue and the compartmentalization effect and strong T_2 contrast are lost. This use of contrast agents is therefore valuable in the visualization of damaged regions of the brain following stroke. Dark regions are those in which the blood–brain barrier is intact, whereas areas that do not suffer signal loss are compromised with a leaky barrier.

Because of the magnetic susceptibility difference between oxy- and deoxyhemoglobin, it is possible to use deoxyhemoglobin as an intrinsic contrast agent. If you hold your breath for about 20 s, the level of the deoxy form builds up sufficiently to be used effectively in T_2-weighted images (special fast imaging methods must be used for this to work). Finally, blood-borne ferrite particle contrast agents are effective in tracing the circulation and in determining the functioning of the **reticuloendothelial system (RES)**. This is the "garbage collection" system of the body and includes components of the liver and spleen. Among other functions, it traps foreign particles such as bacteria and will be labeled with ferrite particle agents when intact. This will be visualized as a signal loss on T_2-weighted scans, since these agents create severe dephasing of protons diffusing within a few micrometers of the particles. The particles are typically tens of nanometers in size (effectively they are colloidal suspensions) and are encapsulated in proteinlike or polymer coatings to control their toxicity, binding, biodistribution, hydrodynamics, and the exchange of water between the bulk and the particle surface.

16.5 HIGHER DIMENSIONAL IMAGING

It is straightforward to extend our two-dimensional imaging pulse sequences to three dimensions. In the case of projection imaging, gradients may be applied in all directions in three-dimensional (3D) space. In the case of spin warp imaging two independently varied phase-encoding gradients along y and z

may be used. The pulse sequence otherwise would be identical to that in Figure 16.9. Multislice spin warp is already a form of 3D acquisition. However, in multislice imaging the in-plane spatial resolution (typically on the order of 1–2 mm) is usually much higher than the slice-to-slice resolution (typically 5–10 mm). Direct 3D imaging, using one frequency-encoded and two phase-encoded dimensions, provides the opportunity to obtain equal spatial resolution in all three axes. There is a price to be paid for such isotropic high resolution in three dimensions. A square 2D image with $256 = 2^8$ pixels on a side and equal resolution in both axes contains $256^2 = 2^{16} = 65,536$ pixels. It requires 256 echo acquisitions, assuming no signal averaging, for each phase encoding step. Assuming a TR of 0.5 s, this takes 128 s, or about 2 min. If the complex data are stored as 8-byte complex floating-point numbers in the computer, the time-domain data set requires 2^{19} bytes of storage, or about 0.5 MB (megabytes). A cubic 3D image with 256 pixels on a side and equal resolution in all axes contains $256^3 = 2^{24}$ pixels, or about 17 megapixels. It requires $256^2 = 65,536$ phase-encoding steps, which will now take a little over 9 hours, and occupy over 134 MB of memory. It is difficult to ask a subject to hold still for 9 hours, and the memory and processing requirements tax even today's computers. For these reasons 3D imaging is always practiced with reduced resolution and fields of view along some axes.

A striking example of a phase-encoded (not multislice) 3D image is shown in Figure 16.15. The image data have been processed by an image-processing computer program that *classifies* each pixel in the data set as either signal or noise (**image segmentation**), discards noise pixels, and presents the signal pixels in the form of a 3D solid object under illumination (**image rendering**). Overlaid on the basic 3D image and inserted into a cutaway of the brain is an oblique (remember the gradient directions may be electronically rotated?) 2D image that highlights the activation of **neurons** (nerve cells) that participate in processing of visual data in the region of the brain known as the **visual cortex**. In the oblique image the contrast is based on the increased capillary blood volume the brain recruits to serve the active neurons and makes use of the magnetic susceptibility contrast mechanism. The brain increases its blood volume in active regions by increasing the diameters of the capillaries; this alters the microscopic field gradients because the same field shift occurs over a larger dimension. By subtracting T_2-weighted images obtained during a rest period and while the brain is performing a particular task (such as perceiving flashing lights in this case), we can highlight only those areas that are experiencing *neuronal activation*. **Functional magnetic resonance imaging (fMRI)** is rapidly becoming an important noninvasive tool in *neuroscience* for studying how the brain functions in health and in organic and psychiatric disorders.

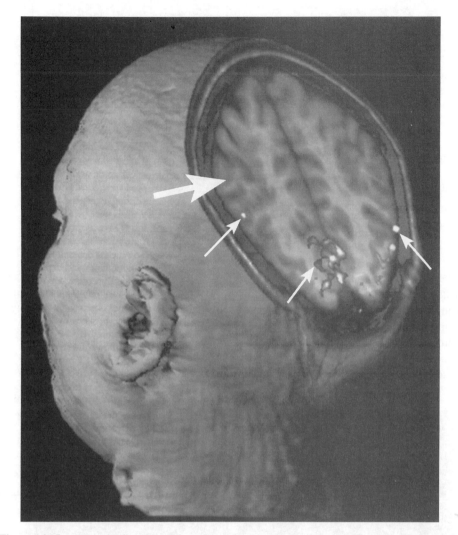

Figure 16.15. Segmented (separated into classes by intensity) and rendered (given solidity by light ray tracing) 3D NMR image data overlaid with a 2D susceptibility contrast image (large arrow) showing activation of brain tissue in the visual cortex (small arrows). (Reprinted with permission from Belliveau JW, Kennedy DN, McKinstry RC, Buchbinder BR, Weisskoff RM, Cohen MS, Vevea JM, Brady TJ, Rosen BR. Science. 1991; 254: 716–719. Copyright 1991 American Association for the Advancement of Science.)

16.6 CHEMICAL SHIFT IMAGING

Up to now, we have discussed MRI contrast based only on relaxation parameters. What if we wish to base the image contrast on specific resonances of the NMR spectrum itself, or perhaps even use the entire spectrum as the contrast parameter? A straightforward alteration of 3D spin warp MRI leads us to a simple form of **chemical shift imaging (CSI)**. To obtain a 2D image with a third dimension containing the chemical shift spectrum, we phase encode the x and y spatial dimensions and then observe the FID or echo in the absence of any field gradient, as shown in Figure 16.16. The two phase encodings constitute two "time" dimensions, and the signal acquisition is the third time dimension. A 3D Fourier transform of this data set yields an image with two spatial dimen-

sions and a third dimension that gives the NMR spectrum at each pixel. This is in effect a highly detailed spatially resolved chemical analysis, but just as in the case of imaging with three spatial dimensions, there may be a severe cost in terms of total data acquisition time and data set size. It is normal in CSI to use much reduced spatial and spectral resolution. Figure 16.17 shows an example of proton CSI in the brain. In this case, solvent suppression was incorporated into the pulse program to reduce the water and fat (lipid methylene) resonances so that other metabolites present in far lower concentration can be seen. Needless to say, the visual presentation of such large amounts of data can be problematic, and only a small part of the data can be represented on the printed page.

Often the spectrum to be resolved is quite simple. For example, we may only be interested in the water and fat

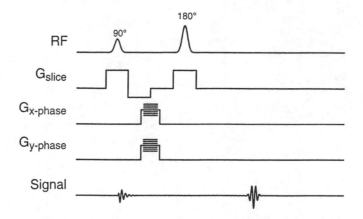

Figure 16.16. Chemical shift imaging pulse sequence.

content. In this case the spectrum consists of only two resonances, water and methylene. Rather than sampling a full-blown FID or echo in order to measure this rather trivial spectrum, we may simply acquire two images. One image is the normal spin warp image. The other is obtained by displacing the Hahn echo (which occurs at the time determined by the timing of the 90° and 180° pulses) from the **gradient echo** (which occurs at the time determined by the amplitudes and durations of the readout and compensation gradients). The displacement is achieved by a missetting of either the compensation gradient or the rf pulse timing. The duration of the interval between the echo positions is set to the time required to achieve a 180° phase shift between the water and methylene resonances. In the resulting images, water and fat will be

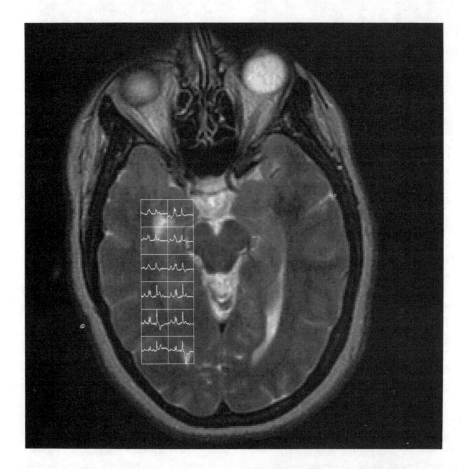

Figure 16.17. Water-suppressed proton chemical shift image of the brain. Proton spectra from six voxels are shown overlaying the normal image at the voxel positions. (Courtesy of Patricia Lani Lee.)

either in phase or 180° out of phase. The sum of the two images yields a pure water image, while the difference of the two images yields a pure fat image. We therefore achieve chemical shift separation with two quick 2D image acquisitions, rather than one very cumbersome 3D acquisition. This method is know as the **phase contrast** or **Dixon** (after its inventor, Thomas Dixon) technique. It and several variations may be extended to multiple-line spectra.

16.7 NMR MOVIES: ECHO PLANAR IMAGING

In the discussion of image contrast we looked at the use of cardiac gating to stroboscopically acquire image data in synchrony with the heart's motion. This reduces motion artifacts and enables us to image at a particular instant of the cardiac cycle. However, this imaging method requires an acquisition time of typically 128 or 256 heart beats, about 1–2 minutes. If some parameter is varying over a shorter time scale, or if the motion is nonperiodic (such as the random contractions of the bowel in the abdomen), this approach will not work. Another method, known as **echo planar imaging** (**EPI**), can be used to acquire images in the order of a few tens of milliseconds, permitting real-time movies to be made.

It is possible to obtain an echo without the use of a 180° pulse, but rather by merely reversing the sign of a frequency-encoding gradient. Initially during a continuously applied gradient, spins evolve in phase according to their position along the frequency-encoding axis. Some spins are fast (far above resonance) compared to the central frequency, while some spins are slow (far below resonance). By reversing the gradient, the roles of fast and slow spins are interchanged, just as the 180° pulse would do to any resonance dispersion. A gradient echo results. Gradient echo imaging may be performed just like spin warp imaging. Because of the absence of the 180° pulse, no chemical shift or $\mathbf{B}_0$ inhomogeneity refocusing occurs, which provides another form of contrast T_2^* weighting. The gradient reversal may be repeated to

produce a train of echoes just as would be produced by a train of 180° pulses in a CPMG spin echo pulse sequence. Now we can add a phase-encoding pulse prior to the readout of each echo and generate enough phase-encoding steps for a complete image acquisition. This is one form of EPI. The pulse sequence is shown in Figure 16.18. The entire image is acquired with the signal elicited by a single 90° pulse and occurs in a small fraction of a second. Special hardware is required to perform the very rapid gradient switching required in EPI. Using EPI, it is possible to obtain a rapid sequence of images of a beating heart without the need for cardiac gating, because the "shutter speed" of the NMR "camera" is sufficiently fast to "stop" the motion. By viewing the images in rapid sequence in real time (**cine** presentation, pronounced "sinnae") we can see a "movie" of the heart.

16.8 NMR MICROSCOPY

The discussion so far has implied that the subjects to be imaged are animal or human sized. There is nothing that prevents us from examining small specimens with MRI of course. The term NMR microscopy refers (rather loosely) to any specimen that fits into a typical 5- or 10-mm NMR tube. (Optical or electron microscopists would certainly scoff at this use of the term "microscopy.") To obtain greater magnification, we merely increase the gradient amplitudes. But remember that magnification is costly in terms of signal-to-noise ratio. Nuclear magnetic resonance microscopy is therefore performed in high-field magnets to get the signal-to-noise advantage of high $\mathbf{B}_0$ The signal-to-noise ratio is one limiting factor when increasing magnification. Another is molecular diffusion. We have already discussed the effects of diffusion in the presence of a field gradient. The dephasing effects limit the spatial resolution obtainable in a liquid specimen to about 1 μm. In most cases, declining signal-to-noise ratio will come into play as a barrier well before reaching the diffusion limit.

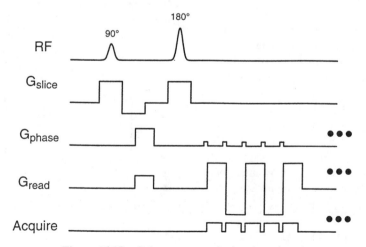

Figure 16.18. Pulse sequence of echo planar imaging.

Often, just as in macrosized imaging, some axes will be acquired at much higher resolution than others, for example, by using a slice thickness of 100 μm and in-plane resolution of 20×20 μm. In exceptional cases, single cells have been imaged (these however were large cells such as frog eggs about 1 mm in diameter). Nuclear magnetic resonance microscopy image acquisitions are typically very long because of the necessity to signal average, often 30 minutes to many hours.

16.9   IN VIVO NMR SPECTROSCOPY

Often the information on NMR relaxation parameters carried by image contrast is insufficient to address a particular problem. We can then look to the rich information content of the spectrum itself. Generally, spectroscopy of the entire body is not of much value, and in vivo spectroscopy is usually carried out as **localized spectroscopy**, that is, over a part of the body. There are various ways of restricting the operation of the spectrometer to a particular region, and they fall into two broad classes: those that depend on the physical dimensions of the rf coil and those that use field gradients in the pulse sequences. Often these approaches are combined. At this time, the use of spectroscopic examinations has not become part of the repertoire of clinical practice, despite a history of in vivo spectroscopy almost as old as MRI itself. In vivo spectroscopy has had a number of landmark successes in solving problems in metabolism research in both animals and humans, but there have been no spectroscopic applications that have been demonstrated to be more effective than other methods for the routine diagnosis of disease.

16.9.1   Surface Coils

There are two reasons for restricting the **receptivity** of the spectrometer to a particular **region of interest** (**ROI**). The first reason has to do with the **filling factor** and its effect on sensitivity. Filling factor represents the fraction of receptive volume within the coil that is actually filled with the specimen. The intensity of $\mathbf{B}_1$ produced by a given current in the rf coil is inversely related to the size of the coil. In comparing two coils of similar physical geometry, the $\mathbf{B}_1$ intensity will vary approximately with the inverse of the cross-sectional area, all other things being equal. By the **reciprocity principle** in electromagnetic theory, the receptivity of the coil (the voltage induced for a given precessing magnetization) is proportional to the magnitude of $\mathbf{B}_1$ generated by a unit current. It is therefore true that smaller coils are more sensitive than larger coils, as long as the smaller coil is capable of containing the specimen or region of interest. We therefore always attempt to maximize the filling factor by using the smallest possible coil that contains the ROI. This principle is used in conventional spectrometers as well, which explains why glass NMR tubes are made with such thin walls (the extra glass of a thick-walled tube wastes filling factor). For in vivo

spectroscopy, this means that we work with the smallest possible ROI and fit the coil to that ROI.

The second reason for maximizing filling factor has to do with the inherent noise properties of the specimen. All materials at temperatures above absolute zero experience random molecular vibrations, which are the thermal fluctuations of a molecular system at absolute temperature T and thermal energy kT (RT for a mole of systems). In these expressions, k is Boltzmann's constant and R is the ideal gas constant. This holds true for electrical conductors as well, but in addition, the thermal fluctuations of the charge carriers—electrons in the case of metals, ions in the case of electrolyte solutions— give rise to fluctuating electric and magnetic fields: this is **thermal noise**, which is also detected by the rf coil. For electrically nonconductive samples, such as solids or solutes in organic solvents, only the electrical noise of the rf coil and the first stage of the preamplifier are significant and determine the ultimate signal-to-noise ratio achieved in the spectrum. In the case of electrolyte solutions, including biological tissues, the specimen noise is also important and in most situations (very small specimens being the prime exclusion) dominates the instrument noise sources. Since the specimen noise power is proportional to its volume, we should minimize the specimen volume "seen" by the coil, which argues for the smallest possible coil that still contains the region in which we are interested.

Working with the smallest possible coil and largest filling factor suggests that the entire body not be enclosed by the coil if we are only interested in a part of the body. If the region of interest is close to the surface, the simplest possible coil we might construct is a simple loop of wire, which is known logically enough as a **surface coil**, pictured in Figure 16.19. A coil surrounding the entire body may still be used as the transmit coil to generate rf pulses, as long as the rf power amplifier is capable of producing enough power to produce the $\mathbf{B}_1$ field we desire. The surface coil will then act as a receiver only. Alternatively, the surface coil may be used as both the transmitter and receiver, just as in a conventional spectrometer. In either case, if the surface coil diameter is tailored to our ROI, it will have a high filling factor and the best possible filling factor and signal-to-noise ratio. The sur-

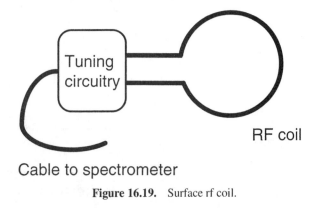

Figure 16.19.   Surface rf coil.

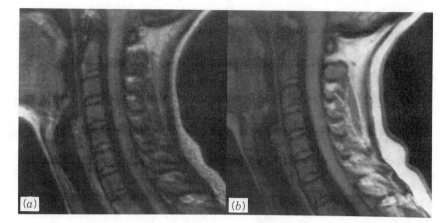

Figure 16.20. Performance comparison of a body coil (left) and a surface coil (right). (Reprinted with permission from Wehrli FW, Shaw D, Kneeland JB, eds. *Biomedical Magnetic Resonance Imaging. Principles, Methodology and Applications.* VCH Publishers, Inc., Weinheim, Germany, 1988. Copyright 1988 VCH Publishers, Inc.)

face coil's receptive volume is roughly a hemisphere of diameter equal to the coil diameter. In practice, this means that a stock of surface coils is usually kept in the laboratory to meet various needs. Most surface coil spectroscopists make their own coils, because it is not practical for commercial vendors of NMR equipment to offer the wide variety of surface coils necessary to fit every possible size and shape requirement. It is even possible to insert a small coil surgically around an organ of interest or to introduce a very small 2–3-mm-diameter coil contained in the tip of a **catheter**, a long tube that may be inserted through a small cut in a surface vein (perhaps the femoral vein in the groin) and snaked all the way up into the right atrium of the heart. This permits localized spectroscopy of a heart wall, which can be used to assess the status of heart muscle that has been damaged by interruption of blood flow in a **myocardial infarction** (heart attack).

A disadvantage of the surface coil compared to the cylindrical coils used in conventional high-field spectroscopy or whole-body MRI is that the B_1 field or (by reciprocity) the receptivity is a strong function of position, as Figure 16.20 shows. The rf field is strongest in the plane of the loop and nearest the wire. The on-axis field drops off rapidly on either side away from the coil plane. The receptive volume is roughly limited to a hemisphere that just encompasses the loop. Figure 16.20 compares two spine images taken with a body coil (left) and with a surface coil (right). The signal intensity with the body coil is more uniform over the entire field of view, while the signal-to-noise ratio is much improved in the region near the surface coil (which has been placed on the right side of the right hand image), and drops off with distance from the coil.

16.9.2 ^{31}P Spectroscopy

One of the most common applications of surface coil spectroscopy is in ^{31}P spectroscopy. It turns out that only a small

number of ^{31}P resonances are visible in tissue, which simplifies the analysis of spectra. Figure 16.21 illustrates the ^{31}P spectrum of skeletal muscle. The three upfield resonances arise from the three phosphates of **adenosine triphosphate** (**ATP**), structure **16-2**, which is one of the compounds used by cells to store energy derived from metabolic processes. The next prominent resonance is that of **phosphocreatine** (**PCr**), structure **16-3**, which is also used to store energy:

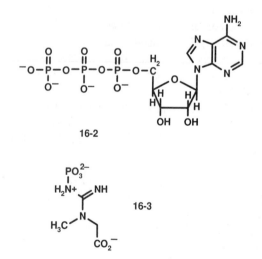

16-2

16-3

The most downfield resonance is that of inorganic phosphate ion, $(PO_4)^{3-}$ (abbreviated P_i). When cells carry out their normal metabolic function, phosphate is cleaved from ATP to produce P_i. The energy status of cells may therefore be gauged by measuring, for instance, the PCr/P_i ratio. Cells with a low value of this ratio are energy depleted, which might result from insufficient blood supply (**ischemia**) or some metabolic disorder.

A useful feature of ^{31}P spectroscopy is that the resonances are pH sensitive. All of the observable phosphates can be

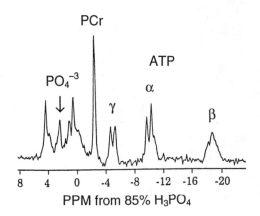

PPM from 85% H₃PO₄

Figure 16.21. In vivo ^{31}P spectrum of skeletal muscle. (Reprinted with permission from Edleman RR, Hesselink JR, Zlatkin MB, eds. *Clinical Magnetic Resonance Imaging*, vol. 1. W. B. Saunders Company, Philadelphia, 1996. Copyright 1996 W. B. Saunders Company.)

protonated. At intracellular physiological pH, about 6.8, P_i is primarily distributed between HPO_4^{2-} and $H_2PO_4^-$ (in fact, the equilibrium between these ions constitutes one of the buffers that maintain our pH within the correct limits). These forms of phosphate each have their own ^{31}P chemical shifts. As the equilibrium shifts with changing pH, the chemical shift, which is the concentration-weighted average of the HPO_4^{2-} and $H_2PO_4^-$ shifts, also changes, making ^{31}P in vivo spectroscopy a means to measure intracellular pH. Because all of the resonances have slightly differing pH dependences, the chemical shift difference between two of the resonances is normally used to determine pH. This eliminates the uncertainty associated with using an external chemical shift reference. A widely used relation for determining the intracellular pH is

$$pH \cong 6.683 + \log \frac{\delta - 3.153}{5.730 - \delta} \qquad (16.8)$$

where δ is the positive chemical shift difference between P_i and PCr.

16.9.3 ^{1}H Spectroscopy

The normal ^{1}H spectrum of tissue consists of the dominant resonance of water at about 4.7 ppm and a weaker resonance from the methylene of lipid (fat) at about 0.9 ppm. This spectrum never changes significantly and is therefore usually uninteresting. However, by suppressing these resonances, many others, much weaker in intensity but far more diagnostically useful, are revealed. There are a number of suppression techniques. A frequency-selective pulse may first be applied to saturate the unwanted resonances. Water, being by far the dominant resonance, is usually suppressed by applying a weak long pulse at the exact water frequency, which saturates the water signal. This pulse, because it is of low amplitude and of long duration, is highly frequency selective and is one form of chemical shift selective, or **CHESS**, pulse. Other CHESS methods rely on **composite pulses** (groups of simple pulses that together have some specified effect) to achieve specially tailored spectral excitation profiles. Fat, because of its short T_2, is easily suppressed if the pulse sequence contains a spin echo with a sufficiently long TE.

Figure 16.22 compares ^{1}H spectra from normal brain tissue. Water and fat signals have been suppressed. Some residual lipid appears as a broad resonance at 0.9 ppm. Creatine (Cr), structure **16-4**, and choline (Cho), structure **16-5**, appear at 3.0 and 3.2 ppm respectively:

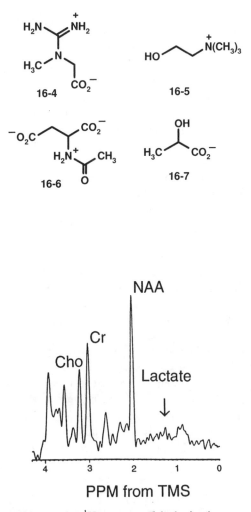

PPM from TMS

Figure 16.22. In vivo ^{1}H spectrum of the brain; the water and fat (lipid CH_2) resonances have been suppressed. (Reprinted with permission from Edleman RR, Hesselink JR, Zlatkin MB, eds. *Clinical Magnetic Resonance Imaging*, vol. 1. W. B. Saunders Company, Philadelphia, 1996. Copyright 1996 W. B. Saunders Company.)

A resonance unique to brain and nerve tissue and believed to be a specific marker for neurons (nerve cells) is N-acetylaspartate (NAA), structure **16-6**, appearing at 2.0 ppm. NAA often decreases in the damaged brain tissue, suggesting that neurons have died. The damaged brain tissue also contains elevated lactate, structure **16-7**, which is an indication that anaerobic metabolism, a pathway forced by oxygen deficit, is taking place. Only the methyl doublet of lactate (1.3 ppm) appears in the spectrum; the methine quartet overlaps the water resonance and is normally never observed in vivo.

16.9.4 Gradient Localized Spectroscopy

In addition to the localization provided by the surface coil itself, the spectroscopic acquisition may be localized by the assistance of field gradient pulses. One simple pulse sequence is known as **PRESS**, for point resolved spectroscopy, which consists of a frequency-selective 90° pulse followed by two frequency-selective 180° pulses. Each pulse is applied in the presence of a gradient pulse along successive coordinate axes. In effect, each pulse selects a slice, so that the final echo arises only from those spins residing in the common volume of three orthogonal slices. The x, y, and z slice widths and positions determine the locations and dimensions of the single spectroscopic voxel analyzed. Standard MRI software permits the specification of the voxel by graphical prescription from a scout image.

Many variations on this scheme exist. A popular method is **STEAM** spectroscopy, for stimulated echo acquisition mode. Three successive 90° pulses generate a **stimulated echo**, which results from dephased transverse magnetization of the initial FID being stored along the z axis by the second pulse and converted back into transverse magnetization by the third pulse. Because the z-stored magnetization retains some phase memory (the z component of a given spin during storage depends upon the extent to which it has precessed during the first interpulse interval), when the stored magnetization is converted back to transverse magnetization by the third pulse, the precession continues (in a phase-coherent but reversed manner), and an echo results. (Actually, there are *five* possible echoes in a three-pulse experiment, but the stimulated echo is maximized by all pulses being 90°.) Each pulse may be applied selectively in the presence of a field gradient, again selecting the x, y, and z dimensions of a spectroscopic voxel.

16.10 NONMEDICAL APPLICATIONS OF MRI

Although MRI grew out of a technique—NMR spectroscopy—that was at the time, and still remains, primarily a tool of chemical analysis, its development and applications have largely been driven by the needs of medicine. Yet there are numerous instances in fields such as chemistry, physics, biology, geology, and many engineering disciplines where MRI could be a useful tool for scientific investigations. It is only in the last few years that imaging capability has become commonly available in high-field research NMR spectrometers, bringing the tools and techniques of MRI into laboratories devoted to these fields. Here, we sample just a few such applications and leave it to your imagination to dream up others.

Essentially any scientific problem in which materials can be beneficially analyzed by NMR spectroscopy and in which spatial variation in the specimen is of interest might be a good candidate for MRI. For example, we might consider the curing of a polymer network by the formation of chemical crosslinks between polymer chains. This is a problem that has been extensively studied by both solid- and fluid-state NMR spectroscopy using many polymer systems. Various parameters of a crosslinked polymeric material may be determined from NMR analysis of spectra and relaxation times of laboratory samples, including the concentration of the crosslinks, their chemical structure, the concentration of residual uncrosslinked polymer chains, and the molecular dynamics of the polymer chains between crosslinks. In the industrial manufacture of a real-world object from polymeric material, however, things often do not proceed as expected on the basis of carefully prepared laboratory samples. The real-world part may be produced by **injection molding**, which involves injecting a softened, uncrosslinked material into a heated mold under high pressure. The raw polymer might contain a **filler**, fine particulate matter that strengthens the material. Silica (SiO_2) and carbon powder are common fillers. The polymer may then be cured to toughen it and to retain the final molded shape. Temperature and pressure gradients often exist during the molding and curing processes, leading to variations in polymer concentration, mixing, cure rate, crosslink concentration, and so forth. All of these are likely to vary from point to point within the specimen. The MRI of the part can be a good way to analyze some of these variables by combining an appropriate spectroscopic technique with an imaging technique. This is a good instance of the use of MRI for nondestructive evaluation, as discussed in Chapter 15.

For example, a rubbery polymer that has a proton T_2 in the range of 10 ms or more may be imaged relatively easily with spin warp imaging. Since the filler is typically a solid with nonspinning T_2^* well below 1 ms (and may not contain protons at all), using a TE in the range of 1–10 ms makes the filler invisible while retaining signal from the polymer. We can choose a frequency-encoding gradient strong enough to ensure that the proton spectrum of the polymer is not resolved in the frequency-encoding axis. If we make the simplifying assumption that the composite polymeric material is made up of just two components, the rubber and the filler, each of which has NMR properties that do not vary with position within the specimen, then the image intensity represents the concentration of rubber, while the complement of the image represents the concentration of filler.

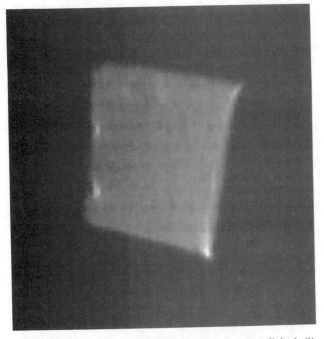

Figure 16.23. Proton image of the polymer in a crosslinked silicone polymeric block that contains in situ precipitated silica filler.

Figure 16.23 illustrates the proton image of a silicone (PDMS, poly-[dimethylsiloxane], structure **16-9**) polymer that contains silica filler produced by the hydrolysis of tetraethylorthosilicate, TEOS, structure **16-10**, in the presence of water. The previously crosslinked polymeric block had been swelled with the TEOS (that is, the TEOS was absorbed by the polymer), and the block was then placed in water, which slowly diffuses in from the outside, reacting with the TEOS and producing silica as it goes. This method for introducing filler is known as in situ **precipitation** and is an excellent way of strengthening silicone polymers. The image, however, shows the nonuniform distribution of polymer, and therefore of filler, which can result by this method.

Another good application of MRI is in the study of fluid distribution in various media. **Porosity**, the volume fraction of an object that is empty space, can be determined by MRI if the pore volume can be filled with an inert fluid (water for instance) that gives a strong MRI signal. The signal intensity of the pure fluid represents 100% porosity, while the signal intensity within the object indicates the object porosity on a proportional basis. The actual pores may or may not be spatially resolved in the image. If they are resolved, an average signal intensity over some appropriate region of interest is used to determine the porosity. This method has been used to determine porosity in rock core samples recovered from oil well drilling operations and in the manufacture of ceramic

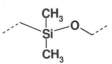

16-9

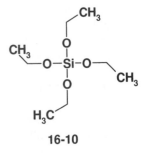

16-10

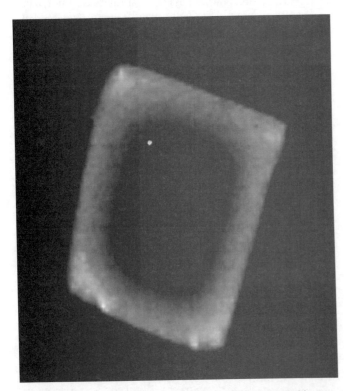

Figure 16.24. Proton image of the penetration of methanol into a square block of poly(methylmethacrylate).

parts (which contain porosity in the early stages of processing). Fluid diffusion into materials may also be studied. Figure 16.24 shows a proton spin warp image of a piece of poly-[methylmethacrylate] that has been immersed in methanol. The image gives the methanol distribution, while the polymer, because of its short T_2, gives no signal itself.

Flow is an important parameter in industrial processes. Magnetic resonance imaging has been used to determine the **velocity profile** in flowing fluids. In the case of **laminar** flow

of a **Newtonian fluid**, it is well established that the velocity along the flow direction is a parabolic function of the radius. In laminar flow in a long tube, the flow profile does not vary with time, and the fluid is directed only along the tube axis. If the flow increases past a certain limit, it becomes **turbulent**. A Newtonian fluid is one in which the viscosity is independent of the **shear rate** (the derivative of axial velocity with respect to position across the tube). Water and other simple liquids are Newtonian fluids, while suspensions and polymer solu-

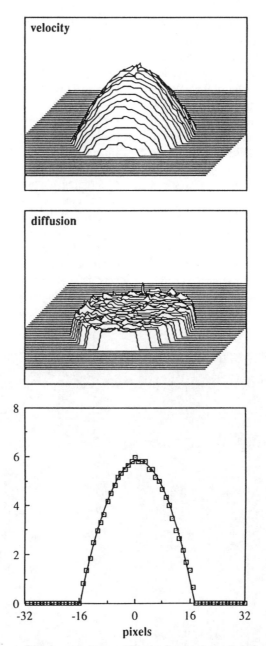

Figure 16.25. Motion-sensitized proton images of water flowing in a tube. The top image gives the velocity (vertical direction) at each point in the tube cross section, the middle image gives the molecular diffusion coefficient, and the bottom shows the expected parabolic velocity profile across the tube diameter. (Reprinted with permission from Callaghan PT. *Principles of Nuclear Magnetic Resonance Microscopy.* Oxford University Press, New York, 1991. Copyright 1991 Oxford University Press.)

tions (paints, pastes, blood, etc.) often exhibit non-Newtonian behavior and nonparabolic velocity profiles. (Ketchup comes out of the bottle more easily if you shake it vigorously first because of its shear-thinning and time-dependent non-Newtonian behavior—but remember to tighten the cap first!) Figure 16.25 illustrates the velocity and diffusion "maps" obtained from water flowing in a tube by using a pulse sequence in which a slice across the tube is selected, dual phase encoding of spatial position within the selected plane is used to resolve the cross-sectional image, and a pair of stepped velocity-encoding gradient pulses is used to induce a phase shift dependent on the velocity along the tube axis. Fourier transformation over the velocity-encoding gradient steps gives the velocity distribution along the flow direction for each point in the cross section. The center of the velocity distribution is the bulk flow velocity, while the width of the velocity distribution is related to the molecular diffusion constant. Since the diffusion constant is an intrinsic property of water, it is constant across the tube cross section, while the velocity is seen to have the expected parabolic profile. This method has been used to determine flow profiles in the stems of live plants.

As a final example, we mention the combination of solid-state spectroscopy (Chapter 15) with imaging. The imaging of solids is difficult because the T_2's are often too short compared with the time it takes to switch and apply field gradients for slice selection and phase and frequency encoding, while in addition extremely large gradient amplitudes are required to overcome the broad line widths. The time to ramp

up the current in a gradient coil is proportional to the inductance of the coil and inversely proportional to the voltage of the power supply. In clinical imagers, voltages of several hundred volts and currents of several hundred amperes are required to produce gradients on the order of 1 G cm^{-1} ramped up in the order of 0.5 ms. A solid with a proton spectral line width of 10 kHz has a T_2^* on the order of 100 μs and would require a gradient around 2.4 G cm^{-1} to just barely resolve 1 cm, and the FID would be completely dephased well before the gradient had reached its full amplitude. Clinical imagers have hardware limitations that preclude imaging solids.

Various approaches have been taken with solid-state MRI, including restricting the measurement to small specimens in small gradient coils (which drastically lowers the coil inductance and the current needed to obtain a given gradient); using line narrowing rf pulse sequences (in combination with imaging gradients) to lengthen the effective T_2; and employing three-dimensional projection reconstruction of FIDs to avoid slice selection and all other gradient switching during data sampling. Imaging has even been performed on samples undergoing magic angle spinning, with the field gradients rotated electronically in synchrony with the spinner motion!

One simple example of a solid-state image is shown in Figure 16.26, which is a two-dimensional ^{31}P FID projection image of a specimen of a 3.5-mm-diameter chicken bone obtained on a 6-T research imager. The gradient of 24 G cm^{-1} was generated by a special small gradient coil producing compensation and readout frequency-encoding gradient pulses. The specimen was reoriented with respect to the

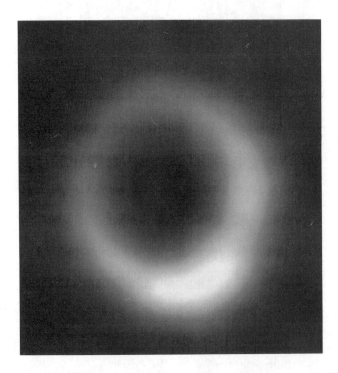

Figure 16.26. Solid-state ^{31}P cross-polarization spin echo projection image of a chicken bone.

frequency-encoding gradient to obtain multiple projection data, instead of using phase encoding. Proton–phosphorus cross polarization followed by a ^{31}P 180° spin echo pulse was used to obtain this image, with a cross-polarization time of 500 μs and an echo time of 200 μs. Despite the large ^{31}P line width of several kilohertz, the cross section of the bone is clearly revealed. Similar images may someday be used to measure **bone mineral density**, which is important in assessing the risk of fracturing a bone in older adults who suffer from bone mineral loss (**osteoporosis**).

CHAPTER SUMMARY

1. In living subjects NMR involves considerations of life support (maintenance of physiological functions, safety, comfort) and subject motion as well as the usual concerns of NMR measurements.

2. Spatial localization in imaging is by means of magnetic field gradients, which impose local Larmor frequencies dependent upon position. The resonance offset frequency (in radians per second) of a spin at position x in a field gradient G_x is equal to $\gamma x G_x$.

3. Sampling (reading out) an FID or echo in the presence of a gradient yields a spectrum that is the projection of the subject along the gradient direction. This type of gradient is known as a frequency-encoding gradient.

4. Applying a pulse of gradient prior to reading out an FID or echo yields a signal with a phase increment proportional to the position of the spin along the gradient direction. This type of gradient is known as a phase-encoding gradient.

5. The earliest MRI method, projection imaging, relied on the acquisition of a series of frequency-encoded projections of the subject at successive angles followed by reconstruction of the set of projections into an image.

6. Slice selection is accomplished by applying a frequency-selective rf pulse in the presence of a gradient.

7. The most commonly used MRI method is spin warp (2DFT) imaging, in which a spin echo, generated by slice-selective 90° and 180° pulses, is read out in the presence of a frequency-encoding gradient following a phase-encoding gradient pulse between the rf pulses. Two-dimensional Fourier transformation over the echo sampling and phase-encoding variables gives the two-dimensional image.

8. The image magnification is proportional to the intensity of the applied gradient, and for a phase encoding gradient it is also proportional to the duration. The minimum resolvable spatial dimension is inversely proportional to the gradient intensity and to gradient (phase-encoding) or sampling (frequency-encoding) duration.

9. By appropriate setting of image parameters such as echo time TE or repetition time TR, we can weight the image pixel intensities (adjust the image contrast) according to the value of the corresponding relaxation time T_2 or T_1.

10. Exogenous contrast agents based on paramagnetic or ferromagnetic substances may be administered to alter the natural relaxation times in order to highlight features of interest.

11. Motion affects magnetic resonance images and may be used to develop image contrast.

12. Chemical shift spectra may be obtained as a third dimension of a two-dimensional imaging technique, by selective reception of a region of interest using a surface rf coil, or by selecting a spectroscopic voxel using field gradients.

ADDITIONAL RESOURCES

1. Mansfield, P., and Morris, P. G., *NMR Imaging in Biomedicine*, Academic, New York, 1982.

2. Wehrli, F. W., Shaw, D., and Kneeland, J. B., Eds., *Biomedical Magnetic Resonance Imaging*, VCH Publishers, New York, 1988.

3. Callaghan, P. T., *Principles of Nuclear Magnetic Resonance Microscopy*, Oxford University Press, New York, 1991.

4. Edelman, R. R., Hesselink, J. R., and Zlatkin, M. B., Eds., *Clinical Magnetic Resonance Imaging*, 2nd ed., W. B. Saunders, Philadelphia, 1996.

5. Mattson, J., and Merrill, S., *The Pioneers of NMR and Magnetic Resonance in Medicine: The Story of MRI*, Bar-Ilan University Press, Ramat Gan, Israel.

6. Ackerman, J. L., and Ellingson, W. A., Eds., *Advanced Tomographic Imaging Methods for the Analysis of Materials*, Materials Research Society, Pittsburgh, 1991.

7. Blümich, B., and Kuhn, W., Eds., *Magnetic Resonance Microscopy. Methods and Applications in Materials Science, Agriculture and Biomedicine*, VCH Publishers, Weinheim, 1992.

8. Brown, M. A., and Semelka, R. C., *MRI: Basic Principles and Applications*, Wiley-Liss, New York, 1995.

Appendix 1

ANSWERS TO REVIEW PROBLEMS

CHAPTER 1

1.1. (a) The same frequency, 4.431598×10^{10} Hz. (b) The microwave region. This is an example of microwave (rotational) spectroscopy.

1.2. Clearly, there has been some *inelastic* scattering. The 1064-nm light corresponds to a frequency of 2.820×10^{14} Hz. The scattered light at 1301 nm corresponds to a *lower* frequency of 2.307×10^{14} Hz. The difference between these is 0.513×10^{14} Hz, which is exactly the infrared absorption frequency of formaldehyde.

1.3. The lifetime of the species must exceed $1/\nu$. Since light of 500 nm has a frequency of 6×10^{14} Hz, the lifetime of the species must exceed 1.7×10^{-15} s.

CHAPTER 2

2.1. $Z = 5$, therefore $N = A - Z = 10 - 5 = 5$; 5 protons and 5 neutrons.

2.2. Group 2: odd atomic number, odd number of neutrons, even mass number. I will be an integer.

2.3. Since $I = 3$, the multiplicity is $2(3) + 1 = 7$; $m = -3, -2, -1, 0, 1, 2, 3$. See Figure A.1.

2.4. From Table 2.1, ν for ^{10}B is 4.5754 MHz T^{-1}. So $\nu = 26.9$ MHz.

2.5. $\Delta E = h\nu\,(\Delta m)$. For neighboring states, $\Delta m = 1$; therefore $\Delta E = 1.78 \times 10^{-26}$ J.

2.6. In this case $\Delta m = 3$, so $\Delta E = 3(1.78 \times 10^{-26}$ J$) = 1.07 \times 10^{-25}$ J. Therefore, $P_{(m=-3)} / P_{(m=+3)} = \exp(-\Delta E/kT) = 0.99997$.

2.7. (a) From Example 2.6 we know that ^{1}H nuclei precess at 250 MHz in a 5.87 T $\mathbf{B}_0$ field. Since $\mathbf{B}_1$ is $10^{-5}\,\mathbf{B}_0$, ^{1}H nuclei will precess around $\mathbf{B}_1$ at 10^{-5} (250 MHz) =

2.5×10^3 Hz. (b) The cycle time $t_0 = 1/\nu = 4.00 \times 10^{-4}$ s = 0.400 ms. (c) Since one complete cycle requires 0.400 ms, $\mathbf{M}$ will rotate 360° in 0.400 ms. The tip angle α will start out at 0° at $t = 0$. After 0.10 ms it will have rotated one-fourth of a cycle (90°) and 180° after 0.20 ms. (d) See Figure A.2. (e) A tip angle of 180° results in a perfectly inverted $\mathbf{M}$, where the population of down spins now outnumbers up spins. No T_2-controlled relaxation is needed since $\mathbf{M}$ already lies along the $-z$ axis. But the normal Boltzmann distribution must be reestablished by longitudinal relaxation (controlled by T_1). See Figure A.3. Note that at the "halfway" point of this relaxation $\mathbf{M}$ momentarily vanishes, since at this point there is an equal (but non-Boltzmann) population

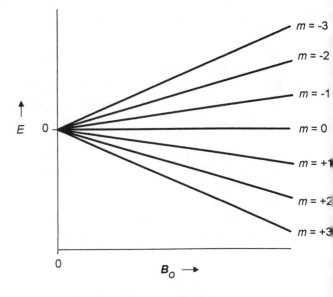

Figure A.1. Spin states of ^{10}B for problem 2.3.

334

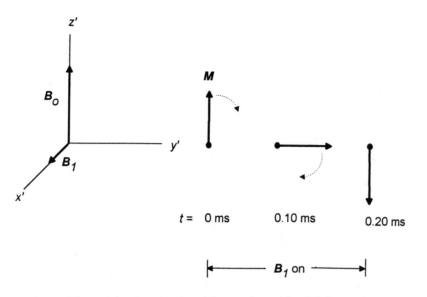

Figure A.2.   Rotating-frame diagram for problem 2.7(d).

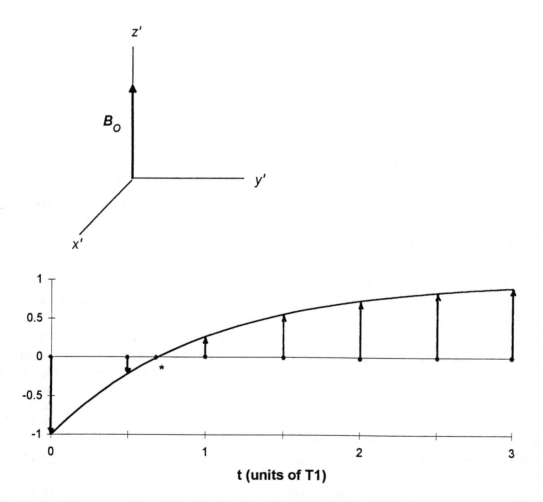

Figure A.3.   Rotating-frame diagram for problem 2.7(e).

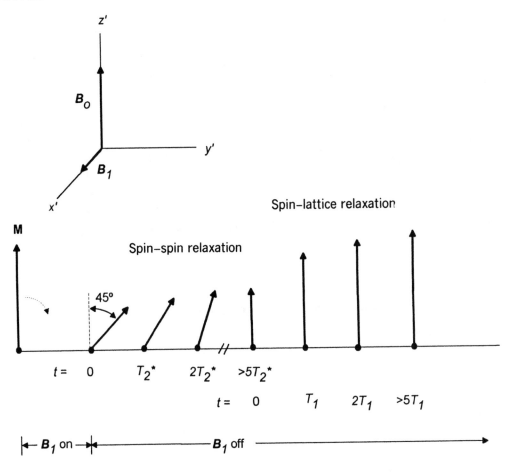

Figure A.4. Rotating-frame diagram for problem 2.8(b).

of spin states. Using Eq. (2.9) we can calculate when this occurs in terms of T_1. By realizing that $P_{eq} = 1 - P_0$ and P_t at this halfway point $= 0.5$, we can write

$$\frac{1 - P_0 - 0.5}{1 - P_0 - P_0} = \frac{0.5 - P_0}{1 - 2P_0} = 0.5 = \exp\left(-\frac{t}{T_1}\right)$$

Solving, $t = (T_1)(-\ln 0.5) = 0.69T_1$.

2.8. (a) If one-fourth of the *excess* up spins are converted to down spins, the number of *excess* up spins will be reduced to half the initial value. Since the magnitude of **M** is directly proportional to the number of *excess* up spins, **M** will also be half its initial magnitude. (b) There will first be rapid spin–spin relaxation as in Figure 2.11 (Example 2.11), followed by slower spin–lattice relaxation like that seen in Figure A.3 in problem 2.7(e). See Figure A.4.

2.9. Since ground-state O_2 has two unpaired electrons with parallel spin, it is a strongly paramagnetic molecule,

with a magnetic moment about 10^3 that of protons. Therefore, O_2 is very efficient at promoting spin–lattice relaxation.

CHAPTER 3

3.1. Look back at Figure 3.5*b* and the associated text.

3.2. (a) Use the ratio of ν values for ^{31}P versus ^1H (in Table 2.1) times 250 MHz to give 101.2 MHz. (b) Follow the procedure in Example 3.9: $t_p = 7.7$ μs, $t_d = 15.5$ μs (64,700 data points per second), $t_{acq} = 1/R = 1.0$ s. The resulting 64,700 data points will just barely fit in 64K of RAM. (c) $n = (15 / 2.4)^2 = 39$. (d) Assuming a total pulse sequence of 3 s ($3T_1$), the total time for FID data acquisition will be 117 s.

3.3. A Gaussian line will be at half height ($a/2$) at two values of $\nu - \nu_0$, whose difference is the halfwidth. Set Eq. (3.16) equal to $a/2$, then solve for the two values of $\nu - \nu_0$:

$$a \exp\left[-\frac{(\nu - \nu_0)^2}{b^2}\right] = \frac{a}{2}$$

$$-\frac{(\nu - \nu_0)^2}{b^2} = -\ln(2)$$

$$\nu - \nu_0 = \pm b\sqrt{\ln(2)} = \pm 0.833b$$

$$\nu_{1/2} = 2(0.833b) = 1.665b$$

If you had trouble with this one, practice by deriving Eq. (3.11).

CHAPTER 4

4.1. The three methyl hydrogens are dynamically homotopic, as are the two ortho hydrogens and the two meta hydrogens.

4.2. Structure **A** has a vertical σ plane bisecting the ring and containing the H–C–H atoms. This renders each pair of like atoms in the two H–C–F groups equivalent. The methylene (CH₂) hydrogens are *not* equivalent; they are diastereotopic. Therefore, we predict two carbons signals and three hydrogen signals. Structure **B** has a C_2 axis through the CH₂ carbon and the opposite ring bond. This renders equivalent not only the two H–C–F groups but also the two methylene hydrogens. This would result in two hydrogen and two carbon signals. Structure **C** has two perpendicular σ planes, as well as a C_2 axis along the intersection of the planes. All four hydrogens are equivalent, as are the carbons bonded to them, giving one hydrogen signal and two carbon signals.

4.3. (a) This conformation is asymmetric; it would exhibit three hydrogen signals. (b) There are two symmetrical conformations that render the CH₂F hydrogens enantiotopic and therefore equivalent:

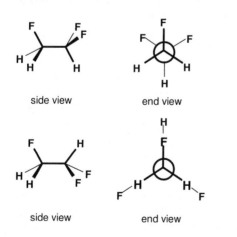

side view   end view

side view   end view

The top one, called the staggered conformation, has a vertical σ plane along the C–C bond. The lower

eclipsed conformation has a vertical σ plane along the C–C bond and another one perpendicular to that bond.

4.4. Let us try the isotope substitution test. The relationship between **4-6A** and **4-6B** is that they are enantiomers!

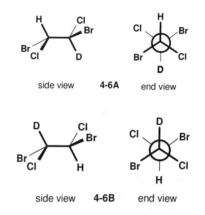

side view   **4-6A**   end view

side view   **4-6B**   end view

Therefore, the two hydrogens related by the center of symmetry are enantiotopic.

CHAPTER 5

5.1. Assuming free rotation around all single bonds, there will be four ^{1}H signals and four ^{13}C signals: one for the three equivalent methyls and one each for the three nonequivalent methylenes. The relative intensities for these signals will be 9 : 2 : 2 : 2, respectively.

5.2. All four hydrogens on the two C_2's are equivalent, as are the four on the two C_3's, and the two on C_4. This will result in three signals in the ratio 4 : 4 : 2 (or 2 : 2 : 1), respectively.

5.3. (a) External 85% phosphoric acid (H₃PO₄, Table 5.1). (b) The first signal is at δ 63.2, the other at δ −38.9. Use Eq. (5.1) and 101.2 MHz for ν_0'; do not forget that signals *downfield* of the reference have positive chemical shifts.

5.4. (a) This signal is due to the 0.1% of CHCl₃ (*undeuteriated* chloroform) present in the solvent. (b) First we calculate the spacing in ppm (Δδ) between the first and second line and between the second and third line: 0.509 and 0.510 ppm in Figure 5.6; 0.31 and 0.32 ppm in Figure 5.7. Then multiply this by ν_0' (62.9 and 100.6 MHz, respectively) to give 32 Hz in all cases. The fact that these separations in hertz are the same regardless of field strength indicates that they cannot be due to chemical shift differences. We will see what causes them in Chapter 8.

5.5. (a) 21739 Hz/62.9 Hz/ppm = 346 ppm. (b) The total number of points is the t_{acq} divided by t_d (both in seconds) = 0.377/2.3 × 10⁻⁵ = 16391. (c) Use Eq. (3.8): $R = 1/t_{acq} = 1/0.377$ s = 2.65 Hz.

CHAPTER 6

6.1. (a) Δν is the difference in hertz between the precession frequency of a nucleus and the operating frequency of the instrument and is the position of an NMR signal as determined by the instrument (Section 5.1). δν is the difference in hertz between the signal of interest and the reference signal (e.g., TMS); the chemical shift δ (ppm) is δν divided by the operating frequency in megahertz (Section 5.2). Δδ is the (de)shielding effect (in ppm) of a substituent on a neighboring nucleus. (b) Since, in ^{1}H NMR, hydrogen is the standard by which the Δδ values are determined, the Δδ value for hydrogen as a substituent is, by definition, zero.

6.2. Use the symmetry concepts from Chapter 4. Ortho, X = Y ≠ H: two signals; X ≠ Y ≠ H: four signals. Meta, X = Y ≠ H: three signals; X ≠ Y ≠ H: four signals. Para, X = Y ≠ H: one signal; X ≠ Y ≠ H: two signals.

6.3. From Table 6.3 we know that a "normal" ring methylene group appears around δ 1.5. Benzylic methylene groups are deshielded by the aromatic ring (Δδ 1.35 ppm, Table 6.2) and appear around δ 2.8, close to the observed value for this compound. However, the circled methylenes are unusually shielded by 0.7 ppm. This is because the phenyl ring is perpendicular to the ring composed of methylene groups, causing the circled methylenes to lie directly above the phenyl ring, in its shielding region. Compare this with Example 6.16.

6.4. (a) The IOU is 1. The presence of two oxygens and a signal at δ 12.05 suggests a carboxylic acid. The remaining nine hydrogens are equivalent, a *tertiary* butyl group. The compound is pivalic acid, $(CH_3)_3CCO_2H$. The predicted chemical shift of the *tert*-butyl hydrogens is δ 0.23 + 0.62 + 2(0.01) + 0.33 = δ 1.20.

(b) The IOU is 1. Both hydrogens are equivalent and appear in the vinyl region. There are three possible isomers: 1,1-dichloroethylene and *cis*- and *trans*-1,2-dichloroethylene:

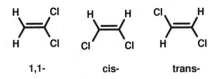

1,1-   cis-   trans-

The predicted (Table 6.4) chemical shift of each compound is

$$\delta_{1,1} = 5.28 + 0.19 + 0.03 = 5.50$$

$$\delta_{cis} = 5.28 + 1.00 + 0.03 = 6.31$$

$$\delta_{trans} = 5.28 + 1.00 + 0.19 = 6.47$$

The trans isomer fits closest and indeed is the correct answer.

(c) The IOU is 4, and all hydrogens are equivalent and lie in the region of benzylic methyl groups. A highly symmetrical aromatic hydrocarbon is suggested. The compound is hexamethylbenzene.

(d) An isomer of the above compound. But this one is less symmetrical, with two methyl groups in a normal position (δ 1.08) and four methyl groups attached to vinyl carbon (Table 6.4 and Figure 6.8). The compound's common name is hexamethyl Dewar benzene:

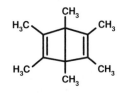

(e) The IOU is 4. Two equivalent aromatic hydrogens at δ 7.00 and a broad signal at δ 4.99 suggest a phenol, with two equivalent *tert*-butyl groups and a benzylic methyl. Two isomers have the requisite symmetry:

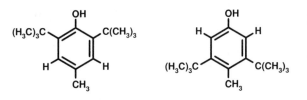

These can be differentiated by the predicted (Table 6.5) chemical shifts of the aromatic hydrogens. Those for the compound on the left should appear at δ 7.28 + (−0.17) + 0.01 + (−0.24) + (−0.14) = 6.74. Those for the compound on the right would appear at δ 7.28 + (−0.50) + 0.01 + (−0.09) + (−0.24) = 6.46. The compound on the left, the antioxidant BHT, is not only a better fit but also the correct answer!

CHAPTER 7

7.1. Refer to Figure 6.5. Notice how the ring carbons are located closer than the hydrogens to the line dividing the shielding zone from the deshielding zone. This means that the carbons will not experience as much of the anisotropic deshielding effect of the aromatic ring current as the hydrogens do.

7.2. We note that both carbons (α and β) in pyrrole are more shielded than those of ethylene, though the β carbons are 10 ppm more shielded. The resonance structures below indicate that resonance of the π electrons with the nitrogen unshared pair accounts for shielding at both positions due to the increased π electron density:

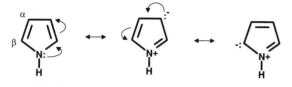

However, the closer proximity of the α carbons to the electron-withdrawing inductive effects of the electronegative oxygen *deshields* them more than it deshields the β carbons.

7.3. Use the symmetry concepts from Chapter 4. Ortho, X ≠ Y ≠ H: six signals; X = Y ≠ H: three signals. Meta, X ≠ Y ≠ H: six signals; X = Y ≠ H: four signals. Para, X ≠ Y ≠ H: four signals; X = Y ≠ H: two signals.

7.4. The chemical shift of CH_3CN is δ 195.5 ppm, while that of CH_3NO_2 is δ 332.5 [relative to $(CH_3)_4N^+I^-$, Figure 7.5]. Equation (5.1) reminds us that signals upfield of (lower frequency than) the reference signal have a negative δ. Therefore, the chemical shift of CH_3CN *relative to CH_3NO_2* is δ 195.5 − 332.5 = −137.0.

7.5. The structure on the left (isopropyl carbenium ion) has the more deshielded charged carbon (δ 320.6). This is because the structure on the right (protonated acetone) has a resonance structure that results from electron donation from the oxygen toward the charged carbon, shielding the latter (to δ 250.3) and resulting in a positive charge on the oxygen:

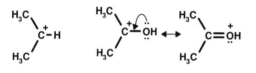

7.6. (a) The IOU is 1, and the spectrum exhibits a signal at δ 214.1: a ketone carbonyl is the only unsaturation. The remaining five carbons give rise to only three signals, so there must be some symmetry in the structure. These carbons must be distributed on both sides of the carbonyl, either one on one side and four on the other or two on one side and three on the other. Only the former could be correct, and then only if three of the four on one side were equivalent. Thus, the only possible structure is pinacolone, which was seen in Example 6.12:

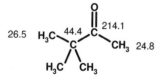

The predicted chemical shifts are

$$δ (CH_3C{=}O) = -2.3 + 30 = 27.7 \text{ ppm}$$

$$δ ((CH_3)_3C) = 15.8 + 8 + 1 = 24.8 \text{ ppm}$$

$$δ ((CH_3)_3C) = 16.3 + 6 + 24 = 46.3 \text{ ppm}$$

(b) Only three signals (not counting the three for $CDCl_3$ in the vicinity of δ 77.4), so there is some symmetry. The IOU is zero, so there is no unsaturation; both oxygens must appear as either alcohols or ethers, and they are probably equivalent. From symmetry considerations, the most likely distribution of seven carbons is in three sets of one, two, and four. The two-carbon set must be the only one attached to oxygen (δ 72.4); the compound is a symmetrical diol. The lone carbon (δ 52.5) is somewhat deshielded and is likely to be on the symmetry axis or plane, between the two hydroxyl-bearing carbons. This reasoning leads to the structure below, seen in Example 6.21:

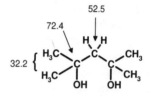

The predicted chemical shifts are

$$δ (CH_3COH) = 13.9 + 8 + 8 = 29.9 \text{ ppm}$$

$$δ (COH) = 22.8 + 6 + 41 = 69.8 \text{ ppm}$$

$$δ (CH_2) = 34.7 + 2(8) + 2(8) = 66.7 \text{ppm}$$

In the case of the methylene carbon the agreement is not as close as we would like. This is probably because the four substituent groups on the pentane backbone do not exert their effects independently.

(c) The IOU is 5. There do not appear to be any vinyl or aromatic carbons, but the signals at δ 85.2 and 92.0 suggest the presence of one or more carbon–carbon triple bonds. The fact that there are five signals for nine carbons suggests a plane or axis of symmetry, with one atom in the plane or axis; this would mean there are two equivalent triple bonds (with ends that are not equivalent) and one ring in the structure. Three structures with nine-membered rings come to mind:

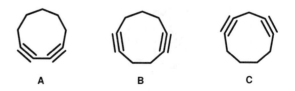

Although these structures are easy to draw, recall that the two atoms connected directly to the triply bonded (formally *sp*-hybridized) carbons prefer to form a linear or just slightly bent arrangement, more like the structures below:

As you can see (especially if you try to construct

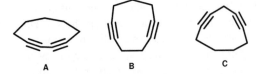

A   B   C

molecular models!), structure **B** requires the least distortion in bond angles and lengths and is therefore expected to be the least strained. In structure **C** the methylene carbon between the two triple bonds should appear around δ 26.1 (the chemical shift of cyclononane) + 2(4.5) = 35.1, while the two equivalent methylenes at the other end of the triple bond should appear around δ 26.1 + 4.5 = 30.6. In both structures **A** and **B**, all four methylenes attached to triply bonded carbons are predicted to have chemical shifts around δ 30.6. So it is difficult to use these chemical shift predictions as an unambiguous criterion to select among the three structures, especially when there is the possibility of *transannular* (across the ring) (de)shielding effects by the triple bonds. The correct structure is, in fact, **B** [see Gleiter, R., et al., *J. Am. Chem. Soc., 113*, 9258 (1991)]. The δ 92 signal is for the more strained triply bonded carbons (those separated by just two methylenes).

(d) From the hint we must have a $Cr(CO)_3$ (δ 233.3) group complexed to an aromatic compound with molecular formula C_7H_8O (IOU = 4). If we assign the signal at δ 55.6 to a CH_3O group, the remaining four signals suggest a monosubstituted benzene ring. Methoxybenzene (anisole) fits the symmetry, so the overall structure of the complex is

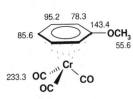

The carbon chemical shifts of *uncomplexed* anisole are shown below:

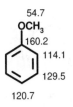

Why are the four aromatic carbon signals upfield of their normal position by an average of 30 ppm? This is due to the proximity of the chromium atom.

7.7. The solvent, $CDCl_3$.

7.8. (a) Phosphorus, like nitrogen, is commonly trivalent, so the IOU is 2. All nine hydrogens are equivalent: a tertiary butyl group [$(CH_3)_3C–$], which would also give rise to the carbon signals at δ 31.3 and 36.4. The only other carbon signal is in the carbonyl region, but there is no oxygen in the molecule! The ^{31}P signal is in the region for a C≡P group (Figure 7.4), and so is the carbon signal, (Section 7.5)! The compound is $(CH_3)_3C–C≡P$.

(b) The IOU is 2. All three hydrogens are equivalent: a methyl group. This leaves only CN, a cyano group (C≡N) with carbons signal at δ 117.9 (Section 7.5) and a nitrogen signal at δ 195.5 (Figure 7.5). The compound is acetonitrile (CH_3CN).

(c) The IOU is 4, and all the carbon signals are in the aromatic region. The fluorine signal is in the correct region for one attached to an aryl ring (Table 7.7). All we need to do is decide whether the F and Cl are ortho or meta. (How can we immediately rule out para?) The predicted chemical shifts for both are shown below:

Clearly the ortho isomer is the better fit.

7.9. The IOU of 5 with four signals in the range δ 110–152 suggests a monosubstituted or para-disubstituted aromatic ring. The signal at δ 189.2 is consistent with an aldehyde carbonyl. The nitrogen must therefore be present as an amine. The remaining four carbons give rise to just two signals in the tetrahedral region, with one (at δ 44.5) attached to nitrogen. The compound is *para*-(diethylamino)benzaldehyde. The signal assignments are given below; the predicted chemical shifts are shown in parentheses:

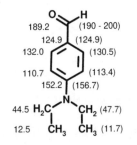

SELF-TEST I

1. The ^1H operating frequency in megahertz.

2. (a) At 250 Hz per δ, a span of 8.0 δ units = 2000 Hz (Section 5.2); radio frequency (rf, Section 1.2); $E = h\nu = (6.63 \times 10^{-34}$ J s) (2.50×10^8 s^{-1}) = 1.66×10^{-25} J (Section 1.2). (b) $\mathbf{B}_0$ = 250.0 MHz/42.58 MHz T^{-1} = 5.87 T (Section 2.2); two, since $I = 2$ (Section 2.1); $\nu_{precession} = \nu_{irradiation}$ = 250 MHz (Section 2.2). (c) The hydrogens of the reference compound TMS define δ 0.00 (Section 5.1); the signal at δ 7.27 is due to the small amount of CHCl$_3$ present in the solvent, CDCl$_3$. (d) δ 1.26 (6H), 2.18 (3H), 2.64 (2H), 3.79 (1H). (e) The IOU is 1. The signal at δ 2.18 suggests a methyl attached to a carbonyl (accounting for the IOU of 1), while the signal at δ 3.65 may suggest an alcohol O–H. Since the remaining hydrogens are all well above δ 3.5, there must be no hydrogens attached to the hydroxyl-bearing carbon. The only structure that fits these criteria is 4-hydroxy-4-methyl-2-pentanone:

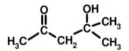

(f) $\delta(CH_3C{=}O) = 0.23 + 1.87 = 2.10$ (obs. 2.18)

$\delta(CH_2C{=}O) = [0.23 + 1.87 + 0.35 + 2(0.01)]$

$= 2.47$ (obs. 2.62)

$\delta[(CH_3)_2] = 0.23 + 0.62 + 0.35 + 2(0.01)$

$= 1.22$ (obs. 1.25)

3. (a) $Z = 6$; $N = 7$ (Section 2.1.1). (b) Since 200 δ corresponds to 5000 Hz (upper scale), the ^{13}C operating frequency must be 5000/200 = 25 MHz (Section 5.2). (c) $I = \frac{1}{2}$ (Table 2.1); this tells us that carbon nuclei can adopt only two ($2I + 1$) spin states when immersed in an external magnetic field (Section 2.2). (d) The signal at δ 0.00 is due to the carbons of reference compound TMS; the signals at δ 76.0, 77.5, and 79.0 are due to the carbon of solvent CDCl$_3$ (Section 5.4.1). (e) Expressed as integers, they are δ 41, 52, 126, 128, 129, 134, and 172. The separation between the signals at δ 134 and 172 is (172 – 134)(25 Hz/δ) = 950 Hz. (f) The IOU value (5) and the four signals in the aromatic region suggest a monosubstituted benzene ring with a relatively nonpolar substituent, while the signal at δ 172 suggests an ester carbonyl that must *not* be directly attached to the ring. The remaining signals are in the correct regions for a methyl group attached to an ester

oxygen ($\delta = -2.3 + 51 = 49$; obs. 52) and a methylene flanked by a phenyl and an ester carbonyl ($\delta = -2.3 + 20 + 23 = 41$; obs. 41). The compound is methyl phenylacetate:

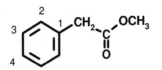

The chemical shifts of the aromatic carbons are predicted to be at the following positions (using the $\Delta\delta$ values for an ethyl substituent):

$\delta_1 = 128.5 + 15.6 = 134.1$ (obs. 134)

$\delta_2 = 128.5 + (-0.5) = 128$ (obs. 128)

$\delta_3 = 128.5 + 0.0 = 128.5$ (obs. 129)

$\delta_4 = 128.5 + (-2.6) = 125.9$ (obs. 126)

4. (a) Because there is no molecular formula, this is a tough one. The 3 : 1 hydrogen ratio indicates there is some integer multiple of 4 hydrogens. There could be 8 (in the ratio 6 : 2), or 12 (9 : 3), and so on. The ^{13}C spectrum exhibits four signals (δ 22, 28, 80, and 170). The last of these most likely represents a carbonyl carbon, perhaps in an ester group. The presence of both a carbonyl carbon and a signal near δ 2.0 in the ^1H spectrum suggests a methyl group attached to the carbonyl carbon (an acetyl group, CH$_3$C$=$O), as in Examples 6.5, 6.12, and 6.17, as well as problem 2 in Self Test I. This would require that the ^1H signal at δ 1.41 represents nine equivalent hydrogens, as in a (CH$_3$)$_3$C– group. We cannot simply connect these two fragments for two reasons. First, the resulting structure would be a ketone (rather than an ester), and its carbonyl carbon would appear farther downfield (Table 7.5). Second, the resulting molecular weight (100) would be 16 units less than the value given. Both of these difficulties can be overcome by inserting an oxygen between the two groups:

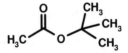

(b) The predicted ^1H chemical shifts are

$\delta(H_3CC{=}O) = 0.23 + 1.77 = 2.00$ (obs. 1.90)

$\delta[(H_3C)_3C] = 0.23 + 0.62 + 2(0.01) + 0.45$

= 1.32 (obs. 1.41)

The predicted ^{13}C chemical shifts are

$$\delta(H_3\mathbf{C}C{=}O) = -2.3 + 20 = 17.7 \text{ (obs. 22)}$$

$$\delta[(H_3\mathbf{C})_3C] = 15.8 + 10 + 6 = 31.8 \text{ (obs. 28)}$$

$$\delta[(H_3C)_3\mathbf{C}] = 16.3 + 9 + 51 = 76.3 \text{ (obs. 80)}$$

Think how much more difficult it would be to deduce this structure *without* knowing the molecular weight of the compound. Fortunately, the technique of **mass spectrometry** readily offers molecular weight and molecular formula information.

5.   (a) The 5H multiplet at δ 7.28 is characteristic of a phenyl (C_6H_5) ring attached to a nonpolar (e.g., alkyl) substituent group. The remaining atoms can be connected in only one way, benzyl alcohol:

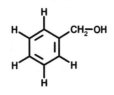

The predicted chemical shifts of the methylene hydrogens is δ 0.9(0.23 + 2.00 + 2.97) = 4.68 (obs. 4.58).
(b) The signal intensities suggest a CH_3 group and a CH_2 group. The chemical shift of the methylene group (δ 4.02) requires that it be attached directly to the deshielding chlorine substituent. Only one structure fits this requirement, 1,2,2-trichloropropane:

The predicted chemical shifts are

$$\delta(H_3C) = 0.23 + 0.62 + 2(0.70) + 0.01$$

$$= 2.26 \text{ (obs. 2.20)}$$

$$\delta(CH_2) = 0.23 + 2.80 + 0.62 + 2(0.70) + 0.01$$

$$= 5.06 \text{ (obs. 4.02)}$$

The latter calculation apparently overestimates the electron-withdrawing effects of the two β chlorines.
(c, d) Because these two compounds have the same molecular formula (and IOU of 9) but different struc-

tures, they are isomers. The close similarity of their ^1H spectra indicates that their structures are probably similar. Each isomer has a 10H multiplet near δ 7.3, suggesting the presence of two equivalent phenyl rings attached to nonpolar substituents (see problem 5a in Self Test I). The two remaining equivalent hydrogens in each spectrum near δ 4 are slightly upfield of the correct range for a hydrogen on carbon connected to both an oxygen and a phenyl, as in problem 5a. The one unaccounted-for IOU must be a ring. There are only two possible structures:

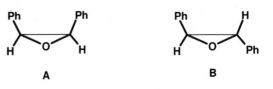

where Ph represents a phenyl group. But which structure belongs to which stereoisomer? Recalling that nuclei located directly above or below an aromatic ring experience a special shielding effect (Section 6.5 and Example 6.20), we note that **B** (the trans isomer) allows the methine hydrogens to occupy just such a position:

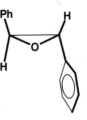

Therefore, problem (c), with the smaller value of $\delta(CH)$, corresponds to **B**; problem (d), with the larger value of $\delta(CH)$, corresponds to **A** (the cis isomer).
(e) The IOU is 1. The 18-hydrogen signal suggests two equivalent $(CH_3)_3C-$ (*t*-Bu) groups, and the remaining signal indicates two equivalent vinyl hydrogens. There are three structures with the appropriate symmetry, each shown with the predicted vinyl hydrogen chemical shift. Only **B** and **C** are close enough to consider:

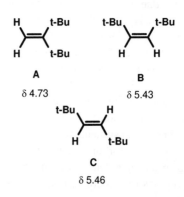

The actual structure is **C**; steric hindrance between the two *t*-Bu groups in **B** make it very unstable.

(f) The IOU is zero, a saturated molecule. The fact that there is an oxygen and a broad signal at δ 1.74 suggests an alcohol, while the nine-hydrogen signal at δ 0.98 suggests a *t*-butyl group. All that remains is a methylene connecting them: $(H_3C)_3CCH_2OH$. The predicted chemical shift of the methylene group is δ 3.46, 0.9[0.23 + 2.97 + 0.62 + 3(0.01)].

(g) An IOU of 4 and a five-hydrogen multiplet at δ 7.23 suggest a phenyl ring with one nonpolar substituent. The remaining atoms can only exist as a CH_2Cl group attached to the ring (benzyl chloride, $PhCH_2Cl$). The predicted methylene chemical shift is δ 4.53 (0.9[0.23 + 2.80 + 2.00]).

(h) The IOU of 4 and four equivalent hydrogens at δ 7.20 suggest only one compound, *para*-dichlorobenzene. The predicted chemical shift of the hydrogens is δ 7.23 [7.27 + 0.02 + (−0.06)].

(i) A very symmetrical molecule with an IOU of 4 but no aromatic or vinyl hydrogens. Since it is very difficult to get so many equivalent methylenes in one structure, we are probably dealing with two sets of multiple equivalent methyl groups, six in one set, four in the other. Further, from the chemical shifts, the environments of both sets must be quite similar. And none of these methyls can be attached to an aromatic ring, because their chemical shifts would be nearer δ 2.4. But everything fits perfectly if there are two equivalent triple bonds in the molecule:

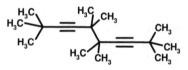

(j) An IOU of 9 and eight equivalent hydrogens in the aromatic region suggest two equivalent *para*-disubstituted benzene rings. The remaining eight equivalent hydrogens must therefore be four equivalent methylene groups, each attached to the aromatic rings. The only way to put all of this together is in [2.2]-paracyclophane:

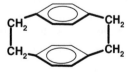

It is interesting to compare the ^1H chemical shifts of paracyclophane with those of *para*-xylene (Example 7.10), δ 2.29 and 6.97. Notice how the methylene hydrogens in paracyclophane are more deshielded by

0.80 ppm, while the aromatic hydrogens in paracyclophane are shielded by 0.45 ppm. This is a direct consequence of the proximity of the two stacked (sandwiched) and slightly bent benzene rings.

(k) To interpret ^{13}C NMR data, remember that the relative intensity of each signal is *not* a simple function of the number of carbons represented by that signal. The IOU is 4. We have 6 signals for 10 carbons, so there is some symmetry to the structure. The 4 signals from δ 114.9 to 154.7 suggest an aromatic ring with 2 different substituents para to each other. The 4 remaining carbons, represented by the signals at δ 31.5 and 33.7 ppm, are too far upfield to be attached to oxygen. Furthermore, their relative intensity suggests that the signal at δ 31.5 probably represents more carbons (with more hydrogens) than the signal at δ 33.7. Here is a structure that fulfills all these criteria:

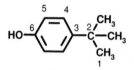

The predicted chemical shift of each carbon can be calculated as follows:

$$\delta_1 = 15.8 + 8 + 7 = 30.8 \text{ (obs. 31.5)}$$

$$\delta_2 = 16.3 + 6 + 17 = 39.3 \text{ (obs. 33.7)}$$

$$\delta_3 = 128.5 + 22.2 + (−7.3) = 143.4 \text{ (obs. 141.6)}$$

$$\delta_4 = 128.5 + (−3.4) + 1.6 = 126.7 \text{ (obs. 125.9)}$$

$$\delta_5 = 128.5 + (−0.4) + (−12.7) = 115.4 \text{ (obs. 114.9)}$$

$$\delta_6 = 128.5 + (−3.1) + 26.6 = 152.0 \text{ (obs. 154.7)}$$

(l) The IOU is 3. Once again, three signals for five carbons indicates a symmetrical structure. The signals at δ 108.3 and 121.6 are in the vinyl or aromatic range, while the δ 35.6 signal could be a methyl carbon attached to an amine nitrogen. If you read Section 7.4.3 carefully, you may remember some examples of heteroaromatic compounds. Note how closely the two downfield signals match those of pyrrole. A methyl group on the nitrogen would preserve the molecule's symmetry and have a predicted chemical shift of δ 39.7 (−2.3 + 42). Thus, the correct structure is *N*-methylpyrrole:

(m) Here we go again: seven carbons and five signals; symmetry. The IOU is 2. There appear to be no aromatic or vinyl carbons, but the signal at δ 183 is a possible carbonyl carbon. None of the other signals is downfield enough to represent a carbon attached to oxygen, though the one at δ 43.1 ppm looks like a carbon attached to a carbonyl carbon. The only way we can have a carbonyl group plus another oxygen, with no other carbons attached to that oxygen, is to have a carboxylic acid (Table 7.5). The remaining IOU suggests that there must be a ring in the structure. Here are the four possibilities:

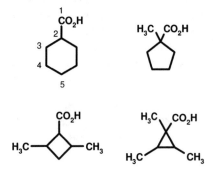

Only the first of these matches the expected chemicals shifts:

$$\delta_1 = 170 - 185$$

$$\delta_2 = 26.9 + 16 = 42.9 \text{ (obs. 43.1)}$$

$$\delta_3 = 26.9 + 2 = 28.9 \text{ (obs. 29.0)}$$

$$\delta_4 = 26.9 + (-2) = 24.9$$

$$\delta_5 = 26.9$$

The last two of these signals are too close to assign with confidence.

(n) Treating P like an N, the IOU is zero. All hydrogens are equivalent, as are the carbons. The ^{31}P chemical shift is in the range for a trialkoxy phosphite [(RO)$_3$P; Figure 7.4]. The only possibility is trimethyl phosphite, (H$_3$CO)$_3$P. In trying to estimate the ^{1}H and ^{13}C chemical shifts for such a compound, you will soon see that there are no Δδ values given for phosphorus functional groups. So, you have to find the nearest analogies from the values given or go to the chemical literature to find better ones. In this case, the methoxy chemical shifts can be estimated to be similar to a methyl ether: ^{1}H, δ 3.30 (0.23 + 2.97); ^{13}C, δ 55.7 (−2.3 + 58).

(o) The IOU is zero. The nitrogen signal suggests a saturated amine (Figure 7.5). As with the above struc-

ture, all carbons are equivalent, as are the hydrogens, so we are dealing with trimethylamine, (H$_3$C)$_3$N. The predicted chemical shifts are ^{1}H, δ 2.23 (0.23 + 2.00); ^{13}C, δ 39.7 (−2.3 + 42).

(p) The IOU is 1. Judging from the nine hydrogen signals at δ 1.59, there must be a *t*-butyl group attached to an electron-withdrawing substituent. The only possibilities are (H$_3$C)$_3$C–NO$_2$ (the nitro derivative) or (H$_3$C)$_3$C–O–N=O (the nitrite ester). Only the latter fits the nitrogen chemical shift (Figure 7.5). The ^{1}H chemical shift can be estimated from the chemical shift of a *t*-butyl carboxylate ester (δ 1.32, problem 4 in Self Test I).

6. First, notice that this structure is aromatic: a planar ring composed of all sp^2 hybridized carbons and containing 18 (4n+2) π electrons in the conjugated network of double bonds. Let us redraw the structure to show the hydrogens:

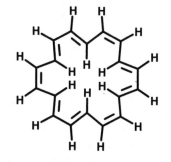

Notice how 12 (equivalent) hydrogens are located outside the aromatic ring, while the remaining 6 lie in the center of the ring. The former hydrogens are subject to an even larger than normal deshielding effect of the aromatic ring (Section 6.5) due to its larger ring current. By contrast, the hydrogens in the center of the ring experience an even more dramatic shielding by the aromatic ring current, because here the induced field is in the opposite direction to the external field (Figure 6.5).

7. The three rings of phenanthrene also comprise an aromatic system (14 π electrons). The indicated hydrogens are about 1.65 ppm more shielded than those of benzene because of the increased deshielding effects of the larger aromatic ring. However, in the ethynyl derivative, the indicated hydrogen also lies in the deshielding region of the triple bond (Figure 6.4), moving its signal even further downfield by 1.71 ppm.

CHAPTER 8

8.1. (a) $L = 2nI + 1 = 2(3)(1) + 1 = 7$.

(b) There will be $[(2)(1)+1]^3 = 27$ spin combinations distributed among the seven states:

M		Degeneracy
−3	↓↓↓	1
−2	↓↓→ ↓→↓ →↓↓	3
−1	↑↓↓ ↓↑↓ ↓↓↑ ↓→→ →↓→ →→↓	6
0	→→→ ↑↓→ ↑→↓ →↑↓ →↓↑ ↓→↑ ↓↑→	7
1	↑↑↓ ↑↓↑ ↓↑↑ ↑→→ →↑→ →→↑	6
2	↑↑→ ↑→↑ →↑↑	3
3	↑↑↑	1

8.2. Symmetry considerations indicate that there are four sets of hydrogens:

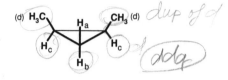

H_a and H_b are not equivalent because they are, respectively, cis and trans to the methyl groups. The two equivalent methyl groups will give rise to a 6H doublet due to three-bond coupling to H_c. The signal for H_a will be split into a doublet (due to two-bond coupling with H_b) of triplets (due to three-bond coupling with the two equivalent H_c nuclei). For the same reasons, H_b will appear as a doublet of triplets. Assuming complete resolution of all the lines, the signal for H_c will be split into a doublet (by H_b) of doublets (by H_a) of quartets (by H_d), 16 lines!

8.3. (a) The IOU is zero, so there are no rings or multiple bonds. At first glance there appear to be four signals, two low-field triplets, a quartet, and an upfield triplet. The integration of these multiplets reveal an intensity ratio of (respectively) 1 : 1 : 4 : 6, yet there are only six hydrogens! Therefore, the two lower field triplets must actually be a doublet of triplet (integrating to 1H), with the quartet integrating to 2H and the triplet to 3H. The J values in the quartet and triplet match (ca. 7 Hz, as determined from the hertz scale at the top of the spectrum). This fact, together with their chemical shifts (and similarity to Figures 8.1 and 8.13), suggests an ethoxy group. The downfield doublet of triplets (with J values of ca. 44 and 5 Hz) indicates a hydrogen split of one nearby (two-bond separation) fluorine and two other equivalent (but more distant, i.e., three-bond separation) fluorines. The two possibilities are shown below:

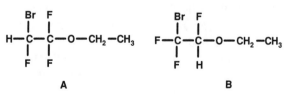

The predicted chemical shift of the lone hydrogen in **A** is δ 8.16 (Table 6.2), while that for **B** is δ 9.15. The observed value, δ 6.25, is somewhat closer to **A**, which is, indeed, the correct structure. Once again, this is a situation where there are too many deshielding substituents to allow for a really accurate prediction of chemical shift (see Example 6.9). (b) δ 1.31 (t, $J = 7$, 3H), 4.02 (q, $J = 7$, 2H), 6.25 (dt, $J = 44$, 5, 1H).

8.4. (a, b) The IOU of 4 and the fact that all the lines are in the region of aromatic carbons suggest that the molecule is one of the isomers of difluorobenzene. That there are 12 lines for 6 carbons indicates the presence of C–F coupling. A careful look at the spacings between lines suggests that the first 3 lines (starting from δ 103.0) comprise a triplet with a 26-Hz [(1.3 ppm)(20 Hz/ppm)] coupling constant; the next two constitute a doublet ($J = 24$ Hz), the next three a triplet ($J = 10$ Hz), followed by two doublets, both with $J = 12$ Hz; a total of 5 signals. But wait! Problem 7.3 showed that whenever a benzene ring is substituted by two identical substituents (in this case, fluorine), the ortho isomer will exhibit three carbon signals, the meta will show four, and the para, just two. How can we reconcile this with this spectrum? Easy! The last two doublets are actually one signal split into a doublet of doublets, with J values of 248 and 12 Hz. Now this spectrum fits our expectations for the meta isomer:

Carbon C_d is predicted (from Table 7.4) to appear at δ 164.5 (the center of the doublet of doublets is at δ 164.9) and be split into a doublet (by the directly connected fluorine, $^1J = 248$ Hz) of doublets (by the other fluorine, $^3J = 12$ Hz). Carbon C_a should appear at δ 99.9 (observed: δ 104.3) and be split into a triplet by the two equivalent fluorines (with $^2J = 26$ Hz). Similarly, C_c will appear at δ 130.3 (observed: δ 131.0) and be split into a triplet by the two equivalent fluorines (with $^3J = 10$ Hz). Finally, C_b will appear at δ 109.7 (observed: δ 111.5) and be split into a doublet (by the closer fluorine, $^2J = 24$ Hz); it will not be coupled to the fluorine four bonds away.

8.5. (a, b) With an IOU of 12 (treat the phosphorus as a nitrogen), relatively few signals all in the aromatic region, and a ^{31}P chemical shift consistent with a triaryl phosphine (Figure 7.4), this must be a highly symmetrical molecule consisting of three phenyl groups attached to the phosphorus, triphenyl phosphine:

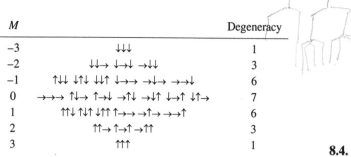

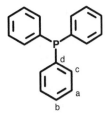

It is going to be difficult to estimate carbon chemical shifts because Table 7.4 does not include data for any phosphorus substituents. So instead, let us see if this structure at least fits the observed carbon–phosphorus couplings. We would predict that three of the carbon signals should be split into doublets, with coupling constants decreasing in the order $C_d > C_c > C_a$ (C_b would involve four-bond coupling). The doublet centered at δ 133.6 has the largest coupling constant (20 Hz), followed by the one centered at δ 137.2 ($J = 12$ Hz) and then the one centered at δ 128.6 ($J = 8$ Hz). Carbon C_b must be the uncoupled signal at δ 128.8. Although the structural assignment is correct, our assignment of ^{13}C signals for C_d and C_c turns out to be reversed. In this case, the two-bond J_{PC} is actually larger (20 Hz) than the one-bond coupling (12 Hz). The reasons for this reversal will be discussed in Chapter 9.

8.6. (a) After reviewing Table 8.1, we see the characteristic trace of $CHCl_3$ at δ 7.24 in Figure 8.18a as well as the $CDCl_3$ triplet at δ 77.0 in Figure 8.18b; so compound **A** was the one dissolved in $CDCl_3$. Figure 8.19b shows a septet at δ 1.24 and a singlet at δ 118.1, both characteristic of CD_3CN; the very weak multiplet at δ 1.94 in Figure 8.19a confirms this as the solvent for compound **B**. (b) This quartet, with a J value of ca. 320 Hz, is the carbon signal for the $CF_3SO_3^-$ (triflate) counterion split into a quartet by the three fluorines. (c) The signals at ca. δ 7.5 and 7.7 are both triplets, while the signal at δ 8.1 is a doublet. This is the pattern expected (Example 8.18) for the meta, para, and ortho hydrogens, respectively, of a monosubstituted benzene ring. (d) In comparing the molecular formulas of **A** and **B** with the generic structure ($R + C_9H_5F_3IO_3S$) of these compounds, the R group in **A** must be $C_2H_3O_2$, while that in **B** must be CH_2Cl (consistent with the ^1H singlet at δ 4.50 and the ^{13}C signal at δ 31). The ^1H signal at δ 3.73 (Figure 8.18a) and the ^{13}C signals at δ 54 and 151 (Figure 8.18b) in the spectra of **A** leave no doubt that the R group is carboxymethyl [$CH_3OC(=O)$]:

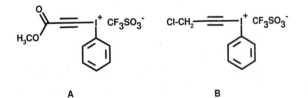

A   B

The table below summarizes the as-yet unassigned ^{13}C signals for each compound:

A	B
135.0	135.8
133.0	134.0
132.5	133.3
116.5	117.4
94.8	102.8
36.0	30.5

We can assign the signals at ca. δ 117 to the aromatic carbon directly attached to (and shielded by) the iodine. Although the remaining aromatic carbons are quite close together, we might guess that their relative chemical shifts parallel those of the attached hydrogens (Section 7.9). Thus, the most downfield carbon (δ 135–136) should correspond to the ortho hydrogen, with the para next (δ 133–134), then the meta (δ 132–133). These assignments are also supported by the relative intensities of the signals and is indeed correct.

We are left with the two signals at δ 95/103 and δ 30/36. Obviously these must be due to the triply bonded carbons. The former signal is slightly deshielded from the position of typical acetylenic carbons (Section 7.5), and this can be ascribed to the deshielding effect of the positively charged iodine. The latter signal, however, is anomalously shielded by 50 ppm! This shows that diamagnetic shielding by the vast electron density around iodine (with its 53 electrons) much more than offsets the effect of its positive charge. [See Williamson, B. L., Stang, P. J., and Arif, A. M., *J. Am. Chem. Soc.*, **115**, 2590 (1993).]

8.7. (a) Careful inspection of this molecular structure should convince you that all four phosphorus atoms are nonequivalent and give rise to the four ^{31}P signals centered at δ 5.1, 9.4, 13.6, and 21.3 (somewhat downfield of uncomplexed triphenyl phosphine; problem 8.5). Each of these signals is split into a complex multiplet (e.g., a doublet of doublets of doublets) by the other three phosphorus nuclei, though the exact multiplicity (and associated coupling constants) of each signal cannot be determined without a further expansion of each signal. The six small, somewhat broad signals (δ −5.5, 4.5, 5.8, 11.5, 31.1, and 32.7) are the ^{195}Pt satellites for three of the four ^{31}P signals. The easiest way to determine which satellites belong to which signals is to take a pair of dividers and find which pair of satellites are equally spaced around each signal. Thus, the satellites at δ −5.5 and 32.7 are equally distant from the signal at δ 13.6, indicating a Pt–P coupling constant of (38.2

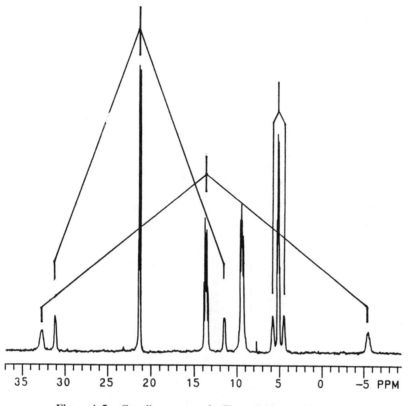

Figure A.5. Coupling patterns for Figure 8.20*a*, problem 8.7(a).

ppm)(121.4 Hz/ppm) = 4640 Hz. In like manner, the satellites at δ 11.5 and 31.1 belong to the signal at δ 21.3 (*J* = 2380 Hz), while those at δ 4.5 and 5.8 belong to the signal at δ 5.1 (*J* = 160 Hz). The Pt–P coupling to the signal at δ 9.4 is too small (< ca. 90 Hz) to be resolved under these conditions. Note that the combined intensity of each pair of satellites corresponds to ca. one-half of the intensity of the main signal, as required by the 33.8% natural abundance of ^{195}Pt. These couplings are shown in Figure A.5. Clearly, the signals with the large Pt–P couplings must correspond to the phosphorus atoms directly bonded to platinum, while the other two signals belong to those attached to the iridium. But we do not yet have a basis for deciding which signal of each pair belongs to which specific phosphorus. We will learn more about how to do this in Chapter 9.

(b) This beautifully symmetrical pattern appears to consist of three main multiplets in the center and their corresponding ^{195}Pt satellites flanking on either side. The center-to-center spacing between the satellite (1.72 ppm, or 515 Hz) is the one-bond Pt–H coupling constant. Careful analysis of the main peaks reveals them to be a doublet (*J* = 0.26 ppm, or 78 Hz) of doublets (*J = 0.19 ppm, or 57 Hz) of triplets (J = 0.03 ppm, or 9 Hz*), due to coupling with two nonequivalent phospho-

rus nuclei and another set of two equivalent (from the standpoint of *J* value) phosphorus nuclei (see Figure A.6). As in part (a), we do not at this point have enough information to decide which phosphorus goes with which coupling. [See Huang, Y.-S., Stang, P. J., and Arif, A. M., *J. Am. Chem. Soc., 112,* 5648 (1990); also *Organometallics, 11,* 845 (1992).]

8.8. (a) Careful consideration of the diol's molecular structure indicates that all hydrogens, except those in each methyl group, are nonequivalent. Even the pairs of methylene hydrogens (H$_d$, H$_e$; H$_f$, H$_g$; and H$_j$, H$_k$) are diastereotopic (Section 4.3) and therefore nonequivalent. Thus, the molecule should give rise to 12 ^{1}H signals:

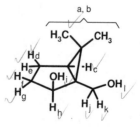

As you can see from Figure 8.21, except for the complex 2H multiplet at δ 1.74–1.96, all 10 of the other signals

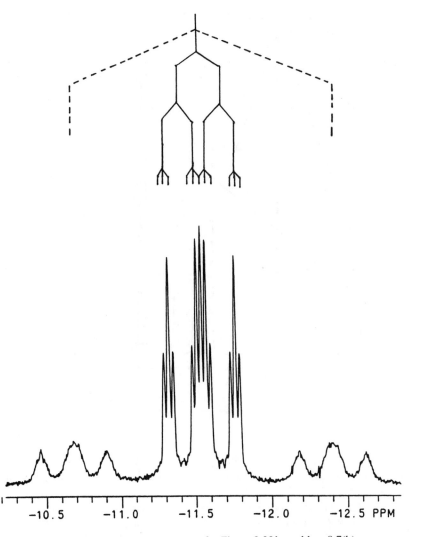

Figure A.6. Coupling patterns for Figure 8.20*b*, problem 8.7(b).

are distinct. Let us see how far we can get with the assignments. First, the methyl groups appear as 3H singlets at δ 1.17 and 1.37, though without the kind of information we will discuss in Chapter 13 we cannot be sure which methyl gives rise to which signal. The two doublet of doublets at δ 3.80 and 3.95 correspond to two hydrogens near an oxygen that are coupled both to each other ($J = 11$ Hz) and essentially equally ($J = 5$ Hz) to one other hydrogen; these must be H_j and H_k. They are both coupled to the hydroxyl hydrogen H_l (no exchange, remember?), so its signal is the triplet at δ 2.03. Hydrogen H_h, also near an oxygen, must give rise to the signal at δ 4.82, which is a doublet of doublets of doublets with J values 10, 7, and 4 Hz. The 4-Hz coupling matches that in the doublet at δ 2.40, which must therefore belong to H_i. The 10-Hz coupling matches one in the multiplet at δ 2.24 and one in the multiplet at δ 1.50; therefore, these latter two signals

must belong to H_f and H_g, which also couple each other with a J of 14 Hz. The complex multiplet at δ 1.74–1.96 must be due to H_d and H_e. Finally, cyclopropyl hydrogen H_c is the doublet at δ 0.97. Based solely on the number of intervening bonds, it should be split into a doublet of doublets (by H_d and H_e), but in fact only one of these couplings is large enough to observe; more on this in Section 9.5.

CHAPTER 9

9.1. The ^{1}H-decoupled ^{31}P spectrum in Figure 8.20*a* is actually two spectra, one for the molecules with a ^{195}Pt nucleus (34% of the molecules) and one for the molecules with other nonmagnetic isotopes of platinum. The latter molecules have a spin system consisting solely of four nonequivalent phosphorus nuclei; the hydrogens can be neglected because of decoupling. Inspection of the spectrum indicates that the differences in chemical shifts are much greater than the coupling constants, so the Pople label for this homonuclear spin system would require four letters in the alphabet as far from each other as possible, something like AJQX. The heteronuclear ^{195}Pt spin system will have four nonequivalent spins of one element and a fifth spin of a different element to give something like an ADGJX system.

Figure 8.20b is also two ^{1}H spectra of the bridging hydrogen, coupled to the four phosphorus nuclei and, in 34% of the molecules, ^{195}Pt. In the former case the Pople label would be AGJMP (where A represents the hydrogen). In the molecules with ^{195}Pt, the label would be AGJMPX.

9.2. In this structure there are three sets of chemically equivalent hydrogens (in the ratio 1 : 2 : 1) and one set of chemically equivalent fluorines. But none of the nuclei are magnetically equivalent because, for example, the ortho H–F coupling is not equal to the para H–F coupling. Therefore, this molecule is an example of an AA′BCXX′ spin system:

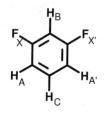

9.3. It may help you to make a model of this compound. There are only two hydrogens (H_d and H_e) within three bonds of H_c. But careful inspection of the three-dimensional relationship between these three nuclei (shown in the diagram below) reveals that H_c makes a dihedral angle of 30° with H_e (for a ^3J value of 8 Hz) and a dihedral angle of 86° with H_d (for a ^3J value of ca. 1 Hz):

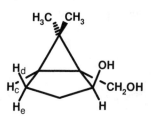

The latter coupling is too small to observe.

9.4. (a, b) The presence of phosphorus should alert us to the possibility of heteronuclear coupling in the ^{1}H spectrum. Thus, the signal centered at δ 5.88 must be a doublet (J_{PH} = 28.5 Hz) of septets (J_{HH} = 3 Hz). The septet must result from coupling to six equivalent hydrogens. Therefore, the signal at δ 1.88 must be a

doublet (J_{PH} = 12 Hz) of doublets (J_{HH} = 3 Hz). The former signal is likely a vinyl hydrogen engaged in two-bond P–H coupling. The latter probably arises from two equivalent methyl groups attached to a vinyl carbon, with perhaps three-bond coupling to the phosphorus and long-range coupling to the other hydrogen. We might therefore suggest this structure:

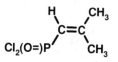

Unfortunately, there are two major problems with this structure. First, the two methyl groups are not equivalent (one is cis, the other trans, to the vinyl H). More importantly, we are still missing a carbon. The correct structure is the corresponding allene:

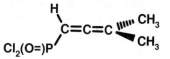

The only seemingly unusual features of this spectrum are the relatively large long-range couplings: the five-bond H–H coupling of 3 Hz and the five-bond P–H coupling of 12 Hz. But as you can see from Table 9.5, these values are just exactly what we should have predicted. (c) The ^{31}P signal would appear as a doublet ($^2J_{PH}$ = 28.5 Hz) of septets ($^5J_{PH}$ = 12 Hz).

9.5. (a, b) The presence of phosphorus should alert us to the possibility of heteronuclear coupling in the ^{13}C spectrum. In fact, judging from the nearly identical intensities, the spectrum seems to consist of five doublets with the following J_{PC} values: δ 27.0 (J = 16 Hz), 52.8 (4 Hz), 85.25 (10 Hz), 114.5 (160 Hz), and 156.8 (16 Hz). The 160-Hz coupling is typical for a one-bond C–P coupling involving a vinyl carbon (Table 9.2), and this carbon must be attached to the only other vinyl carbon at δ 156.8. Since we are apparently dealing with a phosphonate, the signals at δ 52.8 and 85.25 are likely to be the carbons directly bonded to phosphonate oxygens, each with J values in the range typical for two-bond P–O–C coupling. The latter signal is apparently deshielded by more than just the oxygen, perhaps the double bond mentioned earlier. This suggests a five-membered oxaphospholene ring system:

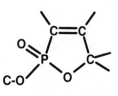

We are left with 2 carbons and 11 hydrogens to complete the structure. The two carbons must either be equivalent and coupled to phosphorus or nonequivalent and not coupled to phosphorus. Thus, they must be methyl groups attached to the same carbon (and therefore slightly nonequivalent; see structure **8-3** in Example 8.14). So, the complete structure of the compound is

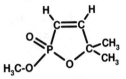

(c) Remember that the multiplicity of the ^{31}P signal (assuming there is no proton decoupling) is dependent on coupling to the neighboring hydrogens, not the ^{13}C's. Since we have established that the two methyls are nonequivalent and not coupled to phosphorus, the ^{31}P signal will be split into a doublet (by the trans H, $^{3}J = 50$ Hz, Table 9.4) of doublets (by the geminal H, $^{2}J = 33$ Hz, Table 9.3) of quartets (by the methyl hydrogens, $^{3}J = 5–15$ Hz, Table 9.4).

CHAPTER 10

10.1. Using the data in Chapter 6, the singlets in the 120°C ^{1}H spectrum of compound **10-15** can be assigned as shown below:

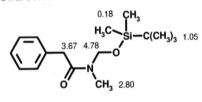

10-15

Compound **10-15**, like DMF (**10-1**), is an amide capable of existing in two relatively stable conformations (**10-15A** and **B**), and each conformation gives rise to a separate set of signals below 40°C:

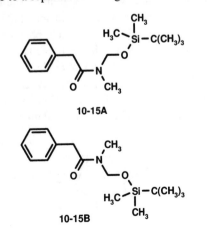

10-15A

10-15B

The fact that the two sets of singlets have essentially identical intensity indicates the two conformations are equally populated. As the temperature increases, the rate of interconversion of the two conformations increases. The pairs of singlets coalesce near 80°C and thereafter exhibit single sharp lines. Note that the pairs of singlets with the larger values of Δν (the methylene and *N*-methyl signals) requires slightly higher temperature to coalesce than the *Si*-methyl and *t*-butyl signals.

10.2. Ester **10-17** is actually a mixture of two *diastereomers* (Section 4.3), the **R,R** and the **S,R**. Diastereomers, unlike enantiomers, do not exhibit identical properties or identical spectra:

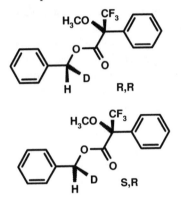

Therefore the benzyl hydrogen resonances, one for each structure, will occur at slightly different positions. The fact that the two signals are of equal intensity confirms that the starting alcohol **10-12** was indeed an equal (racemic) mixture of the two enantiomers.

10.3. The five bonds emanating from phosphorus in **10-18** are directed toward the corners of a trigonal bipyramid (TBP). In a TBP there are two apical (a) positions and three equatorial (e) positions. In **10-18** one of the methoxy groups occupies an apical position, while the other two occupy equivalent equatorial positions:

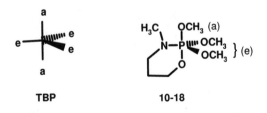

TBP   **10-18**

In addition to small upfield shifts in δ for the other signals, the most significant change in the ^{13}C spectrum of **10-18** going from −90 to 20.4°C is this: The doublets at δ 52.1 and 56.15 coalesce into one doublet at δ 54.5. [Since the doublet at δ 52.1 represents the apical methoxy group, while the signal at δ 56.15 represents two equatorial methoxy groups, the predicted average position is $\frac{1}{3}$ (2 × 56.15 + 52.1) = δ 54.8.] This observation indicates that, at room temperature, the positions of the three methoxy groups are somehow exchanging rapidly, making them all equivalent on the NMR time scale. The process that accomplishes this exchange, **pseudorotation**, involves the simultaneous interconversion of two apical groups (**B**, with a bond angle of 180°) with two equatorial groups (**C**, bond angle 120°). The third equatorial position (**A**) remains equatorial and is called the pivot bond:

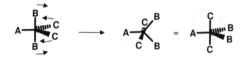

When the phosphorus atom is part of a ring, there are certain restrictions to pseudorotation, the most important being that a small ring cannot span two apical positions. Therefore, when **10-18** undergoes pseudorotation, one of the equatorial methoxy groups serves as the pivot while the other equatorial methoxy exchanges position with the apical methoxy:

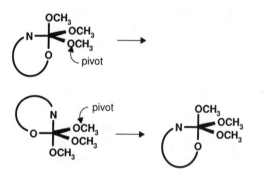

Through a continuing series of these pseudorotations all three methoxy groups become equivalent.

10.4.   Equation (10.12) can be recast in the form

$$\frac{X_C}{K[H]_0} = (1 - X_C)(r - X_C) = r - (r+1)X_C + X_C^2$$

or

$$X_C^2 - \left(1 + r + \frac{1}{K[H]_0}\right) + r = 0$$

The real root of this quadratic equation is given by

$$X_C = \frac{b - \sqrt{b^2 - 4r}}{2}$$

where

$$b = 1 + r + \frac{1}{K[H]_0} \qquad (10.17)$$

Substitution of Eq. (10.17) into Eq. (10.13) gives Eq. (10.15).

10.5.   We recognize that complex **10-20** is a very close analog of chiral shift reagent **10-13**. Furthermore, amine **10-19** is itself chiral, and since it is not otherwise mentioned, we can assume it is a racemic (equal) mixture of the R and S enantiomers. Complexation of racemic **10-19** with **10-20** will first shift downfield all the signals in the 60-MHz ^{1}H spectrum. Since the site of complexation is likely to be the unshared electron pair on nitrogen (not explicitly shown), we note as expected that the downfield shift is greatest for the methine hydrogen (quartet, $\Delta\delta_{obs}$ = 12.9 ppm), followed in order by the methyl hydrogens (doublet, $\Delta\delta_{obs}$ = 9.6 ppm), ortho hydrogens (doublet, $\Delta\delta_{obs}$ = 5.9 ppm), meta hydrogens (triplet, $\Delta\delta_{obs}$ = 1.9 ppm), and para hydrogen (triplet, $\Delta\delta_{obs}$ = 1.6 ppm). (The NH$_2$ hydrogens are too broad to be seen because of exchange processes and quadrupolar coupling.) In addition to these dramatic downfield shifts, the individual signals for each enantiomer are resolved into pairs of identical multiplets by virtue of the asymmetric environment provided by the enantiomerically pure shift reagent.

10.6.   (a) Clearly we are observing the formation of a complex between the host (**10-21**) and the guest resorcinol (**10-22**). The structure of the resulting supramolecular complex must somehow account for the upfield shift of the circled hydrogen in **10-22**. By redrawing the structure of **10-21**, or better yet by making a model, we can appreciate how this complex is formed. Note that the host molecule forms a three-dimensional U-shaped cavity:

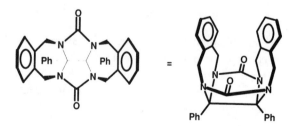

The resorcinol molecule fits inside this cavity, with its two hydroxyl hydrogen bonded to the two carbonyl oxygens of **10-21**. This arrangement also has the three aromatic rings stacked parallel in a "sandwich." The combination of hydrogen bonding and aromatic π-stacking place the circled hydrogen of resorcinol directly in the shielding region of the two upper aromatic rings in **10-21**, accounting for the observed shielding:

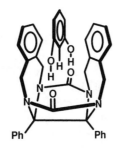

(b) From the data given in part (a), the value of $\Delta\delta = \delta_H - \delta_C = 6.50 - 3.80 = 2.70$ ppm. When the signal is at δ 5.15, the resorcinol is exactly half uncomplexed and half complexed $\{[\textbf{10-22}] = [\text{complex}] = 0.010$ $M/2 = 0.0050\ M$, Eq. (10.14)$\}$. Therefore, the concentration of uncomplexed **10-21** is $0.00538 - 0.0050 = 0.00038\ M$. Now we can use Eq. (10.10): $K = (0.0050)/(0.0050 \times 0.00038) = 2600$.

10.7. The longest way to do this is to take the first derivative of Eq. (10.1), set it equal to zero, and solve the resulting cubic equation for three values of v that correspond to maxima and/or minima. Here is a shortcut. We will adjust our chemical shift scale such that v_{av} is defined as zero, so that $v_A = -v_B$. Now Eq. (10.1) takes the form

$$I(v) = \frac{Ck\ \Delta v_0^2}{-k^2\ v^2 + 4\pi^2\ (v_A^2 - v^2)^2}$$

$$\frac{dI(v)}{dv} = \frac{-(Ck\ \Delta v_0^2)\ [-2k^2 v + 8\pi^2\ (v_A^2 - v^2)(-2v)]}{[-k^2\ v^2 + 4\pi^2\ (v_A^2 - v^2)^2]^2}$$

Setting the simplified numerator to zero, we are left to solve the equation:

$$0 = -k^2 v - 8\pi^2 v\ (v_A^2 - v^2)$$

One solution to this cubic equation is $v = 0$, which corresponds to the minimum at $v = v_{av}$. The other two roots (maxima) can be found by dividing the above equation by v, then solving the resulting quadratic equation:

$$0 = -k^2 - 8\pi^2\ (v_A^2 - v^2)$$

$$v = \pm\ \sqrt{v_A^2 + \frac{k^2}{8\pi^2}}$$

Since Δv, the separation between the two maxima, is simply $2v$, it follows that

$$\Delta v = 2\ \sqrt{v_A^2 + \frac{k^2}{8\pi^2}}\ =\ \sqrt{4v_A^2 + \frac{k^2}{2\pi^2}}$$

Finally, since $\Delta v_0 = v_A - v_B = 2\ v_A$, $4v_A^2$ is nothing more than Δv_0^2.

CHAPTER 11

11.1. (a)

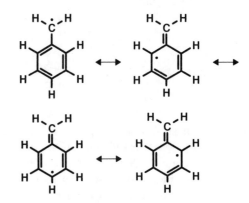

(b) The unpaired electron should be coupled to the two equivalent α hydrogens, two equivalent ortho hydrogens, and one para hydrogen. (On the basis of the above resonance structures, no significant hyperfine coupling is expected to the meta hydrogens.) If all three a values are different, the spectrum will consist of a triplet (due to the α hydrogens) of triplets (due to the ortho hydrogens) of doublets (due to the para hydrogen), a total of 18 lines. All three a values should be negative, a consequence of the one-bond separation between the unpaired electron and the coupling hydrogen in each resonance structure. (c) Recall that the relative magnitude of the a value to each hydrogen is directly proportional to the unpaired electron density at the carbon to which that hydrogen is bonded. Begin by adding up the total (A) of the a values at *each* hydrogen: $A = 2(a_\alpha) + 2(a_{ortho}) + 2(a_{meta}) + (a_{para}) = 4.89$. The fraction of time the unpaired electrons spend on the meta carbons is, therefore, $2(a_{meta})/A = 0.06$. Thus, the unpaired electron spends only 6% of its time on the meta carbon, in complete accord with our qualitative expectations based on the resonance structures above. It is also interesting to note in passing that the relative magnitude of these a values are also very close to predictions of unpaired electron density based on sophisticated molecular orbital calculations.

11.2. (a) The electron will be placed in the lowest (energy) unfilled molecular orbital (LUMO) available in the molecule. In this case, that orbital is ψ_3. (b) The size of the orbital lobes in ψ_3 shows that the unpaired electron will be distributed over all four carbons, spending more time on the terminal carbons (C1 and C4) than on the internal carbons (C2 and C3). This means that the unpaired electron will be coupled more strongly to the four equivalent hydrogens attached to C1 and C4 than to the two equivalent hydrogens attached to C2 and C3. Therefore, the spectrum of the radical anion will appear as a quintet of triplets. (c) Clearly the larger a value (0.76 mT) corresponds to the hyperfine coupling with the C1/C4 hydrogens and the smaller one to the coupling with the C2/C3 hydrogens.

11.3. The formation of PhS–CH$_2$CH$_3$ must involve escape from the cage, exactly analogous to the formation of compound **11-12** from singlet radical pair **11-13** (Example 11.12). Therefore, we expect the same type of multiplet effects observed for the formation of **11-12**, namely *A/E* for both the quartet and the triplet.

11.4. At first glance this appears to be a weird result: One of the *reactants* is exhibiting CIDNP effects! This must indicate that the ethyl iodide (CH$_3$CH$_2$I) is both a reactant *and* a product. That is, there must be some type of process occurring that regenerates ethyl iodide molecules via a radical-pair mechanism. The absence of a net effect suggests that the polarized ethyl iodide arises from a symmetrical radical pair, for example, CH$_3$CH$_2\cdot$ $\cdot$CH$_2$CH$_3$. If this radical pair were born in the singlet state (as it should be under these thermal conditions), escape of an ethyl radical from the cage, followed by abstraction of an iodine atom from another molecule of ethyl iodide, would result in the formation of a new (polarized) ethyl iodide molecule and a new ethyl radical:

$$H_3CCH_2\text{-Li} \longleftrightarrow H_3CCH_2\text{:}^- \text{ Li}^+$$

$$H_3CCH_2\text{:}^- \text{ Li}^+ + H_3CCH_2\text{-I} \xrightarrow[\text{-Li}^+]{\text{SET}} H_3CCH_2{\uparrow} \quad {\downarrow}(\text{I-CH}_2\text{CH}_3)^-$$

$$\text{radical anion of ethyl iodide}$$

$$H_3CCH_2{\uparrow} \quad {\downarrow}(\text{I-CH}_2\text{CH}_3)^- \xrightarrow{\text{-I}^-} H_3CCH_2{\uparrow} \quad {\downarrow}\text{CH}_2\text{CH}_3$$

$$\Big\downarrow \begin{array}{l}\text{1. escape}\\ \text{2. } H_3CCH_2\text{-I}\end{array}$$

SET = single electron transfer

$$H_3CCH_2\text{-I} + H_3CCH_2{\uparrow}$$

$$\text{polarized}$$

The CH$_2$ quartet of the polarized ethyl iodide molecule should exhibit a multiplet effect given by

$$\Gamma_m = \mu\varepsilon a_i a_j J_i \,\sigma = (-)(-)(-)(+)(+)(+) = (-)$$

which corresponds to an *A/E* multiplet effect. The same result is predicted for the CH$_3$ triplet, exactly as experimentally observed.

11.5. All we need to do is take the first derivative of Eq. (3.10):

$$L_\nu = \frac{a\Delta v_i^2}{\Delta v_i^2 + b\,(v - \Delta v_i)^2}$$

$$\frac{dL_\nu}{dv} = a\,\Delta v_i^2 \,\frac{d}{dv}\,[\Delta v_i^2 + b\,(v - \Delta v_i)^2]^{-1}$$

$$= -ab\,\Delta v_i^2\,[\Delta v_i^2 + b\,(v - \Delta v_i)^2]^{-2}\,\frac{d}{dv}\,(v - \Delta v_i)^2$$

$$= \frac{-2ab\,\Delta v_i^2\,(v - \Delta v_i)}{[\Delta v_i^2 + b\,(v - \Delta v_i)^2]^2}$$

Figure 11.2 was drawn using the same parameters as Figure 3.18: $a = 1.00$ intensity units, $\Delta v_i = 5.00$ frequency units, and $b = 200$.

CHAPTER 12

12.1 These NOE results indicate that the signal at δ 0.96 belongs to the *tert*-butyl group *closer* to the C–H hydrogen:

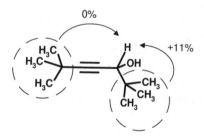

[See Elder, R. C., Florian, L. R., Kennedy, E. R., and Macomber, R. S., *J. Org. Chem.*, **38**, 4177 (1973).]

12.2. Consider the behavior of the three components of a triplet, compared to the two components of a doublet shown in Figure 12.11*b*. The central component of the triplet, comprising half the intensity of the signal, is stationary along the $+y'$ axis. The other two components (totaling the other half of the intensity) precess at $+J$ and $-J$ in the rotating frame (twice as fast as the components of the doublet), as shown below:

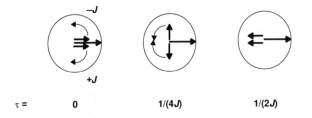

$$\tau = \qquad 0 \qquad\qquad 1/(4J) \qquad\qquad 1/(2J)$$

Thus, half the signal will be independent of τ, while the other half will exhibit cosine dependence at twice the frequency as the doublet did. This results in a signal given by the equation $\frac{1}{2} + \frac{1}{2}[\cos(2\pi J\tau)]$, equivalent to Eq. (12.3b).

12.3. (a) The septet at δ 40 is the carbon signal of deuterated DMSO (Table 8.1). It does not show up in any of the DEPT subspectra because DEPT only shows signals arising from polarization transfer from the ^{1}H nuclei in the molecule. *Deuterated* DMSO has no ^{1}H nuclei!
(b) The generic structure of the polymer is shown below:

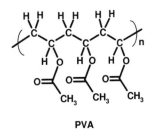

PVA

Based solely on this structure there are three types of carbons expected in the $\delta < 80$ region: the CH_3 at δ 21, the CH_2 at δ 39, and the CH at δ 68; these chemical shifts correspond to expectations based on the data in Section 7.2. (The ester carbonyl carbons show up around δ 170; Table 5.5.) But why are each of these signals composed of several lines and not just singlets? The answer lies in the exact stereochemical structure of the polymer. Recognizing that the CH carbon is a chiral center (Section 4.3), we see that there are two obvious stereoisomers:

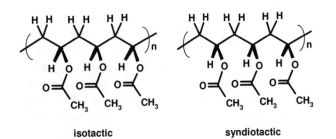

isotactic   **syndiotactic**

Less obvious is that the stereochemical pattern is not limited to only isotactic (same stereochemical configuration at each chiral center) or syndiotactic (alternating stereochemical configuration at each chiral center). The polymer molecule can have sections of *both* stereochemical arrangements in the same molecule as well as random section (atactic). This sample of PVA clearly has at least four different stereochemical regions.

12.4. (a) In addition to the phenyl protons in the region δ 7.1–7.5 there are four signals that can be assigned (from the data in Section 6.2) as follows:

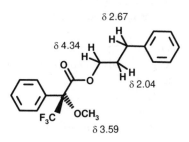

As described in review problem 10.2, the chiral center of the ester renders the two hydrogens in the closest (δ 4.34) CH_2 group diastereotopic (Section 4.3) and hence nonequivalent. Thus, the two hydrogens are coupled not only to the two essentially equivalent hydrogens at δ 2.04 but also to each other, giving two overlapping six-line patterns (triplets of doublets) for A and B of an ABC_2D_2 spin system. The triplet at δ 2.67 simply results from the two equivalent hydrogens being split by the two equivalent δ 2.04 hydrogens. (b) Irradiating the multiplet at δ 2.04 will decouple the center CH_2 group from the other two. The triplet at δ 2.67 collapses to a singlet, while the multiplet at δ 4.34 collapses to an AB quartet (Section 9.11). See Figure A.7; note also the zero beat pattern at the site of irradiation.

CHAPTER 13

13.1. See Figure 13.20.
13.2. (a) Except for the methyl hydrogens there are no other equivalencies, so we expect there will be 18 signals:

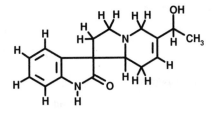

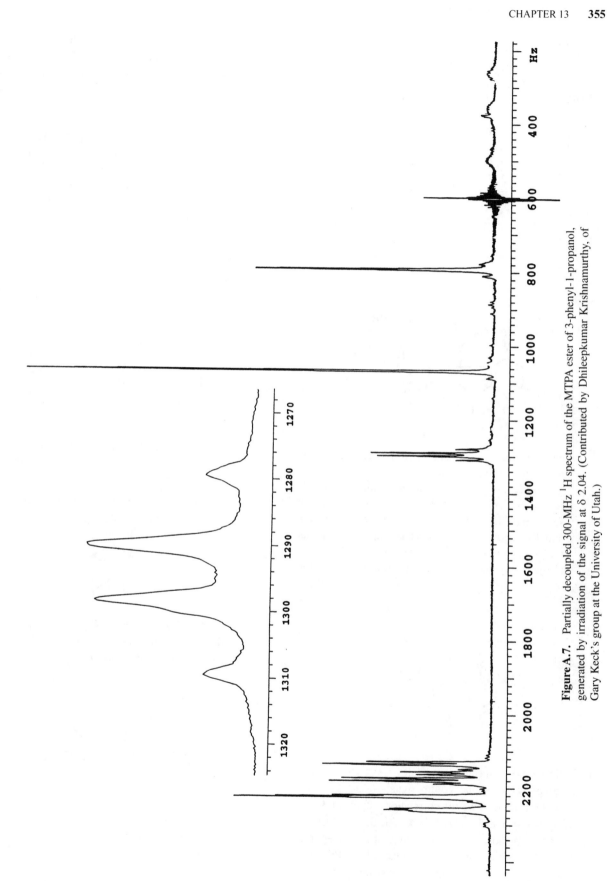

Figure A.7. Partially decoupled 300-MHz ^{1}H spectrum of the MTPA ester of 3-phenyl-1-propanol, generated by irradiation of the signal at δ 2.04. (Contributed by Dhileepkumar Krishnamurthy, of Gary Keck's group at the University of Utah.)

(b)

Hydrogen	Is Coupled to Hydrogen(s)	Hydrogen	Is Coupled to Hydrogen(s)
1	10	9	7
2	6, 7, 11	10	1
3	4, 5, 6	11	2
4	3, 5, 8	12	14
5	3, 4, 8	13	14, 15
6	2	14	12, 13
7	2, 9	15	13
8			

(c)

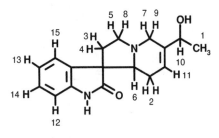

Note that the cross peak between signals **2** and **7** must result from a five-bond long-range interaction.

13.3. Assuming unrestricted rotation around the single bonds, this molecule has nine sets of hydrogens: one set for the CH$_2$ group, three sets on the phenyl group, and five sets on the anthracene ring. From signal integrations it is reasonable to assign signal **1** (δ 5.35) to the methylene hydrogens, signal **2** (δ 7.50) to the unresolved phenyl hydrogens, and signal **6** (δ 8.78) to H$_\epsilon$:

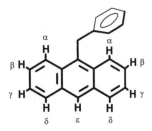

The COSY spectrum indicates weak (long-range) couplings of signal **1** with signals **6** and **2**. In the anthracene ring we see correlations of signal **3** with signals **4** and **5** as well as **4** with **5**. The NOESY spectrum shows all the same correlations except that the **4/5** correlation has been replaced by a **4/6** correlation. The only assignments consistent with these data are: signal **3**, H$_\beta$ and H$_\gamma$; signal **4**, H$_\delta$; and signal

5, H$_\alpha$. Note how the through-bond para coupling between H$_\alpha$ and H$_\delta$ (signals **5** and **4**) in the COSY spectrum is replaced by a through-space NOE interaction between H$_\delta$ and H$_\epsilon$ (signals **4** and **6**) in the NOESY spectrum.

13.4. (a) The IOU for this molecular formula is 4; the three lowest field carbon signals suggest that we are dealing with a conjugated ketone. Integrating the C–H connectivity information from the C,H–HSC and the H–H coupling information from the COSY, we can quickly arrive at the following structural fragments (hydrogen signal assignments are in parentheses):

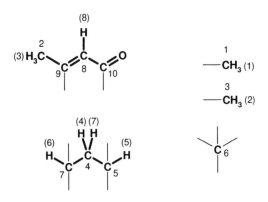

The proximity of H2 to H1, H5, H6, and H7 can be accommodated in two ways, differing in the position of carbons 5 and 7:

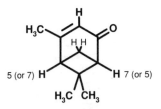

(b) The 2D INADEQUATE makes the final carbon skeleton assignment trivial; hydrogen stereochemistry is based on NOE results:

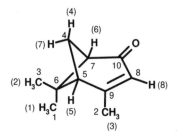

This compound is known as verbenone.

13.5. (a) Note that the IOU value for this molecular formula is 5. Assuming that carbon signal **15** (δ 172) belongs to an ester carbonyl, there seem to be no other unsaturated carbons. This suggests that the structure has four rings.

C	Bears H(s)	Multiplicity	C	Bears H(s)	Multiplicity
1	4 (3H)	CH₃	9	6	CH
2	1 (3H)	CH₃	10	9	CH
3	3, 11	CH₂	11	5	CH
4	8, 12	CH₂	12		C
5	7 (3H)	CH₃	13	16	CH
6	15	CH	14		C
7	2, 10	CH₂	15		C
8	13, 14	CH₂			

(b)

H	Is Coupled to H(s)	H	Is Coupled to H(s)
1	6	9	2, 3
2	9, 10, 11	10	2, 11
3	9, 11	11	2, 3, 10
4	15	12	6, 8, 13
5		13	8, 12, 14
6	1, 12	14	8, 12, 13
7		15	4
8	12, 13, 14	16	

(c) Below is a table of NOE cross peaks not seen in the COSY spectrum.

H	Is Close To	H	H	Is Close To	H
1		5	10		6
3		16	11		15
5		1	15		11
6		10	16		3, 7
7		16			

(d) First we will go through the COSY data, ignoring two-bond couplings already expected on the basis of the C,H–HSC data. For example, the coupling between H6 (attached to C9) and H1 (attached to C2) indicates that C9 is bonded to C2. By like reasoning, C7 is bonded to C3, C3 to C10, C1 to C6, and C4 to C8. Likewise, the NOESY cross peaks further show that C9 is bonded to C11, C7 is bonded to C9, and C13 must be close to C5 and C3. This, plus the chemical shifts of carbons 13 and 14 suggesting at-

tachment of two oxygens, allows us to construct the following carbon skeleton fragments:

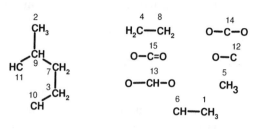

At this point we need some additional connectivity information or we have to start making tentative assignments based on chemical shift.

(e) The 2D INADEQUATE immediately provides the following carbon skeleton:

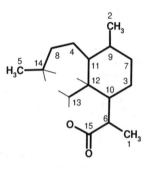

The short bonds are to the three remaining oxygen atoms, and we still need to close three more rings. The final structure of the compound (artemisinin, seen previously in Section 12.12), together with all proton assignments, is given below. Most stereochemical relationships are deducible from the NOESY spectrum:

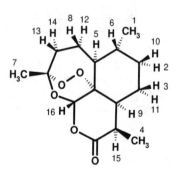

SELF-TEST II

1. (a) The five-hydrogen multiplet at δ 7.3 and the four-carbon signals at δ 124–132 suggest a monosubstituted aromatic ring. The three-hydrogen singlet at δ 1.99 indicates three equivalent hydrogens that are likely part of a methyl group attached to a deshielding

group. The two ^{13}C signals near δ 85 are in the region of triply bonded carbons. The only structure that fulfills all these structural features is 1-phenylpropyne:

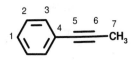

The observed and predicted ^1H and ^{13}C chemical shifts are tabulated below:

Nucleus	δ_{obs}	δ_{calc}	Reference
H1, 2, 3	7.3	7.2–7.4	Table 6.5
H7	1.99	1.73	Table 6.1
C1	127.6	128.9	Table 7.4
C2	128.3	128.6	Table 7.4
C3	131.7	135.4	Table 7.4
C4	124.4	122.7	Table 7.4
C5, 6	80.1, 85.9	range 70–90	Section 7.5
C7	4.1	2.2	Tables 7.1, and 7.2

(b) The signal for C7 will be split by the three methyl hydrogens into a quartet (Section 8.6). To find the value of the residual coupling constant in the off-resonance-decoupled spectrum, use Eq. (12.2) and the value of $^1J_{CH}$ for sp^3-hybridized carbons (125 Hz, Table 9.1):

$$^1J_r = \frac{2\pi J\,\Delta\nu}{\gamma_H\,\mathbf{B}_2}$$

$$= \frac{2\pi\,(125\text{ Hz})\,(2000\text{ Hz})}{267.5 \times 10^6\text{ rad T}^{-1}\text{ Hz})\,(2\times 10^{-4}\text{ T})} = 29\text{ Hz}$$

The signals for C1, C2, and C3 will each be split by one hydrogen into a doublet. The hybridization of carbons in an aromatic ring is sp^2, so use a $^1J_{CH}$ value of 156 Hz (Table 9.1) to calculate the observed residual coupling:

$$^1J_r = \frac{2\pi J\,\Delta\nu}{\gamma_H\,\mathbf{B}_2}$$

$$= \frac{2\pi\,(156\text{ Hz})\,(2000\text{ Hz})}{267.5 \times 10^6\text{ rad T}^{-1}\text{ Hz})\,(2\times 10^{-4}\text{ T})} = 37\text{ Hz}$$

2. (a) First, consider the ^1H spectrum. The triplet at δ 0.95 is likely to be due to a methyl group coupled to two equivalent hydrogens, as in CH_3–CH_2–. The six-

hydrogen doublet (δ 1.22) suggests two *equivalent* methyl groups split by one neighboring hydrogen, as in $(CH_3)_2CH–$ (an isopropyl group). The methine hydrogen of the isopropyl group is found as the expected septet at δ 4.92, indicating that the isopropyl group is attached to an oxygen. The two-hydrogen triplet at δ 2.18 suggests a CH_2 group attached to a carbonyl and split by two equivalent hydrogens, $O=C–CH_2–CH_2–$. The presence of an ester carbonyl is confirmed by the ^{13}C signal at δ 172.7. Putting all this information together in one structure gives isopropyl butyrate:

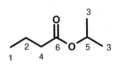

Note how the number and positions of the ^{13}C signals are also consistent with this structure:

Nucleus	δ_{obs}	δ_{calc}	Reference
C1	13.7	13.8	Tables 7.1 and 7.2
C2	18.7	19.3	Tables 7.1 and 7.2
C3	22.0	21.8	Tables 7.1 and 7.2
C4	36.6	35.8	Tables 7.1 and 7.2
C5	67.1	67.3	Tables 7.1 and 7.2
C6	172.7	165–175	Table 7.5
H1	0.95	0.86	Table 6.1
H2	1.70	1.80	Table 6.1
H3	1.22	1.30	Table 6.1
H4	2.18	2.36	Table 6.1
H5	4.92	4.85	Table 6.1

(b) An APT τ value of 8 ms corresponds to a J value of 125 Hz [$1/(8\text{ ms}) = 125$ Hz; Section 12.10]. Such a J value is typical of a one-bond coupling between hydrogen and sp^3 carbon (Table 9.1). Figure 11.12 indicates that, with τ set to $1/J$, methylene and quaternary carbon signals will be positive, while methyl and methine carbon signals will be negative. Thus, the ^{13}C signals at δ 13.7 (C1), 22.0 (C3), and 67.1 (C5) will be negative; the other three will be positive.

3. (a) The two downfield doublets in the ^1H spectrum and the four ^{13}C signals from δ 120–150 suggest a para-disubstituted aromatic ring. The remaining ^1H is most likely a CH_2 group attached to the ring. From the molecular formula, this leaves only $ClNO_2$, likely to occur as a Cl and an NO_2 (nitro) group. So there are two structures to consider:

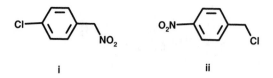

i   ii

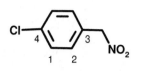

The predicted ¹H and ¹³C chemical shifts of the CH₂ group clearly fit structure **i** much better than **ii**:

Compound	Calculated ¹³C δ^a	Calculated ¹H δ^b
i	52	5.03
ii	84	6.05

^{a}Tables 7.1 and 7.2.
^{b}Table 6.2.

In addition to the methylene signals, all other ¹H and ¹³C signals are well correlated with structure **i**:

Nucleus	δ_{obs}	δ_{calc}	Reference
C1	123.8	123.1	Table 7.4
C2	129.8	130.1	Table 7.4
C3	144.9	143.8	Table 7.4
C4	147.5	145.2	Table 7.4
H1	8.19	8.23	Table 6.5
H2	7.56	7.44	Table 6.5

(b) See Section 13.2. The $^1J_{CH}$ range of 125–160 Hz covers both aliphatic and vinyl/aromatic hydrogens (Table 9.1). Figure A.8 exhibits the following C–H

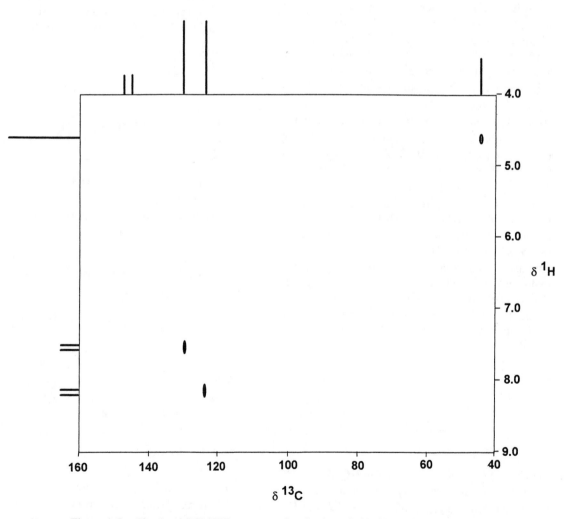

Figure A.8.   Simulated C,H-HSC spectrum of *p*-nitrobenzyl chloride, problem 3.

correlations: C1–H1, C2–H2, and the methylene carbon and hydrogen.

4. (a) Begin by estimating the magnitude of the three possible C–D coupling constants $^1J_{CD}$, $^2J_{CD}$, and $^3J_{CD}$. Use the equation in Example 9.3 (Section 9.2) to estimate J_{CD} values from the J_{CH} values in Tables 9.1, 9.3, and 9.4. Thus,

$$^1J_{CH} = {}^1J_{CD}\left(\frac{\gamma_H}{\gamma_D}\right) = (125\ Hz)\frac{267.512}{41.0648} = 19\ Hz$$

Likewise,

$$^2J_{CH} \approx {}^3J_{CH} = (5\ Hz)\frac{267.512}{41.0648} < 1\ Hz$$

Therefore, the only coupling of any significance is ^1J. The carbonyl carbon should appear as a weak singlet at about δ 210 (Table 7.5). The remaining two carbons are equivalent and will appear at about δ 28 (Table 7.2). From Eq. (8.3), the latter signal will be split by the three ($n = 3$) deuterium nuclei ($I = 1$) into seven lines. The intensity ratio of the lines in this multiplet is 1 : 3 : 6 : 7 : 6 : 3 : 1, which can be deduced from the number of spin combinations that give rise to each M value (Section 8.6.2). The line spacings will all be 19 Hz. See Example 8.16(b).

(b) This spectrum will consist of one signal centered at δ 66 (Table 7.2) that is split into a doublet (by the fluorine, $^1J_{CF} = 165$–350 Hz, Table 9.2) of quartets (by the three equivalent hydrogens, $^1J_{CH} = 125$ Hz.

5. Typical vinyl hydrogens usually appear around δ 5.3 (Section 6.3). But hydrogens attached to aromatic rings are further deshielded to about δ 7.3 by delocalization of the π electrons in the aromatic double bonds (Section 6.5). The ^1H chemical shift of cyclooctatetraene is much closer to that of "normal" vinyl hydrogens, indicating little, if any, aromatic ring current. In fact, unlike the planar molecule benzene (Example 4.3), cyclooctatetraene is tub shaped, making conjugative interaction of the double bonds essentially impossible:

6. (a) The two terminal vinyl hydrogens (H_a) are equivalent (by virtue of a symmetry plane in the page) and experience long-range coupling to H_b ($^4J = 6$ Hz, Table 9.5):

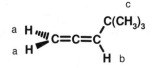

Thus, the H_a signal, which should appear around δ 5.3 (Section 6.3), will be split by H_b into a doublet. The H_c signal will be split by the two H_a's into a triplet centered near δ 5.7 (Table 6.4). The nine equivalent hydrogens of the *tert*-butyl group are not coupled to any other nuclei, so they give rise to a 9H singlet near δ 1.0 (Example 6.12).

(b) Although the 9H singlet is exactly where predicted, the vinyl hydrogens are about 0.6 ppm more shielded than expected. This is a characteristic of the allene (C=C=C) linkage. Note, however, that the "doublet" at δ 4.7 and the "triplet" at δ 5.1 constitute an A_2B system that is rendered complex by second-order effects (Section 9.9). This is because $\Delta\delta\nu/J$ is relatively small (24 Hz/6 Hz = 4). The two small signals flanking the 9H singlet are spinning sidebands (Section 3.3.3). The spacing between them (72 Hz) is too small for them to be ^{13}C satellites (Section 8.6.1).

(c)

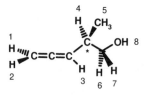

Only the CH_3 hydrogens are equivalent. The chiral center (*) renders the CH_2 hydrogens (H6 and H7), as well as the =CH_2 hydrogens (H1 and H2), diastereotopic (Section 4.3).

(d) Since there are eight sets of hydrogen, there are eight δ values. We can estimate these as follows:

H	δ_{calc}	Reference
1, 2	4.7	Part (b), above
3	5.1	Part (b), above
4	2.96	Table 6.2
5	0.93	Table 6.2
6, 7	3.44	Table 6.2
8	2–4	Section 6.6.1

Note that H1 and H2 are expected to have similar chemical shifts, as are H6 and H7. Neglecting coupling to the OH hydrogen (because of exchange), the following homonuclear H–H couplings will be observable (>1 Hz):

H_a	H_b	n^a	$^nJ_{ab}$	Reference
1	2	2	2.5	Table 9.3
1	3	4	6	Table 9.5
1	4	5	3	Table 9.5
2	3	4	6	Table 9.5
2	4	5	3	Table 9.5
3	4	3	7	Table 9.5
4	5	3	7	Table 9.5
4	6	3	7	Table 9.5
4	7	3	7	Table 9.5

^{a}Number of intervening bonds.

(e)

Signal	Multiplicity	Assignment
1.02	d	H5
2.3	m	H4
3.2	dd(?)	H6, H7
4.1	s	H8
4.7	dd	H1, H2
5.1	dt	H3

From Eq. (8.2), with $n_1 = n_2 = n_3 = n_6 = n_7 = 1$ and $n_5 = 3$, $L = 128$ lines!

(f) Note from part (d) that H4 is coupled to every other hydrogen in the molecule except H8! Irradiation of the H4 signal will decouple it from all other hydrogen nuclei. The decoupled spectrum will consist of the following signals: δ 1.02 (singlet), 3.2 (AB q because of the chiral center), 4.1 (s), 4.7 (d), 5.1 (t). The last two of these will resemble the A_2B pattern in part (b) above.

(g) The ^{13}C signal from the middle carbon in a C=C=C linkage is usually weak and occurs far downfield, in the carbonyl region (δ 200–215; Section 7.7).

(h) The halfwidth ($v_{1/2}$) of the O–H resonance will be broadened by hydrogen exchange (Section 6.6.1), mainly because this exchange accelerates spin–spin relaxation (i.e., decreases the spin–spin relaxation time T_2, which directly determines the halfwidth; Sections 3.5.1). In the case of an N–H signal there is an additional effect operating: The electric quadrupole of nitrogen also decreases T_2.

7. In structure **C** the two OCH_3 groups meta to the $CONH_2$ are symmetry equivalent and distinct from the para one. At +38°C the same is true of compound **D**.

Yet, as the temperature is lowered, it is clear that the two meta OCH_3 groups in **D** become nonequivalent. This indicates that they must be inherently nonequivalent, but some type of process averages them at the higher temperature.

The favored conformation of the molecule would be an all-planar one, allowing a resonance interaction between the phenyl ring and the imidazole ring. In such a conformation the two meta OCH_3 groups are nonequivalent. But this conformation is extremely crowded because of steric interference between the hydrogens and the meta OCH_3 group that are superimposed. Alternatively, the phenyl ring could adopt a conformation perpendicular to the other rings, losing the resonance but relieving the crowding and rendering the meta OCH_3 groups equivalent. The ultimate energy-minimizing compromise is that at temperatures below –67°C the phenyl ring is frozen at an intermediate angle between these extremes, resulting in two enantiomeric conformations (**D1** and **D2**), each with the meta OCH_3 groups nonequivalent (δ 2.76 and 4.13):

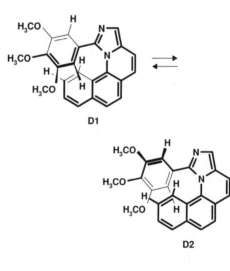

As the temperature increases, restricted rotation of the phenyl ring begins to interconvert **D1** and **D2**. At –23°C the system reaches coalescence of the two meta signals, leaving them buried in the baseline. At +38°C exchange is rapid on the NMR time scale and only one meta signal (at the average chemical shift δ 3.44) appears.

8. First, the magnetogyric ratio of Cl (Table 2.1) is small, and this affects the magnitude of J [Eq. (9.1)]. But more importantly, a chlorine nucleus (as well as any nucleus with $I \geq 1$) has an electric quadrupole moment because of the nonspherical distribution of charge in its nucleus. This fact causes chlorine nuclei to undergo

spin–lattice relaxation extremely rapidly on the NMR time scale. This rapid interconversion and equilibration among its various nuclear spin states effectively decouple the chlorine from spin–spin interactions with neighboring nuclei. However, the EPR time scale is nearly a thousand times faster than the NMR time scale (Section 11.4), so the relaxation of the chlorine nuclei is not fast enough to decouple them from neighboring unpaired electrons.

9. (a) The signal integrations are (from left to right) 5H, 1H, 1H, 1H, 2H, 1H, and 3H. The IOU of 5 and the downfield 5H multiplet suggest a monosubstituted phenyl ring. The doublet at δ 2.13 (that exchanges in D_2O) must belong to an OH hydrogen, with the OH group attached to a methine carbon, whose hydrogen is further coupled to two hydrogens: $-CH_2C(OH)H-$. The two 1H "singlets" near δ 4.9 must belong to geminal vinyl hydrogens on an isopropenyl group, weakly coupled to the vicinal methyl group. There are two ways to assemble these pieces:

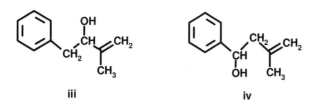

iii   iv

These can be differentiated by estimating the chemical shift of the internal CH_2 group from the data in Table 6.2: **iii**, δ 2.88; **iv**, δ 2.31; observed, δ 2.44. The compound is **iv**.

(b) Eu(hfc)₃ is a chiral chemical shift reagent capable of differentiating enantiotopic nuclei by forming a complex in which these nuclei become diastereotopic. All of the original signals experience pronounced downfield (paramagnetic) shifts ranging from ca. 0.2 ppm for the aromatic hydrogens to 4.0 ppm for the methine hydrogen. More importantly, the original CH_3 singlet (δ 1.80) has now resolved into two closely spaced, equally intense singlets (δ 2.4), the CH_2 doublet (δ 2.44) has split into two complex but equally intense multiplets (δ 4.0 and 4.4), and the methine multiplet (δ 4.82) has become two equally intense multiplets (δ 8.75 and 8.85). The reason for all these "extra" signals is that structure **iv** is actually a racemic (equal) mixture of two enantiomers by virtue of the chiral methine center. In the absence of a chiral shift reagent these two enantiomers exhibit identical NMR spectra. But addition of the reagent causes the signals of the R enantiomer to separate from those of the S.

10. Here we have an IOU of 9. The three oxygens must all be part of carbonyl groups: a carboxylic acid (**C26** and **H30**) and two ketones (**C27** and **C28**). There are also two C=C bonds (**C22–C25**), both of which have chemical shifts suggesting α,β-unsaturation. The four remaining sites of "unsaturation" must be due to rings.

In the absence of 2D INADEQUATE data, let us begin by integrating the HSC and COSY data to establish as much C–C connectivity as we can. The table below lists the hydrogen-bearing carbons, the hydrogen(s) to which each carbon is attached, the hydrogen(s) other than geminal ones to which these hydrogens are coupled, and finally, the carbon(s) to which the original carbon must therefore be connected:

C	H	H	C
1	1 (3H)		
2	9 (3H)		
3	11, 12	2, 7, 17	14, 20
4	3 (3H)	25	8
5	4 (3H)	25	8
6	8, 15	5, 10, 19	7, 21
7	10, 19	8, 14, 15	6, 19
8	25	3, 4	4, 5
9	21, 27	6, 18	10
10	6, 18	13, 21, 27	9, 13
11	22	16, 20	12
12	16, 20	22	11
13	13	2, 5, 6, 18	10, 20, 21
14	7, 17	11, 12	3
15	24, 26	23	17
17	23	14, 24, 26	15, 19
19	14	10, 19, 23	17, 7
20	2	11, 12, 13	3, 13
21	5	8, 13, 15	6, 13
22	28, 29		

Here is how the data in the last column were derived. Take **C3**, for example, which the HSC spectrum indicates is attached to **H11** and **H12**. The COSY spectrum indicates these two hydrogens (in addition to being coupled to each other) are coupled to **H2**, **H7**, and **H17**. These three hydrogens are attached to **C14** and **C20**. Therefore, **C3** must be connected to these two carbons. Extending this reasoning as far as possible, we can deduce the following molecular fragments (where ? represents an unknown nonprotonated carbon):

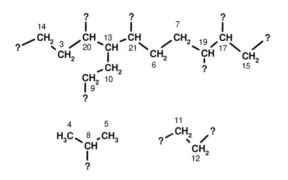

To these we must add the three C=O carbons, the four vinyl carbons, and the two quaternary carbons (C**16** and C**18**) bearing methyl carbons C**1** and C**2**.

The COLOQ results allow us to complete the picture. For example, H**3** and H**4** (attached to C**4** and C**5**, respectively) show long-range coupling to C**24** (as well as C**8**), H**24**/H**26** (attached to C**15**) show long-range coupling to both C**24** and C**28**, and H**28**/H**29** (C**22**) correlate with C**8**, C**24**, and C**28**; so we can add these pieces to the puzzle:

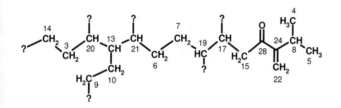

The facts that (a) H**1** (C**1**) correlates with C**14**, C**18**, C**19**, and C**21**, while H**14** (C**19**) correlates with C**7**, C**17**, C**18**, and C**26**, and (b) H**9** (C**2**) correlates with C**12**, C**16**, C**20**, and C**25** lead us to the familiar steroid skeleton:

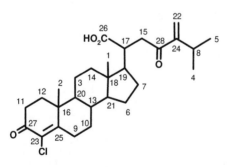

The only carbons left were C**23** and C**27**, which could only occur as shown above.

11. (a) The IOU of 4 is satisfied by the four rings of the steroid skeleton, so the molecule is otherwise fully

saturated. There are eight methyl groups, three more than the standard 27-carbon (cholestane) steroid skeleton. With some very careful analysis you can establish the C–H correlations for about half the carbons. The H–H correlations are difficult to determine from a COSY spectrum of this complex, which had to be reduced in size to fit on one page. You may wish to consult the reference cited at the end of Self Test II to see how the original investigators established not only the connectivity but also the stereochemistry.

(b) The 2D INADEQUATE makes it almost trivial to get the carbon backbone:

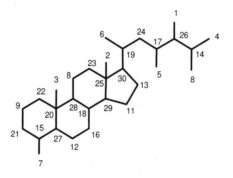

CHAPTER 14

14.1. The expected chemical shifts can be obtained from Figure 14.8. The NH_3 and gaunadinium protons are found, respectively, at approximately δ 7.5 and 7.0 and will probably appear as a slightly broadened singlet due to chemical exchange:

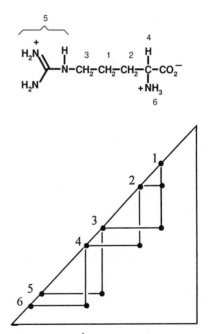

Figure A.9.   Simulated ^{1}H COSY spectrum of arginine.

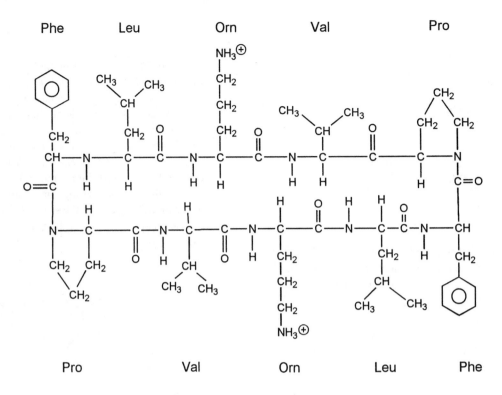

Phe Leu Orn Val Pro

Pro Val Orn Leu Phe

Figure A.10. Structure of gramicidin S.

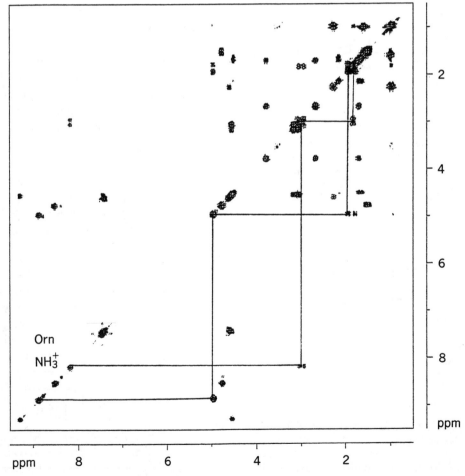

Orn
NH₃⁺

Figure A.11. The ¹H COSY correlations of the ornithine reside in gramicidin S.

364

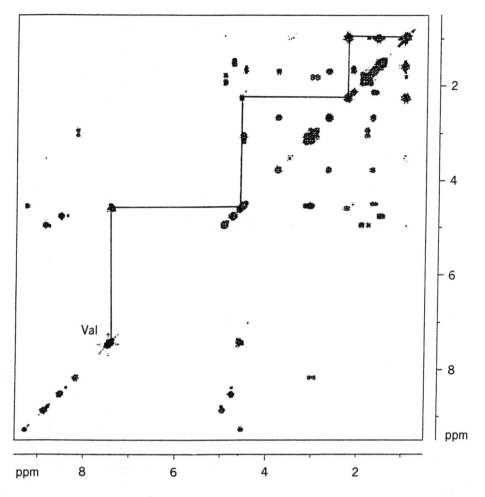

Figure A.12.   The ^{1}H COSY correlations of the valine reside in gramicidin S.

The expected correlations are listed in the table below and are shown in Figure A.9:

H	Is Coupled To	H(s)
1		2, 3
2		1, 4
3		1, 5
4		2, 6
5		3
6		4

14.2.   (a) The structure of Gramicidin S is shown in Figure A.10. There is a twofold axis of symmetry, and therefore the pairs of like amino acids are equivalent and five sets of resonances are expected. (b) Since proline does not have an amide proton, only four pairs of amide protons are expected in the spectral region δ 7–8.7. (c) The amine protons of ornithine are found at δ 8.1 and can be used to map the correlations of the ornithine residue (Figure A.11): δ 8.1 (NH$_3$) → 3.0 (δ) → 1.9 (γ) → 2.0 (β) → 5.0 α) → 8.9 (NH). Valine has the unique one-way stairstep correlation pattern shown in Figure A.12: δ 7.4 (NH) → 4.6 (α) → 2.1 (β) → 1.0 (γ). Compare the patterns of ornithine and valine. It is apparent that each amino acid has its own signature correlation pattern. For the phenylalanine, the aromatic protons are at δ 7.3 and show correlations among themselves but not the other protons. The remaining protons exhibit the correlations shown in Figure A.13: δ 9.2 (NH) → 5.5 (α) → 3.1 and 3.2 (β$_1$ and β$_2$). Leucine's characteristic correlation pattern (also shown in Figure A.13) is δ 8.5 (NH) → 4.8 (α) → 1.7 (β and γ) → 0.9 (δ). Note that the correlation between the β and γ protons (both at δ 1.70) cannot be resolved at the expansion shown. Proline's correlations are more complicated because proline involves a ring system where the geminal β and δ methylene

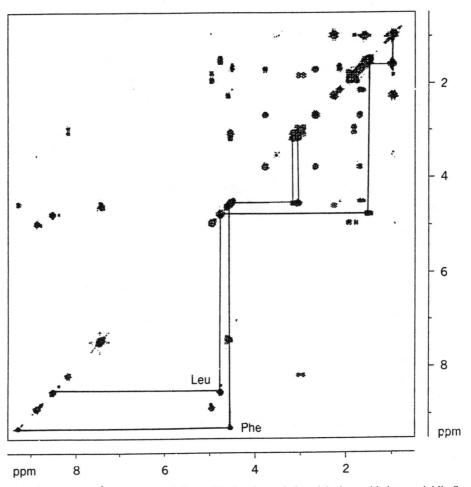

Figure A.13. The ^{1}H COSY correlations of the leucine and phenylalanine reside in gramicidin S.

protons are nonequivalent. Its correlations are shown in Figure A.14. Note that the equivalent γ proton resonances are slightly downfield of the β_2 proton resonance. There are also correlations between the nonequivalent β as well as the nonequivalent δ protons.

Beginning with the proline α proton at δ 4.5, the NOESY/COSY correlations along the peptide backbone can be mapped: pro (α) $\rightarrow$ val (NH) $\rightarrow$ val (α) $\rightarrow$ orn (NH) $\rightarrow$ orn (α) $\rightarrow$ leu (NH) $\rightarrow$ leu (α) $\rightarrow$ phe (NH) $\rightarrow$ phe (α) (Figure A.15, page 368).

14.3. At 30°C, there is little interaction between the adenosine and poly U and the resonances appear at their respective chemical shifts. As the temperature is lowered, the adenosine and poly U begin to reversibly form a large rigid complex. From 30°C to ~20°C, the exchange rate is fast on the NMR time scale, so an averaged spectrum of the two states is observed. Below 20°C, the exchange rate becomes slow on the NMR time scale and only the free excess adenosine

(one-half of the original amount is observed). The spectrum of the rigid 1 : 2 adenosine–poly U complex is broadened beyond detection and is not observed. With a decrease in the exchange rate, the line widths of the free adenosine decrease with decreasing temperature.

14.4. Since there is a concentration dependence to the chemical shift of the benzaldehyde resonances, the exchange rate between the free and bound states is fast on the NMR time scale. The observed chemical shift is a weighted average of the two states. At low concentrations of benzaldehyde, the fraction bound is the largest. Since the average chemical shift is upfield from the free chemical shift, the chemical in the bound state must also be upfield of the free shift. Since the Flavin ring is aromatic and would have ring current magnetic anisotropy, the upfield shift of the benzaldehyde resonances probably arises from this group binding above the Flavin ring in the enzyme–substrate complex.

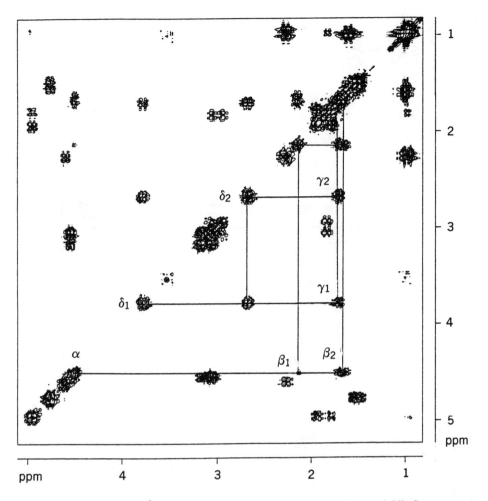

Figure A.14. The ^{1}H COSY correlations of the proline reside in gramicidin S.

CHAPTER 15

15.1. Since the molecular formula of the product is exactly double that of the starting material, we can infer that it is a dimer and, based on the simplicity of the CP/MAS spectrum, a quite symmetrical one at that. If we assume that the three sharp signals each represent a pair of equivalent carbons, their chemical shifts suggest that they belong to (1) simple ketone carbonyls (δ 217.6), (2) carbons bearing an α OH group, an α carbonyl group, and two R groups (δ 83.5, calculated ca. 81), and (3) carbons bearing a β OH group, an α carbonyl group, and one R group (δ 52.5, calculated ca. 50). These facts are best accommodated by the bis-aldol dimer below, with several of the flanking CH$_2$ groups unresolved at this operating frequency. [See Macomber, R. S., and Murphy, P., *Org. Mag. Res.*, **22**, 255 (1984).]

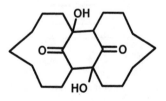

15.2. Since the empirical formulas of both the product and starting material are the same, the product is likely a polymer of the cumulatriene. The two resonances for doubly bonded carbons have been replaced by a signal at δ 88.8, which suggests a symmetrically substituted carbon–carbon triple bond. The signal at δ 48.2 fits for a cyclohexane carbon attached to a triple bond and an alkyl group (calculated ca. 40). These facts are best accommodated by the polymer structure below, rendered rigid and rodlike by the conformational requirements of the six-membered rings on vicinal carbons:

[See Pollack, S. K., Narayanswamy, B., Macomber, R. S., Rardon, D. E., and Constantinides, I., *Macromolecules*, 26 (1993).]

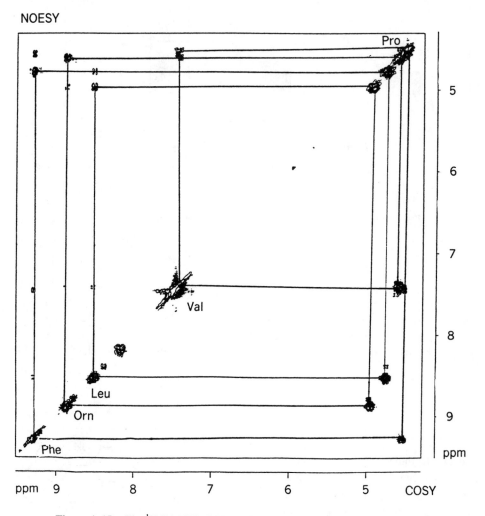

Figure A.15.   The ^{1}H TOCSY (COSY/NOESY) correlations for gramicidin S.

APPENDIX 2: Periodic Table of the Elements

s Orbitals being filled

d Orbitals being filled

p Orbitals being filled

Noble gases

Transition elements

f Orbitals being filled

Period number = n, the highest occupied electron level

Period

Group IA (ns^1)
- H 1 — $1s^1$ — 1.0079
- Li 3 — $2s^1$ — 6.941
- Na 11 — $3s^1$ — 22.9898
- K 19 — $4s^1$ — 39.098
- Rb 37 — $5s^1$ — 85.4678
- Cs 55 — $6s^1$ — 132.905
- Fr 87 — $7s^1$ — (223)

Group IIA (ns^2)
- Be 4 — $2s^2$ — 9.01218
- Mg 12 — $3s^2$ — 24.305
- Ca 20 — $4s^2$ — 40.08
- Sr 38 — $5s^2$ — 87.62
- Ba 56 — $6s^2$ — 137.33
- Ra 88 — $7s^2$ — (226)

Group IIIB $(n-1)d^1ns^2$
- Sc 21 — $3d^14s^2$ — 44.959
- Y 39 — $4d^15s^2$ — 88.9059
- La* 57 — $5d^16s^2$ — 138.905
- Ac⁺ 89 — $6d^17s^2$ — (227)

Group IVB $(n-1)d^2ns^2$
- Ti 22 — $3d^24s^2$ — 47.90
- Zr 40 — $4d^25s^2$ — 91.22
- Hf 72 — $4f^{14}5d^26s^2$ — 178.49
- Rf 104 — (260)

Group VB $(n-1)d^3ns^2$
- V 23 — $3d^34s^2$ — 50.9414
- Nb 41 — $4d^45s^1$ — 92.9064
- Ta 73 — $5d^36s^2$ — 180.948
- Ha 105 — (260)

Group VIB $(n-1)d^5ns^1$
- Cr 24 — $3d^54s^1$ — 51.996
- Mo 42 — $4d^55s^1$ — 95.94
- W 74 — $5d^46s^2$ — 183.85

Group VIIB $(n-1)d^5ns^2$
- Mn 25 — $3d^54s^2$ — 54.938
- Tc 43 — $4d^55s^2$ — (97)
- Re 75 — $5d^56s^2$ — 186.207

Group VIIIB $(n-1)d^6ns^2$ / $(n-1)d^7ns^2$ / $(n-1)d^8ns^2$
- Fe 26 — $3d^64s^2$ — 55.847
- Co 27 — $3d^74s^2$ — 58.9332
- Ni 28 — $3d^84s^2$ — 58.70
- Ru 44 — $4d^75s^1$ — 101.07
- Rh 45 — $4d^85s^1$ — 102.905
- Pd 46 — $4d^{10}$ — 106.4
- Os 76 — $5d^66s^2$ — 190.2
- Ir 77 — $5d^76s^2$ — 192.22
- Pt 78 — $5d^96s^1$ — 195.09

Group IB $(n-1)d^{10}ns^1$
- Cu 29 — $3d^{10}4s^1$ — 63.546
- Ag 47 — $4d^{10}5s^1$ — 107.868
- Au 79 — $5d^{10}6s^1$ — 196.967

Group IIB $(n-1)d^{10}ns^2$
- Zn 30 — $3d^{10}4s^2$ — 65.38
- Cd 48 — $4d^{10}5s^2$ — 112.40
- Hg 80 — $5d^{10}6s^2$ — 200.59

Group IIIA (ns^2np^1)
- B 5 — $2s^22p^1$ — 10.81
- Al 13 — $3s^23p^1$ — 26.9815
- Ga 31 — $4s^24p^1$ — 69.72
- In 49 — $5s^25p^1$ — 114.82
- Tl 81 — $6s^26p^1$ — 204.37

Group IVA (ns^2np^2)
- C 6 — $2s^22p^2$ — 12.011
- Si 14 — $3s^23p^2$ — 28.086
- Ge 32 — $4s^24p^2$ — 72.59
- Sn 50 — $5s^25p^2$ — 118.69
- Pb 82 — $6s^26p^2$ — 207.19

Group VA (ns^2np^3)
- N 7 — $2s^22p^3$ — 14.0067
- P 15 — $3s^23p^3$ — 30.9738
- As 33 — $4s^24p^3$ — 74.9216
- Sb 51 — $5s^25p^3$ — 121.75
- Bi 83 — $6s^26p^3$ — 208.980

Group VIA (ns^2np^4)
- O 8 — $2s^22p^4$ — 15.9994
- S 16 — $3s^23p^4$ — 32.06
- Se 34 — $4s^24p^4$ — 78.96
- Te 52 — $5s^25p^4$ — 127.60
- Po 84 — $6s^26p^4$ — (209)

Group VIIA (ns^2np^5)
- F 9 — $2s^22p^5$ — 18.9984
- Cl 17 — $3s^23p^5$ — 35.453
- Br 35 — $4s^24p^5$ — 79.904
- I 53 — $5s^25p^5$ — 126.904
- At 85 — $6s^26p^5$ — (210)

Group VIIIA (ns^2np^6) — Noble gases
- He 2 — $1s^2$ — 4.0026
- Ne 10 — $2s^22p^6$ — 20.179
- Ar 18 — $3s^23p^6$ — 39.948
- Kr 36 — $4s^24p^6$ — 83.80
- Xe 54 — $5s^25p^6$ — 131.30
- Rn 86 — $6s^26p^6$ — (222)

Lanthanides $4f^{1-14}5d^{0-1}6s^2$
- Ce 58 — $4f^15d^16s^2$ — 140.12
- Pr 59 — $4f^35d^06s^2$ — 140.907
- Nd 60 — $4f^45d^06s^2$ — 144.24
- Pm 61 — $4f^55d^06s^2$ — (145)
- Sm 62 — $4f^65d^06s^2$ — 150.35
- Eu 63 — $4f^75d^06s^2$ — 151.96
- Gd 64 — $4f^75d^16s^2$ — 157.25
- Tb 65 — $4f^95d^06s^2$ — 158.925
- Dy 66 — $4f^{10}5d^06s^2$ — 162.50
- Ho 67 — $4f^{11}5d^06s^2$ — 164.930
- Er 68 — $4f^{12}5d^06s^2$ — 167.26
- Tm 69 — $4f^{13}5d^06s^2$ — 168.934
- Yb 70 — $4f^{14}5d^06s^2$ — 173.04
- Lu 71 — $4f^{14}5d^16s^2$ — 174.97

Actinides $5f^{1-14}6d^{0-1}7s^2$
- Th 90 — $5f^06d^27s^2$ — 232.038
- Pa 91 — $5f^26d^17s^2$ — (231)
- U 92 — $5f^36d^17s^2$ — 238.03
- Np 93 — $5f^46d^17s^2$ — (237)
- Pu 94 — $5f^66d^07s^2$ — (244)
- Am 95 — $5f^76d^07s^2$ — (243)
- Cm 96 — $5f^76d^17s^2$ — (247)
- Bk 97 — $5f^86d^17s^2$ — (247)
- Cf 98 — $5f^{10}6d^07s^2$ — (251)
- Es 99 — $5f^{11}6d^07s^2$ — (254)
- Fm 100 — $5f^{12}6d^07s^2$ — (257)
- Md 101 — $5f^{13}6d^07s^2$ — (258)
- No 102 — $5f^{14}6d^07s^2$ — (255)
- Lr 103 — $5f^{14}6d^17s^2$ — (260)

SUBJECT INDEX